Perturbation Theory for Linear Operators

Aref Jeribi

Perturbation Theory for Linear Operators

Denseness and Bases with Applications

Springer

Aref Jeribi
Department of Mathematics
University of Sfax
Sfax, Tunisia

ISBN 978-981-16-2527-5 ISBN 978-981-16-2528-2 (eBook)
https://doi.org/10.1007/978-981-16-2528-2

This Springer imprint is published by the registered company Springer Nature Singapore Pte Ltd.
The registered company address is: 152 Beach Road, #21-01/04 Gateway East, Singapore 189721, Singapore

To

my mother Sania,

my father Ali,

my wife Fadoua,

my children Adam and Rahma,

my brothers Sofien and Mohamed Amin,

my sister Elhem, and

all members of my extended family

Preface

This book gives a fairly comprehensive review of important aspects of the spectral theory, covering, in particular, the completeness of generalized eigenvectors, the Riesz bases, the semigroup theory, the families of analytic operators, and the Gribov operator acting in the Bargmann space. There is an interesting selection of topics and results, including a number of unfamiliar ones from the literature.

This book is intended for students, researchers in the field of spectral theory of linear non-self-adjoint operators, and mathematicians interested in applications in mathematical physics. Pure analysts will have some new problems to tackle. Readers interested in mathematical physics will have a new theory with practical importance. Most of the material can be understood by a reader with a relatively modest knowledge of functional analysis and operator theory, so I include some chapters connected to this background material that is needed to develop the results of this book, which make the text almost entirely self-contained.

This book covers an important list of topics in applied functional analysis, an excellent choice, and it seems to be well written. The main topics include the following:

- Bases on Hilbert and Banach spaces.
- Carleman-class.
- The evolutionary problem.
- Completeness criteria of the space of generalized eigenvectors of non-self-adjoint operators.
- Spectral theory.
- Analytic operators in Feki-Jeribi-Sfaxi's sense.
- On a Schauder and Riesz bases of eigenvectors of an analytic operator.
- On the basis property of root vectors related to a non-self-adjoint analytic operator.
- On a Riesz basis of finite-dimensional invariant subspaces.
- Perturbation method for sound radiation by a vibrating plate in a light fluid.
- Gribov operator in Bargmann space.
- Applications in mathematical physics and mechanics.

We would like to thank Salma Charfi for the improvement she has made in the introduction of this book. So, we are indebted to her. Moreover, we should mention that the thesis work results, performed under my direction, by my former students and present colleagues, Naouel Ben Ali, Ines Feki, and Hanen Ellouz, have helped us in writing this book. Finally, we apologize in case we have forgotten to quote any author who has contributed, directly or indirectly, to this work.

Sfax, Tunisia Aref Jeribi
April 2021

Introduction

Throughout this book, perturbation theory of linear non-self-adjoint operators in Hilbert and Banach spaces takes an important place and is closely related to many disciplines in mathematics. In general, this theory is based on the idea of perturbation of a normal operator by an unbounded non-self-adjoint operator. Historically, it was developed as an approximation device in quantum field theories since it gives more physical insight and allows the simplification of complex mathematical problems. More precisely, in the perturbation theory, the existence of a basis (possibly with parentheses) of root vectors is an important property and is one of the fundamental properties for vibrating systems.

In this book, we concentrate ourselves exclusively on giving some recent mathematical developments of perturbed non-self-adjoint operators by studying the completeness of the space of generalized eigenvectors, bases on Hilbert and Banach spaces, and asymptotic behavior of the eigenvalues of these operators. Most of the results of this book are motivated by physical problems such as the perturbation method for sound radiation by a vibrating plate in a light fluid, Gribov operator in Bargmann space, and some applications in mathematical physics and mechanics for which we address our study at the end of the book.

Now we describe in more detail the architecture of this book.

We begin this book with six chapters presenting the background material that is needed in our book. Most of this background is drawn from functional analysis and operator theory by giving some definitions, notations, and basic information on functional analysis that underlies most of the concepts presented in this book.

In Chap. 7, we recall some Keldysh results that are not only specific for Hilbert spaces, but they have been formulated for operators in Banach spaces. Indeed, some progress was made in 1951 by M. V. Keldysh to begin the development of a new and profound technique in the theory of non-self-adjoint operators. In fact, M. V. Keldysh established theorems of the completeness of eigenvectors and associated vectors and also theorems concerning the asymptotic properties of the eigenvalues for a wide class of polynomial bundles of non-self-adjoint operators.

On the other hand, we recall the generalization of Keldysh's theorem to operators on a Banach space which was developed by Yu. I. Lyubich, N. K. Nikol'skij, and I. Tweddle in [30].

We turn attention of the reader to the contributions of M. T. Aimar, A. Intissar and J. M. Paoli [1–3], M. T. Aimar, A. Intissar, and A. Jeribi [4], A. Jeribi [26], J. M. Paoli [36], and G. Q. Xu, S. H. Wang [40]. These authors gave a remarkable development in the theory of compact operators or operators with compact resolvent in order to check the denseness of the generalized eigenvectors.

In Chap. 8, we recall some bases such as Hamel and Haar bases given in [23, 24]. An important role is played by those Schauder bases that are orthonormal bases in a separable Hilbert space X and an equivalent characterization of such bases is that they are complete orthonormal sequences which means that every separable Hilbert space has an orthonormal basis. More accurately, a Hilbert basis (named after David Hilbert) [7, 12] is a generalization of Hilbert spaces of the classical notion of orthonormal basis in linear algebra for Euclidean spaces (or Hermitian in the complex case), which are finite dimensional.

We concentrate ourselves in Chap. 9 on the study of the Riesz basis of finite-dimensional invariant subspaces. It is well known that for non-normal operators, there is no analogue of the spectral theorem, so the existence of such Riesz basis of root vectors is an important property since it allows the construction of an invariant non-trivial subspace and yields spectral criteria related to a semigroup generation. Compared to normal operators, a number of new problems arise: first, apart from eigenvectors, the existence of generalized eigenvectors is possible too. Second, in contrast to a normal operator with compact resolvent, which always has an orthonormal basis of eigenvectors, the system of root vectors of a non-normal operator with compact resolvent need not be complete. Third, even if the system is complete, this does not imply that it has additional basis properties. To solve these problems, we find in literature, some pioneering works due to T. Kato [28], N. Dunford and J. T. Schwartz [13], and C. Clark [11]. At the same time, on the basis of the study of the resolvent, A. S. Markus and V. I. Matsaev [31, 32] established some theorems concerning the formation of bases from the root vectors of a non-self-adjoint operator.

This approach was used and extended by C. Wyss, in [39], who established different conditions, for a class of normal operators of non-normal perturbations, in terms of the spectrum which implies the existence of such Riesz bases. Recently, some progress has been made by the authors in [8] to investigate the existence of Riesz bases of eigenvectors of a class of non-normal perturbations of normal operators in the context of the construction of invariant non-trivial subspaces.

Chapter 10 is devoted to introduce analytic operators in Feki-Jeribi-Sfaxi's sense. The idea of the concept of these operators, based on a famous result by B. Sz. Nagy [34] (see also [26]), is that under a suitable powerful method of contour integration, he has been able to transfer most of the theorems for self-adjoint operators to a wider class of closed linear operators in Banach space. More precisely, let $A(\varepsilon)$ be a perturbed operator on a Banach space X, depending on a complex parameter ε as a convergent power series

$$A(\varepsilon) := A_0 + \varepsilon A_1 + \varepsilon^2 A_2 + \cdots + \varepsilon^k A_k + \cdots , \tag{1}$$

where $\varepsilon \in \mathbb{C}$ and $A_0, A_1, A_2, A_3, \cdots$ are linear operators on X, having the same domain $\mathcal{D}$. Under some specific conditions, we focus our attention that B. Sz. Nagy [34] proved the convergence of the series (1) and the closedness of its limit. Moreover, B. Sz. Nagy [34] showed that if the unperturbed operator A_0 has an isolated eigenvalue λ_n with multiplicity one. Then, $A(\varepsilon)$ has a unique eigenvalue $\lambda_n(\varepsilon)$ in the neighborhood of λ_n for sufficiently small $|\varepsilon|$ and this eigenvalue can be expanded into a convergent series

$$\lambda_n(\varepsilon) = \lambda_n + \varepsilon\lambda_{n,1} + \varepsilon^2\lambda_{n,2} + \cdots .$$

The associated eigenvector $\varphi_n(\varepsilon)$ of $A(\varepsilon)$ can also be chosen as a convergent series

$$\varphi_n(\varepsilon) = \varphi_n + \varepsilon\varphi_{n,1} + \varepsilon^2\varphi_{n,2} + \cdots ,$$

where φ_n is an eigenvector of A_0 associated to the eigenvalue λ_n.

Recently, I. Feki, A. Jeribi, and R. Sfaxi ask in [22] what conclusions can be drawn if they consider a more general type of non-analytic perturbation involving more than one perturbation parameter?

Based on rather long and explicit calculations, they investigate under this question and they give new variants on B. Sz. Nagy's results [34] and they consider the operator

$$T(\xi) := T_0 + \xi_1 T_1 + \xi_2 T_2 + \cdots + \xi_k T_k + \cdots , \tag{2}$$

where $\xi = (\xi_k)_{k\geq 1}$ is a sequence of complex numbers such that

$$\tau(\xi) := \sum_{k=1}^{\infty} |\xi_k| < \infty,$$

T_0 is a closed linear operator on, a Banach space, X with domain $\mathcal{D}(T_0)$ and the operators $(T_k)_{k\geq 1}$ are linear on X, having the same domain $\mathcal{D}$ with $\mathcal{D}(T_0) \subset \mathcal{D}$.

The analytic operator in the Kato sense is our central interest in Chaps. 11, 12, and 13. First, we recall the contributions of I. Feki, A. Jeribi, and R. Sfaxi in [20, 21] have tried to formulate some supplements to the results developed in [5] and [34], and they have investigated under sufficient conditions assuring the completeness of the system of eigenvectors of the operator $T(\varepsilon)$. Furthermore, they have extended the main result in [5] to the existence of Schauder basis in any Banach space.

The asymptotic behavior of the eigenvalues of self-adjoint differential operators and their perturbations has long been studied by many authors. Apparently, the first important results for ordinary differential operators were obtained by G. D. Birkhoff [6] and, for partial differential operators, by H. Weyl [38] (see the book by M. A. Naimark [35] for more details). M. V. Keldysh [29] proved the first results for general operators in a Hilbert space using the Tauberian technique. The most complete and the sharpest results for compact perturbations and for the so-called p-subordinate perturbations of self-adjoint operators are due to A. S. Markus and V. I. Matsaev [33]. Later, A. A. Shkalikov [37] obtained an estimate about the distribution of the eigenvalues of G in terms of one of the T. As a tentative approach to grapple with such difficulty, the authors in [15] have the idea to extend the outcomes obtained by A. A. Shkalikov [37] to an abstract analytic operator introduced by B. Sz. Nagy in [34] and considered later in [5, 12, 13, 14, 16, 17, 19, 21] and [20] (see also [9, 10, 25, 27]). Based on some ideas of A. A. Shkalikov [37], the authors in [15] derive a precise description of the localization of the eigenvalues of the analytic operator in the Kato sense.

Recently, H. Ellouz, I. Feki, and A. Jeribi in [18] continue the analysis started in [20], and they focus on the property of bases with parentheses of the analytic operator acting on a separable Hilbert space.

In fact, the authors derive a precise description to the localization of the spectrum of the perturbed operator and they prove an asymptotic relation that allows them to control the jump of the eigenvalue-counting function of some analytic operators where the criteria of A. A. Shkalikov [37] cannot be applied.

Finally, the three last chapters concentrate on a selection of applications to a perturbation method for sound radiation by a vibrating plate in a light fluid, Gribov operator in Bargmann space, and also applications in mathematical physics and mechanics to which the results of the preceding chapters are applied. In Chap. 14, we give some spectral properties about operators governing the perturbation method for sound radiation by a vibrating plate in a light fluid such as compactness and completeness results, existence of Riesz basis, or basis with parentheses. In Chap. 15, we treat the Gribov operator in Bargmann space in the cases of finite and infinite sum on null transverse dimension, and we confirm the existence of Riesz basis of subspaces, Schauder basis, and basis with parentheses. The last chapter contains some applications in mathematical physics and mechanics to investigate the expansion of solution in terms of generalized eigenvectors for a rectilinear transport equation and the behavior of the resolvent in the case of the Lamé system.

References

1. M.T. Aimar, A. Intissar, J.M. Paoli, Densité des vecteurs propres généralisés d'une classe d'opérateurs compacts non auto-adjoints et applications. Comm. Math. Phys. **156**, 169–177 (1993)
2. M.T. Aimar, A. Intissar, J.M. Paoli, Quelques nouvelles propriétés de régularité de l'opérateur de Gribov. Comm. Math. Phys. **172**(3), 461–466 (1995)

3. M.T. Aimar, A. Intissar, J.M. Paoli, Critères de complétude des vecteurs propres généralisés d'une classe d'opérateurs non auto-adjoints compacts ou à résolvante compacte et applications. Publ. Res. Inst. Math. Sci. **32**(2), 191–205 (1996)
4. M.T. Aimar, A. Intissar, A. Jeribi, On an unconditional basis of generalized eigenvectors of the non-self-adjoint Gribov operator in Bargmann space, J. Math. Anal. Appl. **231**, 588–602 (1999)
5. N. Ben Ali, A. Jeribi, On the Riesz basis of a family of analytic operators in the sens of Kato and application to the problem of radiation of a vibriting structure in a light fluid. J. Math. Anal. Appl. **320**(1), 78–94 (2006)
6. G.D. Birkhoff, Boundary Value and Expansion Problems of Ordinary Linear Differential Equations. Trans. Am. Math. Soc. **9**, 373–395 (1908)
7. H. Brezis, Analyse fonctionnelle théorie et applications. Ed. Masson, (1983)
8. S. Charfi, A. Damergi, A. Jeribi, On a Riesz basis of finite-dimensional invariant subspaces and application to Gribov operator in Bargmann space. Line. Mult. Alge. **61**(11), 1577–1591 (2013)
9. S. Charfi, A. Intissar, A. Jeribi, Expansion of solution in terms of generalized eigenvectors for a rectilinear transport equation. Transport Theory Statist. Phys. **38**, 1–16 (2009)
10. S. Charfi, A. Jeribi, I. Walha, Riesz basis property of families of nonharmonic exponential and application to a problem of a radiation of a vibrating structure in a light fluid. Numerical Functional Analysis and Optimization, **32**(4), 370–382 (2011)
11. C. Clark, On relatively bounded perturbations of ordinary differential operators, Pacific J. Math. **25**, 59–70 (1968)
12. J. B. Conway, *A Course in Functional Analysis*, 2nd edn. (Springer, New-York, 1990)
13. N. Dunford, J.T. Schwartz, Linear operators, Part III, (Wiley-Interscience, New-York, 1971)
14. H. Ellouz, I. Feki, A. Jeribi, On a Riesz basis of exponentials related to the eigenvalues of an analytic operator and application to a non-self-adjoint problem deduced from a perturbation method. J. Math. Phys. **54**, 112101 (2013)
15. H. Ellouz, I. Feki, A. Jeribi, On the asymptotic behavior of the eigenvalues of an analytic operator in the sense of Kato and applications. Serdica Math. J. **45**(1), 55–88 (2019)
16. H. Ellouz, I. Feki, A. Jeribi, On a Riesz basis of exponentials related to a family of analytic operators and application. J. Pseudo-Differ. Oper. Appl. **10**(4), 999–1014 (2019)
17. H. Ellouz, I. Feki, A. Jeribi, Non-orthogonal Fusion Frames of an Analytic Operator and Application to a One-Dimensional Wave Control System. Mediterr. J. Math. **16**(2), 16:52 (2019)
18. H. Ellouz, I. Feki, A. Jeribi, On the basis property of root vectors related to a non-self-adjoint analytic operator and applications. (2020)
19. I. Feki, A. Jeribi, R. Sfaxi, On an unconditionl basis of generalized eigenvectors of an analytic operator and application to a problem of radiation of a vibrating structure in a light fluid. J. Math. Anal. Appl. **375**, 261–269 (2011)
20. I. Feki, A. Jeribi, R. Sfaxi, On a Schauder basis related to the eigenvectors of a family of non-self-adjoint analytic operators and applications. Anal. Math. Phys. **3**(4), 311–331 (2013)
21. I. Feki, A. Jeribi, R. Sfaxi, On a Riesz basis of eigenvectors of a non-self-adjoint analytic operator and applications. Line. Mult. Alge. **62**(8), 1049–1068 (2014)
22. I. Feki, A. Jeribi, R. Sfaxi, On a Riesz basis of eigenvectors of a non-self-adjoint perturbation and application to a Gribov operator in Bargmann space. preprint (2021)
23. A. Haar, Zur theorie der orthogonalen Funktionensysteme, Math. Ann. **69**(3), 331–371 (1910)
24. G. Hamel, Eine basis aller zahlen und die unstetigen losungen der funktionalgleichung $f(x+y)=f(x)+f(y)$, Math. Ann. **60**(3), 459–462 (1905)
25. A. Intissar, A. Jeribi, I. Walha, Riesz basis of exponential family for a hyperbolic system. J. Appl. Anal. **25**(1), 13–23 (2019)
26. A.Jeribi, *Denseness, Bases and Frames in Banach Spaces and Applications*, (de Gruyter, Berlin 2018)
27. A. Jeribi, A. Intissar, On an Riesz basis of generalized eigenvectors of the nonself-adjoint problem deduced from a perturbation method for sound radiation by a vibrating plate in a light fluid. J. Math. Anal. Appl. **292**, 1–16 (2004)

28. T. Kato, *Perturbation Theory for Linear Operators*, (Springer, Berlin, 1980)
29. M.V. Keldysh, On eigenvalues and eigenfunctions of some classes of non-self-adjoint equations. Dokl. Akad. Nauk SSSR, **77**, 11–14 (In Russian) (1951)
30. Yu.I. Lyubich, N.K. Nikol'skij, I. Tweddle, Functional analysis I: linear functional analysis, Encycl. Math. Sci. **19**, (Springer 1992)
31. A.S. Markus, Introduction to the spectral theory of polynomial operator pencils, Amer. Math. Soci. Providence, RI, **71** (1988)
32. A.S. Markus, V.I. Matsaev, On the convergence of eigenvector expansions for an operator which is close to being self-adjoint, Mat. Issled. (in Russian) **61**, 104–129 (1981)
33. A.S. Markus, V.I. Matsaev, Comparison Theorems for Spectra of Linear Operators, and spectral asymptotics, Tr. Mosk. Mat. Obshch. **45**, 133–181 (1982) [Trans. Moscow Math. Soc. (1), 139–187 (1984)]
34. B.Sz. Nagy, Perturbations des transformations linéaires fermées, Acta Sci. Math. Szeged. **14**, 125–137 (1951)
35. M.A. Naimark, *Linear Differential Operators*, Nauka, Moscow, (Russian) (1969)
36. J. M. Paoli, Conditions suffisantes de densité du sous-espace engendré par les vecteurs propres généralisés d'un opérateur compact ou à résolvante compacte sur un espace de Hilbert et applications aux opérateurs de Gribov, Thèse de l'université de Corse, France, (1993)
37. A. Shkalikov, On the basis property of root vectors of a perturbed self-adjoint operator, Proc. Steklov Inst. Math. **269**, 284–298 (2010)
38. H. Weyl, Das asymptotische verteilungsgesetz der eigenwerte linearer partieller differentialgleichungen. Math. Ann. **71**, 441–479 (1912)
39. C. Wyss, Riesz bases for p-subordinate perturbations of normal operator. J. Func. Anal. **258**, 208–240 (2010)
40. G. Q. Xu, S. H. Wang, The completeness of systems of generalized eigenfunctions of generators of Riesz semigroups (Chinese). Acta Math. Sinica, **39**(2) 263–267 (1996)

Contents

About the Author

Aref Jeribi is Professor at the Department of Mathematics, University of Sfax, Tunisia. He completed his Habilitation of Mathematics in 2002 at the University of Sfax, Tunisia, and defended his PhD thesis in 1998 at the University of Corsica Pasquale Paoli, France. His research interests include spectral theory, matrix operators, transport theory, Gribov operator, Bargmann space, fixed point theory, Riesz basis, and linear relations.

He is the author of the book *Spectral Theory and Applications of Linear Operators and Block Operator Matrices* (SpringerVerlag, New-York, 2015), co-author of the book *Nonlinear Functional Analysis in Banach Spaces and Banach Algebras: Fixed Point Theory under Weak Topology for Nonlinear Operators and Block Operator Matrices with Applications* (Taylor-Francis, 2015), the author of the book *Denseness, Bases, and Frames in Banach Spaces and Applications* (De Gruyter, Berlin, 2018), the author of the book *Linear Operators and Their Essential Pseudospectra* (Apple Academic Press, CRC Press, Oakville, Boca Raton, 2018), the coauthor of the book *Analyse Numérique Matricielle, Méthodes et Algorithmes, Exercices et Problèmes Corrigés* (Références Sciences. Paris: Ellipses, 2020), the coauthor of the book *Spectral Theory of Multivalued Linear Operators* (Apple Academic Press, CRC Press, Oakville, Boca Raton, 2021), and the author of the book *Résolution de divers problèmes elliptiques par des méthodes d'éléments finis : cours, exercices, et problèmes corrigés* (Références Sciences. Paris: Ellipses, 2021), and 160 research articles published in reputed journals and conference proceedings.

Symbols Description

The most frequently used notations, symbols, and abbreviations are listed below.

$\mathbb{R}, \mathbb{C}$	The fields of real and complex numbers, respectively.
$\mathbb{N}$	The set of natural numbers.
$\mathbb{R}_{\geq 0}$	The set of nonnegative real numbers.
X, Y	Banach spaces.
$\mathscr{C}_{p,q}(X)$	The Schatten-von Neumann ideal of type (p, q).
$\mathscr{C}_p(X)$	The set of operators of Carleman-class.
$\inf(A)$	The infimum of the set A.
$\sup(A)$	The supremum of the set A.
$\chi_A(\cdot)$	The characteristic function of the set A.
$N^{\perp}$	The set of all annihilators of the set N.
${}^{\perp}R$	The set of such annihilators of R.
$\Omega(\varphi_1, \varphi_2)$	The sectors in the complex plane lying between the rays with arguments φ_1, φ_2.
$\mathbb{B}(0, 1)$	The open ball with radius 1 around 0 in X.
$\mathbb{S}_X$	The unit sphere in X.
$\mathbb{D}(\lambda_0, \varepsilon)$	The open disc with radius ε around λ_0 in $\mathbb{C}$.
$\mathscr{D}(A)$	The domain of A.
A^*	The adjoint of A.
A^{α}	The fractional power of the operator A.
$det(I + A)$	The determinant of $I + A$.
$d(\lambda)$	The Fredholm determinant associated to A.
$N(A)$	The null space of the operator A.
$R(A)$	The range of the operator A.
$\alpha(A)$	The dimension of $N(A)$.
$\beta(A)$	The codimension of $R(A)$ in Y.
$G(A)$	The graph of the operator A.
$\mathrm{asc}(A)$	The ascent of the operator A.
$\mathrm{desc}(A)$	The descent of the operator A.
$\rho(A)$	The resolvent set of the operator A.
$\sigma(A)$	The spectrum of the operator A.

$\sigma_p(A)$	The point spectrum of the operator A.
$\sigma_c(A)$	The continuous spectrum of the operator A.
$\sigma_r(A)$	The residual spectrum of the operator A.
$\sigma_j(A)$	The Jeribi essential spectrum of the operator A.
$m_g(\lambda)$	The multiplicity or geometric multiplicity of the eigenvalue λ.
$m_a(\lambda)$	The algebraic multiplicity of λ.
$R(\lambda, A)$ or $(\lambda - A)^{-1}$	The resolvent of A at the point λ.
$W(A)$	The numerical range of A.
$r(A)$	The spectral radius of A.
$a_n(A)$	The nth approximation number of A.
$s_n(A)$	The s-numbers of the operator A.
$n(\Delta, A)$	The sum of multiplicities of all eigenvalues of A contained in Δ.
$\mathscr{L}_A(\lambda)$	The root subspace of A corresponding to λ.
$\text{dist}(x, y)$	The distance between x and y.
X^*, X'	The dual of X.
$\text{dist}(\lambda, \sigma(A))$	The distance between λ and $\sigma(A)$.
$\dim(X)$	The dimension of the space X.
$\|x\|$	The norm of the vector x.
$\|A\|$	The norm of the operator A.
$(X, \|\cdot\|)$	The linear normed space X.
$(x_n)_n$	The sequence $(x_n)_n$.
$G(T)$	The graph of T.
L_p	The L_p space.
$\mathscr{L}(X, Y)$	The space of bounded linear operators from X into Y.
$\mathscr{L}(X)$	The space of bounded linear operators from X into X.
$\mathscr{C}(X, Y)$	The set of all closed, densely defined linear operators from X into Y.
$\mathscr{C}(X)$	The set of all closed, densely defined linear operators from X into X.
$\mathscr{K}(X, Y)$	The set of compact linear operators from X into Y.
$\mathscr{K}(X)$	The ideal of compact linear operators on X.
$F(X)$	The set of finite rank operators on X.
$\mathscr{W}(X, Y)$	The set of weakly compact linear operators from X into Y.
$\mathscr{W}(X)$	The ideal of weakly compact linear operators on X.
$\mathscr{S}(X, Y)$	The set of strictly singular linear operators from X into Y.
$\mathscr{S}(X)$	The ideal of strictly singular linear operators on X.
$\Phi(X, Y)$	The set of Fredholm operators from X into Y.
Φ_A	The Fredholm set of the operator A.
$C[a, b]$	The Banach space of all continuous functions on the closed finite interval $[a, b]$.
$x_n \to x$	x_n converges strongly to x.
$\overline{M}$	The closure of the set M.

Chapter 1
Basic Notations and Results

We begin this chapter by presenting the background material that is needed in our work. Most of this background is drawn from functional analysis and operator theory by giving some definitions, notations, and basic information on functional analysis that underlies most of the concepts presented in this book. The aim of this chapter is to introduce the basic concepts, notations, and elementary results which are used throughout the book.

1.1 Spaces and Operators

1.1.1 Vector and Normed Spaces

We define a linear (or vector) space V as a set $\{x, y, \dots\}$ which is equipped with the following operations:

- summation: $(x, y) \longrightarrow x + y \in V$,
- multiplication by a complex or real number α: $(\alpha, x) \longrightarrow \alpha x \in V$.

These operations satisfy

(i) the summation is commutative, $x + y = y + x$, and associative, $(x + y) + z = x + (y + z)$. Further, there is a zero element $0 \in V$, and an inverse element $-x \in V$, to any $x \in V$ so that $x + 0 = x$ and $x + (-x) = 0$ holds for all $x \in V$,

(ii) $\alpha(\beta x) = (\alpha\beta)x$ and $1x = x$, and

(iii) the summation and multiplication are distributive, $\alpha(x + y) = \alpha x + \alpha y$ and $(\alpha + \beta)x = \alpha x + \beta x$.

A. Jeribi, *Perturbation Theory for Linear Operators*,
https://doi.org/10.1007/978-981-16-2528-2_1

The vectors are the elements of V. Vectors $x_1, \dots, x_n \in V$ are said linearly independent if $\alpha_1 x_1 + \cdots + \alpha_n x_n = 0$ implies that all the numbers $\alpha_1, \dots, \alpha_n$ are zero; otherwise, they are linearly dependent, which means some of them can be expressed as a linear combination of the others. When we replace the set of scalars (or numbers) in the above definition by any algebraic field F, we speak about a vector space over F. In particular, we have a complex and real vector space, respectively, for $F = \mathbb{C}$, $\mathbb{R}$.

A subspace $L \subset V$ is defined as a subset, which is itself a vector space with the same operations. We say that a set $M \subset V$ is linearly independent if all of its finite subsets consist of linearly independent vectors. A map

$$f : V \longrightarrow \mathbb{C}$$

on a vector space V is called a functional; if it maps into the reals we speak about a real functional. A functional $f : V \longrightarrow \mathbb{C}$ is additive if the property

$$f(x + y) = f(x) + f(y)$$

holds for each $x, y \in V$, and homogeneous if

$$f(\alpha x) = \alpha f(x)$$

or antihomogeneous if

$$f(\alpha x) = \overline{\alpha} f(x)$$

for $x \in V, \alpha \in \mathbb{C}$. An additive (anti)homogeneous functional is called (anti)linear.

A real functional p satisfy

$$p(x + y) \leq p(x) + p(y),$$

and

$$p(\alpha x) = |\alpha| p(x)$$

for all $x, y \in V, \alpha \in \mathbb{C}$, is called a seminorm. A map

$$F : V \times \cdots \times V \longrightarrow \mathbb{C}$$

is said a form. In particular, if the range of F is contained in $\mathbb{R}$, the form is real.

We say that a form $F : V \times V \longrightarrow \mathbb{C}$ is bilinear if it is linear in both arguments, and sesquilinear if it is linear in one of them and antilinear in the other. A norm on a vector space V is a seminorm $\|\cdot\|$ such that $\|x\| = 0$ holds iff $x = 0$. We define a normed space as a pair $(V, \|\cdot\|)$. Without loss of generality, we shall speak simply about a normed space V.

An inner (or scalar) product is a strictly positive sesquilinear form on a vector space V. More precisely, it is a map $\langle \cdot, \cdot \rangle$ from $V \times V$ into $\mathbb{C}$ such that the following conditions hold for any $x, y, z \in V$ and $\alpha \in \mathbb{C}$:

(i) $\langle x, \alpha y + z \rangle = \alpha \langle x, y \rangle + \langle x, z \rangle$,
(ii) $\langle x, y \rangle = \langle y, x \rangle$, and
(iii) $\langle x, x \rangle \geq 0$ and $\langle x, x \rangle = 0$ iff $x = 0$.

A linear space with an inner product is said a pre-Hilbert space. Any such space, with the norm

$$\|x\| := \sqrt{\langle x, x \rangle},$$

is at the same time a normed space. A metric on a set X is a map

$$\rho : X \times X \longrightarrow [0, \infty),$$

which is symmetric, $\rho(x, y) = \rho(y, x)$, $\rho(x, y) = 0$ iff $x = y$, and satisfies the following triangle inequality,

$$\rho(x, z) \leq \rho(x, y) + \rho(y, z),$$

for all x, y, and $z \in X$. The pair (X, ρ) is said a metric space and, for more simplicity, we often use the symbol X only. If X is a normed space, then we define a metric on X by

$$\rho(x, y) := \|x - y\|,$$

and we say that is induced by the norm.

Now, let's consider a set X and τ be a family of its subsets which satify the following, topology axioms, conditions:

(i) $X \in \tau$ and $\emptyset \in \tau$,
(ii) if I is each index set and $G_\alpha \in \tau$ for each $\alpha \in I$, then $\bigcup_{\alpha \in I} G_\alpha \in \tau$, and
(iii) $\bigcap_{j=1}^{n} G_j \in \tau$ for each finite subsystem $\{G_1, \ldots, G_n\} \in \tau$.

We say that the family τ is a topology, its elements are open sets, and the set X equipped with a topology is a topological space; when it is suitable we write (X, τ). By definition a family of open sets in a metric space (X, τ) is a topology; we speak about the metric-induced topology τ_ρ, in particular, the norm-induced topology if X is a vector space and ρ is induced by a norm. A Hilbert space is a pre-Hilbert space which is complete with respect to the metric induced by the inner product.

Definition 1.1.1 Let X be a pre-Hilbert space and $\{e_n\}_{n=1}^{\infty}$ be a sequence of mutually orthogonal vectors (i.e., $\langle e_i, e_j \rangle = 0$ for $i \neq j$) of X. The linear span of $\{e_n\}_{n=1}^{\infty}$ is the smallest linear subspace of X (the set of finite linear combinations) containing $\{e_n\}_{n=1}^{\infty}$. ◇

1.1.2 Operators on Quasi-Banach Spaces

By a quasi-norm $\|\cdot\|$ defined on a linear space X we mean a real-valued function with the following properties:

(i) Let $x \in X$. Then, $\|x\| = 0$ if, and only if, $x = 0$.
(ii) $\|\lambda x\| = |\lambda|\|x\|$ for all $x \in X$ and $\lambda \in \mathbb{C}$.
(iii) $\|x + y\| \leq c_X(\|x\| + \|y\|)$ for any $x, y \in X$. Here $c_X > 1$ is a constant.

The assertion (iii) is called the quasi-triangle inequality. If c_X is equal to 1, then $\|\cdot\|$ is said to be a norm. In this case, the assertion (iii) passes into the well-known triangle inequality.

It is easy to see that every quasi-norm $\|\cdot\|$ given on a linear space X induces a metrizable topology such that the algebraic operations are continuous. We recall that a fundamental system of neighborhoods of the zero element is formed by the subsets $\varepsilon\overline{\mathbb{B}}(0, 1)$ with $\varepsilon > 0$, where $\overline{\mathbb{B}}(0, 1)$ is the closure of the open ball with radius 1 around 0 in the linear space X, cf.

$$\mathbb{B}(0, 1) := \{x \in X \text{ such that } \|x\| < 1\}.$$

If a linear space X equipped with a quasi-norm $\|\cdot\|$ which becomes complete with respect to the associated metrizable topology, then X is said a quasi-Banach space. This means that all or any Cauchy sequences are convergent. In the most important case, when $\|\cdot\|$ is a norm on X we call X a Banach space.

1.1.3 Closed and Closable Operators

Let X and Y be two Banach spaces. A transformation (or operator) is a mapping A which assigns to every element x of a set $\mathcal{D}(A) \subset X$ a unique element $y \in Y$, we use, in the sequel, the notation $A(X \longrightarrow Y)$. The set $\mathcal{D}(A)$ on which A acts is said the domain of the operator A. The operator A is called linear, if it satisfies the following

(i) $\mathcal{D}(A)$ is a subspace of X, and
(ii) $A(\lambda x + \beta y) = \lambda Ax + \beta Ay$ for each scalars λ, β, and all elements x, y in $\mathcal{D}(A)$.

The graph $G(A)$, in the product space $X \times Y$, of a linear operator A on a domain $\mathcal{D}(A) \subset X$ into Y is the following set

$$\{(x, Ax) \text{ such that } x \in \mathcal{D}(A)\}.$$

If $\|Ax\| = \|x\|$ for all $x \in X$, then the operator A is called a norm-preserving or isometric. A linear, bijective, and isometric operator defines an isometric isomorphism. The operator A is called bounded if there is a constant M such that

$$\|Ax\| \leq M\|x\|, \quad x \in X.$$

The norm, $\|A\|$, of such operator is given by

$$\|A\| = \sup_{x \neq 0} \frac{\|Ax\|}{\|x\|}.$$

The linear operator A is called closed if its graph $G(A)$ constitutes a closed linear subspace of $X \times Y$. Hence, the notion of a closed linear operator is a natural extension of the notion of a bounded linear operator. We denote by $\mathscr{C}(X, Y)$ (respectively, $\mathscr{L}(X, Y)$), the set of all closed, densely defined linear operators (respectively, the Banach space of all bounded linear operators) from X into Y. If $X = Y$, the sets $\mathscr{L}(X, Y)$ and $\mathscr{C}(X, Y)$ are replaced, respectively, by $\mathscr{L}(X)$ and $\mathscr{C}(X)$. By an operator A from X into Y, we mean a linear operator with a domain $\mathscr{D}(A) \subset X$. We denote by $N(A)$ (respectively, $R(A)$) it is null space (respectively, range). For injective A, the inverse $A^{-1}(Y \longrightarrow X)$ is an operator with $\mathscr{D}(A^{-1}) = R(A)$ and $R(A^{-1}) = \mathscr{D}(A)$. A subspace $U \subset X$ is called A-invariant if $x \in U \bigcap \mathscr{D}(A)$ implies $Ax \in U$. If we speak of a finite-dimensional A-invariant subspace U, we additionally assume that $\dim U < \infty$ and $U \subset \mathscr{D}(A)$. We say that a subspace $D \subset \mathscr{D}(A)$ is a core for A if for each $x \in \mathscr{D}(A)$, there is $(x_n)_n$ in D such that $x_n \to x$ and

$$Ax_n \to Ax.$$

If A is closed, then a subset $D \subset \mathscr{D}(A)$ is said a core for the operator A if $\overline{A|_D} = A$. A closable operator A is a linear operator on $\mathscr{D}(A) \subset X$ into Y which has a closed extension. It is equivalent to the condition that the graph $G(A)$ is a submanifold (or subspace) of a closed linear manifold (or space) which is at the same time a graph. It follows that A is closable if, and only if, the closure $\overline{G(A)}$ of $G(A)$ is a graph. This leads to the criterion: A is closable if, and only if, no element of the form $(0, x)$, $x \neq 0$ is the limit of elements of the form (x, Ax). In other words, A is closable if, and only if, the sequence $(x_n)_n \in \mathscr{D}(A)$, $x_n \to 0$ and $Ax_n \to x$ imply $x = 0$. If the operator A is closable, then there exists a closed operator $\overline{A}$ with

$$G(\overline{A}) = \overline{G(A)}.$$

$\overline{A}$ is called the closure of A. Immediately, we conclude that any closed extension of the operator A is also an extension of the operator $\overline{A}$ that $\overline{A}$ is the smallest closed extension of the operator A. A sequence $(x_n)_n \subset \mathscr{D}(A)$ will be said A-convergent to $x \in X$ if both $(x_n)_n$ and $(Ax_n)_n$ are Cauchy sequences and

$$x_n \to x.$$

The fact that $x \in \mathscr{D}(\overline{A})$ is equivalent to the element $(x, \overline{A}x) \in \overline{G(A)}$, $x \in X$ belongs to $\mathscr{D}(\overline{A})$ if, and only if, there is a sequence $(x_n)_n$ that is A-convergent to x. In this case, we have

$$\overline{A}x = \lim_{n\to\infty} Ax_n.$$

1.1.4 Adjoint Operator

Let X and Y be two Hilbert spaces and $T \in \mathscr{L}(X, Y)$. For each $\varphi \in Y^*$ and $\psi \in X$, the scalar product $\langle \psi, T\varphi \rangle$ is defined and is a semilinear form in φ. Hence, it can be written by Riesz lemma as

$$f[\varphi] = \langle f, \varphi \rangle$$

with an $f \in X^*$. Since f is determined by g, then we can define a function T^* on Y^* into X^* by setting $f = T^*\psi$. So, the defining equation of T^* is

$$\langle T^*\psi, \varphi \rangle = \langle \psi, T\varphi \rangle \text{ for all } \psi \in Y^*, \ \varphi \in X.$$

T^* as defined is a linear operator on Y^* into X^*. Thus, $T^* \in \mathscr{L}(Y^*, X^*)$. Therefore, T^* is called the adjoint operator of T.

For the adjoint of unbounded operator in Hilbert spaces, consider an operator T from X into Y and an operator S from Y^* into X^*. The two operators T and S are said to be adjoint to each other if

$$\langle \psi, T\varphi \rangle = \langle S\psi, \varphi \rangle \text{ for all } \varphi \in \mathscr{D}(T), \ \psi \in \mathscr{D}(S).$$

For each operator T from X into Y, there are, in general, many operators from Y^* into X^* that are adjoint to T. If T is densely defined, then there is a unique maximal operator T^* adjoint to T. This means that T^* is adjoint to T while any other S adjoint to T is a restriction of T^*. Thus, T^* is called the adjoint operator of T. The operator T^* is constructed in the following way. Its domain, $\mathscr{D}(T^*)$, consists of all $\psi \in Y^*$ such that there exists an $f \in X^*$ with the property

$$\langle \psi, T\varphi \rangle = \langle f, \varphi \rangle \text{ for all } \varphi \in \mathscr{D}(T).$$

The $f \in X^*$ is determined uniquely by ψ. Therefore, an operator T^* from Y^* into X^* is defined by setting

$$T^*\psi = f.$$

Let X and Y be two Banach spaces and $A \in \mathscr{L}(X, Y)$. For each $y' \in Y^*$ (the adjoint space of Y), the expression $y'(Ax)$ assigns a scalar to each $x \in X$. Thus, it is a functional $F(x)$. Clearly, F is linear. It is also bounded since

$$\begin{aligned} |F(x)| &= |y'(Ax)| \\ &\leq \|y'\| \|Ax\| \\ &\leq \|y'\| \|A\| \|x\|. \end{aligned}$$

Thus, there is an $x' \in X^*$ (the adjoint space of X) such that

$$y'(Ax) = x'(x), \quad x \in X. \tag{1.1.1}$$

This functional is unique, for any other functional satisfying (1.1.1) would have to coincide with x' on each $x \in X$. Thus, (1.1.1) can be written in the form

$$y'(Ax) = A^* y'(x). \tag{1.1.2}$$

The operator A^* is called the adjoint of A, depending on the mood one is in. If A, B are bounded operators defined everywhere $\mathscr{D}(A) = \mathscr{D}(B) = X$, it is easily checked that

$$(BA)^* = A^* B^*.$$

We just follow the definition of adjoint for bounded operator (see (1.1.2)) for defining the adjoint of unbounded operator. By an operator A from X into Y, we mean a linear operator with a domain $\mathscr{D}(A) \subset X$. We want

$$A^* y'(x) = y'(Ax), \quad x \in \mathscr{D}(A).$$

So, we call that $y' \in \mathscr{D}(A^*)$ if there is an $x' \in X^*$ such that

$$x'(x) = y'(Ax), \; x \in \mathscr{D}(A). \tag{1.1.3}$$

Hence, we define $A^* y'$ to be x'. In order that this definition makes sense, we need x' to be unique, i.e., that $x'(x) = 0$ for all $x \in \mathscr{D}(A)$ should imply that $x' = 0$. Thus, it is true if $\mathscr{D}(A)$ is dense in X. To summarize, we can define A^* for any linear operator from X into Y provided $\mathscr{D}(A)$ is dense in X. We take $\mathscr{D}(A^*)$ to be the set of those $y' \in Y^*$ for which there is an $x' \in X^*$ satisfying (1.1.3). This x' is unique, and we set $A^* y' = x'$. If A, B are only densely defined, $\mathscr{D}(AB)$ need not be dense, and consequently $(BA)^*$ need not exist. If $\mathscr{D}(BA)$ is dense, it follows that $\mathscr{D}(A^* B^*) \subset \mathscr{D}((BA)^*)$ and $(BA)^* z' = A^* B^* z'$, for all $z' \in \mathscr{D}(A^* B^*)$. Let N be a subset of a normed vector space X. A functional $x' \in X^*$ is called an annihilator of N if $x'(x) = 0$, $x \in N$. The set of all annihilators of N is denoted by $N^\perp$. The subspace $N^\perp$ is closed. For any subset R of X^*, we call $x \in X$ an annihilator of R if $x'(x) = 0$, $x' \in R$. We denote the set of such annihilators of R by ${}^\perp R$. The subspace ${}^\perp R$ is closed. If M is a closed subspace of X, then

$${}^\perp(M^\perp) = M.$$

Definition 1.1.2 (i) A densely defined operator A on a Hilbert space is called symmetric, if we have the inclusion $A \subset A^*$, that is, if $\mathscr{D}(A) \subset \mathscr{D}(A^*)$ and $A\varphi = A^*\varphi$ for all $\varphi \in \mathscr{D}(A)$. It is equivalent to A is symmetric if, and only if,

$$\langle A\varphi, \psi \rangle = \langle \varphi, A\psi \rangle \text{ for all } \varphi, \psi \in \mathscr{D}(A).$$

(ii) A is called self-adjoint if $A^* = A$, that is, if, and only if, A is symmetric and $\mathscr{D}(A) = \mathscr{D}(A^*)$.
(iii) A symmetric operator A is said essentially self-adjoint if its closure $\overline{A}$ is self-adjoint.
(iv) A is called skew-symmetric if $\langle A\varphi, \psi\rangle = -\langle \varphi, A\psi\rangle$ for $\varphi, \psi \in \mathscr{D}(A)$, and skew-adjoint if $A = -A^*$. ◇

Definition 1.1.3 Let $A \in \mathscr{L}(X)$ and $r(A) = \sup_{\lambda\in\sigma(A)} |\lambda|$. Then, $r(A)$ is called the spectral radius of A. ◇

Theorem 1.1.1 *Let A be a bounded linear operator on a Banach space X. Then, $\lim_{n\to\infty} \|A^n\|^{\frac{1}{n}}$ exists and*

$$\lim_{n\to\infty} \|A^n\|^{\frac{1}{n}} = r(A).$$

If X is a Hilbert space and A is a self-adjoint operator, then $r(A) = \|A\|$. ◇

1.1.5 *Fredholm Operators*

Let X and Y be two Banach spaces and $A \in \mathscr{C}(X, Y)$. The nullity, $\alpha(A)$, of A is defined as the dimension of $N(A)$ and the deficiency, $\beta(A)$, of A is defined as the codimension of $R(A)$ in Y. An operator $A \in \mathscr{C}(X, Y)$ is called to be a Fredholm operator from X into Y if

(i) $\alpha(A) = \dim N(A)$ is finite,
(ii) $R(A)$ is closed in Y, and
(iii) $\beta(A) := \text{codim} R(A)$ is finite.

The set of Fredholm operators from X into Y is defined by

$$\Phi(X, Y) = \{A \in \mathscr{C}(X, Y),\ \ \alpha(A) < \infty,\ \ R(A) \text{ is closed in } Y \text{ and } \beta(A) < \infty\}.$$

The set of all points λ for which $\lambda - A$ is a Fredholm operator is called the Fredholm set of A and denoted by Φ_A. As is well known, the Fredholm set of an operator A is open (see [60, 91]).

1.2 Some Notions of Spectral Theory

1.2.1 *Closed Graph Theorem*

We first recall the closed graph theorem.

Theorem 1.2.1 *If X, Y are Banach spaces, and A is a closed linear operator from X into Y with $\mathscr{D}(A) = X$, then*

(i) there are positive constants M, r such that $\|Ax\| \leq M$ whenever $\|x\| < r$, and
(ii) $A \in \mathscr{L}(X, Y)$. ◇

Theorem 1.2.2 *If X, Y are Banach spaces, and $A \in \mathscr{L}(X, Y)$ with $R(A) = Y$ and $N(A) = \{0\}$, then $A^{-1} \in \mathscr{L}(Y, X)$.* ◇

1.2.2 Resolvent Set and Spectrum

Definition 1.2.1 Let $A \in \mathscr{C}(X)$. The resolvent set and the spectrum of A are, respectively, defined as

$$\rho(A) = \{\lambda \in \mathbb{C},\ \ \lambda - A \text{ is injective and } R(\lambda, A) := (\lambda - A)^{-1} \in \mathscr{L}(X)\},$$
$$\sigma(A) = \mathbb{C} \backslash \rho(A)$$

and the point spectrum, continuous, and the residual spectrum are defined as

$$\sigma_p(A) = \{\lambda \in \mathbb{C} \text{ such that } \lambda - A \text{ is not injective}\},$$
$$\sigma_c(A) = \{\lambda \in \mathbb{C},\ \ \lambda - A \text{ is injective},\ \overline{R(\lambda - A)} = X,\ R(\lambda - A) \neq X\},$$
$$\sigma_r(A) = \{\lambda \in \mathbb{C} \text{ such that } \lambda - A \text{ is injective},\ \overline{R(\lambda - A)} \neq X\}.$$ ◇

Definition 1.2.2 For an operator A whose point spectrum is symmetric with respect to the imaginary axis and satisfies

$$\sigma_p(A) \bigcap i\mathbb{R} = \emptyset,$$

we say that a partition $\sigma_p(A) = \sigma \bigcup \tau$ separates skew-conjugate points if $\lambda \in \sigma$ if, and only if, $-\overline{\lambda} \in \tau$. ◇

Proposition 1.2.1 *Let A be a closed linear operator with domain $\mathscr{D}(A)$ on a Banach space X. Then,*

(i) $\rho(A)$ is an open subset of $\mathbb{C}$, and
(ii) if $\lambda \in \rho(A)$, then

$$dist(\lambda, \sigma(A)) \geq ||R(\lambda, A)||^{-1},$$

where $dist(\lambda, \sigma(A))$ denotes the distance between λ and $\sigma(A)$. ◇

Proposition 1.2.2 *Let $A \in \mathscr{C}(X)$. For all $\lambda, \mu \in \rho(A)$, we have*

$$(\lambda - A)^{-1} - (\mu - A)^{-1} = (\mu - \lambda)(\lambda - A)^{-1}(\mu - A)^{-1}.$$ ◇

We give spectral mapping theorem for the resolvent.

Theorem 1.2.3 *Let $A \in \mathscr{C}(X)$ and $\lambda \in \rho(A)$. Then,*

(i) $\sigma((\lambda - A)^{-1})\backslash\{0\} = \{(\lambda - \mu)^{-1}$ *such that* $\mu \in \sigma(A)\}$, *and*

(ii) $\sigma_p((\lambda - A)^{-1})\backslash\{0\} = \{(\lambda - \mu)^{-1}$ *such that* $\mu \in \sigma_p(A)\}$. ◇

Proof Let $\mu \in \mathbb{C}$, we have $\mu \in \rho(A)\backslash\{\lambda\}$ if, and only if, $(\lambda - \mu)^{-1} \in \rho((\lambda - A)^{-1})$ and

$$((\lambda - \mu)^{-1} - (\lambda - A)^{-1})^{-1} = (\mu - \lambda)(A - \lambda)(\mu - A)^{-1}. \qquad \text{Q.E.D.}$$

Proposition 1.2.3 *Let $A(X \longrightarrow X)$ be a self-adjoint operator on a Hilbert space X. Then, $\lambda \in \rho(A)$ if, and only if, $\overline{\lambda} \in \rho(A)$.* ◇

Proof Let $\lambda \in \rho(A)$, i.e., $\lambda - A$ is bijective with bounded inverse. Then, the adjoint

$$(\lambda - A)^* = \overline{\lambda} - A^* = \overline{\lambda} - A$$

is also bijective with bounded inverse. Q.E.D.

1.2.3 Bounded Operators

Theorem 1.2.4 *Let X be a reflexive Banach space and let $A \in \mathscr{L}(X)$, one has the alternative*

(i) $X = \overline{R(A) + N(A)}$ *or*

(ii) $\lim\limits_{|z|\to\infty} \|(I - zA)^{-1}\| = \infty$. ◇

Proof Suppose we did not

$$\lim_{|z|\to\infty} \|(I - zA)^{-1}\| = \infty.$$

Then, there is a sequence $(\alpha_n)_n$ in $\mathbb{C}$, with

$$\lim_{n\to\infty} |\alpha_n| = \infty \tag{1.2.1}$$

and $\|(I - \alpha_n A)^{-1}\|$ is bounded. Let $\varphi \in X$ and $\psi_n = (I - \alpha_n A)^{-1}\varphi$. The sequence $(\psi_n)_n$ is then bounded and admits a subsequence which converges weakly to some $\psi \in X$. Hence,

$$\begin{aligned}A\psi_n &= A(I-\alpha_n A)^{-1}\varphi\\ &= \frac{1}{\alpha_n}\alpha_n A(I-\alpha_n A)^{-1}\varphi\\ &= \frac{1}{\alpha_n}[I-(I-\alpha_n A)](I-\alpha_n A)^{-1}\varphi\\ &= \frac{1}{\alpha_n}[(I-\alpha_n A)^{-1}-I]\varphi. \end{aligned} \tag{1.2.2}$$

Since $\|(I-\alpha_n A)^{-1}\|$ is bounded, then it is the same for $\|(I-\alpha_n A)^{-1}-I\|$ and $\|[(I-\alpha_n A)^{-1}-I]\varphi\|$. By using both Eqs. (1.2.1) and (1.2.2), we have

$$\lim_{n\to\infty} A\psi_n = 0.$$

A is continuous so is, for the weak topology. We infer that $A\psi = 0$ and so, $\psi \in N(A)$. Moreover,

$$\begin{aligned}\varphi-\psi_n &= [I-(I-\alpha_n A)^{-1}]\varphi\\ &= -\alpha_n A(I-\alpha_n A)^{-1}\varphi.\end{aligned}$$

Hence, $\varphi-\psi_n \in R(A)$. So, $\varphi-\psi$ is, therefore, in the weak closure of $R(A)$ which, by Hahn-Banach theorem, is also in the strong closure. Thus, $\varphi-\psi \in \overline{R(A)}$ and

$$X = \overline{R(A)} + N(A).$$

The opposite implication is obvious. Q.E.D.

Theorem 1.2.5 *Let X be a reflexive Banach space and let A be a bounded linear operator on X, one has the alternative*

(i) $X = \overline{R(A)} + N(A)$ *or*

(ii) $\lim_{z\to 0} |z|\|(z-A)^{-1}\| = \infty.$ ◇

Proof For $\alpha \neq 0$ and $\frac{1}{\alpha} \in \rho(A)$, we can write

$$\begin{aligned}(I-\alpha A)^{-1} &= \left[\alpha\left(\frac{1}{\alpha}-A\right)\right]^{-1}\\ &= \frac{1}{\alpha}\left(\frac{1}{\alpha}-A\right)^{-1}.\end{aligned} \tag{1.2.3}$$

The result follows from both Eq. (1.2.3) with $z=\frac{1}{\alpha}$ and Theorem 1.2.4. Q.E.D.

1.2.4 Numerical Range

Let X be a complex Hilbert space and let $A \in \mathscr{L}(X)$. The numerical range of such operator A is the set defined by

$$W(A) := \{\langle Ax, x\rangle \ : \ x \in \mathbb{S}_X\},$$

where $\mathbb{S}_X$ is the unit sphere in X given by

$$\mathbb{S}_X := \{x \in X \text{ such that } \|x\| = 1\}.$$

In the beginning of this section, we give some properties of $W(A)$. In fact, we have

$$W(\alpha I + \beta A) = \alpha + \beta W(A) \text{ for } \alpha,\ \beta \in \mathbb{C},$$

$$W(A^*) = \{\overline{\lambda},\ \lambda \in W(A)\}, \text{ and}$$

$$W(U^* AU) = W(A), \text{ for any unitary } U.$$

Most of the results on the well-known Toeplitz-Hausdorff theorem, the numerical range is a convex subset of $\mathbb{C}$ and it satisfies the, so-called, spectral inclusion property

$$\sigma_p(A) \subset W(A) \text{ and } \sigma(A) \subset \overline{W(A)}.$$

Note that if $\dim X < \infty$, then $W(A)$ is closed. Further, the resolvent of the operator A can be estimated in terms of the distance to the numerical range $W(A)$,

$$\|(\lambda - A)^{-1}\| \leq \frac{1}{\operatorname{dist}(\lambda, W(A))}, \quad \lambda \notin \overline{W(A)}, \tag{1.2.4}$$

where $\operatorname{dist}(\lambda, W(A))$ is the distance between λ and the set $W(A)$. If an element $\lambda \in \overline{W(A)}$ is a corner of the numerical range, then $\lambda \in \sigma(A)$. If, in addition, λ is in the numerical range $W(A)$, then λ is in the point spectrum $\sigma_p(A)$. A consequence of the estimate (1.2.4) that if $\lambda \in \sigma_p(A)$ is a boundary point of the numerical range $W(A)$, then there are no associated vectors at λ.

It is useful to consider unbounded operators for many applications. Suppose now that $A \in \mathscr{L}(\mathscr{D}(A), X)$ is a closed linear operator with domain $\mathscr{D}(A)$ densely and continuously embedded in X. Its numerical range is then defined by

$$W(A) := \left\{\langle Ax, x\rangle \ : \ x \in \mathscr{D}(A) \bigcap \mathbb{S}_X\right\}.$$

1.3 Inequalities

In this section, we need to recall some classical inequalities that we will later make use of.

Lemma 1.3.1 (Young's inequality) *Given $a, b,$ and β positive numbers such that $\beta \in [0, 1]$, then*

$$a^{\beta} b^{1-\beta} \leq \beta a + (1 - \beta) b. \qquad \diamond$$

Lemma 1.3.2 (Holder's inequality) *For all $p, q > 1$ such that $\frac{1}{p} + \frac{1}{q} = 1$, and for all $(a_1, \ldots, a_n)$, $(b_1, \ldots, b_n) \in \mathbb{R}^n$, we have*

$$\sum_{k=1}^{n} |a_k||b_k| \leq \left(\sum_{k=1}^{n} |a_k|^p\right)^{\frac{1}{p}} \left(\sum_{k=1}^{n} |b_k|^q\right)^{\frac{1}{q}}. \qquad \diamond$$

Lemma 1.3.3 (Minkowski's inequality) *For all $p \geq 1$, and for all $(a_1, \ldots, a_n)$, $(b_1, \ldots, b_n) \in \mathbb{R}^n$, we have*

$$\left(\sum_{k=1}^{n} |a_k + b_k|^p\right)^{\frac{1}{p}} \leq \left(\sum_{k=1}^{n} |a_k|^p\right)^{\frac{1}{p}} + \left(\sum_{k=1}^{n} |b_k|^p\right)^{\frac{1}{p}}. \qquad \diamond$$

Lemma 1.3.4 (Interpolation inequality) *Given $a,\ b,\ \theta$ and δ positive numbers such that $0 \leq \delta \leq \theta$, then*

$$a^{\delta} b^{\theta-\delta} \leq a^{\theta} + b^{\theta}. \qquad \diamond$$

Let $(x_k)_k$ be a sequence in a Banach space X. We suppose that there exists $C \geq 0$ such that for every reordering $\phi : \mathbb{N} \stackrel{bij}{\longrightarrow} \mathbb{N}$ and every $n \in \mathbb{N}$, we have

$$\left\| \sum_{k=0}^{n} x_{\phi(k)} \right\| \leq C. \tag{1.3.1}$$

Let $\varepsilon_0, \ldots, \varepsilon_n \in \{-1, 1\}$ and consider reorderings ϕ_1 and ϕ_2 that move all $+1$ and all -1 in the sequence $(\varepsilon_0, \ldots, \varepsilon_n)$, respectively, to its beginning. Then, with n_1, n_2 appropriate, and by using (1.3.1), we obtain

$$\left\| \sum_{k=0}^{n} \varepsilon_k x_k \right\| \leq \left\| \sum_{k=0,\ \varepsilon_k=+1}^{n} x_k \right\| + \left\| \sum_{k=0,\ \varepsilon_k=-1}^{n} x_k \right\|.$$

Thus,

$$\begin{aligned} \left\| \sum_{k=0}^{n} \varepsilon_k x_k \right\| &\leq \left\| \sum_{k=0}^{n_1} x_{\phi_1(k)} \right\| + \left\| \sum_{k=0}^{n_2} x_{\phi_2(k)} \right\| \\ &\leq 2C. \end{aligned}$$

So,

$$\sup_{n\in\mathbb{N},\ \varepsilon_k=\pm 1} \left\| \sum_{k=0}^{n} \varepsilon_k x_k \right\| \le 2C. \tag{1.3.2}$$

Let

$$E = \{(\varepsilon_0, \dots, \varepsilon_n) \text{ such that } \varepsilon_k = \pm 1\}, \tag{1.3.3}$$

X be a Hilbert space, and $x_0, \dots, x_n \in X$. We claim that

$$2^{n+1} \sum_{k=0}^{n} \|x_k\|^2 = \sum_{\varepsilon\in E} \|\varepsilon_0 x_0 + \cdots + \varepsilon_n x_n\|^2. \tag{1.3.4}$$

We proceed by induction on n. Since

$$2\|x_0\|^2 = \|x_0\|^2 + \|-x_0\|^2,$$

then the statement is true for the case $n = 0$. Now, we suppose that the statement holds for some $n \ge 0$. Consider

$$\widetilde{E} = \{(\varepsilon_0, \dots, \varepsilon_{n+1}) \text{ such that } \varepsilon_k = \pm 1\}$$

and put

$$x_\varepsilon = \varepsilon_0 x_0 + \cdots + \varepsilon_n x_n.$$

Then,

$$\begin{aligned}\sum_{\varepsilon\in\widetilde{E}} \|\varepsilon_0 x_0 + \cdots + \varepsilon_{n+1} x_{n+1}\|^2 &= \sum_{\varepsilon\in E} (\|x_\varepsilon + x_{n+1}\|^2 + \|x_\varepsilon - x_{n+1}\|^2) \\ &= \sum_{\varepsilon\in E} (2\|x_\varepsilon\|^2 + 2\|x_{n+1}\|^2).\end{aligned}$$

Thus,

$$\begin{aligned}\sum_{\varepsilon\in\widetilde{E}} \|\varepsilon_0 x_0 + \cdots + \varepsilon_{n+1} x_{n+1}\|^2 &= 2\sum_{\varepsilon\in E} \|x_\varepsilon\|^2 + 2\cdot 2^{n+1}\|x_{n+1}\|^2 \\ &= 2^{n+2}\left(\sum_{k=0}^{n} \|x_k\|^2 + \|x_{n+1}\|^2\right),\end{aligned}$$

which prove the claim. Let $\lambda \in \mathbb{C}$. Consider the $n \times n$ Jordan block

$$A = \begin{pmatrix} \lambda & 1 & & \\ & \ddots & \ddots & \\ & & \ddots & 1 \\ & & & \lambda \end{pmatrix}. \tag{1.3.5}$$

The adjoint of the operator A is given by

$$A^* = \begin{pmatrix} \overline{\lambda} & & & \\ 1 & \ddots & & \\ & \ddots & \ddots & \\ & & 1 & \overline{\lambda} \end{pmatrix}.$$

Thus,

$$A^*A = \begin{pmatrix} |\lambda|^2 & \overline{\lambda} & & & \\ \lambda & |\lambda|^2+1 & \ddots & & \\ & \lambda & \ddots & \ddots & \\ & & \ddots & \ddots & \overline{\lambda} \\ & & & \lambda & |\lambda|^2+1 \end{pmatrix}. \tag{1.3.6}$$

Now, consider $x = (\alpha_1, \ldots, \alpha_n) \in \mathbb{C}^n$. In view of Eq. (1.3.6), we get

$$\begin{aligned} \|Ax\|^2 &= \langle Ax, Ax\rangle \\ &= \langle A^*Ax, x\rangle \\ &= |\lambda|^2 \sum_{j=1}^{n} |\alpha_j|^2 + \sum_{j=2}^{n} |\alpha_j|^2 + 2\sum_{j=1}^{n-1} Re(\overline{\alpha_j}\overline{\lambda}\alpha_{j+1}). \end{aligned}$$

Thus,

$$\|Ax\|^2 \geq |\lambda|^2 \sum_{j=1}^{n} |\alpha_j|^2 + \sum_{j=2}^{n} |\alpha_j|^2 - 2|\lambda| \sum_{j=1}^{n-1} |\alpha_j||\alpha_{j+1}|,$$

where $\|\cdot\|$ denotes the Euclidean norm. Without loss of generality, we may assume that $\lambda, \alpha_j \in \mathbb{R}_{\geq 0}$, where $\mathbb{R}_{\geq 0}$ is the set of nonnegative real numbers. If $\lambda \geq 2$, then $\lambda^2 \geq 4$. So, we further estimate

$$\lambda^2 \sum_{j=1}^{n} \alpha_j^2 + \sum_{j=2}^{n} \alpha_j^2 - 2\lambda \sum_{j=1}^{n-1} \alpha_j\alpha_{j+1} = 2\left\| \begin{pmatrix} \alpha_2 \\ \cdot \\ \cdot \\ \alpha_n \end{pmatrix} - \frac{\lambda}{2} \begin{pmatrix} \alpha_1 \\ \cdot \\ \cdot \\ \alpha_{n-1} \end{pmatrix} \right\|^2 + \frac{1}{2}\lambda^2\alpha_1^2 + \left(\frac{1}{2}\lambda^2 - 1\right)(\alpha_2^2 + \cdots + \alpha_{n-1}^2) + (\lambda^2 - 1)\alpha_n^2.$$

Thus,

$$\lambda^2 \sum_{j=1}^{n} \alpha_j^2 + \sum_{j=2}^{n} \alpha_j^2 - 2\lambda \sum_{j=1}^{n-1} \alpha_j \alpha_{j+1} \geq \left(\frac{1}{2}\lambda^2 - 1\right)(\alpha_1^2 + \cdots + \alpha_n^2)$$
$$\geq \frac{1}{4}\lambda^2(\alpha_1^2 + \cdots + \alpha_n^2).$$

Hence, we have the following.

Lemma 1.3.5 *Consider $\lambda \in \mathbb{C}$ with $|\lambda| \geq 2$ and the $n \times n$ Jordan block A, given in (1.3.5). Then,*

$$\|Ax\| \geq \frac{|\lambda|}{2}\|x\|$$

for all $x \in \mathbb{C}^n$. ◇

1.4 Closed Operators

1.4.1 Closed Operator Perturbations

Lemma 1.4.1 *Let A and B be two linear operators from a Banach space X into a Banach space Y, having the same domain $\mathcal{D} \subset X$. Let A be a closed operator. If there exist positive constants θ_1, θ_2 and β satisfying $\beta \in]0, 1]$ and $\theta_2 < \frac{1}{\beta}$ such that*

$$\|(A - B)\varphi\| \leq \theta_1\|\varphi\| + \theta_2\|A\varphi\|^{\beta}\|\varphi\|^{1-\beta} \text{ for all } \varphi \in \mathcal{D}, \tag{1.4.1}$$

then B is also closed. ◇

Proof Let $\varphi \in \mathcal{D}$. Equation (1.4.1) implies that

$$\begin{aligned}\|B\varphi\| &\leq \|A\varphi\| + \|(B - A)\varphi\| \\ &\leq \|A\varphi\| + \theta_1\|\varphi\| + \theta_2\|A\varphi\|^{\beta}\|\varphi\|^{1-\beta},\end{aligned} \tag{1.4.2}$$

and that

$$\begin{aligned}\|A\varphi\| &\leq \|B\varphi\| + \|(A - B)\varphi\| \\ &\leq \|B\varphi\| + \theta_1\|\varphi\| + \theta_2\|A\varphi\|^{\beta}\|\varphi\|^{1-\beta}.\end{aligned} \tag{1.4.3}$$

By using Young's inequality (see Lemma 1.3.1), we have

$$\|A\varphi\|^{\beta}\|\varphi\|^{1-\beta} \leq \beta\|A\varphi\| + (1 - \beta)\|\varphi\|. \tag{1.4.4}$$

So, substituting Eq. (1.4.4) in both Eqs. (1.4.2) and (1.4.3), we get

$$\|B\varphi\| \leq (1+\theta_2\beta)\|A\varphi\| + (\theta_1+\theta_2(1-\beta))\|\varphi\|, \tag{1.4.5}$$

and

$$(1-\theta_2\beta)\|A\varphi\| \leq \|B\varphi\| + (\theta_1+\theta_2(1-\beta))\|\varphi\|. \tag{1.4.6}$$

Since $\theta_2 < \frac{1}{\beta}$, then $1-\theta_2\beta > 0$, so Eq. (1.4.6) implies that

$$\|A\varphi\| \leq \frac{1}{1-\theta_2\beta}\|B\varphi\| + \frac{\theta_1+\theta_2(1-\beta)}{1-\theta_2\beta}\|\varphi\|. \tag{1.4.7}$$

Now, let $(\varphi_n)_{n=1}^{\infty}$ be a sequence in $\mathscr{D}$ such that $\varphi_n \to \varphi$ as $n \to \infty$ and $B\varphi_n \to g$ as $n \to \infty$; and let us prove that

$$\varphi \in \mathscr{D} \text{ and } g = B\varphi.$$

Using Eq. (1.4.7), we have

$$\|A(\varphi_n-\varphi_m)\| \leq \frac{1}{1-\theta_2\beta}\|B(\varphi_n-\varphi_m)\| + \frac{\theta_1+\theta_2(1-\beta)}{1-\theta_2\beta}\|\varphi_n-\varphi_m\| \to 0$$

as $m, n \to \infty$. So, the sequence $(A\varphi_n)_n$ is a Cauchy sequence in Y. Hence, $(A\varphi_n)_n$ is convergent. On the other hand, A is closed, so, $\varphi \in \mathscr{D}$ and $A\varphi_n \to A\varphi$ as $n \to \infty$. By using Eq. (1.4.5), we have

$$\|B(\varphi_n-\varphi)\| \leq (1+\theta_2\beta)\|A(\varphi_n-\varphi)\| + (\theta_1+\theta_2(1-\beta))\|\varphi_n-\varphi\| \to 0 \text{ as } n \to \infty.$$

So, $B\varphi_n \to B\varphi$ as $n \to \infty$ and $g = B\varphi$. Consequently, B is closed. Q.E.D.

1.4.2 A-Bounded, A-Closed, and A-Closable

Let $A \in \mathscr{C}(X)$, then it follows from the closeness of A that $\mathscr{D}(A)$, endowed with the graph norm $\|\cdot\|_A$ (i.e., $\|x\|_A := \|x\| + \|Ax\|$), is a Banach space. Let X_A denote $(\mathscr{D}(A), \|\cdot\|_A)$; in this new space, the operator A satisfies $\|Ax\| \leq \|x\|_A$, and this proves that A is a bounded operator from X_A into X (i.e., $A \in \mathscr{L}(X_A, X)$). Let $J \in \mathscr{L}(X)$. If $\mathscr{D}(A) \subset \mathscr{D}(J)$, then J will be called A-defined. If J is A-defined, we will denote by $\widehat{J}$ its restriction to $\mathscr{D}(A)$. Moreover, if $\widehat{J} \in \mathscr{L}(X_A, X)$, we say that J is A-bounded. One checks easily that if J is closed (or closable) (cf. [83, Remark 1.5, p. 191]), then J is A-bounded.

Definition 1.4.1 (i) An operator J is called A-closed if $x_n \to x$, $Ax_n \to y$ and $Jx_n \to z$ for $(x_n)_n \in \mathscr{D}(A)$ implies that $x \in \mathscr{D}(A)$ and $Jx = z$.

(ii) An operator J is called A-closable if $x_n \to 0$, $Ax_n \to 0$, and $Jx_n \to z$ implies that $z = 0$. $\diamond$

1.5 Lebesgue-Dominated Convergence Theorem

Throughout this section, we use the notation Ω for a closed subset of $\mathbb{R}$, and $(\Omega, \mathscr{B}, \mathrm{d}t)$ for the measure space with measurable subsets $\mathscr{B}$ and the Lebesgue measure $\mathrm{d}t$.

Definition 1.5.1 Suppose that $(\Omega, \mathscr{B}, \mathrm{d}t)$ is the Lebesgue measure space and that $E \in \mathscr{B}$. Let X be a Banach space and let $f : \Omega \longrightarrow X$ be a simple function given by

$$f = \sum_{i=1}^{n} x_i \chi_{E_i},$$

where $\chi_{E_i}(\cdot)$ is the characteristic function of the set and E_i and E_i are disjoint. We define f to be Lebesgue integrable over E if $\|f\|$ is Lebesgue integrable over E, that is,

$$\sum_{i=1}^{n} \|x_i\| \lambda \left(E_i \bigcap E \right) < \infty,$$

where $\lambda(\cdot)$ denotes the Lebesgue measure of the set and we follow the usual convention that $0 \cdot \infty = 0$. The Lebesgue integral of f over E is given by

$$\sum_{i=1}^{n} x_i \lambda \left(E_i \bigcap E \right)$$

and will be denoted by

$$\int_E f(t)\mathrm{d}t. \qquad \diamond$$

Definition 1.5.2 Suppose that $(\Omega, \mathscr{B}, \mathrm{d}t)$ is the Lebesgue measure space and that $E \in \mathscr{B}$. Let Z_1 and Z_2 be two separable Hilbert spaces. The uniformly measurable function

$$F : \Omega \longrightarrow \mathscr{L}(Z_1, Z_2)$$

is Lebesgue integrable over E if there exists a sequence of simple integrable functions F_n converging almost everywhere to F and such that

$$\lim_{n\to\infty} \int_E \|F_n(t) - F(t)\|_{\mathscr{L}(Z_1,Z_2)} \mathrm{d}t = 0.$$

We define the Lebesgue integral by

$$\int_E F(t)\mathrm{d}t = \lim_{n\to\infty} \int_E F_n(t)\mathrm{d}t. \qquad \diamond$$

Definition 1.5.3 Suppose that $(\Omega, \mathscr{B}, \mathrm{d}t)$ is the Lebesgue measure space and that $E \in \mathscr{B}$. Let X be a separable Hilbert space. The strongly measurable function $f : \Omega \longrightarrow X$ is Lebesgue integrable over E if there exists a sequence of simple integrable functions $(f_n)_n$ converging almost everywhere to f and such that

$$\lim_{n\to\infty} \int_E \|f_n(t) - f(t)\|_X \mathrm{d}t = 0.$$

We define the Lebesgue integral by

$$\int_E f(t)dt = \lim_n \to \infty \int_E f_n(t)dt. \qquad \diamond$$

These integrals in the above definition are also called Bochner integrals in the literature.

Definition 1.5.4 Let Z_1 and Z_2 be separable Hilbert spaces and let $F : \Omega \longrightarrow \mathscr{L}(Z_1, Z_2)$. If, for all $z_1 \in Z_1$ and $z_2 \in Z_2$, the function

$$\langle z_2, F(t)z_1\rangle \in L_1(\Omega),$$

then we say that $F(\cdot)$ is Pettis integrable. Furthermore, for all $E \in \mathscr{B}$, we call

$$\int_E F(t)\mathrm{d}t$$

defined by

$$\left\langle z_2, \int_E F(t)\mathrm{d}t z_1 \right\rangle := \langle z_2, z_{F,E} z_1\rangle = \int_E \langle z_2, F(t)z_1\rangle \mathrm{d}t,$$

the Pettis integral of $F(t)$ over E and

$$\int_E F(t)z_1 \mathrm{d}t$$

the Pettis integral of $F(t)z_1$ over E. $\diamond$

We recall the Lebesgue-Dominated convergence theorem.

Theorem 1.5.1 *Let $(f_n)_n$ be a sequence in $L_1(\Omega; X)$, where X is a separable Hilbert space. Suppose that $(f_n)_n$ converges almost everywhere to f, i.e.,*

$$\lim_{n\to\infty} \|f_n(t) - f(t)\| = 0$$

except for t in a set of measure zero. Assume further that there exists a fixed function $g \in L_1(\Omega)$ such that

$$\|f_n(t)\| \leq g(t)$$

for all n and almost all $t \in \Omega$. Then, $f \in L_1(\Omega; X)$ and

$$\lim_{n\to\infty} \int_E f_n(t)\mathrm{d}t = \int_E f(t)\mathrm{d}t$$

for all $E \in \mathscr{B}$. ◇

1.6 Compact, Weakly Compact, Strictly Singular, Discrete, Ascent, Descent, and Riesz Operators

1.6.1 Compact Operator

Definition 1.6.1 Let X and Y be two Banach spaces. A bounded linear operator K from X into Y is called to be compact (or completely continuous) if it transforms every bounded set of X in a relatively compact set of Y. In a similar way, K is called to be compact if for each bounded sequence $(x_n)_n$ of X, the sequence $(Kx_n)_n$ has a convergent sequence in Y. ◇

By $\mathscr{K}(X, Y)$, we denote the set of all compact linear operators from X into Y. If $X = Y$, then the set $\mathscr{K}(X, X)$ is replaced by $\mathscr{K}(X)$.

Theorem 1.6.1 *Let $K \in \mathscr{L}(X, Y)$. Then,*

(i) If $(K_n)_n$ is compact and converge uniformly to K, then K is compact.
(ii) K^* is compact if, and only if, K is compact.
(iii) $\mathscr{K}(X, Y)$ is a two-sided ideal of $\mathscr{L}(X, Y)$. ◇

Definition 1.6.2 An operator $A \in \mathscr{L}(X)$ is called a Volterra operator if it is compact and has no non-zero eigenvalue. ◇

Definition 1.6.3 A bounded operator A on X is said to be finite rank if $\dim R(A) < \infty$. ◇

Definition 1.6.4 The spectrum $\sigma(A)$ of an operator A in a finite-dimensional Banach space is the set of complex numbers λ such that $\lambda - A$ is not one to one. The index $\nu(\lambda)$ of a complex number λ is the smallest nonnegative integer ν such that $(\lambda - A)^{\nu}x = 0$ for every vector x for which $(\lambda - A)^{\nu+1}x = 0$. ◇

1.6.2 Weakly Compact Operator

Definition 1.6.5 An operator $A \in \mathscr{L}(X, Y)$ is said to be weakly compact, if $A(B)$ is relatively weakly compact in Y, for every bounded subset $B \subset X$. ◇

By $\mathscr{W}(X, Y)$, we denote the family of weakly compact operators from X into Y. If $X = Y$, then the family of weakly compact operators on X, $\mathscr{W}(X) := \mathscr{W}(X, X)$, is a closed two-sided ideal of $\mathscr{L}(X)$ containing $\mathscr{K}(X)$ (cf. [50, 56, 60]).

1.6.3 Strictly Singular Operator

Definition 1.6.6 Let X and Y be two Banach spaces. $A \in \mathscr{L}(X, Y)$ is called a strictly singular operator if, for every infinite-dimensional subspace M, the restriction of A to M is not a homeomorphism. ◇

We denote by $\mathscr{S}(X, Y)$, the set of strictly singular operators from X into Y.

For our own use, let us recall the following four facts. The set $\mathscr{S}(X, Y)$ is a closed subspace of $\mathscr{L}(X, Y)$. If $X = Y$, then $\mathscr{S}(X) := \mathscr{S}(X, X)$ is a closed two-sided ideal of $\mathscr{L}(X)$ containing $\mathscr{K}(X)$. In the case where X is a Hilbert space, we have $\mathscr{K}(X) = \mathscr{S}(X)$ and the class of weakly compact operators on L_1-spaces (respectively, $C(K)$-spaces with K being a compact Hausdorff space) is, nothing else but, the family of strictly singular operators on L_1-spaces (respectively, $C(K)$-spaces) (see [90, Theorem 1]).

1.6.4 Discrete Operator

Definition 1.6.7 A closed densely defined operators A on a Banach space X is called either compact resolvent or discrete if

(i) $\rho(A) \neq \emptyset$, and
(ii) for all $\lambda \in \rho(A)$, $(\lambda - A)^{-1} \in \mathscr{K}(X)$. ◇

1.6.5 Ascent and Descent Operators

If $A \in \mathscr{L}(X)$, we define the ascent of A, $\text{asc}(A)$, and the descent of A, $\text{desc}(A)$, respectively, by

$$\text{asc}(A) := \min \left\{ n \in \mathbb{N} \text{ such that } N(A^n) = N(A^{n+1}) \right\},$$

and

$$\text{desc}(A) := \min \left\{ n \in \mathbb{N} \text{ such that } R(A^n) = R(A^{n+1}) \right\}.$$

1.6.6 Riesz Operator

Definition 1.6.8 An operator $A \in \mathscr{L}(X)$ is said a Riesz operator, if it satisfies the following conditions:

(i) for all $\lambda \in \mathbb{C}^*$, $\lambda - A$ is a Fredholm operator on X;
(ii) for all $\lambda \in \mathbb{C}^*$, $\lambda - A$ has a finite ascent and a finite descent; and
(iii) all $\lambda \in \sigma(A)\backslash\{0\}$ are eigenvalues of a finite multiplicity, and have no accumulation points, except possibly zero. ◇

1.7 A-Compact Operators

Definition 1.7.1 Let A and B be two linear operators on a Banach space X. Assume that A is closed. Then, the operator B is said to be A-compact of order $p \in [0, 1]$ if $\mathscr{D}(A) \subset \mathscr{D}(B)$ and if for all $\varepsilon > 0$, there is a constant $c(\varepsilon) > 0$ such that

$$\|B\varphi\| \leq \varepsilon \|A\varphi\|^p \|\varphi\|^{1-p} + c(\varepsilon)\|\varphi\| \text{ for all } \varphi \in \mathscr{D}(A).$$

If $p = 1$, we say that B is A-compact. ◇

Let X be a Banach space and $A \in \mathscr{C}(X)$ such that the resolvent set, $\rho(A)$, of A is not empty ($\lambda_0 \in \rho(A)$). We suppose that B is A-compact. Then, for all $\varepsilon > 0$, there is a constant $c(\varepsilon) > 0$ such that

$$\|B\varphi\| \leq \varepsilon \|A\varphi\| + c(\varepsilon)\|\varphi\| \quad \text{for all} \quad \varphi \in \mathscr{D}(A). \tag{1.7.1}$$

Let $v \in X$ and $\lambda_0 \in \rho(A)$, we have $(\lambda_0 - A)^{-1}v \in \mathscr{D}(A)$. Hence, inequality (1.7.1) implies that for any $\varepsilon > 0$, there exists a constant $c(\varepsilon) > 0$ such that

$$\|B(\lambda_0 - A)^{-1}v\| \leq \varepsilon \|A(\lambda_0 - A)^{-1}v\| + c(\varepsilon)\|(\lambda_0 - A)^{-1}v\|.$$

Furthermore, if we assume that $(\lambda_0 - A)^{-1}$ is compact, then

$$A(\lambda_0 - A)^{-1} = -I + \lambda_0(\lambda_0 - A)^{-1}$$

is bounded. So, there is $M' > 0$ such that

$$\|A(\lambda_0 - A)^{-1}v\| \leq M'\|v\|.$$

Hence, for all $\varepsilon > 0$, there exists a constant $c(\varepsilon) > 0$ such that

$$\|B(\lambda_0 - A)^{-1}v\| \leq \varepsilon M'\|v\| + c(\varepsilon)\|(\lambda_0 - A)^{-1}v\|. \tag{1.7.2}$$

Let $(x_n)_n$ be a bounded sequence of X, i.e.,

$$\|x_n\| \leq M.$$

We may show that $(B(\lambda_0 - A)^{-1}x_n)_n$ has a convergent subsequence. Since the sequence $((\lambda_0 - A)^{-1}x_n)_n$ has a convergent subsequence $((\lambda_0 - A)^{-1}x_{\varphi(n)})_n$, then there is n_0 such that for all $m > n \geq n_0$, we have

$$\|(\lambda_0 - A)^{-1}x_{\varphi(m)} - (\lambda_0 - A)^{-1}x_{\varphi(n)}\| < \varepsilon.$$

Thus, the inequality (1.7.2) implies that for all $\varepsilon > 0$, there exists a constant $c(\varepsilon) > 0$ such that

$$\begin{aligned}\|B(\lambda_0 - A)^{-1}(x_{\varphi(m)} - x_{\varphi(n)})\| &\leq \varepsilon M'\|x_{\varphi(m)} - x_{\varphi(n)}\| \\ &\quad + c(\varepsilon)\|(\lambda_0 - A)^{-1}(x_{\varphi(m)} - x_{\varphi(n)})\| \\ &\leq 2\varepsilon M'M + c(\varepsilon)\varepsilon \\ &= (2M'M + c(\varepsilon))\varepsilon.\end{aligned}$$

Thus, $(B(\lambda_0 - A)^{-1}x_{\varphi(n)})_n$ is a Cauchy sequence in X. So, it is convergent. Hence

$$B(\lambda_0 - A)^{-1} \text{ is compact.} \tag{1.7.3}$$

Conversely, let $\mathscr{D}(B) \supset \mathscr{D}(A)$ and the operator $T = B(\lambda_0 - A)^{-1}$ in X be compact. Under the given $\varepsilon > 0$ let us construct a finite-dimensional operator

$$Q_n = \sum_{k=1}^{n} \langle \cdot, u_k \rangle v_k$$

such that

$$\|T - Q\| < \frac{\varepsilon}{2}.$$

From the lemma condition, it follows that the domain of definition of the conjugate operator A^* is dense in X. Then, the elements u_k can be approximated by elements $w_k \in \mathscr{D}(A^*)$ and the operator

$$P = \sum_{k=1}^{n} \langle \cdot, w_k \rangle v_k$$

can be constructed so that

$$\|Q - P\| < \frac{\varepsilon}{2}.$$

Then, under $u \in \mathscr{D}(A)$ and $\lambda_0 \in \rho(A)$, we have

$$\begin{aligned}
\|Bu\| &= \|T(\lambda_0 - A)u\| \\
&\leq \|P(\lambda_0 - A)u\| + \|(T - P)(\lambda_0 - A)u\| \\
&\leq \sum_{k=1}^{n} |\langle(\lambda_0 - A)u, w_k\rangle| \|v_k\| + \varepsilon(\|Au\| + |\lambda_0| \|u\|) \\
&\leq \sum_{k=1}^{n} (|\langle u, A^* w_k\rangle| + |\lambda_0| |\langle u, w_k\rangle|) \|v_k\| + \varepsilon(\|Au\| + |\lambda_0| \|u\|) \\
&\leq \varepsilon\|Au\| + C(\varepsilon)\|u\|.
\end{aligned}$$

Hence, B is A-compact.

1.8 Dunford-Pettis Property

Definition 1.8.1 Let X be a Banach space. The space X is called to have the Dunford-Pettis property if, for each Banach space Y, every weakly compact operator

$$T : X \longrightarrow Y$$

takes weakly compact sets in X into norm compact sets of Y. ◇

1.9 The Jeribi Essential Spectrum

1.9.1 Definition

Definition 1.9.1 Let X be a Banach space and let $A \in \mathscr{C}(X)$. The Jeribi essential spectrum is defined by

$$\sigma_j(A) := \bigcap_{K \in \mathscr{W}_*(X)} \sigma(A + K),$$

where $\mathscr{W}_*(X)$ stands for each one of the sets $\mathscr{W}(X)$ and $\mathscr{S}(X)$. ◇

For more important contributions concerning the Jeribi essential spectrum and their applications to transport operators were due to A. Jeribi and his collaborators (see the bibliography at the end of the chapter [1, 52, 60, 65, 86, 89]). There is a wealth of literature treating the Jeribi essential spectrum see, for example, References [12, 16–19, 97].

1.9.2 A Characterization of the Jeribi Essential Spectrum

We have the following

Theorem 1.9.1 *Let (Ω, Σ, μ) be an arbitrary positive measure space. Let A be a closed, densely defined, and linear operator on $L_1(\Omega, d\mu)$. Then,*

$$\sigma_j(A) = \bigcap_{K\in\mathcal{K}(L_1(\Omega,d\mu))} \sigma(A+K). \qquad \diamond$$

Corollary 1.9.1 *If X satisfies the Dunford-Pettis property, and if A is a closed, densely defined, and linear operator on X, then*

$$\sigma_j(A) = \bigcap_{K\in\mathcal{K}(X)} \sigma(A+K). \qquad \diamond$$

Theorem 1.9.2 *Let A be a closed, densely defined, and linear operator on $L_p(\Omega, d\mu)$, and let $p \in [1, \infty)$. In the case where $\mathcal{W}_*(L_p(\Omega, d\mu)) = \mathcal{S}(L_p(\Omega, d\mu))$, we have*

$$\sigma_j(A) = \bigcap_{K\in\mathcal{K}(L_p(\Omega,d\mu))} \sigma(A+K). \qquad \diamond$$

1.10 Jordan Chain for an Operator and Multiplicities

A necessary and sufficient condition for the following equation

$$Ax = \lambda_0 x$$

has at least a solution $x \neq 0$ is that $\lambda_0 \in \sigma_p(A)$. Thus, λ_0 is called an eigenvalue of A, and x the corresponding eigenvector. The element $x_0 \neq 0$ that satisfies the following equation

$$Ax_0 = \lambda_0 x_0$$

is called an eigenvector of the operator A, corresponding to the eigenvalue λ_0. If the elements $x_0, x_1, \ldots, x_k$ correlate with the following equations:

$$Ax_p = \lambda_0 x_p + x_{p-1}, \quad p = 1, \ldots, k,$$

then the element x_k is said to be an associated vector of the kth rank to the eigenvector x_0. The number $k+1$ is called the length of the chain $x_0, x_1, \ldots, x_k$. The element x_0 is called an eigenvector of the rth rank if the longest chain corresponding to x_0 has length equal to r. The chain $x_0, x_1, \ldots, x_{r-1}$ is called the Jordan chain. It is easy to see that the elements $x_0, x_1, \ldots, x_{r-1}$ are linearly independent. The null space $N(\lambda_0 - A)$ of the operator $\lambda_0 - A$ is called the eigenspace of A corresponding to the eigenvalue λ_0 of A. It consists of the vector 0 and the totality of eigenvectors corresponding to λ_0. The geometric multiplicity or simply multiplicity of the eigenvalue λ_0 is the dimension of the eigenspace corresponding to the eigenvalue λ_0,

$$m_g(\lambda_0) := \dim N(\lambda_0 - A).$$

When $m_g(\lambda)$ is equal to 1, we called that λ is geometrically simple. The root subspace of A corresponding to an eigenvalue λ of A, is the following set:

$$\mathscr{L}_A(\lambda) := \bigcup_{k=0}^{\infty} N[(A-\lambda)^k]. \tag{1.10.1}$$

A non-zero $x \in \mathscr{D}(A)$ is called a root vector or generalized eigenvector of A, corresponding to an eigenvalue λ of A, if there is a positive integer n, such that

$$(\lambda - A)^n x = 0,$$

i.e., $x \in \mathscr{L}_A(\lambda)$. The algebraic multiplicity of λ is the number

$$m_a(\lambda) := \dim \mathscr{L}_A(\lambda).$$

When $m_a(\lambda)$ is equal to 1, we say that λ is algebraically simple. It is well known that each eigenvalue of a discrete operator must have finite algebraic multiplicity.

1.11 Laurent Series Expansion of the Resolvent

Let λ_0 be an eigenvalue of a closed operator A, acting in a Banach space X that is an isolated point of the spectrum $\sigma(A)$ and a pole of the finite order of the resolvent $(\lambda - A)^{-1}$. Let us expand $(\lambda - A)^{-1}$ in some neighborhood of λ_0 into the Laurent series

$$(\lambda - A)^{-1} = \sum_{n=0}^{\infty} A_n(\lambda - \lambda_0)^n + \frac{A_{-1}}{\lambda - \lambda_0} + \cdots + \frac{A_{-q}}{(\lambda - \lambda_0)^q}, \tag{1.11.1}$$

where A_k, $k = -q, \ldots, \infty$, are bounded operators and series (1.11.1) converges by the operator norm in X. Act from the left in (1.11.1) by the operator $\lambda - A$. Then,

$$I = (\lambda - A) \sum_{n=-q}^{\infty} A_n (\lambda - \lambda_0)^n$$
$$= [(\lambda - \lambda_0) + (\lambda_0 - A)] \sum_{n=-q}^{\infty} A_n (\lambda - \lambda_0)^n.$$

Multiplying this identity by $(\lambda - \lambda_0)^q$, differentiating k times, and passing to the limit under $\lambda \to \lambda_0$, we obtain

$$(\lambda_0 - A)A_n + A_{n-1} = 0, \quad n \neq 0, \ n \neq -q$$
$$(\lambda_0 - A)A_0 + A_{-1} = I,$$
$$(\lambda_0 - A)A_{-q} = 0.$$

Now, act from the right in (1.11.1) by the operator $\lambda - A$. Then, for any $x \in \mathscr{D}(A)$, we get

$$x = \sum_{n=-q}^{\infty} (\lambda - \lambda_0)^n A_n (\lambda - A)x$$
$$= \sum_{n=-q}^{\infty} (\lambda - \lambda_0)^n A_n [(\lambda - \lambda_0) + (\lambda_0 - A)]x.$$

Hence, for $x \in \mathscr{D}(A)$, we have

$$A_n(\lambda_0 - A)x + A_{n-1}x = 0, \quad n \neq 0, \ n \neq q$$
$$A_0(\lambda_0 - A)x + A_{-1}x = x,$$
$$A_{-q}(\lambda_0 - A)x = 0.$$

Definition 1.11.1 If there is a $q > 0$ such that $A_{-q} \neq 0$ (consequently $A_{-n} \neq 0$ for all $n \leq q$), while $A_{-n} = 0$ for all $n > q$, then the point λ_0 is called a pole of the resolvent

$$R(\lambda, A) := (\lambda - A)^{-1}$$

of order q. ◇

Theorem 1.11.1 (Yakubov [98]) *Let λ_0 be an eigenvalue of the operator A and a pole of the finite order of the resolvent $(\lambda - A)^{-1}$. Then, the expansion coefficients of (1.11.1) are correlated as follows:*

(i) $A_{n+1}(\lambda_0 - A) \subset (\lambda_0 - A)A_{n+1} = -A_n$, $n \geq 0$.
(ii) $A_{-p}(\lambda_0 - A) \subset (\lambda_0 - A)A_{-p} = -A_{-(p+1)}$, $p = 1, \ldots, q-1$.
(iii) $A_0(\lambda_0 - A) \subset (\lambda_0 - A)A_0 = -A_{-1} + I$.
(iv) $A_{-q}(\lambda_0 - A) \subset (\lambda_0 - A)A_{-q} = 0$. ◇

Let λ_0 be an eigenvalue of the operator A and a finite-order pole of the resolvent $(\lambda - A)^{-1}$. Then, from Theorem 1.11.1, it follows that

$$(\lambda_0 - A)^{k+1} A_{-q+k} = 0,\ k = 0, \dots, q - 1. \tag{1.11.2}$$

Using Eq. (1.11.2), we have

$$R(A_{-q+k}) \subset N^k,\ k = 0, \dots, q - 1,$$

where

$$N^k := \{x \in \mathscr{D}(A^{k+1}) \text{ such that } (\lambda_0 - A)^{k+1} x = 0\}. \tag{1.11.3}$$

In particular,

$$R(A_{-1}) \subset N^{q-1} \subset \bigcup_{k=0}^{\infty} N^k. \tag{1.11.4}$$

Let x_k be an associated vector of the kth rank. Then, the item (iii) of Theorem 1.11.1 implies that

$$(-A_{-1} + I)x_k = A_0(\lambda_0 - A)x_k = -A_0 x_{k-1}. \tag{1.11.5}$$

Now, the item (i) of Theorem 1.11.1 implies that

$$A_0 x_{k-1} = (-1)^k A_k (\lambda_0 - A)^k x_{k-1} = 0. \tag{1.11.6}$$

It is known, using the residue theory, that in the expansion (1.11.1) the operator A_{-1} plays a special role. Hence, let us establish some elementary properties of the operator A_{-1}, which will be necessary in the sequel. By using (1.11.5) and (1.11.6), we have

$$A_{-1}x = x,\quad x \in \bigcup_{k=0}^{\infty} N^k, \tag{1.11.7}$$

where N^k is given in (1.11.3). From Eq. (1.11.7) in turn, follows

$$\bigcup_{k=0}^{\infty} N^k \subset R(A_{-1}).$$

From this and (1.11.4), we get

$$R(A_{-1}) = \bigcup_{k=0}^{\infty} N^k. \tag{1.11.8}$$

It is easy to see that the coefficients of the principal part of the expansion (1.11.1) satisfy the following correlation

$$R(A_{-q+k}) \subset N^k,\ k = 0, \ldots, q-1.$$

If λ_0 is an eigenvalue of the operator A and a finite-order pole of the resolvent $(\lambda - A)^{-1}$, then from Eqs. (1.11.4) and (1.11.8) follows

$$R(A_{-1}) = N^{q-1} = \bigcup_{k=0}^{\infty} N^k = N^{r-1}. \tag{1.11.9}$$

From Eq. (1.11.9), we have $q \geq r$. On the other hand, since λ_0 is a pole of order q, then

$$A_{-q} = (-1)^{q-1}(\lambda_0 - A)^{q-1} A_{-1} \neq 0.$$

So, there exists $y_0 \in X$ such that

$$(\lambda_0 - A)^{q-1} A_{-1} y_0 \neq 0.$$

Hence,

$$A_{-1} y_0 \notin N^{q-2},$$

and

$$N^{q-2} \neq R(A_{-1}) = \bigcup_{k=0}^{\infty} N^k = N^{r-1}$$

from which follows $r \geq q$. Thus, the rank of the eigenvalue coincides with the pole order of the resolvent, i.e.,

$$r = q.$$

From the general formula on p. 148 of [87], we get the following.

Proposition 1.11.1 *Let X be a Hilbert space, $A \in \mathscr{C}(X)$ and λ_0 be a pole of the resolvent $R(\lambda, A)$ of order p_{λ_0}. Then,*

$$m_a(\lambda_0) \leq p_{\lambda_0} m_g(\lambda_0).$$ ◇

1.12 Bases

1.12.1 Algebraic Bases (Hamel Bases)

Definition 1.12.1 Consider V is a vector space over the field $\mathbb{K}$ ($\mathbb{K} = \mathbb{R}$ or $\mathbb{C}$) and $(e_i)_{i \in I}$ is a vectors family of V, where I is an arbitrary index set.

(i) $(e_i)_{i \in I}$ is said a spanning family of V if each vector of V is a linear combination of a finite subfamily of such a family.

(ii) $(e_i)_{i\in I}$ is said to be a linearly independent family if all its finite subfamilies are linearly independent.
(iii) A basis of V is a linearly independent spanning family.
(iv) Given $(e_i)_{i\in I}$ a basis of V, every element of V can be expressed uniquely as a finite linear combination of basis vectors whose coefficients are referred to as vector components. ◇

1.12.2 On a Schauder Basis

Definition 1.12.2 Let X be an infinite-dimensional Banach space over the field of real or complex numbers. A sequence of vectors $\{e_n\}_{n=1}^{\infty}$ of X is called to be a Schauder basis or a basis for X if to each vector x in the space X, there corresponds a unique sequence of scalars $\{\alpha_n\}_{n=1}^{\infty}$ such that

$$x = \sum_{n=1}^{\infty} \alpha_n e_n.$$

The convergence of the series is understood to be with respect to the strong (norm) topology of X; in other words,

$$\lim_{n\to\infty} \left\| x - \sum_{k=1}^{n} \alpha_k e_k \right\| = 0. \qquad \diamond$$

Example 1.1 Consider the Banach space

$$X = l^p := \left\{ (u_n)_n \text{ such that } \sum_n |u_n|^p < \infty \right\} \quad (1 \le p < \infty).$$

and define sequences

$$e_n = (\delta_{mn})_{m=1}^{\infty} = ((0,\dots,0,1,0,\dots))_{m=1}^{\infty},$$

where the 1 appears in the nth position. Then, $\{e_n\}_{n=1}^{\infty}$ form a Schauder basis for l^p, often called the natural basis for l^p.

Note that $\{e_n\}_{n=1}^{\infty}$ is not a basis for l^{∞}. This is due to the fact that l^{∞} is not separable and hence it has no bases whatsoever. However, using the l^{∞} norm, the sequence $\{e_n\}_{n=1}^{\infty}$ is a basis for the Banach space c_0, and consists of all infinite sequences of scalars which converge to zero.

1.13 Normal Operator

Definition 1.13.1 A closed densely defined operators A on a Banach space X is called normal if it satisfies the following equivalent conditions $AA^* = A^*A$, i.e., $\mathscr{D}(AA^*) = \mathscr{D}(A^*A)$ and $AA^*u = A^*Au$ for all $u \in \mathscr{D}(A^*A)$. ◇

Let A be a normal operator. Then,

$$\|(\lambda - A)^{-1}\| = \frac{1}{\text{dist}(\lambda, \sigma(A))}, \tag{1.13.1}$$

where

$$\text{dist}(\lambda, \sigma(A)) = \min_{z \in \sigma(A)} |\lambda - z|.$$

Theorem 1.13.1 *If A is a closed operator, then*

(i) $\mathscr{D}(A^*)$ *is dense in* Y^*.
(ii) $(A^*)^* = A$. ◇

We denote sectors in the complex plane by

$$\Omega(\varphi_-, \varphi_+) := \{re^{i\theta} \text{ such that } r \geq 0,\ \varphi_- < \theta < \varphi_+\} \text{ and } \Omega(\varphi) := \Omega(-\varphi, \varphi) \tag{1.13.2}$$

for the sectors lying between the rays with arguments φ_-, φ_+ and $-\varphi,\ \varphi$, respectively.

Lemma 1.13.1 (Markus [88, Lemma 6.6]) *Let G be normal with compact resolvent, and*

$$\sigma(G) \bigcap \Omega(2\theta) \subset \mathbb{R}_{\geq 0}$$

with $0 < \theta \leq \frac{\pi}{2}$. Then, for $0 \leq p < 1$, $\alpha > 0$, there exists $r_0 > 0$ such that the contours

$$\Gamma_\pm = \{x + iy \in \mathbb{C} \text{ such that } x \geq r_0,\ y = \pm\alpha x^p\} \tag{1.13.3}$$

satisfy $\Gamma_\pm \subset \rho(G) \bigcap \overline{\Omega(\theta)}$ and we have

$$\int_{\Gamma_\pm} |\lambda|^p \|(\lambda - G)^{-1} f\|^2 |d\lambda| \leq C_1 \|f\|^2,$$

and

$$\int_{\Gamma_\pm} |\lambda|^{p-2} \|G(\lambda - G)^{-1} f\|^2 |d\lambda| \leq C_2 \|f\|^2$$

for all $f \in X$ with some constants C_1, $C_2 \geq 0$. ◇

Lemma 1.13.2 (Markus [88, Lemma 6.7]) *Let G be normal with compact resolvent, and $\sigma(G)\bigcap\Omega(2\theta)\subset\mathbb{R}_{\geq 0}$ with $0<\theta\leq\frac{\pi}{2}$. Let $\{x_k\}_1^\infty$ be a sequence of positive numbers, $0\leq p<1$, and α, c_1, $c_2>0$ such that*

$$\alpha x_1^{p-1}\leq\tan\theta,$$

$$x_n^{1-p}-x_k^{1-p}\geq c_1(n-k)$$

for $n>k$, and

$$dist(x_k,\sigma(G))\geq c_2x_k^p$$

for $k\geq 1$. Then, the lines

$$\gamma_k=\{x_k+iy\in\mathbb{C}\ such\ that\ |y|\leq\alpha x_k^p\}\tag{1.13.4}$$

satisfy $\gamma_k\subset\rho(G)\bigcap\overline{\Omega(\theta)}$, and we have

$$\sum_{k=1}^{\infty}x_k^p\int_{\gamma_k}\|(\lambda-G)^{-1}f\|^2|d\lambda|\leq C_1\|f\|^2,$$

and

$$\sum_{k=1}^{\infty}x_k^{p-2}\int_{\gamma_k}\|G(\lambda-G)^{-1}f\|^2|d\lambda|\leq C_2\|f\|^2,$$

for all $f\in X$ with some constants C_1, $C_2\geq 0$. ◇

In the remaining part of this section, we will state some basic results on the behavior of the spectrum under a relatively compact perturbation (see [88]).

Lemma 1.13.3 (Markus [88, Lemma 3.6]) *Let G be a normal operator with compact resolvent, and let B be G-compact. If*

$$\sigma(G)\bigcap\{\lambda\in\mathbb{C}\ such\ that\ |arg\lambda|<\theta\}\subset\mathbb{R}\ \left(0<\theta\leq\frac{\pi}{2}\right),$$

then the resolvent of the operator $A=G+B$ is also compact, and for any $\delta>0$ only finitely many eigenvalues of A lie in the angles

$$\{\lambda\in\mathbb{C}\ such\ that\ \delta\leq\pm arg\lambda\leq\theta-\delta\}.$$ ◇

1.14 Positive Operators

Definition 1.14.1 Let A be a linear operator with domain $\mathscr{D}(A)$ on a Hilbert space X. A is said to be positive if

$$\langle Af, f\rangle \geq 0 \text{ for all } f \in \mathscr{D}(A).$$

The inequality $A \leq B$ signifies that the operator $B - A$ is nonnegative or positive. ◇

Theorem 1.14.1 (Gohberg and Krein [55, Lemma 1.1. p. 26]) *Let A and B be two positive compact operators on a Hilbert space X. If the operator $0 \leq A \leq B$. Then,*

$$\lambda_j(A) \leq \lambda_j(B) \text{ for all } j \geq 1,$$

where $(\lambda_j(A))_j$ (respectively, $(\lambda_j(B))_j$) denotes the decreasing sequence of eigenvalues of A (respectively, B). ◇

1.15 Spectrum of the Sum of Two Operators

Proposition 1.15.1 (Markus [88, Lemma 8.1, p. 41]) *Let A be an operator with compact resolvent, and suppose that the operator B is A-compact. If Δ is a bounded domain with rectifiable boundary Γ and*

$$\sigma(A + tB) \bigcap \Gamma = \emptyset,$$

for all $t \in [0, 1]$, then the operator $A + B$ also has a compact resolvent, and

$$n(\Delta, A + B) = n(\Delta, A),$$

where $n(\Delta, A)$ (respectively, $n(\Delta, A + B)$) denotes the sum of multiplicities of all eigenvalues of A (respectively, $A + B$) contained in Δ. ◇

We give the Kato-Rellich theorem.

Theorem 1.15.1 *If T_α is a family of analytic operators in the sense of Kato and σ_0 is a simple eigenvalue of T_0 (T_α with $\alpha = 0$), then for α small enough, T_α has only one eigenvalue $\sigma(\alpha)$ in the neighborhood of σ_0 and is simple. Moreover, $\sigma(\alpha)$ and the eigenvector corresponding $\varphi(\alpha)$ can be developed in entire series of α.* ◇

Let $A = A^*$ be a self-adjoint operator in a separable Hilbert space X with domain $\mathscr{D}(A)$. Below, we will assume that A is lower semibounded; moreover, we will assume (without loss of generality) that A is uniformly positive. Suppose that A has a discrete spectrum, and let $\{\mu_k\}_{k=1}^{\infty}$ be its eigenvalues, each taken with its geometric multiplicity. Denote a complete orthonormal system of eigenvectors of the operator A that correspond to these eigenvalues by $\{\varphi_k\}_{k=1}^{\infty}$. We have

$$A^{-1} = \sum_{k=1}^{\infty} \frac{\langle \cdot, \varphi_k \rangle \varphi_k}{\mu_k}.$$

If we suppose that $\mathscr{D}(B) \supset \mathscr{D}(A)$, then

$$BA^{-1} = \sum_{k=1}^{\infty} \frac{\langle \cdot, \varphi_k \rangle B\varphi_k}{\mu_k}.$$

Consequently,

$$\left\| BA^{-1} - \sum_{k=1}^{N} \frac{\langle f, \varphi_k \rangle B\varphi_k}{\mu_k} \right\|^2 \le \|f\|^2 \sum_{k=N+1}^{\infty} \frac{\|B\varphi_k\|^2}{\mu_k^2}. \tag{1.15.1}$$

If, further, we suppose that the condition

$$\sum_{k=1}^{\infty} \frac{\|B\varphi_k\|^2}{\mu_k^2} < \infty \tag{1.15.2}$$

holds, then by using (1.15.1), we have

$$\left\| BA^{-1} - \sum_{k=1}^{N} \frac{\langle f, \varphi_k \rangle B\varphi_k}{\mu_k} \right\|^2 \le \varepsilon \|f\|^2,$$

where $\varepsilon \to 0$ as $N \to \infty$ (since the series (1.15.2) converges). Therefore, the operator BA^{-1} can be approximated with arbitrary accuracy by finite-dimensional operators. Thus, the operator

$$BA^{-1} \tag{1.15.3}$$

is compact.

If we suppose that $\mathscr{D}(B) \supset \mathscr{D}(A)$ and the condition (1.15.2) holds, then from Lemma 1.13.3, it follows that the spectrum of the perturbed operator

$$A + B \text{ is discrete.} \tag{1.15.4}$$

1.16 Notes and Remarks

- For an introduction to the theory of bounded or unbounded linear operators, we refer to the books of Davies [47], Dunford and Schwartz [50], I. C. Gohberg, S. Goldberg and Kaashoek [54], Jeribi [60–62, 64] and Kato [83].
- In 1905, Hamel [59] proved in a particular case, but via a general method, the existence of a basis for a vector space of infinite dimension. Precisely, he demonstrated, using the axiom of choice, the existence of a basis of $\mathbb{R}$, seen as a vector space over the rational. This is the first demonstration of the existence of such a basis for a vector space of infinite dimension.
- Schauder bases were described by Juliusz Schauder in 1927 [92, 93], although such bases were discussed earlier. For example, the Haar basis was given in 1909 [58], and Georg Faber discussed in 1910 [53] a basis for continuous functions on an interval, sometimes called a Faber Schauder system.
- The concept of strictly singular operators was introduced by T. Kato in his pioneering paper [83] as a generalization of the notion of compact operators. For a detailed study of the properties of a strictly singular operators, we can refer to the references [56, 83].
- The Dunford-Pettis property, as defined above, was explicitly defined by Grothendieck [57] who undertook an extensive study about it and also about some related properties. It is well known that any L_1-space has the Dunford-Pettis property [49]. Moreover, if Ω is a compact Hausdorff space, then $C(\Omega)$ has the Dunford-Pettis property [57]. For further examples, we may refer to [48] or [50, pp. 494, 479, 508, and 511]. Let us notice that the Dunford-Pettis property is not conserved under conjugation. However, if X is a Banach space whose dual has the Dunford-Pettis property, then X has also the Dunford-Pettis property (see, e.g., [57]). For more information, we may refer to Diestel's paper [48] which contains a real survey of the Dunford-Pettis property, as well as some related topics.

(i) Note that, if $\rho(A) \neq \emptyset$, then A is closed. In fact, if $\lambda \in \rho(A)$, then $(\lambda - A)^{-1}$ is closed, which is also valid for $\lambda - A$. Then, according to the closed graph theorem (see Theorem 1.2.1), we deduce that

$$\rho(A) = \{\lambda \in \mathbb{C} \text{ such that } \lambda - A \text{ is bijective}\},$$

and hence

$$\sigma(A) = \sigma_p(A) \bigcup \sigma_c(A) \bigcup \sigma_r(A).$$

(ii) Let $\lambda \in \rho(A)$, then by using the closed graph theorem the inverse

$$R(\lambda, A) := (\lambda - A)^{-1}$$

is a bounded operator on X and will be called the resolvent of A at the point λ.
(a) If J is bounded, then J is A-bounded.
(b) If J is closed, then J is A-closed.

(c) If J is closable, then J is A-closable.

(d) If A is closed then, by using [94, Lemma 2.1], we have J is A-closed if, and only if, J is A-closable if, and only if, J is A-bounded.

(iii) If $\dim X = +\infty$, then the operators with compact resolvent are necessarily unbounded. In fact, let $\lambda \in \rho(A)$ such that $(\lambda - A)^{-1} \in \mathscr{K}(X)$. If $\lambda - A \in \mathscr{L}(X)$, then

$$I = (\lambda - A)(\lambda - A)^{-1}$$

is compact by the ideal property. Thus, absurd.

(iv) If X satisfies the Dunford-Pettis property, $\mathscr{W}_*(X) = \mathscr{W}(X)$ and if A is a densely defined, and closed linear operator on X. Then,

(a) $\sigma_j(A + K) = \sigma_j(A)$ for all $K \in \mathscr{W}(X)$,

(b) $\sigma_c(A) \bigcup \sigma_r(A) \subset \sigma_j(A)$.

(v) Let $p \in [1, \infty)$ and let A be a closed, densely defined, and linear operator on $L_p(\Omega, d\mu)$. Then,

$$\sigma_c(A) \bigcup \sigma_r(A) \subset \sigma_j(A).$$

References

1. B. Abdelmoumen, A. Dehici, A. Jeribi, M. Mnif, Some new properties of Fredholm theory, Schechter essential spectrum, and application to transport theory. J. Inequal. Appl. Art. ID 852676, 1–14 (2008)
2. B. Abdelmoumen, A. Jeribi, M. Mnif, Invariance of the Schechter essential spectrum under polynomially compact operators perturabation. Extr. Math. **26**(1), 61–73 (2011)
3. B. Abdelmoumen, A. Jeribi, M. Mnif, Measure of weak noncompactness, some new properties in Fredholm theory, characterization of the Schechter essential spectrum and application to transport operators. Ricerche Mat. **61**, 321–340 (2012)
4. B. Abdelmoumen, A. Jeribi, M. Mnif, On graph measures in Banach spaces and description of essential spectra of multidimensional transport equation. Acta Math. Sci. Ser. B Engl. Ed. **32**(5), 2050-2064 (2012)
5. F. Abdmouleh, A. Ammar, A. Jeribi, Stability of the S-essential spectra on a Banach space. Math. Slovaca **63**(2), 299–320 (2013)
6. F. Abdmouleh, A. Ammar, A. Jeribi, A Characterization of the pseudo-Browder essential spectra of linear operators and application to a transport equation. J. Comput. Theor. Transp. **44**(3), 141–153 (2015)
7. F. Abdmouleh, S. Charfi, A. Jeribi, On a characterization of the essential spectra of the sum and the product of two operators. J. Math. Anal. Appl. **386**(1), 83–90 (2012)
8. F. Abdmouleh, A. Jeribi, Symmetric family of Fredholm operators of indices zero, stability of essential spectra and application to transport operators. J. Math. Anal. Appl. **364**, 414–423 (2010)
9. F. Abdmouleh, A. Jeribi, Gustafson, Weidmann, Kato, Wolf, Schechter, Browder, Rakočević and Schmoger essential spectra of the sum of two bounded operators and application to a transport operator. Math. Nachr. **284**(2–3), 166–176 (2011)
10. A. Ammar, B. Boukattaya, A. Jeribi, Stability of the S-left and S-right essential spectra of a linear operator, Acta Math. Sci. Ser. B Engl. Ed. **34**(6), 1922–1934 (2014)

11. A. Ammar, B. Boukettaya, A. Jeribi, A note on the essential pseudospectra and application. Linear Multilinear Algebra **64**(8), 1474–1483 (2016)
12. A. Ammar, B. Boukettaya, A. Jeribi, Relationship between the left-right Jeribi essential spectra and the left-right Weyl essential spectra, Libertas Mathematica (new series) **39**(2), 13–26 (2019)
13. A. Ammar, A. Jeribi, K. Mahfoudhi, Browder essential approximate pseudospectrum and defect pseudospectrum on a Banach space, Extr. Math. **34**(1), 29–40 (2019)
14. A. Ammar, A. Jeribi, K. Mahfoudhi, The condition pseudospectrum of a operator pencil. Asian-Eur. J. Math. **14**(4), 2150057, (2021) p. 12
15. A. Ammar, A. Jeribi, K. Mahfoudhi, Measure of noncompactness, essential approxi- mation and defect pseudospectrum. Methods Funct. Anal. Topol. **25**(1), 1–11 (2019)
16. C. Belabbaci, Jeribi essential spectrum, Lib. Math. (N.S.) **37**(1), 65–73 (2017)
17. C. Belabbaci, M. Aissani, M. Terbeche, S-essential spectra and measure of noncompactness, Math. Slovaca **67**(5), 1203–1212 (2017)
18. C. Belabbaci, Spectral radius of S-essential spectra, Mat. Stud. **54**(1), 91–97 (2020)
19. C. Belabbaci, The S-Jeribi essential spectrum, Ukrains'kyi Matematychnyi Zhurnal, **73**(3), (2021)
20. A. Ammar, F. Bouzayeni, A. Jeribi, Perturbation of unbounded linear operators by γ-relative boundedness, *Ric. Mat.* **67**(2), 433–445 (2018)
21. A. Ammar, M.Z. Dhahri, A. Jeribi, Stability of essential approximate point spectrum and essential defect spectrum of linear operator. Filomat **29**(9), 1983–1994 (2015)
22. A. Ammar, M.Z. Dhahri, A. Jeribi, Some properties of the M-essential spectra of closed linear operator on a Banach space. Funct. Anal. Approx. Comput. **7**(1), 15–28 (2015)
23. A. Ammar, M.Z. Dhahri, A. Jeribi, A Characterization of S-essential spectrum by mean of measure of non-strict-singularity and application. Azerbaijan J. Math. **5**(1), 2218–6816 (2015)
24. A. Ammar, N. Djaidja, A. Jeribi, The essential spectrum of a sequence of linear operators in Banach spaces. Int. J. Anal. Appl. **1–7** (2017)
25. A. Ammar, A. Jeribi, A characterization of the essential pseudospectra on a Banach space. Arab. J. Math. **2**(2), 139–145 (2013)
26. A. Ammar, A. Jeribi, A characterization of the essential pseudospectra and application to a transport equation. Extr. Math. **28**, 95–112 (2013)
27. A. Ammar, A. Jeribi, Measures of noncompactness and essential pseudo-spectra on Banach Space. Math. Methods Appl. Sci. **37**(3), 447–452 (2014)
28. A. Ammar, A. Jeribi, The Weyl essential spectrum of a sequence of linear operators in Banach spaces. Indag. Math. (N.S.) **27**(1), 282–295 (2016)
29. A. Ammar, A. Jeribi, The essential pseudo-spectra of a sequence of linear operators. Complex Anal. Oper. Theory **12**(3), 835–848 (2018)
30. A. Ammar, A. Jeribi, K. Mahfoudhi, A characterization of the essential approximation pseudospectrum on a Banach space. Filomat **31**(11), 3599–3610 (2017)
31. A. Ammar, A. Jeribi, K. Mahfoudhi, A characterization of the condition pseudospectrum, *Funct. Anal. Approx. Comput.* **10**(2), 13–21 (2018)
32. A. Ammar, A. Jeribi, N. Moalla, On a characterization of the essential spectra of a 3×3 operator matrix and application to three-group transport operators. Ann. Funct. Anal. AFA **4**(2), 153–170, electronic only (2013)
33. A. Ammar, A. Jeribi, B. Saadaoui, Frobenius-Schur factorization for multivalued 2×2 matrices linear operator. Mediterr. J. Math. **14**(1), 29, Paper No. 29 (2017)
34. H. Baloudi, A. Jeribi, Left-Right Fredholm and Weyl spectra of the sum of two bounded operators and application. Mediterr. J. Math. **11**, 939–953 (2014)
35. H. Baloudi, A. Jeribi, Holomorphically Weyl-decomposably regular. (English) Funct. Anal. Approx. Comput. **8**(2), 13–22 (2016)
36. N. Ben Ali, A. Jeribi, N. Moalla, On some subsets of Schechter's essential spectrum of a matrix operator and application to transport operator. Math. Nachr. **283**(9), 1245–1256 (2010)
37. A. Ben Amar, A. Jeribi, M. Mnif, Some applications of the regularity and irreducibility on transport theory. Acta Appl. Math. **110**, 431–448 (2010)

38. A. Ben Amar, A. Jeribi, B. Krichen, Essential spectra of a 3×3 operator matrix and application to three-group transport equation. Inte. Equa. Oper. Theo. **68**, 1–21 (2010)
39. A. Ben Amar, A. Jeribi, M. Mnif, Some results on Fredholm and semi-Fredholm operators. Arab. J. Math. (Springer) **3**(3), 313–323 (2014)
40. M. Benharrat, A. Ammar, A. Jeribi, B. Messirdi, On the Kato, semi-regular and essentially semi-regular spectra. Func. Anal. Appr. Comp. **6**(2), 9–22 (2014)
41. W. Chaker, A. Jeribi, B. Krichen, Demicompact linear operators, essential spectrum and some perturbation results. Math. Nachr. **288**(13), 1476–1486 (2015)
42. S. Charfi, A. Jeribi, On a characterization of the essential spectra of some matrix operators and applications to two-group transport operators. Math. Z. **262**(4), 775–794 (2009)
43. S. Charfi, A. Jeribi, N. Moalla, Time asymptotic behavior of the solution of an abstract Cauchy problem given by a one-velocity transport operator with Maxwell boundary condition. Collectanea Mathematica **64**, 97–109 (2013)
44. S. Charfi, A. Jeribi, R. Moalla, Essential spectra of operator matrices and applications. Methods Appl. Sci. **37**(4), 597–608 (2014)
45. S. Charfi, A. Jeribi, I. Walha, Essential spectra, matrix operator and applications. Acta Appl. Math. **111**(3), 319–337 (2010)
46. M. Damak, A. Jeribi, On the essential spectra of some matrix operators and applications. Electron. J. Differ. Equ. **11**, 1–16 (2007)
47. E.B. Davies, *Linear Operators and Their Spectra* (Cambridge University Press, Cambridge, 2007)
48. J. Diestel, A survey of results related to Dunford-Pettis property, in Contemporary Mathematics, vol. 2, Amer. Math. Soc. of Conference on Integration, Topology and Geometry in Linear Spaces (1980), pp. 15–60
49. N. Dunford, B.J. Pettis, Linear operations on summable functions. Trans. Amer. Math. Soc. **47**, 323–392 (1940)
50. N. Dunford, J.T. Schwartz, *Linear Operators, Part III* (Wiley-Interscience, New-York, 1971)
51. A. Elleuch, A. Jeribi, On a characterization of the structured Wolf, Schechter and Browder essential pseudospectra. Indag. Math. **27**, 212–224 (2016)
52. A. Elleuch, A. Jeribi, New description of the structured essential pseudospectra. Indag. Math. **27**, 368–382 (2016)
53. G. Faber, Uber die orthogonalfunktionen des herrn haar. Deutsche Math. Ver (in German) **19**, 104–112 (1910)
54. I.C. Gohberg, S. Goldberg, M.A. Kaashoek, Classes of linear operators, Vol. I, Operator Theory: Advances and Applications, Vol. 29, Birkhauser, Basel (1990)
55. I.C. Gohberg, M.G. Krein, *Introduction to the Theory of Linear Non-self-Adjoint Operators in Hilbert Space* (Amer. Math. Soci, Providence, 1969)
56. S. Goldberg, *Unbounded Linear Operators* (McGraw-Hill, New-York, 1966)
57. A. Grothendieck, Produits tensoriels topologiques et espaces nucleaires. Mem. Am. Math. Soc. **16**, 13–30 (1955)
58. A. Haar, Zur theorie der orthogonalen Funktionensysteme. Mathematische Annalen **69**(3), 331–371 (1910)
59. G. Hamel, Eine basis aller zahlen und die unstetigen losungen der funktionalgleichung $f(x+y) = f(x) + f(y)$. Math. Ann. **60**(3), 459–462 (1905)
60. A. Jeribi, *Spectral Theory and Applications of Linear Operators and Block Operator Matrices* (Springer, New-York, 2015)
61. A. Jeribi, *Denseness, Bases and Frames in Banach Spaces and Applications* (de Gruyter, Berlin, 2018)
62. A. Jeribi, *Linear Operators and Their Essential Pseudospectra* (CRC Press, Boca Raton, 2018)
63. A. Jeribi (ed.), Operator theory. Proceedings of the international conference on operator theory, Hammamet, Tunisia, April 30 - May 3, 2018. De Gruyter Proceedings in Mathematics. Berlin: De Gruyter p. 216 (2021) (ISBN 978-3-11-059686-1/hbk; 978-3-11-059819-3/ebook). x

64. A. Jeribi, M.A. Hammami, A. Masmoudi (eds.), Applied mathematics in Tunisia, in *International conference on advances in applied mathematics (ICAAM)* (Hammamet, Tunisia, 2013); *Springer Proceedings in Mathematics and Statistics*, vol. 131 (Springer, Cham, 2015), p. 397. (ISBN 978-3-319-18040-3/hbk; 978-3-319-18041-0/ebook). xix
65. A. Jeribi, Quelques remarques sur les oprateurs de Fredholm et application l'quation de transport. C. R. Acad. Sci. Paris Srie I **325**, 43–48 (1997)
66. A. Jeribi, Quelques remarques sur le spectre de Weyl et applications. C. R. Acad. Sci., Paris Srİ Math. **327**(5), 485–490 (1998)
67. A. Jeribi, Développement de certaines propriétés fines de la théorie spectrale et applications à des modèles monocinétiques et à des modèles de Reggeons, Thesis of Mathematics. University of Corsica, Frensh (16 Janvier 1998)
68. A. Jeribi, Une nouvelle caractrisation du spectre essentiel et application, C. R. Acad. Sci. Paris, t. 331, Srie I, 525–530 (2000)
69. A. Jeribi, A characterization of the essential spectrum and applications. Boll. dell. Unio. Mate. Ital., **8**(B–5), 805–825 (2002)
70. A. Jeribi, A characterization of the Schechter essential spectrum on Banach spaces and applications. J. Math. Anal. Appl. **271**, 343–358 (2002)
71. A. Jeribi, Some remarks on the Schechter essential spectrum and applications to transport equations. J. Math. Anal. Appl. **275**, 222–237 (2002)
72. A. Jeribi, On the Schechter essential spectrum on Banach spaces and applications. Ser. Math. Info. **17**, 35–55 (2002)
73. A. Jeribi, Fredholm operators and essential spectra. Arch. Ineq. Appl. **2**(2–3), 123–140 (2004)
74. A. Jeribi, B. Krichen, M. Zarai Dhahri, Essential spectra of some of matrix operators involving γ-relatively bounded entries and an applicatons. Linear Multilinear Algebra 1654–1668 (2016)
75. A. Jeribi, K. Latrach, Quelques remarques sur le spectre essentiel et application l'quation de transport, C. R. Acad. Sci. Paris, t. 323, Srie I, 469–474 (1996)
76. A. Jeribi, M. Mnif, Fredholm operators, essential spectra and application to transport equation. Acta Appl. Math. **89**, 155–176 (2005)
77. A. Jeribi, N. Moalla, Fredholm operators and Riesz theory for polynomially compact operators. Acta Appl. Math. **90**(3), 227–245 (2006)
78. A. Jeribi, N. Moalla, A characterization of some subsets of Schechter's essential spectrum and application to singular transport equation. J. Math. Anal. Appl. **358**, 434–444 (2009)
79. A. Jeribi, N. Moalla, I. Walha, Spectra of some block operator matrices and application to transport operators. J. Math. Anal. Appl. **351**(1), 315–325 (2009)
80. A. Jeribi, N. Moalla, S. Yengui, S-essential spectra and application to an example of transport operators. Math. Methods Appl. Sci. **37**(16), 2341–2353 (2014)
81. A. Jeribi, N. Moalla, S. Yengui, Some results on perturbation theory of matrix operators, M-essential spectera of matrix operators and application to an example of transport operators. J. Math. Appl. 44 (2021)
82. A. Jeribi, I. Walha, Gustafson, Weidmann, Kato, Wolf, Schechter and Browder Essential Spectra of Some Matrix Operator and Application to Two-Group Transport Equation. Math. Nachr. **284**(1), 67–86 (2011)
83. T. Kato, *Perturbation Theory for Linear Operators* (Springer, Berlin, 1980)
84. K. Latrach, A. Jeribi, On the essential spectrum of transport operators on L_1-spaces. J. Math. Phys. **37**(12), 6486–6494 (1996)
85. K. Latrach, A. Jeribi, Sur une quation de transport intervenant en dynamique des populations, C, R. Acad. Sci. Paris, t. 325, Série I, 1087–1090 (1997)
86. K. Latrach, A. Jeribi, Some results on Fredholm operators, essential spectra and application. J. Math. Anal. Appl. **225**(2), 461–485 (1998)
87. Z.H. Luo, B.Z. Guo, Ö. Morgül, *Stability and Stabilization of Linear Infinite Dimensional Systems with Applications* (Springer, London, 1999)
88. A. S. Markus, Introduction to the spectral theory of polynomial operator pencils, Amer. Math. Soci., Providence, RI, 71 (1988)

89. N. Moalla, M. Damak, A. Jeribi, Essential spectra of some matrix operators and application to two-group transport operators with general boundary conditions. J. Math. Anal. Appl. **323**(2), 1071–1090 (2006)
90. A. Pelczynski, On strictly singular and strictly cosingular operators. I. Strictly singular and strictly cosingular operators in $C(X)$-spaces. II. Strictly singular and strictly cosingular operators in $L(\mu)$-spaces, Bull. Acad. Polon. Sci., 13, 13–36 and 37–41 (1965)
91. M. Schechter, *Principls of functional analysis* (Academic, 1971)
92. J. Schauder, Zur theorie stetiger abbildungen in funktionalraumen. Math. Z. **26**, 47–65 (1927)
93. J. Schauder, Eine eigenschaft des haarschen orthogonalsystems. Math. Z. **28**, 317–320 (1928)
94. M. Schechter, On the essential spectrum of an arbitrary operator. J. Math. Anal. Appl. **13**, 205–215 (1966)
95. C. Wyss, Perturbation theory for Hamiltonian operator matrices and Riccati equations, PhD thesis, University of Bern, (2008)
96. C. Wyss, Riesz bases for p-subordinate perturbations of normal operator. J. Func. Anal. **258**, 208–240 (2010)
97. Snežana Č. Živković-Zlatanović, M. Berkani, Topological uniform descent, quasi-Fredholmness and operators originated from semi-B-Fredholm theory, Complex Anal. Oper. Theory **13**(8), 3595–3622 (2019)
98. S. Yakubov, *Completeness of Root Functions of Regular Differential Operators* (Longman, Harlow, 1994)

Chapter 2
Analysis with Operators

In this chapter, we give some fundamentals and elementary results about projections, particular functions such as the function of finite-order and the sine-type function, which are used in the sequel. We also recall Phragmén-Lindelöf theorems, spectral properties of holomorphic operator functions. After that, we introduce some definitions about the theory of semigroup. Some concepts of comparison of operators such as concepts of subordination and fully subordination are given.

2.1 Projections

This section is devoted to the study of the concept of projection, in general, as well as the orthogonal and spectral projection.

2.1.1 Generalities

Definition 2.1.1 A bounded operator P on X is said projection if $P^2 = P$. ◇

Proposition 2.1.1 *Let P and Q be two projections from X on the subspaces F and D, respectively. If $\|P - Q\| < 1$, then*

$$\dim F = \dim D.$$ ◇

Definition 2.1.2 Let X, Y be normed linear spaces.

(i) A linear operator $T : X \longrightarrow Y$ is a topological isomorphism if T is a bijection and both T and T^{-1} are continuous.
(ii) We say that X and Y are topologically isomorphic if there exists a topological isomorphism $T : X \longrightarrow Y$. ◇

A. Jeribi, *Perturbation Theory for Linear Operators*,
https://doi.org/10.1007/978-981-16-2528-2_2

Every isometric isomorphism is a topological isomorphism, but the converse need not hold. Let P and Q be two projections in a Banach space X with

$$\|P - Q\| < 1.$$

Let $u \in N(Q) \bigcap R(P)$. Then,

$$\|u\| = \|(P - Q)u\|;$$

as $\|P - Q\| < 1$, this is only possible for $u = 0$. Hence,

$$N(Q) \bigcap R(P) = \{0\}$$

and $Q_{|R(P)}$ is injective. A Neumann series argument shows that $I - Q + P$ is an isomorphism in X. Consequently, for every $v \in X$, there exists $u \in X$ such that

$$v = (I - Q + P)u = (I - Q)u + Pu.$$

This implies

$$X = R(I - Q) + R(P) = N(Q) + R(P).$$

Moreover, if $v \in R(Q)$, then

$$v = Qv = QPu.$$

Hence, $Q_{|R(P)}$ maps onto $R(Q)$. So,

$$X = N(Q) \oplus R(P),$$

and Q induces an isomorphism $Q_{|R(P)} : R(P) \longrightarrow R(Q)$.

2.1.2 Orthogonal Projection

Definition 2.1.3 A projection P on a Hilbert space X is called orthogonal projection if it satisfies $P^* = P$. ◇

Proposition 2.1.2 (Gohberg and Krein [18, Lemma I.4.2 p. 17]) *Let A be a compact operator on a Hilbert space X such that the subspace generated by its generalized eigenvectors $F_A \neq X$, and let Q_A be the orthogonal projection which projects X on $F_A^{\perp}$. Then, $Q_A A Q_A$ is a Volterra operator.* ◇

Proposition 2.1.3 (Gohberg and Krein [18, p. 2]) *Let A be a bounded operator on a Hilbert space X, F be a subspace invariant by A, and P be the orthogonal projection which projects X on F. Then, if both the operators A and $PAP + Q$ are invertible,* *then the operator* $QAQ + P$ *is also invertible, and we have*

$$(QAQ + P)^{-1} = QA^{-1}Q + P,$$

where $Q = I - P$. ◇

Proposition 2.1.4 *Let* G *be a normal operator with domain* $\mathscr{D}(G)$ *dense in* X *and* F *be a closed subspace of* X *invariant by* G *and* G^**, then the orthogonal projection* P_F *of* F *commutes with* G *and* G^*. ◇

2.1.3 Spectral Projection

Let A be a closed linear operator on X. Suppose that the spectrum $\sigma(A)$ can be decomposed into two isolated parts σ and σ' so that we can draw a closed curve Γ passing on the resolvent set $\rho(A)$ and having σ in its interior and σ' in its exterior. Define

$$P := E(\lambda, A) = \frac{1}{2\pi i} \oint_{\Gamma} R(\lambda, A) \mathrm{d}\lambda \tag{2.1.1}$$

the spectral (or Riesz) projection of X on F in the direction F', where one traverses Γ in the positive direction. We have $F = PX$ and $F' = (I - P)X$. F and F' are called the spectral spaces corresponding, respectively, to the isolated parts σ and σ' of the spectrum. F includes, in particular, all own elements corresponding to eigenvalues λ interior to σ, i.e., the solutions f of the equation

$$(A - \lambda)f = 0.$$

Proposition 2.1.5 *P is a projection.* ◇

Proposition 2.1.6 *Let* A *be a closed linear operator on* X *and* P*, defined in (2.1.1), be the corresponding spectral projection. Then,*

$$AP = \frac{1}{2\pi i} \oint_{\Gamma} \lambda R(\lambda, A) \mathrm{d}\lambda.$$ ◇

We recall some facts about direct sums of subspaces. By a subspace of a Banach space X we understand a linear subspace in the algebraic sense, i.e., it need not be topologically closed. For a finite system $U_1, \ldots, U_n \subset X$ of subspaces, the sum $U_1 + \cdots + U_n$ is called algebraic direct, denoted

$$U_1 \dotplus \cdots \dotplus U_n,$$

if $x_1 + \cdots + x_n = 0$ with $x_j \in U_j$ implies $x_1 = \cdots = x_n = 0$. The corresponding projections

$$P_j : U_1 \dotplus \cdots \dotplus U_n \longrightarrow U_j$$

are not necessarily bounded and we shall use the term algebraic projection in this context. The sum is called topological direct, denoted

$$U_1 \oplus \cdots \oplus U_n,$$

if it is algebraic direct and the algebraic projections $P_1, \ldots, P_n$ are bounded. In this case, the sum is closed (and thus a Banach space) if and only if every U_j is closed. The notion "projection on a Banach space X" will always refer to a bounded operator $P : X \longrightarrow X$ satisfying

$$P^2 = P;$$

such a projection gives rise to the topological direct

$$X = N(P) \oplus R(P).$$

Let $(V_\lambda)_{\lambda\in\Lambda}$ be a family of subspaces of a Banach space X with Λ an arbitrary index set. We will denote by

$$\sum_{\lambda\in\Lambda} V_\lambda = \left\{x_{\lambda_1} + \cdots + x_{\lambda_n},\ n \in \mathbb{N},\ x_{\lambda_j} \in V_{\lambda_j}\right\}$$

the sum of the family $(V_\lambda)_{\lambda\in\Lambda}$ in the algebraic sense. There is an obvious generalization of algebraic direct sums to the case of infinitely many subspaces.

Definition 2.1.4 The family $(V_\lambda)_{\lambda\in\Lambda}$ of subspaces of a Banach space X is called finitely linearly independent if $x_{\lambda_1} + \cdots + x_{\lambda_n} = 0$, $x_{\lambda_j} \in V_{\lambda_j}$, then

$$x_{\lambda_1} = \cdots = x_{\lambda_n} = 0$$

for every finite subset $\{\lambda_1, \ldots, \lambda_n\} \subset \Lambda$. ◇

Lemma 2.1.1 *For a family $(V_\lambda)_{\lambda\in\Lambda}$ of subspaces of a Banach space X, the following properties are equivalent:*

(i) *$(V_\lambda)_{\lambda\in\Lambda}$ is finitely linearly independent.*
(ii) *Every $x \in \sum_{\lambda\in\Lambda} V_\lambda$ has a unique representation*

$$x = \sum_{\lambda\in\Lambda} x_\lambda$$

with $x_\lambda \in V_\lambda$ and almost all x_λ zero.
(iii) *There is a family of algebraic projections $(P_\lambda)_{\lambda\in\Lambda}$ corresponding to $(V_\lambda)_{\lambda\in\Lambda}$ with domain $\mathscr{D}(P_\lambda) = \sum_{\mu\in\Lambda} V_\mu$, range $R(P_\lambda) = V_\lambda$, and the property that $P_\mu P_\lambda = 0$ whenever $\mu \neq \lambda$.* ◇

Proof The implication $(i) \Rightarrow (ii)$ is clear; for $(ii) \Rightarrow (iii)$ define

$$P_\lambda\left(\sum_\mu x_\mu\right) = x_\lambda$$

for each $\lambda \in \Lambda$. $(iii) \Rightarrow (i)$ From $R(P_\lambda) = V_\lambda$ and $P_\mu P_\lambda = 0$ for $\mu \neq \lambda$, we obtain $P_\lambda x_\mu = 0$ for $x_\mu \in V_\mu$ and $\mu = \lambda$. Hence $x_{\lambda_1} + \cdots + x_{\lambda_n} = 0$ implies

$$x_{\lambda_j} = P_{\lambda_j}(x_{\lambda_1} + \cdots + x_{\lambda_n}) = 0.$$

This completes the proof. Q.E.D.

Because of the uniqueness of the expansion

$$x = \sum_{\lambda\in\Lambda} x_\lambda,$$

we call the sum of the finitely linearly independent family $(V_\lambda)_{\lambda\in\Lambda}$ algebraic direct and use the notation

$$\sum_{\lambda\in\Lambda}^{\dot{+}} V_\lambda.$$

2.1.4 Sum of Spectral Projection

Let $P_0, \ldots, P_n$ be projections in a Hilbert space X with $P_j P_k = 0$ for $j \neq k$ and $x \in X$. Let $x_k = P_k x$ and use Eq. (1.3.4) considering that $\varepsilon \in E$ for which

$$\|\varepsilon_0 x_0 + \cdots + \varepsilon_n x_n\|$$

becomes maximal, where E is defined in (1.3.3). Then, we obtain

$$\begin{aligned}\sum_{k=0}^{n} \|P_k x\|^2 &\leq \|\varepsilon_0 x_0 + \cdots + \varepsilon_n x_n\|^2 \\ &= \left\|\left(\sum_{k=0}^{n} \varepsilon_k P_k\right)\left(\sum_{k=0}^{n} x_k\right)\right\|^2 \\ &\leq C^2 \left\|\sum_{k=0}^{n} P_k x\right\|^2.\end{aligned}$$

Furthermore, if we choose $\varepsilon \in E$ such that $\|\varepsilon_0 x_0 + \cdots + \varepsilon_n x_n\|$ is minimal, we find

$$\left\|\sum_{k=0}^{n} P_k x\right\|^2 = \left\|\left(\sum_{k=0}^{n} \varepsilon_k P_k\right)\left(\sum_{k=0}^{n} \varepsilon_k x_k\right)\right\|^2$$
$$\leq C^2 \|\varepsilon_0 x_0 + \cdots + \varepsilon_n x_n\|^2$$
$$\leq C^2 \sum_{k=0}^{n} \|P_k x\|^2.$$

So,

$$C^{-2} \sum_{k=0}^{n} \|P_k x\|^2 \leq \left\|\sum_{k=0}^{n} P_k x\right\|^2 \leq C^2 \sum_{k=0}^{n} \|P_k x\|^2, \tag{2.1.2}$$

where

$$C = \max\left\{\left\|\sum_{k=0}^{n} \varepsilon_k P_k\right\| \text{ with } \varepsilon_k = \pm 1\right\}.$$

2.1.5 l^2-Decomposition

2.1.5.1 Definitions

Definition 2.1.5 We say that a family $(V_\lambda)_{\lambda\in\Lambda}$ of closed subspaces of a Banach space X forms an l^2-decomposition of X if

(i) the sum $\sum_{\lambda\in\Lambda} V_\lambda \subset X$ is dense, and
(ii) there exists $c \geq 1$ such that

$$c^{-1} \sum_{\lambda\in F} \|x_\lambda\|^2 \leq \left\|\sum_{\lambda\in F} x_\lambda\right\|^2 \leq c \sum_{\lambda\in F} \|x_\lambda\|^2$$

for all finite subsets $F \subset \Lambda$ and $x_\lambda \in V_\lambda$, where Λ is an arbitrary index set. ◇

Proposition 2.1.7 (Wyss [33, Proposition 2.1.6]) *If the family $(V_\lambda)_{\lambda\in\Lambda}$ forms an l^2-decomposition of a Banach space X, then every $x \in X$ has a unique expansion*

$$x = \sum_{\lambda\in\Lambda} x_\lambda$$

with $x_\lambda \in V_\lambda$. Its members are given by

$$x_\lambda = P_\lambda x,$$

where P_λ is the projection onto V_λ defined by the projection corresponding to the subset $\{\lambda\} \subset \Lambda$. ◇

Because of the uniqueness of the expansion

$$x = \sum_{\lambda \in \Lambda} x_\lambda,$$

we use the notation

$$X = \bigoplus_{\lambda \in \Lambda}^{2} V_\lambda$$

for an l^2-decomposition.

Definition 2.1.6 Let $T(X \longrightarrow X)$ be a closed operator on a separable Hilbert space X. We say that an l^2-decomposition

$$X = \bigoplus_{k \in \mathbb{N}}^{2} V_k$$

is finitely determining for T if $\dim V_k < \infty$, $V_k \subset \mathscr{D}(T)$, $T(V_k) \subset V_k$, and

$$\sum_{k=0}^{\infty} V_k$$

is a core for T. ◇

A finitely determining l^2-decomposition is not uniquely determined since any finite collection of the subspaces V_k can be replaced by its sum. Note that the restrictions

$$T_{|V_k} : V_k \longrightarrow V_k$$

are bounded since the V_k are finite dimensional. The assumption of $\sum_k V_k$ being a core for T will then enable us to carry over results for the finite-dimensional parts $T_{|V_k}$ to the whole operator T. Without the core property, the theory still applies to an operator generated by the parts $T_{|V_k}$.

2.1.5.2 General Operator

Let $T(X \longrightarrow X)$ be an operator and

$$X = \bigoplus_{k \in \mathbb{N}}^{2} V_k$$

be an l^2-decomposition with $\dim V_k < \infty$, $V_k \subset \mathscr{D}(T)$, and $T(V_k) \subset V_k$. Let P_k be the projection onto V_k corresponding to the given l^2-decomposition. We claim that the restriction

$$T_0 = T_{|\sum_{k=0}^{\infty} V_k} \text{ is closable.} \tag{2.1.3}$$

In fact, let $y_n \in \mathscr{D}(T_0)$ with $y_n \to 0$ and $T_0 y_n \to z$. We can write y_n in the following form

$$y_n = \sum_{j \in \mathbb{N}} P_j y_n,$$

where the sum is actually finite since

$$\mathscr{D}(T_0) = \sum_k V_k.$$

Hence, the T-invariance of the V_k's gives

$$P_k T_0 y_n = P_k \sum_j T P_j y_n = T P_k y_n. \tag{2.1.4}$$

Thus, since $T_{|V_k}$ is a bounded operator because V_k is finite dimensional, then

$$\begin{aligned} P_k z &= P_k \lim_{n \to \infty} T_0 y_n \\ &= \lim_{n \to \infty} P_k T_0 y_n \\ &= \lim_{n \to \infty} T_{|V_k} P_k y_n \\ &= T_{|V_k} \lim_{n \to \infty} P_k y_n \\ &= T_{|V_k}(0) = 0. \end{aligned}$$

Since $P_k z = 0$ for all k, then we deduce that $z = 0$, i.e., T_0 is closable. This completes the proof of the claim. On the other hand, $\sum_k V_k$ is then a core for $\overline{T_0}$. So,

$$\bigoplus_{k \in \mathbb{N}}^{2} V_k$$

is finitely determining for the closure $\overline{T_0}$.

2.1.5.3 Closed Operator

Let $T(X \longrightarrow X)$ be a closed operator with finitely determining l^2-decomposition

$$X = \bigoplus_{k \in \mathbb{N}}^{2} V_k.$$

Let P_k be the projections onto V_k corresponding to the l^2-decomposition. Let $y \in \mathscr{D}(T)$. Since $\sum_k V_k$ is a core for T, then there is a sequence $y_n \in \sum_k V_k$ with $y_n \to y$, and $Ty_n \to Ty$. Using (2.1.4), we have $P_k T y_n = T P_k y_n$ and

$$\begin{aligned} P_k T y &= P_k \lim_{n\to\infty} T y_n \\ &= \lim_{n\to\infty} P_k T y_n \\ &= T_{|V_k} \lim_{n\to\infty} P_k y_n \\ &= T P_k y. \end{aligned}$$

So,

$$\begin{aligned} \sum_k \|T P_k y\|^2 &= \sum_k \|P_k T y\|^2 \\ &\leq c\|Ty\|^2 < \infty \end{aligned}$$

and

$$y = \sum_k P_k y \in \left\{ \sum_{k=0}^{\infty} x_k \in \bigoplus_{k\in\mathbb{N}}^{2} V_k \text{ such that } \sum_{k=0}^{\infty} \|T x_k\|^2 < \infty \right\}$$

with

$$\begin{aligned} Ty &= \sum_k P_k T y \\ &= \sum_k T P_k y. \end{aligned}$$

Furthermore, if $\sum_k x_k \in \bigoplus_{k\in\mathbb{N}}^{2} V_k$ with $\sum_k \|T x_k\|^2 < \infty$, then

$$\mathscr{D}(T) \ni \sum_{k=0}^{n} x_k \to \sum_{k=0}^{\infty} x_k$$

and

$$T \sum_{k=0}^{n} x_k = \sum_{k=0}^{n} T x_k \to \sum_{k=0}^{\infty} T x_k.$$

Thus, since T is closed, then

$$\sum_k x_k \in \mathscr{D}(T).$$

So,

$$\mathscr{D}(T) = \left\{ \sum_{k=0}^{\infty} x_k \in \bigoplus_{k\in\mathbb{N}}^{2} V_k \text{ such that } \sum_{k=0}^{\infty} \|T x_k\|^2 < \infty \right\}, \tag{2.1.5}$$

$$Tx = \sum_{k=0}^{\infty} Tx_k \text{ for } x = \sum_{k=0}^{\infty} x_k \in \mathscr{D}(T). \tag{2.1.6}$$

Hence, T is bounded if, and only if, the restrictions $T_{|V_k}$ are uniformly bounded and, in this case,

$$\|T\| \leq c \sup_{k \in \mathbb{N}} \|T_{|V_k}\|.$$

In fact, if $\sup_k \|T_{|V_k}\| < \infty$, then for $x = \sum_k x_k \in \mathscr{D}(T)$, we have

$$\begin{aligned}
\|Tx\|^2 &= \left\| \sum_k T_{|V_k} x_k \right\|^2 \\
&\leq c \sum_k \|T_{|V_k} x_k\|^2 \\
&\leq c \left(\sup_k \|T_{|V_k}\| \right)^2 \sum_k \|x_k\|^2 \\
&\leq c^2 \left(\sup_k \|T_{|V_k}\| \right)^2 \|x\|^2.
\end{aligned}$$

Therefore, T is bounded with norm

$$\|T\| \leq c \sup_k \|T_{|V_k}\|.$$

Furthermore, the point spectrum of T is given by

$$\sigma_p(T) = \bigcup_{k=0}^{\infty} \sigma(T_{|V_k}). \tag{2.1.7}$$

In fact, we have

$$\sigma(T_{|V_k}) \subset \sigma_p(T)$$

for all $k \in \mathbb{N}$. Now, If $\lambda \in \sigma_p(T)$, then there exists an element $\sum_k x_k \in \mathscr{D}(T) \backslash \{0\}$ such that

$$0 = (T - \lambda) \sum_{k=0}^{\infty} x_k = \sum_{k=0}^{\infty} (T - \lambda) x_k.$$

Thus,

$$(T_{|V_k} - \lambda) x_k = 0$$

for all k. Since $x_{k_0} \neq 0$ for some k_0, we find $\lambda \in \sigma(T_{|V_{k_0}})$. Also, the residual spectrum of T is given by

$$\sigma_r(T) = \emptyset.$$

In fact, let $z \notin \sigma_p(T)$. Then, the injective operator $T - z$ maps each finite-dimensional T-invariant subspace V_k onto itself. Hence, we deduce that

$$\sum_k V_k \subset R(T - z).$$

So, the range is thus dense. Now, we want to derive the formula for the resolvent set. In fact, the resolvent set is given by

$$\rho(T) = \left\{ z \in \mathbb{C} \backslash \sigma_p(T) \text{ such that } \sup_{k \in \mathbb{N}} \|(T_{|V_k} - z)^{-1}\| < \infty \right\}. \tag{2.1.8}$$

Indeed, for one inclusion, let $z \in \mathbb{C} \backslash \bigcup_k \sigma(T_{|V_k})$ such that

$$\sup_k \|T_{|V_k}\| = \sup_k \|(T_{|V_k} - z)^{-1}\| < \infty.$$

Using (2.1.5), (2.1.6), and their properties, we have

$$S : \sum_{k=0}^{\infty} x_k \longrightarrow \sum_{k=0}^{\infty} (T_{|V_k} - z)^{-1} x_k$$

defining a bounded operator $S : X \longrightarrow X$, which has the finitely determining decomposition $\bigoplus_{k \in \mathbb{N}}^{2} V_k$ and $R(S) \subset \mathscr{D}(T - z)$. It is easy to see

$$(T - z)Sx = x$$

for all $x \in X$. Since $z \notin \sigma_p(T)$, i.e., $T - z$ is injective, then $z \in \rho(T)$ with

$$(T - z)^{-1} = S.$$

For the other inclusion. Let $z \in \rho(T)$, then $z \notin \sigma(T_{|V_k})$ for all k. Since $T_{|V_k} \subset T$, then

$$(T_{|V_k} - z)^{-1} \subset (T - z)^{-1}.$$

Hence,

$$\|(T_{|V_k} - z)^{-1}\| \leq \|(T - z)^{-1}\| \text{ for all } k.$$

So, if T is closed with a finitely determining decomposition $\bigoplus_{k \in \mathbb{N}}^{2} V_k$, then the point spectrum of T is non-empty and at most countably infinite. For

$$x = \sum_k x_k \in \bigoplus_{k\in\mathbb{N}}^2 V_k,$$

we have $x \in \mathscr{L}_T(\lambda)$ if, and only if, $x_k \in \mathscr{L}_T(\lambda)$ for all $k \in \mathbb{N}$. Moreover, $\bigoplus_{k\in\mathbb{N}}^2 V_k$ is finitely determining for $(T-z)^{-1}$, $z \in \rho(T)$, and

$$(T-z)^{-1}x = \sum_k (T_{|V_k} - z)^{-1} x_k$$

for

$$x = \sum_k x_k \in \bigoplus_{k\in\mathbb{N}}^2 V_k.$$

Example 2.1 Consider an orthogonal decomposition

$$X = \bigoplus_{k\in\mathbb{N}} V_k$$

of a Hilbert space into finite-dimensional subspaces V_k and

$$T_k : V_k \longrightarrow V_k$$

is a linear operator. Let $T(X \longrightarrow X)$ be the operator defined by

$$\mathscr{D}(T) = \sum_{k\in\mathbb{N}}^{\dotplus} V_k, \quad T_{|V_k} = T_k.$$

By using (2.1.3), we have that the operator T is closable and that $\bigoplus_{k\in\mathbb{N}} V_k$ is a finitely determining l^2-decomposition for $\overline{T}$. Now, Eq. (2.1.7) gives

$$\sigma_p(\overline{T}) = \bigcup_{k=0}^{\infty} \sigma(T_{|V_k}) = \bigcup_{k=0}^{\infty} \sigma(T_k).$$

In particular, for any given non-empty subset $\sigma \subset \mathbb{C}$ which is at most countable, we may choose the operators T_k such that

$$\sigma_p(\overline{T}) = \sigma.$$

Example 2.2 Consider an orthogonal decomposition

$$X = \bigoplus_{k \in \mathbb{N}} V_k$$

such that $\dim V_k = 2$ and an operator $T_0(X \longrightarrow X)$ with

$$\mathscr{D}(T_0) = \sum_k V_k$$

such that all V_k are invariant. Suppose that the restriction $T_{0|V_k}$ have eigenvalues k and $k+i$. By using (2.1.3), we have T_0 is closable and by using (2.1.7), we have

$$\sigma_p(\overline{T_0}) = \bigcup_k \{k, k+i\}.$$

Moreover, if there are unit length eigenvectors $v_k,\ w_k \in V_k$ corresponding to k and $k+i$, respectively, which satisfy

$$\langle v_k, w_k \rangle = 1 - k^{-q}$$

with $q > 6$, then

$$\sup_{k \in \mathbb{N}} \|(T_{0|V_k} - z)^{-1}\| = \infty,$$

for all $z := x + iy \in \mathbb{C} \backslash \bigcup_k \{k, k+i\}$ with $y \neq \frac{1}{2}$. In fact, it is easy to see that

$$\|(T_{0|V_k} - z)^{-1}\| \geq \frac{|(k-z)^{-1}| - |(k+i-z)^{-1}|}{\sqrt{2(1 - \langle v_k, w_k \rangle)}} \to \infty \text{ as } k \to \infty.$$

By using the characterization of the resolvent set, given in (2.1.8), we see that

$$\left\{ z \in \mathbb{C} \text{ such that } Im(z) \neq \frac{1}{2} \right\} \subset \sigma(\overline{T_0}).$$

Since the spectrum is a closed set, this implies that

$$\sigma(\overline{T_0}) = \mathbb{C}.$$

2.2 Spectral Theory of Compact and Discrete Operators

2.2.1 Riesz-Schauder Theorem

Theorem 2.2.1 *Let $A = I - K$ with $K \in \mathscr{K}(X)$, then there is a unique $r \in \mathbb{N}$ such that*

$$\{0\} = N(A^0) \subsetneq N(A) \subsetneq N(A^2) \ldots \subsetneq N(A^r) = N(A^{r+1}),$$

and

$$X = A^0(X) \supsetneq A(X) \supsetneq A^2(X) \supsetneq \cdots \supsetneq A^r(X) = A^{r+1}(X). \quad \diamond$$

We can see the proof of Theorem 2.2.1 in [20].

Corollary 2.2.1 *Let $A = I - K$ with $K \in \mathscr{K}(X)$. If A is one to one, then A is invertible with inverse, A^{-1} is bounded.* $\diamond$

Proof If A is one to one, then $N(A) = \{0\}$. So, $r = 0$ (see Theorem 2.2.1). Hence, $A(X) = X$. Thus, A is onto. So, A is invertible. Since A is bounded, then by using Theorem 1.2.2, A^{-1} is bounded. Q.E.D.

The fundamental results of the spectral theory of compact operators given in all the books of functional analysis are summarized here by the following theorem:

Theorem 2.2.2 (Riesz-Schauder theorem) *Let A be a compact operator in $\mathscr{L}(X)$. Then,*

(i) if X is infinite dimensional, then $0 \in \sigma(A)$;

(ii) $\sigma(A)$ is a discrete set having no limit points except perhaps $\lambda = 0$, i.e.,

$$\sigma(A)\backslash\{0\} = \sigma_p(A)\backslash\{0\}.$$

Further, any non-zero $\lambda \in \sigma(A)$ is an eigenvalue of finite multiplicity (i.e., the corresponding space of eigenvectors is finite dimensional);

(iii) $\sigma(A)$ is countable with 0 as only possible accumulation point, i.e., either $\sigma(A)$ is finite or there exists a sequence $(\lambda_n)_{n=1}^{\infty} \subset \mathbb{C}$ such that $\lim\limits_{n\to+\infty} \lambda_n = 0$ and

$$\sigma(A) = \{\lambda_n,\ n \in \mathbb{N}^*\} \bigcup \{0\};$$

(iv) every non-zero number in the spectrum of A is an eigenvalue of A of finite multiplicity;

(v) for each $\lambda \in \sigma_p(A)\backslash\{0\}$, the eigenspace $N(\lambda - A)$ has a finite dimensional;

(vi) a non-zero number is an eigenvalue of A if, and only if, it is an eigenvalue of A^*, and the multiplicities of this eigenvalue are the same for A and A^*;

(vii) every non-zero number in $\sigma(A)$ is a pole of A and has finite positive index. For such a number λ, the projection P_λ has a non-zero finite-dimensional range given by the formula

$$P_\lambda(X) = \{x \in X \text{ such that } (\lambda - A)^{\nu} x = 0\},$$

where ν is the order of the pole;

(viii) for each $\lambda \in \sigma_p(A)\backslash\{0\}$, there is $n_0(\lambda) \in \mathbb{N}$ (the index of the eigenvalue λ) such that

$$\begin{cases} \forall n < n_0(\lambda),\ N(\lambda - A)^n \neq N(\lambda - A)^{n_0(\lambda)}, \\ \forall n \geq n_0(\lambda),\ N(\lambda - A)^n = N(\lambda - A)^{n_0(\lambda)}; \end{cases}$$

(ix) *for each* $\lambda \in \sigma_p(A)\backslash\{0\}$, $N[(\lambda - A)^{n_0(\lambda)}]$ *is finite dimensional (this dimension is called the algebraic multiplicity of* λ*); and*

(x) *we have the topological direct sum*

$$X = N[(\lambda - A)^{n_0(\lambda)}] \oplus R[(\lambda - A)^{n_0(\lambda)}]. \qquad \diamond$$

Proof We will prove the statement (ii), and the rest are assigned as exercise for the reader.

(ii) Let $f(z) = zA$. Then, $f(z)$ is an analytic compact operator-valued function on the entire plane. Thus,

$$\{z \in \mathbb{C} \text{ such that } zA\psi = \psi \text{ has a solution } \psi \neq 0\}$$

is a discrete set (it is not the entire plane since it does not contain $z = 0$) and if $\frac{1}{\lambda}$ is not in this discrete set, then

$$(\lambda - A)^{-1} = \frac{1}{\lambda}\left(I - \frac{1}{\lambda}A\right)^{-1}$$

exists. The fact that the non-zero eigenvalues have finite multiplicity follows immediately from the compactness of A. Q.E.D.

Lemma 2.2.1 (B. Russo [31]) *Let* a, b, c, *and* $d \in \mathbb{R}$ *such that* $a < b$ *and* $c < d$. *We suppose that the following conditions are satisfied:*

(i) $\left(\int_a^b |N(x,y)|^r \mathrm{d}y\right)^{\frac{1}{r}} \leq C_1$ *a.e.* $x \in [c,d]$, *where* $r > 0$;

(ii) $\left(\int_c^d |N(x,y)|^\sigma \mathrm{d}x\right)^{\frac{1}{\sigma}} \leq C_2$ *a.e.* $y \in [a,b]$, *where* $\sigma > 0$; *and*

(iii) $q \geq p, q > \sigma$ *and* $\left(1 - \frac{\sigma}{q}\right) p' < r$, *where* p' *is the conjugate of* p $(p' = \frac{p}{p-1})$.

Then, the operator

$$K\psi(x) = \int_a^b N(x,y)\psi(y)\mathrm{d}y, \quad x \in [c,d]$$

is compact from $L_p([a,b])$ *into* $L_q([c,d])$. $\diamond$

2.2.2 Discrete Operators

Proposition 2.2.1 *Let* A *be a closed operator such that* $\rho(A) \neq \emptyset$. *Then,* $(\lambda - A)^{-1}$ *is compact if, and only if, the canonical injection from* $\mathcal{D}(A)$ *into* X *is compact.* $\diamond$

The proof of Proposition 2.2.1 may be found in [20].

Proposition 2.2.2 *Let A be a closed densely defined linear operator with compact resolvent on a Hilbert space X. Then,*

(i) $\sigma(A) = \sigma_p(A)$.
(ii) Either $\sigma(A)$ is finite or there exists a sequence $(\lambda_n)_{n=1}^{\infty} \subset \mathbb{C}$ such that

$$\lim_{n\to+\infty} |\lambda_n| = +\infty$$

and

$$\sigma(A) = \{\lambda_n,\ n \in \mathbb{N}^*\}.$$

(iii) For each $\lambda \in \sigma(A)$, the eigenspace $N(\lambda - A)$ is finite dimensional.
(iv) For each $\lambda \in \sigma(A)$, there is $n_0(\lambda) \in \mathbb{N}$ (the index of the eigenvalue λ) such that

$$\begin{cases} \forall n < n_0(\lambda),\ N[(\lambda - A)^n] \neq N[(\lambda - A)^{n_0(\lambda)}] \\ \forall n \geq n_0(\lambda),\ N[(\lambda - A)^n] = N[(\lambda - A)^{n_0(\lambda)}]. \end{cases}$$

(v) For each $\lambda \in \sigma(A)$, $N[(\lambda - A)^{n_0(\lambda)}]$ is finite dimensional (this dimension is called the algebraic multiplicity of λ), $R[(\lambda - A)^{n_0(\lambda)}]$ is closed in X and for all $n \geq n_0(\lambda)$,

$$R[(\lambda - A)^n] = R[(\lambda - A)^{n_0(\lambda)}].$$

(vi) We have the topological direct sum

$$X = N[(\lambda - A)^{n_0(\lambda)}] \oplus R[(\lambda - A)^{n_0(\lambda)}].$$ ◇

Let X be a Hilbert space and let $A \in \mathscr{C}(X)$. Then,

$$\rho(A^*) = \overline{\rho(A)} = \{\overline{\lambda} \in \mathbb{C} \text{ such that } \lambda \in \rho(A)\}$$

and

$$\sigma(A^*) = \overline{\sigma(A)}.$$

Moreover,

$$(\overline{\lambda} - A^*)^{-1} = [(\lambda - A)^{-1}]^*.$$

In particular, if

$$\sigma(A) = \{\lambda_k,\ k \in \mathbb{N}\},$$

then

$$\sigma(A^*) = \{\overline{\lambda_k},\ k \in \mathbb{N}\}.$$

It is easy to deduce the following result.

Proposition 2.2.3 *Let X be a Hilbert space and let A be a closed and densely defined linear operator with compact resolvent on X. Then,*

(i) $n_0(\overline{\lambda_k}) = n_0(\lambda_k)$ *($n_0(\lambda_k)$ is the index of the eigenvalue λ_k of A),*
(ii) $N[(\overline{\lambda_k} - A^*)^{n_0(\lambda_k)}] = (R[(\lambda_k - A)^{n_0(\lambda_k)}])^\perp$, *and*
(iii) $R[(\overline{\lambda_k} - A^*)^{n_0(\lambda_k)}] = (N[(\lambda_k - A)^{n_0(\lambda_k)}])^\perp$. ◇

2.3 Functions

2.3.1 Function of Finite Order

Definition 2.3.1 An entire function $f(z)$ is said to be of exponential type if the inequality

$$|f(z)| \le Ae^{B|z|}, \quad \forall z \in \mathbb{C} \tag{2.3.1}$$

holds for some positive constants A and B. The smallest of constants B such that (2.3.1) holds is said to be exponential type of f. ◇

Definition 2.3.2 (i) An entire function $f(z)$ is said to be of finite order if there exist $M > 0$ and $p > 0$ such that

$$M_f(r) \le Me^{r^p}$$

for r sufficiently large, where

$$M_f(r) := \max_{|z|=r} |f(z)|.$$

The greatest lower bound ρ of such numbers p is called the order of f, that is,

$$\rho = \overline{\lim_{r\to\infty}} \frac{\log\log M_f(r)}{\log r}.$$

Similarly, we define the type σ of an entire function $f(z)$ of order ρ to be the greatest lower bound of the positive numbers L such that

$$M_f(r) \le Me^{Lr^p}$$

for r sufficiently large, that is,

$$\sigma = \overline{\lim_{r\to\infty}} \frac{\log M_f(r)}{r^p}.$$

(ii) An entire function of finite order p is said to be of minimum type if for any $\varepsilon > 0$, there exists $M > 0$ such that

$$M_f(r) \leq M e^{\varepsilon r^p}. \qquad \diamond$$

By $f(x) = O(\phi(x))$ we mean generally that

$$|f(x)| < A\phi(x)$$

if x is sufficiently near to some given limit. In particular, $O(1)$ means a bounded function. By $f(x) = o(\phi(x))$ we mean that

$$\frac{f(x)}{\phi(x)} \to 0$$

as x tends to a given limit. In particular, $o(1)$ means a function which tends to zero.

An integral function $f(z)$ is said to be of finite order if there is a positive number A such that, as $|z| = r \to \infty$,

$$f(z) = O(e^{r^A}).$$

The lower bound ρ of numbers A for which this is true is called the order of the function. Thus, if $f(z)$ is of order ρ,

$$f(z) = O(e^{r^\rho + \varepsilon})$$

for every positive value of ε, but not for any negative value. Let $m(r)$ denote the minimum of $|f(z)|$ on the circle $|z| = r$.

Theorem 2.3.1 (Titchmarsh [32]) *If $f(z)$ is a function of order ρ, then*

$$m(r) > e^{-r^{\rho+\varepsilon}}$$

on circles of an arbitrarily large radius. $\diamond$

Theorem 2.3.2 (Young [36, Theorem 5, p. 64]) *If $f(z)$ is an entire function of finite order ρ, then*

$$n(r) = O(r^{\rho+\xi})$$

for every positive number ξ, where $n(r)$ denotes the number of zeros of $f(z)$ contained in the closure of the open disc $\mathbb{D}(0, r) := \{z \in \mathbb{C}$ such that $|z| < r\}$. $\diamond$

As a corollary, we have the following important result. If $f(z)$ is an entire function of finite order ρ, then we infer from Theorem 2.3.2 that

$$n(r) \leq Ar^\beta$$

for some constant A and all values of r. Let $z_1, z_2, z_3, \dots$ are zeros of f, other than $z = 0$. Without loss of generality, we may suppose that the zeros of $f(z)$ have been numbered so that $0 < |z_1| \leq |z_2| \leq \cdots$. Let $\alpha > \rho$, and consider β such that $\rho < \beta < \alpha$. Let $r = |z_n|$, then $n(r) = n$. So,

$$n \leq A|z_n|^{\beta} \text{ for } n = 1, 2, 3, \dots.$$

Since

$$\sum_{n=1}^{\infty} \frac{1}{n^{\frac{\alpha}{\beta}}} < \infty,$$

then the series

$$\sum_{n=1}^{\infty} \frac{1}{|z_n|^{\alpha}}$$

is convergent whenever $\alpha > \rho$.

2.3.2 *Function of Sine Type*

Definition 2.3.3 The growth indicator of an exponential-type function f is a 2π-periodic function on $\mathbb{R}$ defined by the equality

$$h_f(\phi) = \lim_{r\to\infty} \sup \frac{1}{r} \log|f(re^{i\phi})|, \ \phi \in [-\pi, \pi].$$

The indicator diagram of f is a convex set G_f such that

$$h_f(\phi) = \sup_{k\in G_f} Re(ke^{-i\phi}), \ \phi \in [-\pi, \pi]. \qquad \diamond$$

Definition 2.3.4 An entire function f of exponential type is said to be of sine type if

(i) the zeros of f lie in a strip $\{z \in \mathbb{C}$ such that $|Imz| \leq H\}$ for some $H > 0$, and
(ii) there exist $h \in \mathbb{R}$, and positive constants c_1, c_2 such that

$$c_1 \leq |f(x + ih)| \leq c_2,$$

for all $x \in \mathbb{R}$. $\diamond$

Definition 2.3.5 Let $\Lambda := \{\lambda_n, \ n \in \mathbb{N}\}$ be a complex sequence. The set Λ is said to beseparated if the condition

$$\delta(\Lambda) := \inf_{k \neq n} |\lambda_k - \lambda_n| > 0$$

is fulfilled. ◇

2.3.3 *Generating Function in $L_2(0, T)$*

Definition 2.3.6 An entire function f of exponential type is said to be a function of the Cartwright class if

$$\int_{\mathbb{R}} \frac{\max(\log|f(x)|, 0)}{1+x^2}\,dx < +\infty. \qquad ◇$$

In particular, the function f of exponential type satisfying the condition

$$\int_{\mathbb{R}} \frac{|f(x)|^2}{1+x^2}\,dx < \infty$$

belongs to the Cartwright class. Furthermore, the indicator diagram of a Cartwright class function is an interval $[i\alpha, i\beta]$, $\alpha \leq \beta$, of the imaginary axis. Its length is the width of indicator diagram (see [2, pp. 59–60]).

Definition 2.3.7 An entire function of exponential type with simple zeros $\{\lambda_n\}_1^\infty$ and with the width of indicator diagram T is called a generating function of exponential family $\{e^{i\lambda_n t}\}_1^\infty$ in $L_2(0, T)$. ◇

2.4 Phragmén-Lindelöf Theorems

The following important extension of the maximum-modulus theorem was first given by Phragmén and Lindelöf.

Theorem 2.4.1 (Phragmén-Lindelöf) *Let $f(z)$ be an analytic function of $z = re^{i\theta}$, regular in the region D between two straight lines making an angle $\frac{\pi}{\alpha}$ at the origin, and on the lines themselves. Suppose that*

$$|f(z)| \leq M \tag{2.4.1}$$

on the lines, and that, as $r \to \infty$,

$$f(z) = O(e^{r^\beta}),$$

where $\beta < \alpha$, uniformly in the angle. Then, the inequality (2.4.1) holds throughout the region D. ◇

Theorem 2.4.2 (Phragmén-Lindelöf) *Let $f : z \longrightarrow f(z)$ be a holomorphic function on the angular sector between two half-line pheasants an angle $\frac{\pi}{\alpha}$ at the origin and holomorphic on these half-lines. Suppose that $|f(z)| \leq M$ on the half-lines and for all $\varepsilon > 0$, there is $A > 0$ such that on the sector, we have*

$$|f(z)| \leq Ae^{\varepsilon|\lambda|^{\alpha}}.$$

Then,

$$|f(z)| \leq M$$

on all the sectors. ◇

The following theorem is deduced from Theorem 2.4.2 by the change of variable

$$z \longrightarrow \frac{1}{z}.$$

Theorem 2.4.3 (Phragmén-Lindelöf) *Let g be a function with complex variables and analytic inside of a sector S of the form $\{z \in \mathbb{C}$ such that $|z| \leq R$ and $\theta_1 \leq arg(z) \leq \theta_2\}$, with*

$$|\theta_2 - \theta_1| \leq \frac{\pi}{p}.$$

Suppose again, g analytic in a neighborhood of the points $\{z \in \mathbb{C}$ such that $arg(z) = \theta_1$ and $z \neq 0\}$ and $\{z \in \mathbb{C}$ such that $arg(z) = \theta_2$ and $z \neq 0\}$ such that for all $\varepsilon > 0$, there is $A > 0$ such that for all $z \in S\backslash\{0\}$,

$$|g(z)| \leq Ae^{\varepsilon|\lambda|^{-p}}$$

and $|g(z)|$ is bounded on both $\{z \in \mathbb{C}$ such that $arg(z) = \theta_1$ and $z \neq 0\}$ and $\{z \in \mathbb{C}$ such that $arg(z) = \theta_2$ and $z \neq 0\}$. Then, $|g(z)|$ is bounded in a neighborhood of 0 in S. ◇

2.5 Holomorphic Operator Functions

2.5.1 Spectrum and Multiplicities

By an operator-valued function, or simply operator function, we will mean a function W defined on Ω whose values are arbitrary closed linear operators in a separable

Hilbert space X, where Ω is an open, connected subset of $\mathbb{C}$. The resolvent set of W is the set of $\lambda \in \Omega$ such that $W(\lambda)$ has a bounded inverse. The spectrum of W, denoted $\Sigma(W)$, is just the complement of the resolvent set in Ω (this notation is used to make a clear distinction between the operator-function spectrum and $\sigma(W(\lambda))$, the spectrum of the fixed operator $W(\lambda)$; $\lambda \in \Sigma(W)$ if, and only if, $0 \in \sigma(W(\lambda))$. The point spectrum of W, denoted $\Sigma_p(W)$, is the set of $\lambda \in \Omega$ such that

$$N(W(\lambda)) \neq \{0\}.$$

As usual, these points are called eigenvalues, non-zero vectors in the kernel are corresponding eigenvectors, and the dimension of the kernel is the corresponding geometric multiplicity. If the domain of $W(\lambda)$ is independent of λ and if for each x in this constant domain the function

$$\lambda \longrightarrow W(\lambda)x$$

is analytic in the usual strong sense, then we can define so-called chains of and canonical systems of eigen and associated vectors as well as a concept of algebraic multiplicity of an eigenvalue λ_0, denoted in the following by $m(\lambda_0; W)$, in a natural way which extends the usual definition via Jordan chains for an operator A, i.e., for the linear operator function

$$W(\lambda) = A - \lambda$$

in which case we may also write $m_a(\lambda_0)$ for the algebraic multiplicity. As we will be concerned with simple eigenvalues in the following, we will not discuss this definition further here, but refer rather to the details in [26, Sect. 11.2].

Definition 2.5.1 Let U be a domain in $\mathbb{C}$, an operator-valued function $A(z)$ $(z \in U)$ with values in $\mathscr{L}(X)$ is said to be uniformly holomorphic if for any $z_0 \in U$ the uniform limit

$$\lim_{z \to z_0} \frac{A(z) - A(z_0)}{z - z_0} \quad (= A'(z_0))$$

exists. ◇

Definition 2.5.2 Let U be a domain in $\mathbb{C}$, an operator-valued function $A(z)$ $(z \in U)$ with values in $\mathscr{L}(X)$ is said to be meromorphic if it is holomorphic in U except in certain points which are poles for it. ◇

Theorem 2.5.1 (Gohberg et al. [17, p. 203]) *If $W : \Omega \longrightarrow \mathscr{L}(X)$ is analytic and Fredholm valued and if $W(\lambda)$ has a bounded inverse for some $\lambda \in \Omega$, then $\Sigma(W)$ is discrete, i.e., consists only of at most countably many eigenvalues, each of finite algebraic multiplicity, with no accumulation point in Ω. Furthermore,*

$$\lambda \longrightarrow W(\lambda)^{-1}$$

is meromorphic on Ω, its poles are exactly the eigenvalues of W, and the principal part of its Laurent expansion around each pole has only finitely many non-zero terms, the coefficients in these being of finite rank. ◇

If Δ is a Cauchy domain with $\overline{\Delta} \subset \Omega$ and if $W : \Omega \longrightarrow \mathscr{L}(X)$ is analytic, then W is said to be normal with respect to Δ if $W(\lambda)$ has a bounded inverse for every $\lambda \in \partial\Delta$ and $W(\lambda)$ is Fredholm for every $\lambda \in \Delta$ (see [17, p. 205]). We have the following Rouché's theorem for operator functions.

Theorem 2.5.2 (Gohberg et al. [17, p. 206]) *If $W, S : \Omega \longrightarrow \mathscr{L}(X)$ are analytic, if W is normal with respect to a Cauchy domain Δ ($\overline{\Delta} \subset \Omega$), and if $\|S(\lambda)W(\lambda)^{-1}\| < 1$ for all $\lambda \in \partial\Delta$, then*

$$V := W + S$$

is also normal with respect to Δ and

$$\sum_{\lambda \in \Delta} m(\lambda; V) = \sum_{\lambda \in \Delta} m(\lambda; W).$$ ◇

Theorem 2.5.3 (Yosida [35, Theorem 1, p. 211]) *Let A be a closed linear operator with domain and range both in a complex Banach space X. Then, the resolvent set $\rho(A)$ is an open set of the complex plane. In each component (the maximal connected sets) of $\rho(A)$, $(\lambda - A)^{-1}$ is a holomorphic function of λ.* ◇

2.5.2 Zeros of a Holomorphic Function

Theorem 2.5.4 *The zeros of a function f that is holomorphic on the domain Γ have the following properties:*

(i) If s_0 is a zero of f, then $f(s)$ is either identically zero on Γ or the order of the zero is finite, that is, there exists an m such that

$$f(s) = (s - s_0)^m g(s),$$

where g is holomorphic on Γ and $g(s_0) \neq 0$.

(ii) If f is not identically zero on Γ, then the zeros of f are isolated, that is, for every zero s_0 of f, there exists a $\delta > 0$ such that $f(s) \neq 0$ for all s satisfying

$$0 < |s - s_0| < \delta.$$

(iii) If the zeros of f have a limit point in Γ, then f is identically zero.

(iv) In every compact subset V of $\mathbb{C}$ with $V \subset \Gamma$, there are only finitely many zeros, provided that f is not identically zero. ◇

We also use the following auxiliary result from complex analysis, cf. [26, Lemma 1.6] and [24, Theorem I.11].

Lemma 2.5.1 (Wyss [34]) *Let $U \subset \mathbb{C}$ be a bounded, simply connected domain, $F \subset U$ be compact, z_0 be an interior point of F, and $\eta > 0$. Then, there exists a constant $C > 0$ such that the following holds:*

If $a,\ b \in \mathbb{C}$ and $f : aU + b \longrightarrow \mathbb{C}$ with $f(az_0 + b) \neq 0$ is holomorphic and bounded, then there is a set $E \subset \mathbb{C}$ being the union of finitely many discs with radii summing up to at most $|a|\eta$ such that

$$|f(z)| \geq \frac{|f(az_0 + b)|^{1+C}}{\|f\|^C_{aU+b,\infty}}$$

for all $z \in (aF + b)\backslash E$, where

$$\|f\|_{aU+b,\infty} = \sup_{z \in U} |f(az + b)|. \qquad \diamondsuit$$

Lemma 2.5.2 (Shkalikov [30, Lemma 2]) *Let $f(\lambda)$ be a bounded holomorphic function in the rectangle*

$$\Pi = \{\lambda \in \mathbb{C} \text{ such that } |Re\lambda - r| < a \text{ and } |Im\lambda| < c\}.$$

For $\delta \in (0, 1)$, we set $a' = a(1 - \delta)$ and $c' = c(1 - \delta)$, and we denote by Π' the rectangle defined by

$$\Pi' = \{\lambda \in \mathbb{C} \text{ such that } |Re\lambda - r| < a' \text{ and } |Im\lambda| < c'\}.$$

Let $M := \sup_{\lambda \in \Pi} f(\lambda)$ and $M' := \sup_{\lambda \in \Pi'} f(\lambda)$. Then,

(i) the number, $n_f(\Pi')$, of zeros of the function $f(\lambda)$ in the rectangle Π' satisfies the estimate

$$n_f(\Pi') \leq C(\log M - \log M'),$$

(ii) for any $\varepsilon \in (0, 1)$, there exist discs such that the total sum of their radii is $\leq \varepsilon$; and

$$\log|f(\lambda)| \geq C \log \varepsilon (\log M - \log M') + \log M', \ \lambda \in \Pi' \backslash \mathscr{E},$$

where $\mathscr{E}$ is the union of these discs,

(iii) if γ is a straight line segment contained in Π' that does not pass through the zeros of f in Π', then the variation of the argument of the function f along γ is estimated as

$$|[\arg f(\lambda)]_\gamma| \leq C(\log M - \log M').$$

In all the estimates, the constant C depends only on δ and the ratio $\frac{a}{c}$ and do not depend on r and f. $\diamondsuit$

2.5.3 *Determinant of Operator*

An important role in the comparison of spectra of operators is played by the concept of determinant of operators having the form $I + K$, where K is a finite-rank operator on X.

Definition 2.5.3 The determinant of the operator $I + K$ is the product of all its eigenvalues with multiplicity taken into account. The determinant of $I + K$ is denoted by $\det(I + K)$. ◇

Proposition 2.5.1 (Markus [26, p. 14]) *Let K be a finite-rank operator on X. Then, the following assertions hold:*

(i) $|\det(I + K)| \leq (1 + \|K\|)^{rk(K)}$, *where* $rk(K) := \dim R(K)$ *designates the rank of* K.
(ii) If $A(z)$ is a holomorphic operator-valued function with values in $\mathscr{L}(X)$, then the function $\det(I + KA(z))$ *is holomorphic.* ◇

Lemma 2.5.3 (Markus and Matsaev [27, p. 149]) *Let γ be a Jordan curve that is not closed (possibly unbounded) and $K(z)$ $(z \in \gamma)$ be a continuous operator-valued function with $\|K(z)\| < 1$ and* $\dim K(z) \leq n$. *Then, we have*

$$|[\arg\det(I + K(z))]_\gamma| \leq n\pi.$$ ◇

Lemma 2.5.4 (Markus [26, Lemma 7.1, p. 38]) *Suppose that A and $A + K$ are operators on a Hilbert space X with compact resolvent, Δ is a bounded domain with rectifiable boundary Γ, and*

$$\sigma(A) \bigcap \Gamma = \sigma(A + K) \bigcap \Gamma = \emptyset.$$

Then, $\det(I - KR(\lambda, A)) \neq 0$ *for $\lambda \in \Gamma$, and*

$$n(\Delta, A + K) - n(\Delta, A) = \frac{1}{2\pi}[\arg\det(I - KR(\lambda, A))]_\Gamma.$$ ◇

2.6 Semigroup Theory

2.6.1 *Definitions*

Definition 2.6.1 Let X be a Banach space. A strongly continuous semigroup is an operator-valued function $T(t)$ from $\mathbb{R}_{\geq 0}$ into $\mathscr{L}(X)$ that satisfies the following properties:

$$T(t + s) = T(t)T(s) \text{ for } t,\ s \geq 0 \tag{2.6.1}$$

$$T(0) = I \tag{2.6.2}$$

$$\|T(t)z_0 - z_0\| \to 0 \text{ as } t \to 0^+ \text{ for all } z_0 \in X. \tag{2.6.3}$$

We shall subsequently use the standard abbreviation C_0-semigroup for a strongly continuous semigroup. ◇

Definition 2.6.2 The infinitesimal generator A of a C_0-semigroup on a Banach space X is defined by

$$Az = \lim_{t \to 0^+} \frac{1}{t}(T(t) - I)z$$

whenever the limit exists; the domain of A, $\mathscr{D}(A)$, being the set of elements in X for which the limit exists. ◇

2.6.2 *Example*

Consider $A \in \mathscr{L}(X)$, and put

$$e^{tA} = \sum_{n=0}^{\infty} \frac{(tA)^n}{n!}. \tag{2.6.4}$$

Proposition 2.6.1 *Let $A \in \mathscr{L}(X)$. Then, the infinite sum*

$$\sum_{n=0}^{\infty} \frac{(tA)^n}{n!}$$

converges in the uniform topology. ◇

Proof For $M > N$, consider the finite sums, for which we get

$$\begin{aligned} \left\| \sum_{n=0}^{M} \frac{(tA)^n}{n!} - \sum_{n=0}^{N} \frac{(tA)^n}{n!} \right\| &= \left\| \sum_{n=N+1}^{M} \frac{(tA)^n}{n!} \right\| \\ &\leq \sum_{n=N+1}^{M} \left\| \frac{(tA)^n}{n!} \right\| \\ &\leq \sum_{n=N+1}^{M} \frac{t^n \|A\|^n}{n!}. \end{aligned}$$

In view of

$$\sum_{n=N+1}^{M} \frac{t^n \|A\|^n}{n!} \to 0 \text{ as } M,\ N \to \infty,$$

we conclude that the last series is a Cauchy series, and so the right-hand side of (2.6.4) converges in the uniform topology. Q.E.D.

It is easy to verify the condition (2.6.2). We now verify the conditions (2.6.1) and (2.6.3).

Proposition 2.6.2 *Let $A \in \mathscr{L}(X)$. Then,*

$$\|e^{tA}z_0 - z_0\| \to 0 \text{ as } t \to 0^+ \text{ for all } z_0 \in X. \qquad \diamond$$

Proof It is easy to show that

$$\begin{aligned}\|e^{tA}z_0 - z_0\| &= \left\| \sum_{n=1}^{\infty} \frac{(tA)^n}{n!} z_0 \right\| \\ &\le \sum_{n=1}^{\infty} \frac{t^n \|A\|^n}{n!} \|z_0\| \\ &= [e^{t\|A\|} - 1]\|z_0\|.\end{aligned}$$

The continuity of $e^{t\|A\|}$ implies that (2.6.3) holds. Q.E.D.

Proposition 2.6.3 *Let $A \in \mathscr{L}(X)$. Then,*

$$e^{(t+s)A} = e^{tA}e^{sA} \text{ for } t,\ s \ge 0. \qquad \diamond$$

Proof Let $t,\ s \ge 0$, we have

$$\begin{aligned}\sum_{n=0}^{\infty} \frac{(t+s)^n A^n}{n!} &= \sum_{n=0}^{\infty} \frac{A^n}{n!} \sum_{k=0}^{n} C_n^k t^k s^{n-k} \\ &= \sum_{n=0}^{\infty} \sum_{k=0}^{n} \frac{t^k A^k}{k!} \frac{s^{n-k} A^{n-k}}{(n-k)!} \\ &= \sum_{k=0}^{\infty} \sum_{n=k}^{\infty} \frac{t^k A^k}{k!} \frac{s^{n-k} A^{n-k}}{(n-k)!} \\ &= \sum_{k=0}^{\infty} \frac{t^k A^k}{k!} \sum_{l=0}^{\infty} \frac{s^l A^l}{l!},\end{aligned}$$

which completes the proof. Q.E.D.

Corollary 2.6.1 *Let $A \in \mathscr{L}(X)$. Then, e^{tA} is a C_0-semigroup.* $\diamond$

Proposition 2.6.4 *Let $A \in \mathscr{L}(X)$. Then, the infinitesimal generator of the C_0-semigroup e^{tA} is A.* $\diamond$

Proof We have

$$\begin{aligned}\frac{1}{t}\|e^{tA}z - z - tAz\| &= \frac{1}{t}\left\|\sum_{n=2}^{\infty}\frac{t^n A^n}{n!}z\right\| \\ &\leq \frac{1}{t}\sum_{n=2}^{\infty}\frac{t^n\|A^n\|}{n!}\|z\| \\ &\leq \frac{1}{t}\sum_{n=2}^{\infty}\frac{t^n\|A\|^n}{n!}\|z\| \\ &= \left[\frac{e^{t\|A\|}-1}{t} - \|A\|\right]\|z\|.\end{aligned}$$

From this, we see that the infinitesimal generator of the C_0-semigroup e^{tA} is A. This completes the proof. Q.E.D.

We have actually shown that e^{tA} is differentiable at zero. Similarly, one can show that it is differentiable for all t and that

$$\frac{de^{tA}}{dt} = Ae^{tA}.$$

We recall the uniform boundedness theorem.

Theorem 2.6.1 (Banach-Steinhaus theorem) *Let $(T_n)_n$ be a family of bounded linear operators in $\mathscr{L}(X, Y)$, where X is a Banach space and Y is a normed linear space. If the family $\{T_n x\}$ is bounded for each x (that is, $\|T_n x\|_Y \leq M_x$, where M_x depends on x, but is independent of n), then $\{\|T_n\|\}$ is uniformly bounded in n.* ◇

2.7 Concepts of Subordination and Fully Subordination

2.7.1 Concepts of Subordination

Definition 2.7.1 Let G and S be two linear operators on a Banach space X. We call that S is G-bounded with order $p \in [0, 1]$, if $\mathscr{D}(G) \subset \mathscr{D}(S)$ and there exist strictly positive constants a, b such that

$$\|S\varphi\| \leq b\|G\varphi\|^p\|\varphi\|^{1-p} + a\|\varphi\|, \text{ for every } \varphi \in \mathscr{D}(G).$$

For $p = 1$, we call that S is G-bounded. ◇

Definition 2.7.2 Let G and S be two linear operators on a Banach space X. The operator S is said subordinate to G with order $p \in [0, 1]$ (or p-subordinate) if

$\mathscr{D}(G) \subset \mathscr{D}(S)$ and there is a constant $c > 0$ such that

$$\|S\varphi\| \leq c\|G\varphi\|^p \|\varphi\|^{1-p} \text{ for all } \varphi \in \mathscr{D}(G).$$

If $p = 1$, then S is said to be subordinate to G. ◇

In the case $p = 0$, the notion of subordination is simply reduced to the boundedness of S. First note that, for $\lambda > 0$, we get

$$\lambda^p + \lambda^{p-1} \geq 1. \tag{2.7.1}$$

In fact, we have $\lambda^p \geq 1$ for $\lambda \geq 1$ and $\lambda^{p-1} \geq 1$ for $0 < \lambda \leq 1$. Let G and S be linear operators on a Banach space X with $\mathscr{D}(G) \subset \mathscr{D}(S)$ and $0 < p \leq 1$. We suppose that there exists a constant $C > 0$ such that

$$\|Su\| \leq C(\varepsilon^{-p}\|u\| + \varepsilon^{1-p}\|Gu\|) \text{ for all } u \in \mathscr{D}(G),\ \varepsilon > 0. \tag{2.7.2}$$

Let S be p-subordinate to G. If $\|Gu\| \neq 0$, then by using (2.7.1) with $\lambda = \|u\|(\varepsilon\|Gu\|)^{-1}$, we obtain

$$\begin{aligned}\|Su\| &\leq b\|u\|^{1-p}\|Gu\|^p \left(\left(\frac{\|u\|}{\varepsilon\|Gu\|}\right)^p + \left(\frac{\|u\|}{\varepsilon\|Gu\|}\right)^{p-1}\right)\\ &= b(\varepsilon^{-p}\|u\| + \varepsilon^{1-p}\|Gu\|).\end{aligned} \tag{2.7.3}$$

Example 2.3 Consider the operators

$$\begin{cases} G : \mathscr{D}(G) \subset L_2(]a,b[) \longrightarrow L_2(]a,b[) \\ \qquad\qquad \psi \longrightarrow G\psi(x) = \dfrac{\mathrm{d}^2\psi}{\mathrm{d}x^2}(x) \\ \mathscr{D}(G) = \{\psi \in H^2(]a,b[) \text{ such that } \psi(a) = \psi(b) = 0\}, \end{cases}$$

and

$$\begin{cases} S : \mathscr{D}(S) \subset L_2(]a,b[) \longrightarrow L_2(]a,b[) \\ \qquad\qquad \psi \longrightarrow G\psi(x) = \dfrac{\mathrm{d}\psi}{\mathrm{d}x}(x) \\ \mathscr{D}(S) = C^1([a,b]). \end{cases}$$

The operator S is $\frac{1}{2}$-subordinate to G. In fact, let $\psi \in \mathscr{D}(G)$, we have

$$\int_a^b |\psi'(x)|^2 \mathrm{d}x = \int_a^b \psi'(x)\overline{\psi'(x)}\mathrm{d}x. \tag{2.7.4}$$

By integrating by parts (2.7.4) and by using the boundary condition, we get

$$\int_a^b |\psi'(x)|^2 \mathrm{d}x = -\int_a^b \psi(x)\overline{\psi''(x)}\mathrm{d}x. \tag{2.7.5}$$

Using both Cauchy-Schwartz inequality and Eq. (2.7.5), it is easy to deduce that

$$\int_a^b |\psi'(x)|^2 \mathrm{d}x \le \|\psi\|_{L_2(]a,b[)} \|\psi''\|_{L_2(]a,b[)}.$$

Hence,

$$\|S\psi\|_{L_2(]a,b[)} \le \|G\psi\|^{\frac{1}{2}}_{L_2(]a,b[)} \|\psi\|^{\frac{1}{2}}_{L_2(]a,b[)} \text{ for all } \psi \in \mathscr{D}(G).$$

Example 2.4 Consider the Laplacian on a domain $\Omega \subset \mathbb{R}^m$ with Dirichlet boundary conditions

$$\begin{cases} G : \mathscr{D}(G) \subset L_2(\Omega) \longrightarrow L_2(\Omega) \\ \qquad\qquad\qquad\quad \psi \longrightarrow G\psi(x) = \Delta\psi \\ \\ \mathscr{D}(G) = \{\psi \in H^2(\Omega) \text{ such that } \psi = 0 \text{ on } \partial\Omega\}. \end{cases}$$

Let ∂_k be the operator of taking the kth partial derivative with domain $H^1(\Omega)$. Then, for $\psi \in \mathscr{D}(G)$, we have

$$\begin{aligned} \int_\Omega |\partial_k \psi(x)|^2 \mathrm{d}x &\le \sum_{j=1}^m \int_\Omega |\partial_j \psi(x)|^2 \mathrm{d}x \\ &= \sum_{j=1}^m \int_\Omega \partial_j \psi(x) \overline{\partial_j \psi(x)} \mathrm{d}x \\ &= -\sum_{j=1}^m \int_\Omega \psi(x) \overline{\partial_j^2 \psi(x)} \mathrm{d}x \\ &= -\int_\Omega \psi(x) \overline{\Delta\psi(x)} \mathrm{d}x \\ &\le \|\psi\|_{L_2(\Omega)} \|G\psi\|_{L_2(\Omega)}. \end{aligned}$$

So, ∂_k is $\frac{1}{2}$-subordinate to G for $j = 1, \dots, m$.

Example 2.5 We consider a class of operators arising from Reggeon field theory. This theory was invented by the theoretical physicist, Vladimir Naumovich Gribov in 1967 and constitutes one of the most important attempts to understand strong interactions, i.e., the interaction between protons and neutrons among other less stable particles. It is indeed governed by the Gribov operator

$$H_{\mu,\lambda} = \mu A^* A + i\lambda A^*(A^* + A)A$$

and, in Reggeon field theory, the real parameters μ is Pomeron intercept, λ is the triple coupling of Pomeron and $i^2 = 1$. It is convenient to regard the above operators as acting on the following Bargmann space:

$$E_0 = \left\{ \varphi : \mathbb{C} \longrightarrow \mathbb{C} \text{ entire such that } \int_{\mathbb{C}} e^{-|z|^2} |\varphi(z)|^2 \mathrm{d}x\mathrm{d}y < \infty \text{ and } \varphi(0) = 0 \right\}.$$

The Bargmann space E_0 with the paring

$$\langle \varphi, \psi \rangle = \frac{1}{\pi} \int_{\mathbb{C}} e^{-|z|^2} \varphi(z)\overline{\psi}(z)\mathrm{d}x\mathrm{d}y$$

is a Hilbert space and

$$e_n(z) = \frac{z^n}{\sqrt{n!}}, \quad n = 0, 1, \ldots$$

is an orthonormal basis in E_0. The space E_0 was used in [3] as representation space for the canonical computation rules of quantum mechanics. Since then it had occurred in many different contents, e.g., in representation theory of nilpotent Lie groups and as a class of symbols in the theory of Hankel and Toeplitz operators. In this representation, the standard annihilation and creation operators are given by the following:

$$\begin{cases} A : \mathscr{D}(A) \subset E_0 \longrightarrow E_0 \\ \qquad\qquad \varphi \longrightarrow A\varphi(z) = \dfrac{d\varphi}{dz}(z) \\ \mathscr{D}(A) = \{\varphi \in E_0 \text{ such that } A\varphi \in E_0\} \end{cases}$$

and

$$\begin{cases} A^* : \mathscr{D}(A^*) \subset E_0 \longrightarrow E_0 \\ \qquad\qquad \varphi \longrightarrow A^*\varphi(z) = z\varphi(z) \\ \mathscr{D}(A^*) = \{\varphi \in E_0 \text{ such that } A^*\varphi \in E_0\}. \end{cases}$$

Let $U = A^*A$ be the harmonic oscillator with domain

$$\mathscr{D}(U) = \{\varphi \in E_0 \text{ such that } U\varphi \in E_0\}$$

and

$$\widetilde{U} = \mu U,$$

where μ is the intercept pomeron; $\widetilde{U}$ is a self-adjoint operator with compact resolvent in the Bargmann space E_0, and A^*A is a generator of an analytic semigroup. Let

$$\widetilde{U}\Phi = \sum_{k=1}^{\infty} \lambda_k \langle \Phi, e_k \rangle e_k$$

be its spectral decomposition, where $\lambda_k = \mu k$ is the kth eigenvalue of $\widetilde{U}$ associated to the eigenvector

$$e_k(z) = \frac{z^k}{\sqrt{k!}}.$$

For $\beta > 0$, we define the operator $\widetilde{U}^\beta$ by

$$\widetilde{U}^\beta \Phi = \sum_{k=1}^{\infty} \lambda_k^\beta \langle \Phi, e_k \rangle e_k$$

with

$$\mathscr{D}(\widetilde{U}^\beta) = \left\{ \Phi \in E_0 \text{ such that } \sum_{k=1}^{\infty} \lambda_k^{2\beta} |\langle \Phi, e_k \rangle|^2 < \infty \right\}.$$

From both

$$\mathscr{D}(A^{*3}A^3) = \mathscr{D}(U^3)$$

and

$$\|A^{*3}A^3\Phi\| \approx \|U^3\Phi\|,$$

it is easy to deduce that

$$\mathscr{D}(U^{\frac{3}{2}}) \subset \mathscr{D}(H_{\mu,\lambda})$$

and there exists $C_{\mu,\lambda} > 0$ such that

$$\|H_{\mu,\lambda}\Phi\| \le C_{\mu,\lambda} \|U^{\frac{3}{2}}\Phi\|$$

for any Φ in $\mathscr{D}(U^{\frac{3}{2}})$. But

$$\begin{aligned} \|U^{\frac{3}{2}}\Phi\|^2 &= \langle U^{\frac{3}{2}}\Phi, U^{\frac{3}{2}}\Phi \rangle \\ &= \langle U^3\Phi, \Phi \rangle \\ &\le \|U^3\Phi\| \|\Phi\| \end{aligned}$$

consequently, for every Φ in $\mathscr{D}(U^3)$, we get

$$\|H_{\mu,\lambda}\Phi\| \leq C_{\mu,\lambda}\|U^3\Phi\|^{\frac{1}{2}}\|\Phi\|^{\frac{1}{2}},$$

and for every Φ in $\mathscr{D}(A^{*3}A^3) := \{\varphi \in E_0$ such that $A^{*3}A^3\varphi \in E_0\}$, we have

$$\|H_{\mu,\lambda}\Phi\| \leq C_{\mu,\lambda}\|A^{*3}A^3\Phi\|^{\frac{1}{2}}\|\Phi\|^{\frac{1}{2}}.$$

Hence, $H_{\mu,\lambda}$ is $\frac{1}{2}$-subordinate to $A^{*3}A^3$.

Example 2.6 Let G and S be operators on $L_2([0, 1])$ defined by

$$\begin{cases} G : \mathscr{D}(G) \subset L_2(]0, 1[) \longrightarrow L_2(]0, 1[) \\ \qquad\qquad \psi \longrightarrow G\psi(x) = \dfrac{d^2\psi}{dx^2}(x) \\ \\ \mathscr{D}(G) = H^2(\Omega) \end{cases}$$

and

$$\begin{cases} S : \mathscr{D}(S) \subset L_2(]0, 1[) \longrightarrow L_2(]0, 1[) \\ \qquad\qquad \psi \longrightarrow G\psi(x) = \dfrac{d\psi}{dx}(x) \\ \\ \mathscr{D}(S) = H^1(\Omega). \end{cases}$$

Consider the function

$$u_\lambda(x) = \frac{\lambda}{2}x^2 + x$$

with $\lambda \in \mathbb{C}$. It is easy to see that $u_\lambda \in \mathscr{D}(G)$, $u'_\lambda(x) = \lambda x + 1$ and $u''_\lambda(x) = \lambda$. So,

$$(G - \lambda)u_\lambda = 0$$

and $Su_\lambda \neq 0$. Hence, S is not p-subordinate to $G - \lambda$ for $0 < p \leq 1$. Since $\mathscr{D}(S) = H^1(\Omega)$, then S is not bounded. So, S is not p-subordinate to $G - \lambda$ for any $p \in [0, 1]$ and $\lambda \in \mathbb{C}$.

Example 2.7 Consider the operator

$$\begin{cases} G_n : \mathscr{D}(G_n) \subset L_2(]0, 1[) \longrightarrow L_2(]0, 1[) \\ \qquad\qquad \psi \longrightarrow G_n\psi(x) = \dfrac{d^n\psi}{dx^n}(x) \\ \\ \mathscr{D}(G_n) = \{\psi \in H^n(]0, 1[) \text{ such that } \psi^{(k)}(0) = \psi^{(k)}(1) \text{ for } k = 0, \ldots, n-1\}. \end{cases}$$

We claim that for $0 \leq k \leq n$ and $n \geq 1$, G_k is $\frac{k}{n}$-subordinate to G_n, i.e.,

$$\|G_k\psi\| \leq \|\psi\|^{1-\frac{k}{n}}\|G_n\psi\|^{\frac{k}{n}} \text{ for } \psi \in \mathscr{D}(G_n). \tag{2.7.6}$$

The cases $k = 0$ and $k = n$ are trivial. Let $0 < k < n$, we show the result (2.7.6) by induction on n. The calculation in Example 2.3 proves that the assertion (2.7.6) is true for $n = 2$. Now, if (2.7.6) holds for some $n \geq 2$ and let $\psi \in \mathscr{D}(G_{n+1})$, then by using (2.7.6) twice, one time with $n = 2$, we find

$$\begin{aligned}\|G_1\psi^{(n-1)}\|^2 &\leq \|\psi^{(n-1)}\|\|G_2\psi^{(n-1)}\| \\ &\leq \|\psi\|^{\frac{1}{n}}\|G_n\psi\|^{\frac{n-1}{n}}\|G_{n+1}\psi\|.\end{aligned}$$

This implies that

$$\begin{aligned}\|G_n\psi\|^{2-\frac{n-1}{n}} &= \|G_n\psi\|^{\frac{n+1}{n}} \\ &\leq \|\psi\|^{\frac{1}{n}}\|G_{n+1}\psi\|.\end{aligned}$$

Hence

$$\|G_n\psi\| \leq \|\psi\|^{\frac{1}{n+1}}\|G_{n+1}\psi\|^{\frac{n}{n+1}}.$$

Using again (2.7.6), we deduce for $k \leq n-1$

$$\begin{aligned}\|G_k\psi\| &\leq \|\psi\|^{1-\frac{k}{n}}\|G_n\psi\|^{\frac{k}{n}} \\ &\leq \|\psi\|^{1-\frac{k}{n}}\left(\|\psi\|^{\frac{1}{n+1}}\|G_{n+1}\psi\|^{\frac{n}{n+1}}\right)^{\frac{k}{n}}.\end{aligned}$$

Thus,

$$\begin{aligned}\|G_k\psi\| &\leq \|\psi\|^{1-\frac{k}{n}+\frac{k}{n(n+1)}}\|G_{n+1}\psi\|^{\frac{k}{n+1}} \\ &= \|\psi\|^{1-\frac{k}{n+1}}\|G_{n+1}\psi\|^{\frac{k}{n+1}}.\end{aligned}$$

This proves the claim.

Example 2.8 Consider the operators A, Q_1, Q_2 on $L_2(0, 1)$ given by

$$Au = u''', \quad Q_1u = -u'' + u, \quad \text{and} \quad Q_2u = -u'' + 2u,$$

$$\mathscr{D}(A) = \{u \in H^3(0, 1) \text{ such that } u(0) = u(1) = 0, \ u'(0) = u'(1)\},$$

$$\mathscr{D}(Q_1) = \mathscr{D}(Q_2) = \{u \in C^2[0, 1] \text{ such that } u(0) = u(1) = 0\}.$$

Then, A is skew-adjoint with compact resolvent, $0 \in \rho(A)$, and

$$\sup_{r>0} \frac{n(\overline{\mathbb{D}}(0, r), A)}{r^{\frac{1}{3}}} < \infty.$$

The operators Q_1 and Q_2 are symmetric nonnegative and $\frac{2}{3}$-subordinate to A. Indeed, for $u \in \mathscr{D}(Q_1)$, we can find constants $b_0 \geq 0$ such that

$$\begin{aligned}\|Q_1 u\|_{L_2} &= \|-u''+u\|_{L_2}\\ &\le \|-u''\|_{L_2}+\|u\|_{L_2}\\ &\le b_0\|u\|_{L_2}^{1-\frac{2}{3}}\|u\|_{H^3(0,1)}^{\frac{2}{3}}.\end{aligned} \tag{2.7.7}$$

We can prove that for every $\lambda \in \mathbb{C}\backslash\sigma_p(A)$, there is a constant $C \geq 0$ such that

$$\|u\|_{H^3(0,1)} \leq C\|(A-\lambda)u\|_{L_2} \text{ for } u \in \mathscr{D}(A). \tag{2.7.8}$$

Using (2.7.7) and (2.7.8), we deduce that Q_1 is $\frac{2}{3}$-subordinate to A. By using the same reasoning as before we can prove that Q_2 is $\frac{2}{3}$-subordinate to A.

For $p > 0$, the following result gives a connection to relative boundedness, cf. S. G. Krein [23, p. 146]. Let G and S be two operators in a Banach space X with $\mathscr{D}(G) \subset \mathscr{D}(S)$ and $0 < p \leq 1$. Then, S is p-subordinate to G if, and only if, (2.7.2) holds true. In fact, as the case $u = 0$ is trivial, we may assume $u \neq 0$. Suppose first that S is p-subordinate to G. If $\|Gu\| = 0$, then $\|Su\| = 0$ and (2.7.2) holds. If $\|Gu\| \neq 0$, the proof is contained in (2.7.3). Conversely, suppose that (2.7.2) holds. If $\|Gu\| = 0$, then

$$\|Su\| \leq C\varepsilon^{-p}\|u\| \to 0 \text{ as } \varepsilon \to \infty,$$

that is $\|Su\| = 0$. If $\|Gu\| \neq 0$, then by using (2.7.2) with $\varepsilon = \|u\|/\|Gu\|$, we get

$$\begin{aligned}\|Su\| &\leq C\left(\left(\frac{\|u\|}{\|Gu\|}\right)^{-p}\|u\| + \left(\frac{\|u\|}{\|Gu\|}\right)^{1-p}\|Gu\|\right)\\ &= 2C\|u\|^{1-p}\|Gu\|^p.\end{aligned}$$

If the operator S is p-subordinate to G with bound b, then S is G-bounded with G-bound 0 for $0 \leq p < 1$ and G-bound $\leq b$ for $p = 1$. While boundedness implies relative boundedness, there is, in general, no relation between p-subordination for different p. For example, if $N(G) \neq \{0\}$, then the condition

$$N(G) \subset N(S)$$

is necessary for a bounded (i.e., 0-subordinate) operator S to be p-subordinate to G with $p > 0$. The situation is different for $0 \in \rho(G)$.

Proposition 2.7.1 *If S is G-bounded with order $p \in [0, 1]$ and $0 \in \rho(G)$, then S is G-bounded with order q for all $q \in [p, 1]$.* ◇

Proof For $u \in \mathscr{D}(G)$, we have

$$\begin{aligned}\|Su\| &\leq b\|u\|^{1-p}\|Gu\|^p + a\|\varphi\|\\ &= b\|u\|^{1-q}\|G^{-1}Gu\|^{q-p}\|Gu\|^p + a\|\varphi\|\\ &\leq b\|G^{-1}\|^{q-p}\|u\|^{1-q}\|Gu\|^q + a\|\varphi\|.\end{aligned}$$

This completes the proof. Q.E.D.

Corollary 2.7.1 *If S is p-subordinate to G and $0 \in \rho(G)$, then S is q-subordinate to G for all $q \in [p, 1]$.* ◇

Lemma 2.7.1 *Let X be a Hilbert space and G and S be two linear operators on X with $\mathscr{D}(G) \subset \mathscr{D}(S)$. If the operator G has compact resolvent and S is G-bounded with order p, ($p \in [0, 1]$), then*

(i) If $N(G) = \{0\}$, then the operator S is p-subordinate to G.
(ii) If $N(G) \neq \{0\}$, then the operator $S - P$ is p-subordinate to $G + P$, where P denotes the orthogonal projection onto $N(G)$. ◇

Proof Consider G is an operator with compact resolvent and S is a G-bounded perturbation of order p, i.e.,

$$\|S\varphi\| \leq b\|G\varphi\|^p\|\varphi\|^{1-p} + a\|\varphi\|, \text{ for every } \varphi \in \mathscr{D}(G),$$

where $p \in [0, 1]$, a and b are strictly positive constants.

Let us begin by the proof of the item (i):

(i) The fact that G has compact resolvent and $N(G) = \{0\}$, then $0 \in \rho(G)$ and we can estimate

$$a\|\varphi\| = a\|G^{-1}G\varphi\|^p\|\varphi\|^{1-p} \leq a\|G^{-1}\|^p\|G\varphi\|^p\|\varphi\|^{1-p}.$$

Thus,

$$\|S\varphi\| \leq (b + a\|G^{-1}\|^p)(\|G\varphi\|^p\|\varphi\|^{1-p}),$$

for every $\varphi \in \mathscr{D}(G)$. So, according to Definition 2.7.1, we infer that S is p-subordinate to G. This achieves the proof of the item (i). For the second item, we have

(ii) Suppose that $N(G) \neq \{0\}$ and consider P is the orthogonal projection onto $N(G)$ and G_0 the operator $G_0 := G + P$. Thus, since $p \in [0, 1]$, we have

$$\begin{aligned}\|(S - P)\varphi\| &\leq b\|G\varphi\|^p\|\varphi\|^{1-p} + (a + 1)\|\varphi\| \\ &\leq b(\|G_0\varphi\| + \|P\varphi\|)^p\|\varphi\|^{1-p} + (a + 1)\|\varphi\|.\end{aligned}$$

So,

$$\begin{aligned}\|(S - P)\varphi\| &\leq b(\|G_0\varphi\|^p + \|P\varphi\|^p)\|\varphi\|^{1-p} + (a + 1)\|\varphi\| \\ &\leq b\|G_0\varphi\|^p\|\varphi\|^{1-p} + (a + b + 1)\|\varphi\|,\end{aligned}$$

this proves that $S - P$ is G_0-bounded of order p. By using Proposition 2.1.4, we get $0 \in \rho(G_0)$. Thus, $S - P$ is p-subordinate to G_0. This completes the proof. Q.E.D.

2.7.2 Concepts of Fully Subordination

Definition 2.7.3 A functional $\Phi : X \longrightarrow \mathbb{R}$ is called sublinear functional if it satisfies the following:

(i) $\Phi(\lambda\varphi) = \lambda\Phi(\varphi)$, for all $\lambda \in \mathbb{R}_{\geq 0}$ and all $\varphi \in X$, and
(ii) $\Phi(\varphi + \psi) \leq \Phi(\varphi) + \Phi(\psi)$, for all $\varphi,\ \psi \in X$. ◇

Definition 2.7.4 Let G and S be two linear operators on X such that $\mathscr{D}(G) \subset \mathscr{D}(S)$ and $\beta \in [0, 1]$. We say that S is β-fully subordinate to G if for every sufficiently small η, there exist $\Phi_\eta : X \longrightarrow \mathbb{R}$ continuous and sublinear functional, such that

$$\|S\varphi\| \leq \eta\|G\varphi\|^\beta\|\varphi\|^{1-\beta} + \Phi_\eta(\varphi), \quad \text{for every } \varphi \in \mathscr{D}(G).$$

For $\beta = 1$, we call that S is fully subordinate to G (see the reference [23] for more details). ◇

A comparison between the two concepts of subordination and fully subordination follows from the coming proposition:

Proposition 2.7.2 *Let G and S be two linear operators on X with $\mathscr{D}(G) \subset \mathscr{D}(S)$. If the operator G has compact resolvent and S is β-fully subordinate to G, ($\beta \in [0, 1]$), then*

(i) If $N(G) = \{0\}$, then the operator S is β-subordinate to G.
(ii) If $N(G) \neq \{0\}$, then the operator $S - P$ is β-subordinate to $G + P$, where P denotes the orthogonal projection onto $N(G)$. ◇

Proof Let G be an operator with compact resolvent and let S be a β-fully subordinate to G. We begin by the proof of the first item:

(i) Since G has compact resolvent and $N(G) = \{0\}$, then $0 \in \rho(G)$, and we can estimate

$$\begin{aligned}\|\varphi\| &= \|G^{-1}G\varphi\|^\beta\|\varphi\|^{1-\beta}\\ &\leq \|G^{-1}\|^\beta\|G\varphi\|^\beta\|\varphi\|^{1-\beta}.\end{aligned}$$

So, for sufficiently small η, we get

$$\begin{aligned}\Phi_\eta(\varphi) &\leq \|\Phi_\eta\|\|\varphi\|\\ &\leq \|\Phi_\eta\|\|G^{-1}\|^\beta\|G\varphi\|^\beta\|\varphi\|^{1-\beta},\end{aligned}$$

for every $\varphi \in \mathscr{D}(G)$. Hence,

$$\|S\varphi\| \leq (\eta + \|\Phi_\eta\|\|G^{-1}\|^\beta)\|G\varphi\|^\beta\|\varphi\|^{1-\beta},$$

for each $\varphi \in \mathscr{D}(G)$. In view of Definition 2.7.4, we infer that S is β-subordinate to G. This completes the proof of the first item.

(ii) Consider P is the orthogonal projection onto $N(G)$ and G_0 is the operator $G_0 := G + P$. We claim that $N(G_0) = \{0\}$. Furthermore, since $\beta \in [0, 1]$, we get

$$\begin{aligned}\|(S-P)\varphi\| &\le \eta\|G\varphi\|^{\beta}\|\varphi\|^{1-\beta} + (\|\Phi_\eta\| + 1)\|\varphi\| \\ &\le \eta(\|G_0\varphi\| + \|P\varphi\|)^{\beta}\|\varphi\|^{1-\beta} + (\|\Phi_\eta\| + 1)\|\varphi\| \\ &\le \eta(\|G_0\varphi\|^{\beta} + \|P\varphi\|^{\beta})\|\varphi\|^{1-\beta} + (\|\Phi_\eta\| + 1)\|\varphi\| \\ &\le \eta\|G_0\varphi\|^{\beta}\|\varphi\|^{1-\beta} + (\|\Phi_\eta\| + \eta + 1)\|\varphi\|,\end{aligned}$$

this shows that $S - P$ is G_0-fully subordinate with order β. Finally, using the first item (i), we deduce that $S - P$ is β-subordinate to G_0. This achieves the proof. Q.E.D.

Lemma 2.7.2 *Let X be a Hilbert space and G and S be two linear operators on X, having the same domain $\mathscr{D}$ and S be closed. If for every sufficiently small η, there exist a continuous and sublinear functional $\Phi_\eta : X \longrightarrow \mathbb{R}$ such that the following inequality*

$$\|(G-S)\varphi\| \le \eta\|S\varphi\|^{\beta}\|\varphi\|^{1-\beta} + \Phi_\eta(\varphi) \text{ for all } \varphi \in \mathscr{D} \tag{2.7.9}$$

is satisfied for some $\beta \in]0, 1]$, then G is closed. ◇

Proof Let $\eta > 0$ and $\varphi \in \mathscr{D}$. Then, Eq. (2.7.9) implies that

$$\begin{aligned}\|G\varphi\| &\le \|(G-S)\varphi\| + \|S\varphi\| \\ &\le \|S\varphi\| + \eta\|S\varphi\|^{\beta}\|\varphi\|^{1-\beta} + \|\Phi_\eta\|\|\varphi\|\end{aligned} \tag{2.7.10}$$

and

$$\begin{aligned}\|S\varphi\| &\le \|G\varphi\| + \|(S-G)\varphi\| \\ &\le \|G\varphi\| + \|\Phi_\eta\|\|\varphi\| + \eta\|S\varphi\|^{\beta}\|\varphi\|^{1-\beta}.\end{aligned} \tag{2.7.11}$$

Using Young's inequality (Lemma 1.3.1), we have

$$\|S\varphi\|^{\beta}\|\varphi\|^{1-\beta} \le \beta\|S\varphi\| + (1-\beta)\|\varphi\|. \tag{2.7.12}$$

Hence, substituting Eq. (2.7.12) in Eqs. (2.7.10) and (2.7.11), we obtain

$$\|G\varphi\| \le (1+\eta\beta)\|S\varphi\| + (\|\Phi_\eta\| + \eta(1-\beta))\|\varphi\|, \tag{2.7.13}$$

and

$$(1-\eta\beta)\|S\varphi\| \le \|G\varphi\| + (\|\Phi_\eta\| + \eta(1-\beta))\|\varphi\|.$$

Since η is sufficiently small, we can suppose that $\eta < \frac{1}{\beta}$, then $1 - \eta\beta > 0$, and we get

$$\|S\varphi\| \leq \frac{1}{1-\eta\beta}\|G\varphi\| + \frac{\eta(1-\beta)+\|\Phi_\eta\|}{1-\eta\beta}\|\varphi\|. \tag{2.7.14}$$

Now, let $(\varphi_n)_{n=1}^{\infty}$ be a sequence in $\mathscr{D}$ such that $\varphi_n \to \varphi$ as $n \to \infty$ and $G\varphi_n \to g$ as $n \to \infty$. In the next, we will prove that $\varphi \in \mathscr{D}$ and

$$g = G\varphi.$$

Using Eq. (2.7.14), we have

$$\|S(\varphi_n - \varphi_m)\| \leq \frac{1}{1-\eta\beta}\|G(\varphi_n - \varphi_m)\| + \frac{\eta(1-\beta)+\|\Phi_\eta\|}{1-\eta\beta}\|\varphi_n - \varphi_m\| \to 0$$

as $m, n \to \infty$. So, the sequence $(S\varphi_n)_n$ is a Cauchy sequence in X. Hence, $(S\varphi_n)_n$ is convergent. On the other hand, S is closed, so $\varphi \in \mathscr{D}$ and $S\varphi_n \to S\varphi$ as $n \to \infty$. Equation (2.7.13) enables us to get

$$\|G(\varphi_n - \varphi)\| \leq (1+\eta\beta)\|S(\varphi_n - \varphi)\| + \big(\eta(1-\beta) + \|\Phi_\eta\|\big)\|\varphi_n - \varphi\|.$$

So, $G\varphi_n \to G\varphi$ as $n \to \infty$ and $g = G\varphi$. Consequently, G is closed. Q.E.D.

2.8 Notes and Remarks

- The result of Theorem 2.5.4 can be found in Theorem 10.18 of Rudin [28] and in Theorems 7.2 and 7.4 of Levison and Redheffer [25].
- A proof of Lemma 2.5.1 for the special case of U and F being discs, $z_0 = 0, a = 1$, $b = 0$, and $f(0) = 1$ can be found in Levin [24, Theorem I.11]. The general form stated in Lemma 2.5.1 is obtained from this particular case by means of a conformal mapping.
- Variants of the assertions expressed in a different form in Lemma 2.5.2 can be found, for example, in the monograph [24, Chap. 1]. In the form presented in Lemma 2.5.2, the first and third assertions are contained in [27, Lemmas 1.1, 1.3], and the second assertion in [29, Lemma 2.1].

(i) In the case where A has compact resolvent, its spectrum consists of isolated eigenvalues only, so for every eigenvalue λ, there is the associated Riesz projection P_λ, which satisfies $R(P_\lambda) = \mathscr{L}_A(\lambda)$, where $\mathscr{L}_A(\lambda)$ is the root subspace of A corresponding to λ defined in (1.10.1).
(ii) An entire function of exponential type is of finite order at most 1.
(iii) If a sequence $(\lambda_n)_{n=1}^{\infty}$ is separated, then all the points λ_n are entirely distinct.
(iv) Let K be a finite-rank operator on X. Then,

$$|\det(I+K)| \geq |1 - \|K\||^{rk(K)}.$$

Indeed,

$$
\begin{aligned}
|\det(I+K)| &= \left|\prod_{n=1}^{rk(K)} (1+\lambda_n(K))\right| \\
&= \prod_{n=1}^{rk(K)} |1+\lambda_n(K)| \\
&\geq \prod_{n=1}^{rk(K)} |1-|\lambda_n(K)||\,, \qquad (2.8.1)
\end{aligned}
$$

where $\lambda_n(K)$ is the nth eigenvalue of K. Since K is a finite-rank operator, then we have

$$|\lambda_n(K)| \leq \|K\|.$$

So, Eq. (2.8.1) implies that

$$|\det(I+K)| \geq |1-\|K\||^{rk(K)}.$$

(v) $I+K$ is invertible if, and only if, $\det(I+K) \neq 0$, and in this case

$$\|(I+K)^{-1}\| \leq \frac{(1+\|K\|)^{rk(K)}}{|\det(I+K)|}.$$

(vi) The notion of G-boundedness with order p is a natural generalization of the notion of p-subordination. Our initial line of inquiry was suggested by a comparison between the two concepts. The concept of p-subordinate perturbations was studied by the references [18, I.7.1] and [20, 26], see also [1, 4–16, 19, 21, 22] and [34, 3.2].

References

1. M. T. Aimar, A. Intissar, A. Jeribi, On an unconditional basis of generalized eigenvectors of the non-self-adjoint Gribov operator in Bargmann space, J. Math. Anal. Appl. **231**, 588–602 (1999)
2. S.A. Avdonin, S.A. Ivanov, *Families of Exponentials: The Method of Moments in Controlability Problems for Distributed Systems* (Cambridge University Press, Cambridge, UK, 1995)
3. V. Bargmann, On a Hilbert space of analytic functions and associated integral transform. Commun. Pure Appl. Math. **14**, 187–214 (1961)
4. N. Ben Ali, A. Jeribi, On the Riesz basis of a family of analytic operators in the sens of Kato and application to the problem of radiation of a vibriting structure in a light fluid, J. Math. Anal. Appl. **320**(1), 78–94 (2006)

5. S. Charfi, A. Damergi, A. Jeribi, On a Riesz basis of finite-dimensional invariant subspaces and application to Gribov operator in Bargmann space. Line. Mult. Alge. **61**(11), 1577–1591 (2013)
6. S. Charfi, A. Intissar, A. Jeribi, Expansion of solution in terms of generalized eigenvectors for a rectilinear transport equation. Transport Theory Statist. Phys. **38**, 1–16 (2009)
7. S. Charfi, A. Jeribi, I. Walha, Riesz basis property of families of nonharmonic exponential and application to a problem of a radiation of a vibrating structure in a light fluid. Numerical Functional Analysis and Optimization, **32**(4), 370–382 (2011)
8. H. Ellouz, I. Feki, A. Jeribi, On a Riesz basis of exponentials related to the eigenvalues of an analytic operator and application to a non-self-adjoint problem deduced from a perturbation method, J. Math. Phys. **54**, 112101 (2013)
9. H. Ellouz, I. Feki, A. Jeribi, On the asymptotic behavior of the eigenvalues of an analytic operator in the sense of Kato and applications, Serdica Math. J. **45**(1), 55–88 (2019)
10. H. Ellouz, I. Feki, A. Jeribi, On a Riesz basis of exponentials related to a family of analytic operators and application. J. Pseudo-Differ. Oper. Appl. **10**(4), 999–1014 (2019)
11. H. Ellouz, I. Feki, A. Jeribi, Non-orthogonal Fusion Frames of an Analytic Operator and Application to a One-Dimensional Wave Control System. Mediterr. J. Math. **16**(2), 16:52 (2019)
12. H. Ellouz, I. Feki, A. Jeribi, On the basis property of root vectors related to a non-self-adjoint analytic operator and applications, (2021)
13. I. Feki, A. Jeribi, R. Sfaxi, On an unconditionl basis of generalized eigenvectors of an analytic operator and application to a problem of radiation of a vibrating structure in a light fluid, J. Math. Anal. Appl. **375**, 261–269 (2011)
14. I. Feki, A. Jeribi, R. Sfaxi, On a Schauder basis related to the eigenvectors of a family of non-self-adjoint analytic operators and applications, Anal. Math. Phys. **3**(4), 311–331 (2013)
15. I. Feki, A. Jeribi, R. Sfaxi, On a Riesz basis of eigenvectors of a non-self-adjoint analytic operator and applications. Line. Mult. Alge. **62**(8), 1049–1068 (2014)
16. I. Feki, A. Jeribi, R. Sfaxi, On a Riesz basis of eigenvectors of a non-self-adjoint perturbation and application to a Gribov operator in Bargmann space, preprint (2021)
17. I.C. Gohberg, S. Goldberg, M.A. Kaashoek, Classes of linear operators, Vol. I, Operator Theory: Advances and Applications, Vol. 29, Birkhauser, Basel, (1990)
18. I.C. Gohberg, M.G. Krein, *Introduction to the Theory of Linear Non-self-Adjoint Operators in Hilbert Space* (Amer. Math. Soci, Providence, 1969)
19. A. Intissar, A. Jeribi, I. Walha, Riesz basis of exponential family for a hyperbolic system. J. Appl. Anal. **25**(1), 13–23 (2019)
20. A. Jeribi, *Spectral Theory and Applications of Linear Operators and Block Operator Matrices* (Springer, New-York, 2015)
21. A. Jeribi, *Denseness, Bases and Frames in Banach Spaces and Applications* (de Gruyter, Berlin, 2018)
22. A. Jeribi, A. Intissar, On an Riesz basis of generalized eigenvectors of the nonself-adjoint problem deduced from a perturbation method for sound radiation by a vibrating plate in a light fluid. J. Math. Anal. Appl. **292**, 1–16 (2004)
23. S.G. Krein, *Linear Differential Equations in Banach Space* (Amer. Math. Soci., 1971)
24. B.J. Levin, *Distribution of Zeros of Entire Functions* (Amer. Math. Soc, Providence, 1980)
25. N. Levison, R.M. Redheffer, *Complex Variables* (Holden-Day Inc, San Francisco, 1970)
26. A.S. Markus, *Introduction to the Spectral Theory of Polynomial Operator Pencils*, vol. 71 (Amer. Math. Soci., Providence, RI, 1988)
27. A.S. Markus, V.I. Matsaev, Comparison Theorems for Spectra of Linear Operators, and spectral asymptotics. Tr. Mosk. Mat. Obshch. **45**, 133–181 (1982) [Trans. Moscow Math. Soc. (1), 139–187 (1984)]
28. W. Rudin, *Real and Complex Analysis*, 2nd edn. (McGraw-Hill Book Company, New-York, 1974)
29. A.A. Shkalikov, The basis problem of the eigenfunctions of ordinary differential operators with Integral boundary conditions. Vestn. Mosk. Univ., Ser. 1: Mat., Mekh. (6), 12–21 (1982) [Moscow Univ. Math. Bull. **37**(6), 10–20 (1982)]

30. A.A. Shkalikov, On the basis property of root vectors of a perturbed self-adjoint operator. Proc. Steklov Inst. Math. **269**, 284–298 (2010)
31. B. Russo, On the Hausdorff-Young theorem for integral operators. Pac. J. Math. **68**, 241–253 (1977)
32. E.C. Titchmarsh, *The Theory of Functions* (Oxford University Press, 1932)
33. C. Wyss, Perturbation theory for Hamiltonian operator matrices and Riccati equations, Ph.D. thesis, University of Bern (2008)
34. C. Wyss, Riesz bases for p-subordinate perturbations of normal operator. J. Funct. Anal. **258**, 208–240 (2010)
35. K. Yosida, *Functional Analysis* (Springer, Berlin, 1980)
36. R.M. Young, *An Introduction to Nonharmonic Fourier Series* (Academic, London, 1980)

Chapter 3
Series of Complex Terms

This chapter is devoted to study the convergence of series of complex terms. In this chapter, some operators with Carleman-class, the spectral theory of compact operators, and the objective of this chapter are to establish the equality

$$\sum_{n=1}^{+\infty} a_n \left(\sum_{k=1}^{+\infty} s_k\right)^n = \sum_{n=1}^{+\infty} a_n \sum_{l=n}^{+\infty} \sum_{i_1+i_2+\cdots+i_n=l} s_{i_1} s_{i_2} \ldots s_{i_n}, \tag{3.0.1}$$

where $(s_k)_{k\geq 1}$ and $(a_k)_{k\geq 1}$ are given complex sequences satisfying some convergence conditions. Our purpose here is to establish some results required in the sequel given by Feki et al. [1].

3.1 Identity Results

To reach this goal (prove Eq. (3.0.1)), we shall first introduce the following technical results for future use.

3.1.1 Technical Results

Lemma 3.1.1 *Let $s = (s_k)_{k\geq 0}$ be a sequence in $\mathbb{C}$ and $s^* = (s_k^*)_{k\geq 0}$ be the transformed sequence given by*

$$s_0^* = 1, \quad s_k^* = \sum_{\nu=0}^{k-1} s_\nu^* \, s_{k-\nu}, \quad \text{for every } k \geq 1. \tag{3.1.1}$$

A. Jeribi, *Perturbation Theory for Linear Operators*,
https://doi.org/10.1007/978-981-16-2528-2_3

Then,

$$s_k^* = \sum_{l=1}^{k} \sum_{i_1+i_2+\cdots+i_l=k} s_{i_1}s_{i_2}s_{i_3}\dots s_{i_l}, \quad \text{for every} \quad k \geq 1. \tag{3.1.2}$$

(Here $1 \leq i_l \leq k$*, for every* $l = 1, 2, \dots, k$ *and* $k \geq 1$*).* ◇

Proof Let us proceed by induction on $k \geq 1$. For $k = 1$, we have

$$s_1^* = s_0^* s_1 = s_1 = \sum_{l=1}^{1} \sum_{i_1=1} s_{i_1}.$$

So, the recurrence property is valid. Let $k \in \mathbb{N}^*$. Suppose that the property is true up to order k and let us prove that

$$s_{k+1}^* = \sum_{l=1}^{k+1} \sum_{i_1+i_2+\cdots+i_l=k+1} s_{i_1}s_{i_2}s_{i_3}\cdots s_{i_l}.$$

We have

$$\begin{aligned} s_{k+1}^* &= \sum_{\nu=0}^{k} s_\nu^* s_{k+1-\nu} \\ &= s_0^* s_{k+1} + \sum_{\nu=1}^{k} s_\nu^* s_{k+1-\nu}. \end{aligned}$$

Taking into account the fact that $s_0^* = 1$ and the induction hypothesis, we infer that

$$\begin{aligned} s_{k+1}^* &= s_{k+1} + \sum_{\nu=1}^{k} \left(\sum_{l=1}^{\nu} \sum_{i_1+i_2+\cdots+i_l=\nu} s_{i_1}s_{i_2}s_{i_3}\cdots s_{i_l} \right) s_{k+1-\nu} \\ &= s_{k+1} + \sum_{l=1}^{k} \left(\sum_{\nu=l}^{k} \sum_{i_1+i_2+\cdots+i_l=\nu} s_{i_1}s_{i_2}\dots s_{i_l} \right) s_{k+1-\nu} \\ &= s_{k+1} + \sum_{l=2}^{k+1} \left(\sum_{\nu=l-1}^{k} \sum_{i_1+i_2+\cdots+i_{l-1}=\nu} s_{i_1}s_{i_2}\dots s_{i_{l-1}} \right) s_{k+1-\nu} \\ &= s_{k+1} + \sum_{l=2}^{k+1} \sum_{i_1+i_2+\cdots+i_l=k+1} s_{i_1}s_{i_2}\dots s_{i_l} \\ &= \sum_{l=1}^{k+1} \sum_{i_1+i_2+\cdots+i_l=k+1} s_{i_1}s_{i_2}\dots s_{i_l}. \end{aligned}$$

This finishes the proof of the lemma. Q.E.D.

Lemma 3.1.2 *Let x and y in $\mathbb{C}$. If we set $t_k = yx^k$ for all $k \geq 1$, then*

$$t_k^* = yx^k(1+y)^{k-1},$$

for all $k \geq 1$. ◇

Proof If $t_k = 0$, for all integer $k \geq 0$, it is clear that $t_k^* = 0$ for all integer $k \geq 1$. Assume that x and y are non-zero complex numbers and let $t_k = yx^k$ for all integer $k \geq 1$. By induction on the integer $k \geq 1$, let us show that

$$t_k^* = yx^k(1+y)^{k-1}$$

for all integer $k \geq 1$. For $k = 1$, the recurrence property is true since $t_1^* = t_1 = yx$. Suppose that the recurrence property is valid to the order k ($k \geq 1$), and let us show that it remains valid to the order $k+1$. By the recurrence hypothesis and Eq. (3.1.1), we obtain

$$\begin{aligned} t_{k+1}^* &= \sum_{\nu=0}^{k} t_\nu^* \, t_{k+1-\nu} \\ &= t_{k+1} + \sum_{\nu=1}^{k} yx^\nu(1+y)^{\nu-1} yx^{k+1-\nu}. \end{aligned}$$

Thus,

$$\begin{aligned} t_{k+1}^* &= yx^{k+1} + x^{k+1}y^2 \sum_{\nu=1}^{k} (1+y)^{\nu-1} \\ &= yx^{k+1} - x^{k+1}y^2 \frac{1-(1+y)^k}{y} \\ &= yx^{k+1}(1+y)^{k-1}, \end{aligned}$$

which completes the proof. Q.E.D.

Lemma 3.1.3 *Let y in $\mathbb{C}$. For every integer $k \geq 1$, we have the following:*

$$\sum_{l=1}^{k} y^l \sum_{i_1+i_2+\cdots+i_l=k} 1 = y(1+y)^{k-1}.$$ ◇

Proof Due to Lemmas 3.1.1 and 3.1.2, for every integer $k \geq 1$ and every y in $\mathbb{C}$, we get

$$y(1+y)^{k-1} = \sum_{l=1}^{k} y^l \sum_{i_1+i_2+\cdots+i_l=k} 1.$$ Q.E.D.

Lemma 3.1.4 *For every integer $k \geq 1$ and $1 \leq l \leq k$, we have the following:*

$$\sum_{i_1+i_2+\cdots+i_l=k} 1 = C_{k-1}^{l-1}. \qquad \diamond$$

Proof Let y in $\mathbb{C}$. From Lemma 3.1.3 and the binomial expansion formula, we have

$$\begin{aligned}\sum_{l=1}^{k} y^l \sum_{i_1+i_2+\cdots+i_l=k} 1 &= \sum_{l=0}^{k-1} C_{k-1}^{l} y^{l+1} \\ &= \sum_{l=1}^{k} C_{k-1}^{l-1} y^l.\end{aligned}$$

By identification, for every integer $k \geq 1$ and $1 \leq l \leq k$, it follows that

$$\sum_{i_1+i_2+\cdots+i_l=k} 1 = C_{k-1}^{l-1}.$$

Q.E.D.

Lemma 3.1.5 *Let* $s = (s_k)_{k\geq 1}$ *be a sequence in* $\mathbb{C}$. *Suppose there exist* $\lambda > 0$ *and* $0 < q < \frac{1}{\lambda+1}$ *such that*

$$|s_k| \leq \lambda q^k,$$

for every integer $k \geq 1$, *then*

$$\sum_{k=1}^{\infty} |s_k| \leq \frac{\lambda q}{1-q},$$

and

$$\sum_{k=0}^{\infty} |s_k^*| \leq \frac{1-q}{1-q(1+\lambda)}. \qquad \diamond$$

Proof Let $(s_k)_{k\geq 1}$ be a sequence in $\mathbb{C}$ satisfying

$$|s_k| \leq \lambda q^k,$$

for every integer $k \geq 1$, where $\lambda > 0$ and $0 < q < \frac{1}{1+\lambda}$. Clearly,

$$\begin{aligned}\sum_{k=1}^{\infty} |s_k| &\leq \lambda \sum_{k=1}^{\infty} q^k \\ &= \frac{\lambda q}{1-q} < +\infty.\end{aligned}$$

Besides,

$$|s_k| \leq |t_k|,$$

for every integer $k \geq 1$, where $t_k = \lambda q^k$, for every integer $k \geq 1$. By using both Lemma 3.1.2 and Notes and Remarks 3.3 (Eq. (3.3.1)), it follows that

$$\begin{aligned}|s_k^*| &\leq \widetilde{t}_k^* \\ &= \lambda q^k (1+\lambda)^{k-1},\end{aligned}$$

for every integer $k \geq 1$. Thus,

$$\begin{aligned}\sum_{k=0}^{\infty} |s_k^*| &\leq 1 + \sum_{k=1}^{\infty} \widetilde{t}_k^* \\ &\leq 1 + \frac{\lambda}{1+\lambda} \sum_{k=1}^{\infty} (q(1+\lambda))^k \\ &\leq 1 + \frac{\lambda q}{1 - q(1+\lambda)} \\ &\leq \frac{1-q}{1-q(1+\lambda)},\end{aligned}$$

which completes the proof. Q.E.D.

Now, let us recall Fubini's absolute convergence inversion criterion.

Lemma 3.1.6 *Let $(U_{n,l})_{n,l\geq 1}$ be a double sequence in $\mathbb{C}$ satisfying the following:*

(i) for every fixed integer $n \geq 1$, the series

$$\sum_{l\geq 1} |U_{n,l}|$$

converge, and

(ii) the series

$$\sum_{n\geq 1} \sum_{l\geq 1} |U_{n,l}|$$

converge.

Then, we can interchange the order of summation in a doubly indexed infinite series

$$\sum_{n=1}^{+\infty} \sum_{l=1}^{+\infty} U_{n,l} = \sum_{l=1}^{+\infty} \sum_{n=1}^{+\infty} U_{n,l}. \qquad \diamond$$

3.1.2 Proof of Eq. (3.0.1) When $(a_k)_k \equiv 1$

Having obtained these results, we are ready to prove Eq. (3.0.1) when $(a_k)_k \equiv 1$.

Proposition 3.1.1 *For any sequence $s = (s_k)_{k\geq 1}$ in $\mathbb{C}$ such that*

$$\tau(s) = \sum_{k=1}^{+\infty} |s_k| < 1$$

and

$$\sum_{k=0}^{+\infty} |s_k^*| < +\infty,$$

where $s^* = (s_k^*)_{k\geq 0}$ is given by (3.1.1), we have the following:

$$\sum_{n=1}^{+\infty}\left(\sum_{k=1}^{+\infty} s_k\right)^n = \sum_{n=1}^{+\infty}\sum_{l=1}^{n}\sum_{i_1+i_2+\cdots+i_l=n} s_{i_1}s_{i_2}\ldots s_{i_l}. \qquad \diamond$$

Proof Setting

$$\mathscr{S} = \sum_{n=1}^{+\infty}\left(\sum_{k=1}^{+\infty} s_k\right)^n - \sum_{n=1}^{+\infty}\sum_{l=1}^{n}\sum_{i_1+i_2+\cdots+i_l=n} s_{i_1}s_{i_2}\ldots s_{i_l}.$$

By Eq. (3.1.2) and the fact that $s_0^* = 1$, we can write

$$\mathscr{S} = \sum_{n=0}^{+\infty}\left(\sum_{k=1}^{+\infty} s_k\right)^n - \sum_{n=0}^{+\infty} s_n^*.$$

Putting $s_0 = 0$ and applying the Cauchy product formula of two absolutely convergent series, we obtain after taking (3.0.1) into account,

$$\begin{aligned}
\mathscr{S}\left(\sum_{k=1}^{+\infty} s_k\right) &= \sum_{n=0}^{+\infty}\left(\sum_{k=1}^{+\infty} s_k\right)^{n+1} - \sum_{n=0}^{+\infty} s_n^* \sum_{k=0}^{+\infty} s_k \\
&= \sum_{n=0}^{+\infty}\left(\sum_{k=1}^{+\infty} s_k\right)^{n} - 1 - \sum_{n=0}^{+\infty}\left(\sum_{\nu=0}^{n} s_\nu^*\, s_{n-\nu}\right) \\
&= \sum_{n=0}^{+\infty}\left(\sum_{k=1}^{+\infty} s_k\right)^{n} - 1 - \sum_{n=1}^{+\infty}\left(\sum_{\nu=0}^{n-1} s_\nu^*\, s_{n-\nu}\right) \\
&= \sum_{n=0}^{+\infty}\left(\sum_{k=1}^{+\infty} s_k\right)^{n} - 1 - \sum_{n=1}^{+\infty} s_n^* \\
&= \sum_{n=0}^{+\infty}\left(\sum_{k=1}^{+\infty} s_k\right)^{n} - \sum_{n=0}^{+\infty} s_n^* = \mathscr{S}.
\end{aligned}$$

Hence,

$$\mathscr{S}\left(\sum_{k=1}^{+\infty} s_k - 1\right) = 0.$$

Since

$$\left|\sum_{k=1}^{+\infty} s_k\right| < \sum_{k=1}^{+\infty} |s_k| < 1,$$

it comes that $\mathscr{S} = 0$. Accordingly, we get

$$\sum_{n=1}^{+\infty}\left(\sum_{k=1}^{+\infty} s_k\right)^n = \sum_{n=1}^{+\infty}\sum_{l=1}^{n}\sum_{i_1+i_2+\cdots+i_l=n} s_{i_1}s_{i_2}\ldots s_{i_l}.$$ Q.E.D.

Proposition 3.1.2 *Let $s = (s_k)_{k\geq 1}$ be a sequence in $\mathbb{C}$ such that*

$$\tau(s) = \sum_{k=1}^{+\infty} |s_k| < 1$$

and

$$\sum_{k=0}^{+\infty} |s_k^*| < +\infty,$$

where $s^* = (s_k^*)_{k\geq 0}$ is given by (3.1.1). If

$$\sum_{k=0}^{+\infty} \widetilde{s}_k^* < +\infty,$$

where $\widetilde{s}_k = |s_k|$, for every integer $k \geq 1$, then we have

$$\sum_{n=1}^{+\infty}\left(\sum_{k=1}^{+\infty} s_k\right)^n = \sum_{l=1}^{+\infty}\sum_{k=l}^{+\infty}\sum_{i_1+i_2+\cdots+i_l=k} s_{i_1}s_{i_2}\cdots s_{i_l}.$$ ◇

Proof Assume that

$$\sum_{k=0}^{+\infty} \widetilde{s}_k^* < +\infty.$$

Due to Lemma 3.1.1, we have

$$\begin{aligned}\sum_{n=1}^{+\infty}\left(\sum_{k=1}^{+\infty} s_k\right)^n &= \sum_{n=1}^{+\infty}\sum_{l=1}^{n}\sum_{i_1+i_2+\cdots+i_l=n} s_{i_1}s_{i_2}\ldots s_{i_l}\\ &= \sum_{n=1}^{+\infty}\sum_{l=1}^{+\infty} U_{n,l},\end{aligned}$$

where

$$U_{n,l} = \chi_{[|1,n|]}(l) \sum_{i_1+i_2+\cdots+i_l=n} s_{i_1} s_{i_2} \dots s_{i_l},$$

for every $n, l \geq 1$ and $\chi_{[|1,n|]}$ is the characteristic function of $[|\ 1, n\ |] = \{1, 2, \dots, n\}$. For any fixed integer $n \geq 1$, we have

$$\sum_{l=1}^{+\infty} \left|U_{n,l}\right| = \sum_{l=1}^{n} \left| \sum_{i_1+i_2+\cdots+i_l=n} s_{i_1} s_{i_2} \dots s_{i_l} \right| < +\infty.$$

Besides,

$$\begin{aligned}\sum_{n=1}^{+\infty}\sum_{l=1}^{+\infty} \left|U_{n,l}\right| &\leq \sum_{n=1}^{+\infty}\sum_{l=1}^{n} \sum_{i_1+i_2+\cdots+i_l=n} |s_{i_1}||s_{i_2}| \dots |s_{i_l}| \\ &\leq \sum_{n=1}^{+\infty} \widetilde{s}_n^* < +\infty.\end{aligned}$$

Using Lemma 3.1.6, it follows that

$$\begin{aligned}\sum_{n=1}^{+\infty} \left(\sum_{k=1}^{+\infty} s_k\right)^n &= \sum_{n=1}^{+\infty}\sum_{l=1}^{+\infty} U_{n,l} \\ &= \sum_{l=1}^{+\infty}\sum_{n=1}^{+\infty} U_{n,l} \\ &= \sum_{l=1}^{+\infty}\sum_{n=l}^{+\infty} \sum_{i_1+i_2+\cdots+i_l=n} s_{i_1} s_{i_2} \dots s_{i_l}.\end{aligned}$$

Hence, the proof is complete. Q.E.D.

Proposition 3.1.3 *Let $s = (s_k)_{k\geq 1}$ be a sequence in $\mathbb{C}$ such that*

$$\tau(s) = \sum_{k=1}^{+\infty} |s_k| < 1$$

and

$$\sum_{k=0}^{+\infty} |s_k^*| < +\infty,$$

where $s^* = (s_k^*)_{k\geq 0}$ is given by (3.1.1). If

$$\sum_{k=0}^{+\infty} \widetilde{s}_k^* < +\infty,$$

where $\widetilde{s}_k = |s_k|$, *for every integer* $k \geq 1$, *then for every integer* $l \geq 1$, *we have*

$$\left(\sum_{k=1}^{+\infty} s_k\right)^l = \sum_{k=l}^{+\infty} \sum_{i_1+i_2+\cdots+i_l=k} s_{i_1} s_{i_2} \ldots s_{i_l}. \qquad \diamond$$

Proof Let z be a fixed complex number such that $|z| < 1$. Setting $v_k = z s_k$, for all $k \geq 1$. Clearly, $|v_k| \leq |s_k|$ for every $k \geq 1$. From Eq. (3.1.2), we have

$$\widetilde{v}_k^* < \widetilde{s}_k^*, \quad \forall\, k \geq 1.$$

By the assumption, it comes that

$$\sum_{k=1}^{+\infty} |v_k| \leq \sum_{k=1}^{+\infty} |s_k| < 1$$

and

$$\sum_{k=0}^{+\infty} \widetilde{v}_k^* \leq \sum_{k=0}^{+\infty} \widetilde{s}_k^* < +\infty.$$

So, while using Proposition 3.1.1, we obtain

$$\begin{aligned}\sum_{n=1}^{+\infty} \left(\sum_{k=1}^{+\infty} v_k\right)^n &= \sum_{n=1}^{+\infty} \sum_{l=1}^{n} \sum_{i_1+i_2+\cdots+i_l=n} v_{i_1} v_{i_2} \ldots v_{i_l} \\ &= \sum_{l=1}^{+\infty} \sum_{k=l}^{+\infty} \sum_{i_1+i_2+\cdots+i_l=k} v_{i_1} v_{i_2} \ldots v_{i_l},\end{aligned}$$

i.e.,

$$\begin{aligned}\sum_{l=1}^{+\infty} z^l \left(\sum_{k=1}^{+\infty} s_k\right)^l &= \sum_{n=1}^{+\infty} \sum_{l=1}^{n} z^l \sum_{i_1+i_2+\cdots+i_l=n} s_{i_1} s_{i_2} \ldots s_{i_l} \\ &= \sum_{l=1}^{+\infty} z^l \left(\sum_{k=l}^{+\infty} \sum_{i_1+i_2+\cdots+i_l=k} s_{i_1} s_{i_2} \ldots s_{i_l}\right),\end{aligned}$$

for every z in $\mathbb{C}$ with $|z| < 1$. By identification, we infer that, for every integer $l \geq 1$

$$\left(\sum_{k=1}^{+\infty} s_k\right)^l = \sum_{k=l}^{+\infty} \sum_{i_1+i_2+\cdots+i_l=k} s_{i_1} s_{i_2} \ldots s_{i_l}.$$

Hence, the proof of proposition is completed. Q.E.D.

Proposition 3.1.4 *Let* $s = (s_k)_{k \geq 1}$ *be a sequence in* $\mathbb{C}$ *such that*

$$\tau(s) = \sum_{k=1}^{+\infty} |s_k| < 1$$

and

$$\sum_{k=0}^{+\infty} |s_k^*| < +\infty,$$

where $s^* = (s_k^*)_{k\geq 0}$ *is given by (3.1.1). If*

$$\sum_{k=0}^{+\infty} \widetilde{s}_k^* < +\infty,$$

where $\widetilde{s}_k = |s_k|$, *for every integer* $k \geq 1$, *then for* z *in* $\mathbb{C}$ *such that* $|z| < 1$ *and every integer* $l \geq 1$, *we have*

$$\left(\sum_{k=1}^{+\infty} s_k z^k\right)^l = \sum_{k=l}^{+\infty} \left(\sum_{i_1+i_2+\cdots+i_l=k} s_{i_1} s_{i_2} \ldots s_{i_l}\right) z^k. \qquad \diamond$$

Proof For a fixed z in $\mathbb{C}$, with $|z| < 1$, let $(w_k)_{k\geq 1}$ be the sequence in $\mathbb{C}$, given by

$$w_k = z^k s_k,$$

for every $k \geq 1$. Clearly,

$$|w_k| \leq |s_k|$$

for all $k \geq 1$. By Eq. (3.1.2), we get

$$\widetilde{w}_k^* \leq \widetilde{s}_k^*, \quad \forall\, k \geq 1.$$

By the assumption, it comes that

$$\sum_{k=1}^{+\infty} |w_k| \leq \sum_{k=1}^{+\infty} |s_k| < 1$$

and

$$\sum_{k=0}^{+\infty} \widetilde{w}_k^* \leq \sum_{k=0}^{+\infty} \widetilde{s}_k^* < +\infty.$$

Due to Proposition 3.1.3, it follows that

$$\left(\sum_{k=1}^{+\infty} s_k z^k\right)^l = \sum_{k=l}^{+\infty} \left(\sum_{i_1+i_2+\cdots+i_l=k} s_{i_1} s_{i_2} \ldots s_{i_l}\right) z^k,$$

for every $z \in \mathbb{C}$ with $|z| < 1$ and every integer $l \geq 1$. Hence, the proof is complete. Q.E.D.

3.1.3 General Case

More general, we have the following result.

Theorem 3.1.1 *For any sequence $a = (a_k)_{k\geq 1}$ and $s = (s_k)_{k\geq 1}$ in $\mathbb{C}$ such that*

$$\tau(s) = \sum_{k=1}^{+\infty} |s_k| < 1$$

and

$$\sum_{k=0}^{+\infty} \widetilde{s}_k^* < +\infty,$$

where $\widetilde{s}^* = (\widetilde{s}_k^*)_{k\geq 0}$ is given by

$$\begin{cases} \widetilde{s}_0^* = 1 \\ \widetilde{s}_k^* = \sum_{l=1}^{k} \sum_{i_1+i_2+\cdots+i_l=k} |s_{i_1}||s_{i_2}|\ldots|s_{i_l}|, \quad k \geq 1. \end{cases}$$

Then,

$$\sum_{n=1}^{+\infty} a_n \left(\sum_{k=1}^{+\infty} s_k\right)^n = \sum_{n=1}^{+\infty} a_n \sum_{l=n}^{+\infty} \sum_{i_1+i_2+\cdots+i_n=l} s_{i_1}s_{i_2}\ldots s_{i_n}. \qquad \diamond$$

Proof From the assumption and Proposition 3.1.3, it is clear that

$$\sum_{n=1}^{+\infty} a_n \left(\sum_{k=1}^{+\infty} s_k\right)^n = \sum_{n=1}^{+\infty} a_n \sum_{l=n}^{+\infty} \sum_{i_1+i_2+\cdots+i_n=l} s_{i_1}s_{i_2}\ldots s_{i_n}.$$

Hence, the proof is complete. Q.E.D.

Theorem 3.1.2 *For any sequence $a = (a_k)_{k\geq 1}$ and $s = (s_k)_{k\geq 1}$ in $\mathbb{C}$ such that*

$$\tau(s) = \sum_{k=1}^{+\infty} |s_k| < 1$$

and

$$\sum_{k=0}^{+\infty} \widetilde{s}_k^* < +\infty,$$

where $\widetilde{s}^* = (\widetilde{s}_k^*)_{k\geq 0}$ *is given by*

$$\begin{cases} \widetilde{s}_0^* = 1 \\ \widetilde{s}_k^* = \sum_{l=1}^{k} \sum_{i_1+i_2+\cdots+i_l=k} |s_{i_1}||s_{i_2}|\ldots|s_{i_l}|, \quad k \geq 1. \end{cases}$$

If $a = (a_k)_{k\geq 1}$ *is a bounded sequence, then we have*

$$\sum_{n=1}^{+\infty} a_n \left(\sum_{k=1}^{+\infty} s_k\right)^n = \sum_{l=1}^{+\infty}\sum_{n=1}^{l} a_n \sum_{i_1+i_2+\cdots+i_n=l} s_{i_1}s_{i_2}\ldots s_{i_n}.$$ ◇

Proof Assume that $a = (a_k)_{k\geq 1}$ is a bounded sequence. We can write

$$\sum_{n=1}^{+\infty} a_n \left(\sum_{k=1}^{+\infty} s_k\right)^n = \sum_{n=1}^{+\infty}\sum_{l=1}^{+\infty} V_{n,l},$$

with

$$V_{n,l} = a_n\, \chi_{[|n,+\infty[|}(l) \sum_{i_1+i_2+\cdots+i_n=l} s_{i_1}s_{i_2}\ldots s_{i_n},$$

for every integer $n, l \geq 1$ and where $\chi_{[|n,+\infty[|}$ is the characteristic function of $[|n, +\infty[|$. From the assumption and Proposition 3.1.3, we have

$$\begin{aligned} \sum_{l=1}^{+\infty} \left|V_{n,l}\right| &\leq |a_n| \sum_{l=n}^{+\infty} \sum_{i_1+i_2+\cdots+i_n=l} |s_{i_1}||s_{i_2}|\ldots|s_{i_n}| \\ &= |a_n|\,(\tau(s))^n < +\infty, \end{aligned}$$

for every integer $n \geq 1$. Besides,

$$\sum_{n=1}^{+\infty}\sum_{l=1}^{+\infty} \left|V_{n,l}\right| \leq \sum_{n=1}^{+\infty} |a_n|\,(\tau(s))^n.$$

From the assumption $a = (a_n)_{n\geq 1}$ is a bounded sequence, i.e., there exists $M > 0$ such that $|a_n| \leq M$, for every $n \geq 1$ and since $\tau(s) < 1$, we obtain

$$\begin{aligned} \sum_{n=1}^{+\infty}\sum_{l=1}^{+\infty} \left|V_{n,l}\right| &\leq M \sum_{n=1}^{+\infty} (\tau(s))^n \\ &= \frac{M\tau(s)}{1-\tau(s)} < +\infty. \end{aligned}$$

By virtue of Lemma 3.1.6, this implies

$$\sum_{n=1}^{+\infty}\left(\sum_{k=1}^{+\infty} s_k\right)^n = \sum_{n=1}^{+\infty}\sum_{l=1}^{+\infty} V_{n,l}$$
$$= \sum_{l=1}^{+\infty}\sum_{n=1}^{+\infty} V_{n,l}$$
$$= \sum_{l=1}^{+\infty}\sum_{n=1}^{l} a_n \sum_{i_1+i_2+\cdots+i_n=l} s_{i_1}s_{i_2}\ldots s_{i_l}.$$

Hence, the proof is complete. Q.E.D.

Proposition 3.1.5 *Let $s = (s_k)_{k\geq 0}$ be a sequence in $\mathbb{C}$ and $(s_k^*)_{k\geq 0}$ be the transformed sequence given by $s_0^* = 1$ and for all $k \geq 1$,*

$$s_k^* = \sum_{\nu=0}^{k-1} s_\nu^* s_{k-\nu}.$$

Then, for all $k \geq 1$, we have

$$\sum_{\nu=0}^{k-1} s_\nu^* s_{k-\nu} = \sum_{\nu=k}^{\infty}\left(\sum_{i_1+i_2+\cdots+i_k=\nu} s_{i_1}s_{i_2}\ldots s_{i_k}\right). \quad \diamond$$

Proof Due to Lemma 3.1.1, we have

$$\sum_{k=1}^{\infty} s_k^* = \sum_{k=1}^{\infty}\sum_{l=1}^{k}\sum_{i_1+i_2+\cdots+i_l=k} s_{i_1}s_{i_2}s_{i_3}\ldots s_{i_l}.$$

Interchanging the two sums of the right-hand side of the last equation, we get

$$\sum_{k=1}^{\infty} s_k^* = \sum_{l=1}^{\infty}\sum_{k=l}^{\infty}\left(\sum_{i_1+i_2+\cdots+i_l=k} s_{i_1}s_{i_2}\ldots s_{i_l}\right)$$
$$= \sum_{k=1}^{\infty}\sum_{\nu=k}^{\infty}\left(\sum_{i_1+i_2+\cdots+i_k=\nu} s_{i_1}s_{i_2}\ldots s_{i_k}\right). \tag{3.1.3}$$

On the other hand, we have

$$\sum_{k=1}^{\infty} s_k^* = \sum_{k=1}^{\infty}\sum_{\nu=0}^{k-1} s_\nu^* s_{k-\nu}. \tag{3.1.4}$$

Hence, Eqs. (3.1.3) and (3.1.4) imply that

$$\sum_{\nu=k}^{\infty}\left(\sum_{i_1+i_2+\cdots+i_k=\nu} s_{i_1}s_{i_2}\ldots s_{i_k}\right)=\sum_{\nu=0}^{k-1} s_\nu^* s_{k-\nu},\quad \forall k\geq 1.\qquad \text{Q.E.D.}$$

3.2 Duality Bracket

In the sequel, our objective is to establish the equality

$$\sum_{n=0}^{+\infty} a_n\left(\sum_{k=1}^{+\infty} s_k\right)^n=\sum_{k=0}^{+\infty} s_k^{**}, \tag{3.2.1}$$

where $(s_k)_{k\geq 1}$ and $(a_k)_{k\geq 0}$ are given complex sequences and $(s_k^{**})_{k\geq 0}$ is to be defined. To reach this goal, we shall first prove the result when $(a_k)_k\equiv 1$.

3.2.1 *Proof of Eq. (3.2.1) When* $(a_k)_k\equiv 1$

Proposition 3.2.1 *Let* $s=(s_k)_{k\geq 0}$ *be a sequence in* $\mathbb{C}$ *such that* $s_0=0$ *and*

$$\sum_{k=1}^{+\infty}|s_k|<1,$$

then

$$\sum_{n=1}^{+\infty}\left(\sum_{k=1}^{+\infty} s_k\right)^n=\sum_{k=1}^{+\infty}\sum_{l=1}^{k}\sum_{i_1+i_2+\cdots+i_l=k} s_{i_1}s_{i_2}\ldots s_{i_l}$$
$$=\sum_{l=1}^{+\infty}\sum_{k=l}^{+\infty}\sum_{i_1+i_2+\cdots+i_l=k} s_{i_1}s_{i_2}\ldots s_{i_l}.$$

◇

Proof Setting

$$\mathscr{S}=\sum_{n=0}^{+\infty}\left(\sum_{k=1}^{+\infty} s_k\right)^n-\sum_{n=0}^{+\infty} s_n^*,$$

we get

$$\mathscr{S}\left(\sum_{k=1}^{+\infty} s_k\right) = \sum_{n=0}^{+\infty}\left(\sum_{k=1}^{+\infty} s_k\right)^{n+1} - \sum_{n=0}^{+\infty} s_n^* \sum_{k=0}^{+\infty} s_k$$
$$= \sum_{n=0}^{+\infty}\left(\sum_{k=1}^{+\infty} s_k\right)^{n} - 1 - \sum_{n=0}^{+\infty}\left(\sum_{\nu=0}^{n} s_\nu^*\, s_{n-\nu}\right). \tag{3.2.2}$$

Since $s_0 = 0$, then

$$\sum_{n=0}^{+\infty}\left(\sum_{\nu=0}^{n} s_\nu^*\, s_{n-\nu}\right) = \sum_{n=1}^{+\infty}\left(\sum_{\nu=0}^{n-1} s_\nu^*\, s_{n-\nu}\right). \tag{3.2.3}$$

Hence, both Eqs. (3.2.2) and (3.2.3) imply that

$$\mathscr{S}\left(\sum_{k=1}^{+\infty} s_k\right) = \sum_{n=0}^{+\infty}\left(\sum_{k=1}^{+\infty} s_k\right)^{n} - 1 - \sum_{n=1}^{+\infty}\left(\sum_{\nu=0}^{n-1} s_\nu^*\, s_{n-\nu}\right)$$
$$= \sum_{n=0}^{+\infty}\left(\sum_{k=1}^{+\infty} s_k\right)^{n} - 1 - \sum_{n=1}^{+\infty} s_n^*$$
$$= \sum_{n=0}^{+\infty}\left(\sum_{k=1}^{+\infty} s_k\right)^{n} - \sum_{n=0}^{+\infty} s_n^* = \mathscr{S}.$$

Hence,

$$\mathscr{S}\left(\sum_{k=1}^{+\infty} s_k - 1\right) = 0.$$

Viewing in mind the fact that

$$\left|\sum_{k=1}^{+\infty} s_k\right| < \sum_{k=1}^{+\infty} |s_k| < 1,$$

we deduce that $\mathscr{S} = 0$. Q.E.D.

As a consequence of Proposition 3.2.1, we have the following corollaries.

Corollary 3.2.1 *For every $k \geq 1$ and $x \in \mathbb{C}$, we have the following:*

$$\sum_{l=1}^{k} x^k \sum_{i_1+i_2+\cdots+i_l=k} 1 = x(1+x)^{k-1}. \qquad \diamond$$

Proof Let x and y in $\mathbb{C}$. In view of Proposition 3.2.1, we have

$$
\begin{aligned}
\sum_{n=1}^{+\infty}\left(\sum_{k=1}^{+\infty} yx^k\right)^n &= \sum_{k=1}^{+\infty}\sum_{l=1}^{k}\sum_{i_1+i_2+\cdots+i_l=k}(yx^{i_1})(yx^{i_2})\ldots(yx^{i_l}) \\
&= \sum_{k=1}^{+\infty} x^k \sum_{l=1}^{k} y^l \sum_{i_1+i_2+\cdots+i_l=k} 1. \qquad (3.2.4)
\end{aligned}
$$

On the other hand, we have

$$
\begin{aligned}
\sum_{n=1}^{+\infty}\left(\sum_{k=1}^{+\infty} yx^k\right)^n &= \sum_{n=1}^{+\infty} y^n \left(\sum_{k=1}^{+\infty} x^k\right)^n \\
&= \sum_{n=1}^{+\infty}\left(\frac{yx}{1-x}\right)^n .
\end{aligned}
$$

Thus,

$$
\begin{aligned}
\sum_{n=1}^{+\infty}\left(\sum_{k=1}^{+\infty} yx^k\right)^n &= \frac{yx}{1-x(1+y)} \\
&= yx\sum_{n=0}^{+\infty}(x(1+y))^n \\
&= \sum_{n=1}^{+\infty} x^n(1+y)^{n-1}y. \qquad (3.2.5)
\end{aligned}
$$

Hence, by identification, Eqs. (3.2.4) and (3.2.5) imply that for every $k \geq 1$

$$
\sum_{l=1}^{k} x^k \sum_{i_1+i_2+\cdots+i_l=k} 1 = x(1+x)^{k-1}. \qquad \text{Q.E.D.}
$$

Corollary 3.2.2 *Let $(\xi_k)_{k\geq 1}$ be a sequence in $\mathbb{C}$ such that*

$$
0 < \tau(\xi) = \sum_{l=1}^{k} |\xi_k| < +\infty,
$$

then for every $l \geq 1$, we have

$$
\sum_{k=l}^{+\infty}\sum_{i_1+i_2+\cdots+i_l=k} |\xi_{i_1}||\xi_{i_2}|\cdots|\xi_{i_l}| = (\tau(\xi))^l . \qquad \diamond
$$

Proof Let $0 < x < \frac{1}{\tau(\xi)}$. By Proposition 3.2.1, we have

$$\sum_{l=1}^{+\infty} x^l \sum_{k=l}^{+\infty} \sum_{i_1+i_2+\cdots+i_l=k} |\xi_{i_1}||\xi_{i_2}|\cdots|\xi_{i_2}| = \sum_{l=1}^{+\infty} (x\tau(\xi))^l .$$

Hence, for every $l \geq 1$

$$\sum_{k=l}^{+\infty} \sum_{i_1+i_2+\cdots+i_l=k} |\xi_{i_1}||\xi_{i_2}|\cdots|\xi_{i_l}| = (\tau(\xi))^l . \qquad \text{Q.E.D.}$$

3.2.2 *Proof of Eq. (3.2.1) When $(a_k)_{k\geq 1}$ is Any Sequence in $\mathbb{C}$*

Having obtained these results, we are now ready to establish Eq. (3.2.1) when $(a_k)_{k\geq 1}$ is any sequence in $\mathbb{C}$.

Proposition 3.2.2 *For any sequences $(a_k)_{k\geq 1}$ and $(s_k)_{k\geq 1}$, we have formerly*

$$\begin{aligned}\sum_{n=1}^{+\infty} a_n \left(\sum_{k=1}^{+\infty} s_k\right)^n &= \sum_{k=1}^{+\infty} a_k \sum_{l=1}^{k} \sum_{i_1+i_2+\cdots+i_l=k} s_{i_1}s_{i_2}\ldots s_{i_l} \\ &= \sum_{l=1}^{+\infty}\sum_{k=l}^{+\infty} a_k \sum_{i_1+i_2+\cdots+i_l=k} s_{i_1}s_{i_2}\ldots s_{i_l}.\end{aligned} \qquad \diamond$$

Proof Let $P_{\mathbb{C}}$ denote the space of polynomials with coefficients in $\mathbb{C}$ and with one variable and $P'_{\mathbb{C}}$ be its algebraic dual space. For the sequence $(a_n)_{n\geq 0}$, we can associate a linear functional on polynomes, $U \in P'_{\mathbb{C}}$ given by its moments as follows:

$$a_n = (U)_n = \langle U, x^n\rangle \text{ for all } n \geq 0,$$

where $\langle\cdot,\cdot\rangle$ is the duality bracket (see [2]). Then,

$$\begin{aligned}\sum_{n=0}^{+\infty} a_n \left(\sum_{k=1}^{+\infty} s_k\right)^n &= \sum_{n=0}^{+\infty} \langle U, x^n\rangle \left(\sum_{k=1}^{+\infty} s_k\right)^n \\ &= \sum_{n=0}^{+\infty} \left\langle U, \left(x\sum_{k=1}^{+\infty} s_k\right)^n\right\rangle \\ &= \left\langle U, \sum_{n=0}^{+\infty}\left(\sum_{k=1}^{+\infty} x s_k\right)^n\right\rangle.\end{aligned} \qquad (3.2.6)$$

Setting $s_k(x) = xs_k$ for all $k \geq 0$, we get in view of both Proposition 3.2.1 and Lemma 3.1.1(i), we infer that

$$\sum_{n=0}^{+\infty}\left(\sum_{k=1}^{+\infty}s_k(x)\right)^n=\sum_{n=0}^{+\infty}s_n^*(x), \tag{3.2.7}$$

where $s_0^*(x)=1$ and

$$s_n^*(x)=\sum_{l=1}^{n}\sum_{i_1+i_2+\cdots+i_l=n}s_{i_1}s_{i_2}s_{i_3}\ldots s_{i_l},\quad \forall n\geq 1.$$

Hence, Eqs. (3.2.6) and (3.2.7) imply that

$$\begin{aligned}
\sum_{n=0}^{+\infty}a_n\left(\sum_{k=1}^{+\infty}s_k\right)^n &= \left\langle U, s_0^*(x)+\sum_{k=1}^{+\infty}s_k^*(x)\right\rangle\\
&= \langle U,1\rangle+\sum_{k=1}^{+\infty}\left\langle U, s_k^*(x)\right\rangle\\
&= a_0+\sum_{k=1}^{+\infty}\left\langle U,\sum_{l=1}^{k}\sum_{i_1+i_2+\cdots+i_l=k}s_{i_1}(x)s_{i_2}(x)\ldots s_{i_l}(x)\right\rangle\\
&= a_0+\sum_{k=1}^{+\infty}\left\langle U,\sum_{l=1}^{k}x^l\sum_{i_1+i_2+\cdots+i_l=k}s_{i_1}s_{i_2}\ldots s_{i_l}\right\rangle\\
&= a_0+\sum_{k=1}^{+\infty}\sum_{l=1}^{k}a_l\sum_{i_1+i_2+\cdots+i_l=k}s_{i_1}s_{i_2}\ldots s_{i_l}\\
&= a_0+\sum_{l=1}^{+\infty}\sum_{k=l}^{+\infty}a_l\sum_{i_1+i_2+\cdots+i_l=k}s_{i_1}s_{i_2}\ldots s_{i_l}\\
&= a_0+\sum_{l=1}^{+\infty}a_l\sum_{k=l}^{+\infty}\sum_{i_1+i_2+\cdots+i_l=k}s_{i_1}s_{i_2}\ldots s_{i_l}.
\end{aligned}$$

Due to Lemma 3.1.1(ii), we have

$$\sum_{k=l}^{\infty}\left(\sum_{i_1+i_2+\cdots+i_l=k}s_{i_1}s_{i_2}\ldots s_{i_l}\right)=\sum_{k=0}^{l-1}s_k^*s_{l-k},$$

where $s_0^*=1$ and

$$s_k^*=\sum_{\nu=0}^{k-1}s_\nu^*s_{k-\nu}\quad \forall\, k\geq 1.$$

Consequently, we obtain

$$\begin{aligned}\sum_{n=0}^{+\infty} a_n \left(\sum_{k=1}^{+\infty} s_k\right)^n &= a_0 + \sum_{l=1}^{+\infty} a_l \sum_{k=0}^{l-1} s_k^* s_{l-k} \\ &= \sum_{k=0}^{+\infty} s_k^{**},\end{aligned}$$

where the sequence $(s_k^{**})_{k\geq 0}$ is given by

$$\begin{cases} s_0^{**} = a_0 \\ s_k^{**} = a_k s_k^*, \quad \forall k \geq 1 \\ s_k^* = \displaystyle\sum_{\nu=0}^{k-1} s_\nu^* s_{k-\nu}, \quad \forall k \geq 1, \end{cases}$$

which completes the proof. Q.E.D.

3.3 Notes and Remarks

(i) For any fixed λ in $\mathbb{C}$, let $s_k = \lambda$, for all $k \geq 1$. By an easy induction, we can show that $s_k^* = \lambda(1+\lambda)^{k-1}$, for all $k \geq 1$.

(ii) Due to (3.1.1) for any sequences $s = (s_k)_{k\geq 1}$ and $t = (t_k)_{k\geq 1}$ in $\mathbb{C}$ satisfying $|s_k| \leq |t_k|$, for all $k \geq 1$, we have

$$|s_k^*| \leq \widetilde{t}_k^*, \quad \text{for every } \ k \geq 0, \tag{3.3.1}$$

where $\widetilde{t}_k = |t_k|$, for every $k \geq 1$.

References

1. I. Feki, A. Jeribi, R. Sfaxi, Nagy's perturbation of a non-selfadjoint operator and application to a Gribov operator in Bargmann space, preprint (2021)
2. F. Marcellán, R. Sfaxi, Second structure relation for semiclassical orthogonal polynomials. J. Comput. Appl. Math. **200**(2), 537–554 (2007)

Chapter 4
Carleman-Class

This chapter is concerned with operators with Carleman-class and the spectral theory of compact operators. The focus of this part aims at defining the singular values of a compact operator and at introducing the Carleman-class of operators $\mathscr{C}_p$.

4.1 Singular Values

4.1.1 Singular Values of a Compact Operator

Let K be a compact operator acting on a separable Hilbert space X. Note that K^*K is compact, positive, and self-adjoint. Let

$$\lambda_1(K^*K) \geq \lambda_2(K^*K) \geq \lambda_3(K^*K) \geq \cdots$$

be the sequence of non-zero eigenvalues of K^*K, where every eigenvalue is repeated as many times as its multiplicity.

Definition 4.1.1 We say that a sequence of singular values of compact operator K is the sequence of eigenvalues of the compact positive operators $\sqrt{K^*K}$, ranked in descending order, where each eigenvalue is repeated as many times as its multiplicity. The eigenvalues of the operator $\sqrt{K^*K}$ are called the s-numbers of the operator K. ◇

Note that $s_n(K)$ is the s-numbers of the operator K and we have

$$s_n(K) = \sqrt{\lambda_n(K^*K)}.$$

A. Jeribi, *Perturbation Theory for Linear Operators*,
https://doi.org/10.1007/978-981-16-2528-2_4

4.1.2 Polar Representation of a Bounded Operator

An operator $U \in \mathscr{L}(X)$ is said to be partially isometric if it maps the subspaces $N(U)^{\perp}$ isometrically onto $R(U)$. For a partially isometric operator U, the linear $R(U)$ is a closed subspace. It is easily seen that if U is a partially isometric operator, so is the operator U^*. The operator U defines an isometric mapping of the subspace $R(U^*)$ onto $R(U)$, and the operator U^* defines the inverse mapping of $R(U)$ onto $R(U^*)$, so that

$$U^*U = P_1$$

and

$$UU^* = P_2,$$

where P_1 and P_2 are the orthoprojectors (orthogonal projector, i.e., $P = P^*$) which project X onto the subspaces $R(U^*)$ and $R(U)$, respectively.

Every bounded linear operator A admits a representation (see [2]) in the form

$$A = UH, \tag{4.1.1}$$

where $H = \sqrt{A^*A}$ and U is a partially isometric operator which maps the subspace $R(A^*)$ isometrically onto $R(A)$. The representation (4.1.1) is called the polar representation of the operator A. One can easily prove the following properties of the operators which appear in the polar representation of the operator A:

(i) $U^*A = H$,
(ii) $H_1 = UHU^*$, $H = U^*H_1U$, where $H_1 = \sqrt{AA^*}$, and
(iii) $A = H_1U$, $H_1 = AU^*$.

4.1.3 The Dimension of an Operator

By the dimension of the operator A is meant the number $r(A)$ $(\leq \infty)$ equal to the dimension of the subspace $\overline{R(A)}$. It is easily seen that

$$\begin{aligned} r(A) &= r(\sqrt{A^*A}) \\ &= r(\sqrt{AA^*}) \\ &= r(A^*). \end{aligned}$$

An operator $A \in \mathscr{L}(X)$ is said to be finite dimensional if $r(A) < \infty$.

Obviously, every operator K defined by

$$Kf = \sum_{j=1}^{n} \langle f, \phi_j \rangle \psi_j, \quad (f \in X)$$

where $\{\phi_j\}_1^n$ and $\{\psi_j\}_1^n$ are arbitrary systems of vectors from X, is a finite-dimensional operator, and

$$r(K) \leq n.$$

It is easy to show that

$$r(K) = \min(\dim X_1, \dim X_2),$$

where X_1 and X_2 are the linear hulls of the systems $\{\phi_j\}_1^n$ and $\{\psi_j\}_1^n$, respectively.

4.1.4 The Schmidt Expansion of a Compact Operator

Let A be a compact operator and let $A = UH$ be its polar representation. We denote by ϕ_j ($j = 1, \ldots, \dim \overline{R(H)}$) an orthonormal system (i.e., elements $\{\phi_j\}_{j=1}^{\infty}$ which have norm 1 and are pairwise orthogonal) of eigenvectors of the operator H which is dense in $R(H)$. Then,

$$H = \sum_{j=1}^{\dim \overline{R(H)}} s_j(A) \langle \cdot, \phi_j \rangle \phi_j, \tag{4.1.2}$$

where the series on the right side converges in the uniform operator norm. Applying the operator U to both sides of (4.1.2), we obtain

$$A = \sum_{j=1}^{\dim \overline{R(A)}} s_j(A) \langle \cdot, \phi_j \rangle U\phi_j.$$

Since $\phi_j \in R(H)$, the system $U\phi_j$ ($j = 1, \ldots, \dim \overline{R(A)}$) is orthonormal. Thus, every linear compact operator A admits a Schmidt expansion

$$A = \sum_{j=1}^{\dim \overline{R(A)}} s_j(A) \langle \cdot, \phi_j \rangle \psi_j, \tag{4.1.3}$$

where $\{\phi_j\}$ and $\{\psi_j\}$ are certain orthonormal systems, and the series in (4.1.3), just as does the series in (4.1.2), converges in the uniform norm. It follows from (4.1.3) that

$$A^* = \sum_{j=1}^{\dim \overline{R(A)}} s_j(A) \langle \cdot, \psi_j \rangle \phi_j. \tag{4.1.4}$$

From both (4.1.3) and (4.1.4), we obtain

$$A^*A\phi_j = [s_j(A)]^2\phi_j, \quad j = 1, \ldots, \dim \overline{R(A)}$$

and

$$AA^*\psi_j = [s_j(A)]^2\psi_j, \quad j = 1, \ldots, \dim \overline{R(A)}.$$

Hence, we conclude that

$$s_j(A) = s_j(A^*), \quad j = 1, 2, \ldots.$$

So, we have the following.

Proposition 4.1.1 *Let A be a compact operator on a Hilbert space X. Then,*

$$s_j(A) = s_j(A^*) \text{ for all } j \geq 1. \qquad \diamond$$

4.1.5 *Some Properties of Singular Values*

Proposition 4.1.2 (Principe of min-max) *Let K be a compact operator on a Hilbert space X and $s_n(K)$ be the sequence of its singular values. Then,*

$$s_n(K) = \min_{\dim M = n-1} \max_{\varphi \in M^\perp} \frac{\|K\varphi\|}{\|\varphi\|}. \qquad \diamond$$

4.1.5.1 Characterization of Singular Values Operator

Let A be a compact operator on a separable Hilbert space X and K be an operator of rank $m \leq n$. We may prove that

$$s_{n+1}(A) \leq \|A - K\|. \tag{4.1.5}$$

Notice that

$$\dim R(K) = \dim N(K)^\perp = m.$$

Now, let $M = N(K)^\perp$. Since K is finite rank, then K is compact. Hence, K is closed, and therefore $N(K)$ is closed. So,

$$\begin{aligned} M^\perp &=^\perp (N(K)^\perp) \\ &= \overline{N(K)}. \end{aligned}$$

By using Proposition 4.1.2, we have

$$s_{n+1}(A) \le s_{m+1}(A) \le \max_{\varphi \in N(K)} \frac{\|A\varphi\|}{\|\varphi\|}.$$

Moreover, for all $\varphi \in N(K)$,

$$\begin{aligned} \|A\varphi\| &= \|(A-K)\varphi\| \\ &\le \|A-K\|\|\varphi\|. \end{aligned}$$

So,

$$s_{n+1}(A) \le \|A-K\|.$$

Thus, $s_{n+1}(A)$ bounds the set

$$\{\|A-K\| \text{ such that } \dim R(K) \le n\}$$

from below. It remains to show that the minimum is achieved and is equal to $s_n(A)$. For this, we consider the Schmidt representation of the operator A (see (4.1.3)),

$$A = \sum_{j=1}^{\infty} s_j(A)\langle \cdot, \varphi_j\rangle \psi_j, \tag{4.1.6}$$

where $\{\varphi_j\}$ and $\{\psi_j\}$ are certain orthonormal systems. Let

$$K_n = \sum_{j=1}^{n} s_j(A)\langle \cdot, \varphi_j\rangle \psi_j, \tag{4.1.7}$$

we have K_n is an operator with rank n. According to the foregoing (see Eq. (4.1.5)), we have

$$s_{n+1}(A) \le \|A-K_n\|.$$

However, both Eqs. (4.1.6) and (4.1.7) imply that, for all $\varphi \in X$,

$$\begin{aligned} (A-K_n)\varphi &= A\varphi - K_n\varphi \\ &= \sum_{j=n+1}^{\infty} s_j(A)\langle \varphi, \varphi_j\rangle \psi_j. \end{aligned}$$

Therefore,

$$\begin{aligned} \|(A-K_n)\varphi\|^2 &= \sum_{j=n+1}^{\infty} \left[s_j(A)\right]^2 |\langle \varphi, \varphi_j\rangle|^2 \\ &\le \left(\sup_{j\ge n+1} s_j(A)\right)^2 \sum_{j=n+1}^{\infty} |\langle \varphi, \varphi_j\rangle|^2 \\ &\le (s_{n+1}(A))^2 \,\|\varphi\|^2. \end{aligned}$$

Hence, for all $\varphi \in X$,

$$\|(A - K_n)\varphi\| \leq s_{n+1}(A)\|\varphi\|,$$

and so

$$\|A - K_n\| \leq s_{n+1}(A).$$

Thus,

$$s_{n+1}(A) = \|A - K_n\|. \tag{4.1.8}$$

Hence, the minimum is achieved in $K = K_n$ and is equal to $s_{n+1}(A)$. So,

$$s_{n+1}(A) = \min_{\dim R(K) \leq n} \|A - K\|. \tag{4.1.9}$$

4.1.5.2 *A*-Compact Operator

Let A be a closed linear operator with dense domain $\mathcal{D}(A)$ in a separable Hilbert space X. Let $\lambda_0 \in \rho(A)$, $\varepsilon > 0$, and B be a closed linear operator with domain $\mathcal{D}(B)$ such that $\mathcal{D}(A) \subset \mathcal{D}(B)$ and

$$T := B(\lambda_0 - A)^{-1}$$

is compact in X. By using (4.1.8), there is a finite-rank operator

$$Q_n = \sum_{k=1}^{n} s_k(T)\langle \cdot, u_k \rangle v_k$$

such that

$$\lim_{n \to \infty} \|T - Q_n\| = 0,$$

where $u_k, v_k \in X$. Hence, for $\frac{\varepsilon}{2} > 0$, there is n_0 such that for all $n \geq n_0$, we have

$$\|T - Q_n\| < \frac{\varepsilon}{2}. \tag{4.1.10}$$

Since A is closed and $\mathcal{D}(A)$ is dense in X, and using Theorem 1.13.1(i), we will have $\mathcal{D}(A^*)$ is dense in X. This allows us to deduce, from $u_k \in X$, there is a sequence $(w_{k_l})_l \in \mathcal{D}(A^*)$ such that

$$\lim_{l \to \infty} \|u_k - w_{k_l}\| = 0.$$

So, for

$$\frac{\varepsilon}{2\sum_{k=1}^{n}\|v_k\|} > 0,$$

there is l_0 such that, for all $l \geq l_0$, we have

$$\|u_k - w_{k_l}\| < \frac{\varepsilon}{2\sum_{k=1}^{n}\|v_k\|}.$$

Let

$$P_{n,l} = \sum_{k=1}^{n}\langle\cdot, w_{k_l}\rangle v_k.$$

In what follows, we will show that, for all $l \geq l_0$,

$$\|Q_n - P_{n,l}\| \leq \frac{\varepsilon}{2}. \tag{4.1.11}$$

In fact, let $x \in X\backslash\{0\}$, we have

$$\begin{aligned}\big\|(Q_n - P_{n,l})x\big\| &= \left\|\sum_{k=1}^{n}\langle x, u_k\rangle v_k - \sum_{k=1}^{n}\langle x, w_{k_l}\rangle v_k\right\| \\ &= \left\|\sum_{k=1}^{n}\langle x, u_k - w_{k_l}\rangle v_k\right\|.\end{aligned}$$

Thus,

$$\begin{aligned}\big\|(Q_n - P_{n,l})x\big\| &\leq \sum_{k=1}^{n}|\langle x, u_k - w_{k_l}\rangle|\|v_k\| \\ &\leq \left(\sum_{k=1}^{n}\|u_k - w_{k_l}\|\|v_k\|\right)\|x\|.\end{aligned}$$

Hence,

$$\begin{aligned}\frac{\|(Q_n - P_{n,l})x\|}{\|x\|} &\leq \sum_{k=1}^{n}\|u_k - w_{k_l}\|\|v_k\| \\ &< \frac{\varepsilon}{2}.\end{aligned}$$

So,

$$\|Q_n - P_{n,l}\| \leq \frac{\varepsilon}{2}.$$

Thus, by using both Eqs. (4.1.10) and (4.1.11), we have for all $n \geq n_0$ and for all $l \geq l_0$

$$\begin{aligned} \|T - P_{n,l}\| &\leq \|T - Q_n\| + \|Q_n - P_{n,l}\| \\ &< \frac{\varepsilon}{2} + \frac{\varepsilon}{2} \\ &= \varepsilon. \end{aligned} \tag{4.1.12}$$

Let $\varphi \in \mathscr{D}(A)$, we have

$$\begin{aligned} \|B\varphi\| &= \|T(A - \lambda_0)\varphi\| \\ &\leq \|P_{n,l}(A - \lambda_0)\varphi\| + \|(T - P_{n,l})(A - \lambda_0)\varphi\| \\ &\leq \left\| \sum_{k=1}^{n} \langle (A - \lambda_0)\varphi, w_{k_l}\rangle v_k \right\| + \|T - P_{n,l}\| \|(A - \lambda_0)\varphi\|. \end{aligned}$$

Now, using the last inequalities and Eq. (4.1.12), we get for all $n \geq n_0$ and for all $l \geq l_0$

$$\begin{aligned} \|B\varphi\| &\leq \sum_{k=1}^{n} \|\langle (A - \lambda_0)\varphi, w_{k_l}\rangle v_k\| + \varepsilon \left(\|A\varphi\| + |\lambda_0|\|\varphi\|\right) \\ &\leq \sum_{k=1}^{n} |\langle A\varphi, w_{k_l}\rangle - \lambda_0\langle \varphi, w_{k_l}\rangle| \|v_k\| + \varepsilon \left(\|A\varphi\| + |\lambda_0|\|\varphi\|\right) \\ &\leq \sum_{k=1}^{n} \left(|\langle A\varphi, w_{k_l}\rangle| + |\lambda_0||\langle \varphi, w_{k_l}\rangle|\right) \|v_k\| + \varepsilon \left(\|A\varphi\| + |\lambda_0|\|\varphi\|\right) \\ &\leq \sum_{k=1}^{n} \left(|\langle \varphi, A^* w_{k_l}\rangle| + |\lambda_0||\langle \varphi, w_{k_l}\rangle|\right) \|v_k\| + \varepsilon \left(\|A\varphi\| + |\lambda_0|\|\varphi\|\right) \\ &\leq \varepsilon\|A\varphi\| + \varepsilon|\lambda_0|\|\varphi\| + \sum_{k=1}^{n} \left(\|\varphi\|\|A^* w_{k_l}\| + |\lambda_0|\|\varphi\|\|w_{k_l}\|\right) \|v_k\| \\ &\leq \varepsilon\|A\varphi\| + \left(\varepsilon|\lambda_0| + \sum_{k=1}^{n} \left(\|A^* w_{k_l}\| + |\lambda_0|\|w_{k_l}\|\right) \|v_k\| \right) \|\varphi\|. \end{aligned}$$

This proves that B is A-compact.

4.1.5.3 Some Results Concerning Singular Values Operator

Proposition 4.1.3 *Let A and B be two compact operators on a separable Hilbert space X. Then,*

$$s_{n+m-1}(A + B) \leq s_n(A) + s_m(B) \textit{ for all } n,\ m = 1, 2, \ldots. \qquad \diamond$$

Proof Let K_1 be the operator of rank $n - 1$ and K_2 be the operator of rank $m - 1$ such that

$$s_n(A) = \|A - K_1\|$$

and

$$s_m(B) = \|B - K_2\|.$$

We have

$$R(K_1 + K_2) \subset R(K_1) + R(K_2).$$

Indeed, let $y \in R(K_1 + K_2)$, there exists $x \in \mathscr{D}(K_1) \bigcap \mathscr{D}(K_2)$ such that

$$\begin{aligned} y &= (K_1 + K_2)x \\ &= K_1 x + K_2 x \in R(K_1) + R(K_2). \end{aligned}$$

Thus,

$$\begin{aligned} \dim R(K_1 + K_2) &\leq \dim(R(K_1) + R(K_2)) \\ &\leq \dim R(K_1) + \dim R(K_2). \end{aligned}$$

Hence,

$$\dim R(K_1 + K_2) \leq m + n - 2.$$

By using Theorem 4.1.9, we obtain

$$s_{n+m-1}(A + B) = \min_{\dim R(K) \leq n+m-2} \|A + B - K\|.$$

So,

$$\begin{aligned} s_{n+m-1}(A + B) &\leq \|A + B - (K_1 + K_2)\| \\ &\leq \|A - K_1\| + \|B - K_2)\| \\ &\leq s_n(A) + s_m(B), \end{aligned}$$

which completes the proof. Q.E.D.

Proposition 4.1.4 *Let A be a compact operator on a Hilbert space X and B be a bounded operator on X. Then,*

(i) $s_j(BA) \leq \|B\| s_j(A)$ for all $j = 1, 2, \ldots$.
(ii) $s_j(AB) \leq \|B\| s_j(A)$ for all $j = 1, 2, \ldots$. ◇

Proof (i) We have the following identity

$$\begin{aligned} \left[s_j(BA)\right]^2 &= \left[\lambda_j\left(\sqrt{(BA)^* BA}\right)\right]^2 \\ &= \lambda_j((BA)^* BA) \\ &= \lambda_j(A^* B^* BA), \end{aligned}$$

and

$$\begin{aligned} \left[s_j(A)\right]^2 &= \left[\lambda_j\left(\sqrt{A^* A}\right)\right]^2 \\ &= \lambda_j(A^* A). \end{aligned}$$

Moreover, for all $f \in X$, we have

$$\begin{aligned}\langle A^*B^*BAf, f\rangle &= \langle B^*BAf, Af\rangle \\ &= \langle BAf, BAf\rangle \\ &= \|BAf\|^2 \\ &\leq \|B\|^2\|Af\|^2 \\ &= \|B\|^2\langle A^*Af, f\rangle.\end{aligned}$$

Hence, for all $f \in X$,

$$\langle \big(\|B\|^2A^*A - A^*B^*BA\big) f, f\rangle \geq 0.$$

So, by using Theorem 1.14.1, we have

$$\lambda_j(\|B\|^2A^*A) \geq \lambda_j(A^*B^*BA).$$

Thus,

$$\big[s_j(BA)\big]^2 \leq \|B\|^2\big[s_j(A)\big]^2.$$

So,

$$s_j(BA) \leq \|B\|s_j(A).$$

(ii) By using Proposition 4.1.1, we have

$$s_j(AB) = s_j(B^*A^*).$$

Thus, (i) implies that

$$s_j(B^*A^*) \leq \|B^*\|s_j(A^*) = \|B\|s_j(A).$$ Q.E.D.

Let us consider the linear function defined by

$$\begin{aligned}\Phi : \mathscr{L}(X) &\longrightarrow \mathscr{L}(X), \\ A &\longrightarrow P\circ A,\end{aligned}$$

where P is the orthogonal projection on a closed subspace F of X. We have

$$\|P\| \leq 1$$

and

$$\|A_{|F}\| \leq \|A\|.$$

So,

$$\|\Phi(A)\| \leq \|A\|.$$

Thus,

$$\|\Phi\| \leq 1.$$

Moreover,

$$\Phi(F_{n-1}(X)) \subset F_{n-1}(X),$$

where $F_n(X)$ is a subspace of $\mathscr{L}(X)$ of operators of rank less than or equal to n. Let $A' \in F_{n-1}(F)$ and let A be the extension of A' to X obtained, for example, taking $A = 0$ on $F^{\perp}$. So, $A \in F_{n-1}(X)$ and

$$\Phi(A) = A'.$$

Hence,

$$\Phi(F_{n-1}(X)) = F_{n-1}(F).$$

Thus,

$$\operatorname{dist}(\Phi(A), F_{n-1}(F)) \leq \operatorname{dist}(A, F_{n-1}(X))$$

for all $A \in \mathscr{L}(X)$. Moreover, if F is invariant by A, then

$$A_{|F} = \Phi(A)$$

and

$$\begin{aligned} s_n(A_{|F}) &= \operatorname{dist}(A_{|F}, F_{n-1}(F)) \\ &\leq \operatorname{dist}(A, F_{n-1}(X)) \\ &= s_n(A). \end{aligned}$$

So,

$$s_n(A_{|F}) \leq s_n(A) \text{ for all } n \geq 1. \tag{4.1.13}$$

At the moment, we have two two-sided ideals of $\mathscr{L}(X)$:

- finite rank operators:

$$F(X) := \{A \in \mathscr{L}(X) \text{ such that } (s_n(A))_{n\geq 1} \text{ zero from a certain rank}\}$$

- compact operators:

$$\mathscr{K}(X) := \left\{A \in \mathscr{L}(X) \text{ such that } \lim_{n\to\infty} s_n(A) = 0\right\}.$$

4.1.6 *Intermediate Ideals Between $F(X)$ and $\mathscr{K}(X)$*

We will now introduce intermediate ideals between $F(X)$ and $\mathscr{K}(X)$.

Definition 4.1.2 Let $p > 0$ and $q > 0$. We say that a Schatten-von Neumann ideal of type (p, q), the set of operators defined by

$$\mathscr{C}_{p,q}(X) = \left\{ A \in \mathscr{L}(X) \text{ such that } (n^{\frac{1}{p}-\frac{1}{q}} s_n(A))_n \in l^q \right\}.$$

In the case where $q = \infty$,

$$\begin{aligned}\mathscr{C}_{p,\infty}(X) &= \{A \in \mathscr{L}(X) \text{ such that } (n^{\frac{1}{p}} s_n(A))_n \text{ is bounded}\} \\ &= \{A \in \mathscr{L}(X) \text{ such that } s_n(A) = O(n^{-\frac{1}{p}}) \text{ as } n \to \infty\}.\end{aligned}$$ ◇

The sets $\mathscr{C}_{p,q}(X)$ $q \in (0, \infty]$ are two-sided ideals of $\mathscr{L}(X)$. They all contain $F(X)$ and are contained in $\mathscr{K}(X)$. None is closed for the uniform norm since

$$\overline{F(X)} = \mathscr{K}(X),$$

but we may introduce on $\mathscr{C}_{p,q}(X)$ a metrizable topologies for which they are complete.

Let $A \in \mathscr{C}_{p,q}(X)$. For $q < \infty$, we define

$$N_{p,q}(A) = \left(\sum_{n=1}^{\infty} n^{\frac{p}{q}-1} (s_n(A))^q \right)^{\frac{1}{q}}.$$

For $q = \infty$, we define

$$N_{p,\infty}(A) = \sup_{n \geq 1} \left[n^{\frac{1}{p}} s_n(A) \right].$$

The $N_{p,q}(\cdot)$ are quasi-norm on $\mathscr{C}_{p,q}(X)$. The quasi-norm $N_{p,q}(\cdot)$ on $\mathscr{C}_{p,q}(X)$ induces a metrizable topology of vector space. A fundamental system of neighborhoods of the zero element is formed by the subsets

$$B_{p,q}(\varepsilon) = \{A \in \mathscr{C}_{p,q}(X) \text{ such that } N_{p,q}(A) < \varepsilon\}$$

for which $\mathscr{C}_{p,q}(X)$ is complete (quasi-Banach).

Theorem 4.1.1 (Pietsch [6])

(i) $\mathscr{C}_{p_0,q_0}(X) \subset \mathscr{C}_{p_1,q_1}(X)$ *for* $0 < p_0 < p_1 < \infty$ *and* q_0, q_1 *arbitrary.*
(ii) $\mathscr{C}_{p,q_0}(X) \subset \mathscr{C}_{p,q_1}(X)$ *for all* $p > 0$ *and* $0 < q_0 < q_1 \leq \infty$.
(iii) The $\mathscr{C}_{p,q}(X)$ *are ordered according to the lexicographical inclusion of* (p, q). ◇

Let $A \in \mathscr{C}_{p,\infty}(X)$ such that

$$\lim_{n\to\infty} n^{\frac{1}{p}} s_n(A) \neq 0.$$

Then, there is $a > 0$ and a subsequence $(n_k^{\frac{1}{p}} s_{n_k}(A))_k$ such that for all $k \geq 1$, we have

$$n_k^{\frac{1}{p}} s_{n_k}(A) \geq a.$$

We infer that, for all finite rank F, we have

$$N_{p,\infty}(A - F) = \sup_{n\geq 1} \left[n^{\frac{1}{p}} s_n(A - F) \right] \geq a$$

and so A is not in the closeness of $F(X)$ in $\mathscr{C}_{p,\infty}(X)$. If, by against,

$$\lim_{n\to\infty} n^{\frac{1}{p}} s_n(A) = 0,$$

taking a Schmidt decomposition of A, and its partial sums, we easily see that A is in the closeness which is denoted by $\mathscr{C}^{(0)}_{p,\infty}(X)$. Hence,

$$\begin{aligned} \mathscr{C}^{(0)}_{p,\infty}(X) &= \left\{ A \in \mathscr{L}(X) \text{ such that } \lim_{n\to\infty} n^{\frac{1}{p}} s_n(A) = 0 \right\} \\ &= \{A \in \mathscr{L}(X) \text{ such that } s_n(A) = o(n^{-\frac{1}{p}}) \text{ as } n \to \infty\}. \end{aligned}$$

We know that any compact operator can be approximated to the uniform norm by finite-rank operator. It requires the approximation problem in $\mathscr{C}_{p,q}(X)$ for finite-rank operators for the topology of $\mathscr{C}_{p,q}(X)$. In the case $0 < q < \infty$, the answer is provided by the following theorem.

Theorem 4.1.2 (Pietsch [6]) *For* $0 < q < \infty$, *the quasi-normed ideal* $\mathscr{C}_{p,q}(X)$ *satisfies the following:*

$$\overline{F(X)}^{\mathscr{C}_{p,q}(X)} = \mathscr{C}_{p,q}(X),$$

where $\overline{F(X)}^{\mathscr{C}_{p,q}(X)}$ *stands for the closeness of* $F(X)$ *for the topology of* $\mathscr{C}_{p,q}(X)$. ◇

Theorem 4.1.2 is false for $q = \infty$.

Lemma 4.1.1 *Let X be a Hilbert space, $K \in \mathscr{C}_{p,\infty}(X)$ and let P' be the space generated by the generalized eigenvectors associated with the non-zero eigenvalues of K. Then,*

$$K^*_{|(P')^\perp} \in \mathscr{C}_{p,\infty}(X^*). \qquad \diamond$$

Proof It suffices to note that the singular values, $s_n(K^*_{|F})$, of $K^*_{|F}$ are less than or equal to $s_n(K)$ for every subspace F of X (see Proposition 4.1.1 and Eq. (4.1.13)). This completes the proof. Q.E.D.

Let A be a closed and linear operator with a dense domain $\mathscr{D}(A)$ on a Hilbert space X. We assume that

(i) there exists $\lambda \in \rho(A)$ such that $(\lambda - A)^{-1}$ is a compact operator and
(ii) there exists $\xi_0 \in \rho(A)$ such that $(\xi_0 - A)^{-1} \in \mathscr{C}^{(0)}_{p,\infty}(X)$ for some $p > 0$.

We have

$$(\xi_0 - A)^{-1}_{|M_\infty} = (\xi_0 - A_{|M_\infty})^{-1},$$

where

$$M_\infty := \bigcap_{k=0}^{\infty} R[(\lambda_k - A)^{n_0(\lambda_k)}]$$

with $n_0(\lambda_k)$ is the index of the eigenvalue λ_k of A. By combining both hypothesis (ii) and (4.1.13), we have

$$s_n((\xi_0 - A)^{-1}_{|M_\infty}) = o(n^{-\frac{1}{p}})$$

as $n \to \infty$. Let

$$K_1 = (\xi_0 - A)^{-1}_{|M_\infty}$$

and let $\lambda \in \mathbb{C}\backslash\{\xi_0\}$. We have

$$\begin{aligned}(\lambda - A)_{|M_\infty} &= \lambda - A_{|M_\infty} \\ &= \lambda - \xi_0 + \xi_0 - A_{|M_\infty}.\end{aligned}$$

This implies that

$$\begin{aligned}(\lambda - A)_{|M_\infty} K_1 &= [(\lambda - \xi_0) + (\xi_0 - A_{|M_\infty})]K_1 \\ &= (\lambda - \xi_0)K_1 + I.\end{aligned}$$

Hence,

$$\begin{aligned}\frac{1}{\xi_0 - \lambda}(\lambda - A)_{|M_\infty} K_1 &= -K_1 + \frac{1}{\xi_0 - \lambda} \\ &= \frac{1}{\xi_0 - \lambda} - K_1.\end{aligned}$$

So,

$$\frac{1}{\xi_0-\lambda}(\lambda-A)_{|M_\infty}K_1\left(\frac{1}{\xi_0-\lambda}-K_1\right)^{-1}=I$$

on M_∞. We can show the same

$$\frac{1}{\xi_0-\lambda}K_1\left(\frac{1}{\xi_0-\lambda}-K_1\right)^{-1}(\lambda-A)_{|M_\infty}=I$$

on $\mathscr{D}(A)\bigcap M_\infty$. Thus,

$$((\lambda-A)_{|M_\infty})^{-1}=\frac{1}{\xi_0-\lambda}K_1\left(\frac{1}{\xi_0-\lambda}-K_1\right)^{-1}$$

which is closed and all defined on M_∞, and therefore bounded (by the closed graph theorem). By referring to Proposition 4.1.4 (i), we deduce that

$$s_n((\lambda-A)^{-1}_{|M_\infty})=o(n^{-\frac{1}{p}})\text{ as } n\to\infty.$$

4.2 Spectral Theory of Compact Operators

Let A be a closed linear operator on a complex Banach space X and let λ_0 be an isolated singular point of the resolvent $(\lambda-A)^{-1}$. Then, $(\lambda-A)^{-1}$ can be expanded into Laurent series

$$(\lambda-A)^{-1}=\sum_{n=-\infty}^{+\infty}(\lambda-\lambda_0)^nA_n,$$

with

$$A_n=\frac{1}{2\pi i}\int\limits_C(\lambda-\lambda_0)^{-n-1}(\lambda-A)^{-1}d\lambda,$$

where C is a circumference of sufficiently small radius: $|\lambda-\lambda_0|=\varepsilon$ such that the circle $|\lambda-\lambda_0|\leq\varepsilon$ does not contain other singularities than $\lambda=\lambda_0$, and the integration is performed counter-clockwise.

Theorem 4.2.1 *Let A be a compact operator or with compact resolvent and $\lambda_0\in\sigma(A)$ (with $\lambda_0\neq 0$ in the compact case). Then,*

(i) λ_0 is a pole of order $n_0(\lambda_0)$ of the resolvent.

(ii) $A_{-1}=\frac{1}{2\pi i}\int_C(\lambda-A)^{-1}d\lambda$ is the spectral projection associated to λ_0 (projection on $N[(\lambda_0-A)^{n_0(\lambda_0)}]$ with the direction $R[(\lambda_0-A)^{n_0(\lambda_0)}]$).

(iii) $(A-\lambda_0)A_0 = A_{-1} - I$,
$(A-\lambda_0)^n A_n = A_0$ *for all* $n \in \mathbb{N}$, *and*
$A_{n-1} = (A-\lambda_0)^{-n} A_{-1}$ *for all* $n \leq -1$.

If $\varphi \in X$, *we have in a neighborhood of* λ_0

$$(\lambda - A)^{-1}\varphi = \sum_{n=-\infty}^{+\infty} (\lambda-\lambda_0)^n A_n \varphi.$$

In particular, if $\varphi \in \bigcap_{n \leq -1} N(A_n)$, $(\lambda - A)^{-1}\varphi$ *extends to a holomorphic function in* λ_0 *with values in* X. ◇

We recall the Macaev theorem.

Theorem 4.2.2 *Let* A *be a quasi-nilpotent operator, i.e.,* $\sigma(A) = \{0\}$ *and let* $M(r)$ *be the function defined by*

$$M(r) = \max_{|\lambda|=r} \|(I - \lambda A)^{-1}\|.$$

If $s_n(A) = O(n^{-\frac{1}{p}})$ *(respectively,* $o(n^{-\frac{1}{p}})$*) for some* $p > 0$, *then* $\log(M(r)) = O(r^p)$ *(respectively,* $o(r^p)$*). Moreover, if* $A \in \mathscr{C}_p(X)$, *then*

$$\int_0^\infty \frac{\log(M(r))}{r^{p+1}} dr < \infty.$$ ◇

4.2.1 Quasi-Nilpotent Operator

Lemma 4.2.1 *Let* X *be a Hilbert space,* $K \in \mathscr{C}^{(0)}_{p,\infty}(X)$ *for some* $p > 0$ *and let* P' *be the space generated by the generalized eigenvectors associated with the non-zero eigenvalues of* K. *Then,* $(P')^\perp$ *is invariant by* K^*. ◇

Proof Since P' is invariant by K, then $(P')^\perp$ is invariant by K^*. Q.E.D.

Let X be a Hilbert space, K be a compact operator on X, and P' be the space generated by the generalized eigenvectors associated with the non-zero eigenvalues of K. We claim that the operator $K^*_{|P'^\perp}$ is quasi-nilpotent, i.e.,

$$\sigma(K^*_{|P'^\perp}) = \{0\}. \tag{4.2.1}$$

Suppose that $\sigma(K^*_{|P'^\perp}) \neq \{0\}$. Then, there is an $\lambda_1 \in \sigma(K^*_{|P'^\perp})$ such that $\lambda_1 \neq 0$. $K^*_{|P'^\perp}$ is compact as a restriction of a compact operator, and so λ_1 is an eigenvalue of $K^*_{|P'^\perp}$. Thus, there is $\psi_1 \in P'^\perp$, $\psi_1 \neq 0$ such that

$$K^*_{|P'^{\perp}}\psi_1 = K^*\psi_1 = \lambda_1\psi_1.$$

Hence,

$$\psi_1 \in N(\lambda_1 - K^*) = [R(\overline{\lambda_1} - K)]^{\perp}$$

and $\overline{\lambda_1} \in \sigma(K)$. Let $n_0(\overline{\lambda_1})$ be the index of the eigenvalue $\overline{\lambda_1}$ of K. Since

$$R[(\overline{\lambda_1} - K)^{n_0(\overline{\lambda_1})}] \subset R(\overline{\lambda_1} - K),$$

we have

$$\psi_1 \in [R(\overline{\lambda_1} - K)^{n_0(\overline{\lambda_1})}]^{\perp}.$$

Moreover, $\psi_1 \in P'^{\perp}$ and $\psi_1 \in [N(\overline{\lambda_1} - K)^{n_0(\overline{\lambda_1})}]^{\perp}$. In view of Theorem 2.2.2(x), we have

$$X = N[(\lambda_1 - K^*)^{n_0(\lambda_1)}] \oplus R[(\lambda_1 - K^*)^{n_0(\lambda_1)}].$$

So, $\psi_1 = 0$ and $K^*_{|P'^{\perp}}$ cannot have non-zero eigenvalues. This proves the claim.

4.2.2 Entire Function

Let X be a Hilbert space and $K \in \mathscr{C}^{(0)}_{p,\infty}(X)$ for some $p > 0$. Suppose that there are m rays $\gamma_1, \gamma_2, \ldots, \gamma_m$ centered at the origin divided on the whole complex plan, such that

(i) the angle between two adjacent half-lines is less than or equal to $\frac{\pi}{p}$, and
(ii) for each of these half-lines there is a positive integer N such that

$$\|(\lambda - K)^{-1}\| = O\left(\frac{1}{|\lambda|^N}\right) \quad \text{when } |\lambda| \to 0.$$

The singularities of

$$\lambda \longrightarrow (\lambda - K^*)^{-1}\varphi$$

are located at the points of $\sigma(K^*)$, i.e., in 0 and in a sequence of isolated points λ_n. For $\lambda_n \in \sigma(K^*)$, $\lambda_n \neq 0$, consider the Laurent series expansion of $(\lambda - K^*)^{-1}$ in the neighborhood of λ_n,

$$(\lambda - K^*)^{-1} = \sum_{k=-n_0(\lambda_n)}^{\infty} A_k(\lambda - \lambda_n)^k.$$

Let P' be the space generated by the generalized eigenvectors associated with the non-zero eigenvalues of K and $\varphi \in (P')^{\perp}$, we have

$$\varphi \in [N(\overline{\lambda_n} - K)^{n_0(\lambda_n)})]^{\perp} = R[(\lambda_n - K^*)^{n_0(\lambda_n)}] = N(A_{-1})$$

and, $\varphi \in N(A_k)$ for all $k < 0$. Hence,

$$\lambda \longrightarrow (\lambda - K^*)^{-1}\varphi$$

extends to a holomorphic function in λ_n. So, $(\lambda - K^*)^{-1}\varphi$ extends to a holomorphic function on $\mathbb{C}\backslash\{0\}$, and therefore also for $\lambda^N(\lambda - K^*)^{-1}\varphi$. Now, let us show that $\lambda^N(\lambda - K^*)^{-1}\varphi$ extends analytically at 0. For this, let $\widetilde{f_\varphi} : \mathbb{C}\backslash\{0\} \longrightarrow X$ be the map extending $\lambda^N(\lambda - K^*)^{-1}\varphi$ to $\mathbb{C}\backslash\{0\}$. Let us write the Laurent series expansion of $\widetilde{f_\varphi}$ in the neighborhood of 0,

$$\widetilde{f_\varphi}(\lambda) = \sum_{k=-\infty}^{+\infty} \lambda^k v_k,$$

where $v_k \in X$. Let $\psi \in X$, the Laurent development of

$$\lambda \longrightarrow \langle \widetilde{f_\varphi}(\lambda), \psi\rangle$$

in the neighborhood of 0 is

$$\sum_{k=-\infty}^{+\infty} \lambda^k \langle v_k, \psi\rangle.$$

If, for all $\psi \in X$, $\langle \widetilde{f_\varphi}(\lambda), \psi\rangle$ extends to a holomorphic function in 0, we have for all $k < 0$ and for all $\psi \in X$,

$$\langle v_k, \psi\rangle = 0$$

which gives for all $k < 0$, $v_k = 0$, and so $\widetilde{f_\varphi}$ extends to an entire function. Therefore, let us show that for all $\psi \in X$, $\langle \widetilde{f_\varphi}(\lambda), \psi\rangle$ extends analytically to 0. It suffices to show that the function $\langle \widetilde{f_\varphi}(\lambda), \psi\rangle$ is bounded in the neighborhood of 0. The fact that P' is invariant by K implies that $P'^{\perp}$ is invariant by K^* (see Lemma 4.2.1). Let K_1 be the restriction of K^* to $P'^{\perp}$. Using (4.2.1), we have K_1 is quasi-nilpotent, i.e.,

$$\sigma(K_1) = \{0\}.$$

For all $\varphi \in P'^{\perp}$ and for all $\lambda \in \rho(K^*)$, we have

$$(\lambda - K^*)^{-1}\varphi = (\lambda - K_1)^{-1}\varphi$$

and

$$(\lambda - K_1)^{-1}\varphi = \frac{1}{\lambda}\left(I - \frac{1}{\lambda}K_1\right)^{-1}\varphi.$$

By using the Macaev theorem for the quasi-nilpotent operator K_1, we have $K \in \mathscr{C}_{p,\infty}^{(0)}(X)$ if, and only if, $K^* \in \mathscr{C}_{p,\infty}^{(0)}(X^*)$. This implies that

$$K_1 = K^*_{|P'^{\perp}} \in \mathscr{C}_{p,\infty}^{(0)}(P'^{\perp})$$

and

$$\log\left(\left\|\left(I - \frac{1}{\lambda}K_1\right)^{-1}\right\|\right) = o(|\lambda|^{-p}),$$

when λ tends to 0, i.e., for all $\varepsilon > 0$, there is $A > 0$, such that for all $\lambda \in \rho(K^*)$

$$\left\|\left(I - \frac{1}{\lambda}K_1\right)^{-1}\right\| \leq Ae^{\varepsilon|\lambda|^{-p}}.$$

By using an easy argument, and for all $\varepsilon > 0$, there is $A > 0$, such that for all $\lambda \in \rho(K^*)$

$$\frac{1}{|\lambda|}\left\|\left(I - \frac{1}{\lambda}K_1\right)^{-1}\varphi\right\| \leq Ae^{\varepsilon|\lambda|^{-p}}.$$

By using a density argument we deduce that, for all $\varepsilon > 0$, there is $A > 0$ such that, for all $\lambda \neq 0$,

$$|\langle \widetilde{f}_{\varphi}(\lambda), \psi\rangle| \leq Ae^{\varepsilon|\lambda|^{-p}}.$$

Moreover, $|\langle \widetilde{f}_{\varphi}(\lambda), \psi\rangle|$ is bounded on m half-lines range by conjugation in $\mathbb{C}$, and those in which

$$\|(\lambda - K)^{-1}\| = O(|\lambda|^{-N})$$

in the neighborhood of 0. Using Theorem 2.4.3, we have for all $\varphi \in (P')^{\perp}$, the function

$$\begin{aligned} \rho(K^*) &\longrightarrow X, \\ \lambda &\longrightarrow \lambda^N(\lambda - K^*)^{-1}\varphi \end{aligned}$$

extends to an entire function.

4.3 Generalized Eigenvectors Associated with the Non-zero Eigenvalues

4.3.1 Holomorphic Function

Let X be a Hilbert space, $K \in \mathscr{C}_{p,\infty}^{(0)}(X)$ for some $p > 0$ and let P' be the space generated by the generalized eigenvectors associated with the non-zero eigenvalues of K. Let $\varphi \in (P')^{\perp}$, and consider the function

$$\begin{aligned} f : \rho(K^*) &\longrightarrow X, \\ \lambda &\longrightarrow (\lambda - K^*)^{-1}\varphi. \end{aligned}$$

Let $\lambda_0 \in \sigma(K^*)$, $\lambda_0 \neq 0$ with an index $m(\lambda_0)$. There is $\varepsilon > 0$ such that, for $0 < |\lambda - \lambda_0| \leq \varepsilon$, we have

$$(\lambda - K^*)^{-1}\varphi = \sum_{n=0}^{\infty}(\lambda - \lambda_0)^n A_n\varphi + \sum_{n=1}^{m(\lambda_0)}(\lambda - \lambda_0)^{-n} A_{-n}\varphi,$$

where

$$A_n = \frac{1}{2i\pi}\int_C \frac{(\lambda - K^*)^{-1}}{(\lambda - \lambda_0)^{n+1}}d\lambda,$$

C denotes the circle of center λ_0 containing no other spectral value, A_{-1} is the projection of X onto $N[(\lambda_0 - K^*)^{m(\lambda_0)}]$ along $R[(\lambda_0 - K^*)^{m(\lambda_0)}]$, and

$$A_{n-1} = (K^* - \lambda_0)^{-n}A_{-1}$$

for all $n \leq -1$. Since $N[(\overline{\lambda_0} - K)^{m(\lambda_0)}]$ is included in P', we have

$$(P')^{\perp} \subset [N(\overline{\lambda_0} - K)^{m(\lambda_0)}]^{\perp} = R[(\lambda_0 - K^*)^{m(\lambda_0)}] = N(A_{-1}) \subset N(A_n)$$

for all $n \leq -1$. Therefore, for every φ belonging to $(P')^{\perp}$, the singular part of the Laurent series expansion of $(\lambda - K^*)^{-1}\varphi$ in the neighborhood of a non-zero eigenvalue vanishes. Thus, $(\lambda - K^*)^{-1}\varphi$ extends analytically to $\mathbb{C}\backslash\{0\}$ into X.

4.3.2 *Norm of the Resolvent*

Let X be a Hilbert space, $A \in \mathscr{C}(X)$, and $1 \leq p < \infty$. We assume that

(i) there exists $\xi_0 \in \rho(A)$, such that $K := (\xi_0 - A)^{-1}$ is a compact operator and $s_n(K) = o(n^{-\frac{1}{p}})$, where n tends to infinity, and
(ii) there is a constant $\varepsilon > 0$ such that the numerical range of the operator A is included in the sector

$$S_p = \left\{\lambda \in \mathbb{C} \text{ such that } -\frac{\pi}{2p} + \varepsilon \leq arg(\lambda) \leq \frac{\pi}{2p} - \varepsilon\right\}.$$

Without loss of generality, we can assume that $0 \in \rho(A)$. Hence,

$$\frac{\langle A^{-1}\psi, \psi\rangle}{\|\psi\|^2} \in S_p$$

for all $\psi \in X$. Let γ an all half-lines from the origin of the complex plane and not belonging to the sector S_p and $\lambda \in \gamma$ with $|\lambda|$ large enough and not belonging to the sector S_p. We have

$$\left|\frac{1}{\lambda} - \frac{\langle A^{-1}\psi, \psi\rangle}{\|\psi\|^2}\right| \geq \frac{|\sin(\theta_0)|}{|\lambda|},$$

where θ_0 denotes the smaller of the two angles formed by the ray γ and the edges of sectors S_p. When we simplify by $|\lambda|$, we obtain

$$|\langle (I - \lambda A^{-1})\psi, \psi\rangle| \geq \|\psi\|^2 |\sin(\theta_0)|.$$

By using the Cauchy inequality, we deduce that

$$\|(I - \lambda A^{-1})^{-1}\| \leq \frac{1}{|\sin(\theta_0)|}.$$

Since

$$(\lambda - A)^{-1} = -A^{-1}(I - \lambda A^{-1})^{-1},$$

then we obtain

$$\|(\lambda - A)^{-1}\| \leq \frac{\|A^{-1}\|}{|\sin(\theta_0)|}.$$

This proves that the resolvent $\|(\lambda - A)^{-1}\|$ is uniformly bounded on γ independently of λ.

4.4 $\mathscr{C}_p$ Carleman-Class

4.4.1 Definition

Definition 4.4.1 Let K be a compact operator in a Hilbert space X. Then, K is said to belong to the Carleman-class $\mathscr{C}_p$, $p > 0$, with order p, if the series

$$\sum_{n=1}^{\infty} s_n^p(K)$$

converges, where $(s_n(K))_n$ are the eigenvalues of the operator $\sqrt{K^*K}$. ◇

We denote by $\mathscr{C}_p(X)$ $(p > 0)$ the set of operators of Carleman-class $\mathscr{C}_p$. In particular case $p = 2$, $\mathscr{C}_2$ is exactly the space of Hilbert-Schmidt operators and for $p = 1$, $\mathscr{C}_1$ is called the space of nuclear operators or the space of trace class operators on X. For a systematic treatment of the operators of Carleman-class, we refer to the Gohberg and Krein's book [2] and the Jeribi's book [4].

4.4.2 The Resolvent Representation

Theorem 4.4.1 (Lidskii [5]) *If K belongs to the Carleman-class $\mathscr{C}_p$ for some $p > 0$, then $(I - \lambda K)^{-1}$ can be represented as the ratio of two entire functions*

$$(I - \lambda K)^{-1} = \frac{D_\lambda}{d(\lambda)} \tag{4.4.1}$$

for which the inequalities

$$|d(\lambda)| \leq e^{c_1|\lambda|^p}$$

and

$$\|D_\lambda\| \leq e^{c_2|\lambda|^p}$$

are valid for all $\lambda \in \mathbb{C}$ (c_1 and c_2 being positive constants). ◇

Theorem 4.4.2 (Lidskii [5]) *Under the conditions of Theorem 4.4.1, the resolvent $(I - \lambda K)^{-1}$ can be represented as a ratio of entire functions*

$$(I - \lambda K)^{-1} = \frac{D_K(\lambda)}{\Delta_K(\lambda)}$$

such that

$$|D_K(\lambda)| \leq e^{\varepsilon(\lambda)|\lambda|^p}$$

and

$$|\Delta_K(\lambda)| \leq e^{\varepsilon(\lambda)|\lambda|^p},$$

where $\varepsilon(\lambda) \to 0$ as $\lambda \to \infty$. ◇

Lemma 4.4.1 (Shkalikov [7, Lemma 8]) *Let $F(\lambda)$ be a scalar meromorphic function belonging to the Carleman-class $\mathscr{C}_p$ $(p > 0)$ in an angle*

$$\Lambda_\alpha = \{\lambda \in \mathbb{C} \text{ such that } |\arg \lambda| < \alpha\}$$

and the poles of $F(\lambda)$ in this angle lie inside the strip $|Im\lambda| \leq h,\ h > 0$. Suppose that $|F(\lambda)| \leq M$ on the half-lines $Im\lambda = \pm(h + \delta),\ \delta > 0$, inside the angle Λ_α. Then, the following estimate holds inside the strip $|Im\lambda| \leq h + \delta$ as $Re\lambda \to \infty$

outside an exceptional set of discs $\mathscr{D}$:

$$\log|F(\lambda)| \leq C\left(M + \sup_{|t-r|\leq r^\eta}(n(t+1,F) - n(t,F))\right), \qquad r = |\lambda|, \qquad (4.4.2)$$

where $n(t, F)$ is the pole-counting function for F and the number η can be taken arbitrarily small. For any $d > 0$, the exceptional set of discs $\mathscr{D}$ can be chosen in such a way that the total sum of the radii of the discs from $\mathscr{D}$ inside the rectangle

$$|Im\lambda| \leq h, \ \ t \leq Re\lambda \leq t+1$$

does not exceed d for any sufficiently large t. The constant C in estimate (4.4.2) depends on δ, η, and d (the dependence on d is proportional to $\log d$) but does not depend on r and F. ◇

4.4.3 Some Properties of $\mathscr{C}_p$ Carleman-Class

Proposition 4.4.1 *For all $p \geq 1$, if $A, B \in \mathscr{C}_p(X)$, then*

$$A + B \in \mathscr{C}_p(X).$$

◇

Proof By using Proposition 4.1.3, we have

$$\begin{aligned} s_{2j}(A+B) &= s_{(j+1)+j-1}(A+B) \\ &\leq s_{j+1}(A) + s_j(B) \ \ (j = 1, 2, \ldots), \end{aligned}$$

and

$$\begin{aligned} s_{2j+1}(A+B) &= s_{(j+1)+(j+1)-1}(A+B) \\ &\leq s_{j+1}(A) + s_{j+1}(B) \ \ (j = 0, 1, 2, \ldots). \end{aligned}$$

So,

$$\begin{aligned} \sum_{j=1}^{\infty}\left[s_j(A+B)\right]^p &= \sum_{j=1}^{\infty}\left[s_{2j}(A+B)\right]^p + \sum_{j=0}^{\infty}\left[s_{2j+1}(A+B)\right]^p \\ &\leq \sum_{j=1}^{\infty}\left[s_{j+1}(A) + s_j(B)\right]^p + \sum_{j=0}^{\infty}\left[s_{j+1}(A) + s_{j+1}(B)\right]^p \\ &\leq \sum_{j=1}^{\infty}\left[s_{j+1}(A) + s_j(B)\right]^p + \sum_{j=1}^{\infty}\left[s_j(A) + s_j(B)\right]^p. \end{aligned}$$

In view of $s_{j+1}(A) \leq s_j(A)$, we get

$$\sum_{j=1}^{\infty}\left[s_j(A+B)\right]^p \leq 2\sum_{j=1}^{\infty}\left[s_j(A)+s_j(B)\right]^p.$$

Now, according to Minkowski inequality (Lemma 1.3.3), for $p \geq 1$, we have

$$\left(\sum_{j=1}^{\infty}\left[s_j(A)+s_j(B)\right]^p\right)^{\frac{1}{p}} \leq \left(\sum_{j=1}^{\infty}\left[s_j(A)\right]^p\right)^{\frac{1}{p}} + \left(\sum_{j=1}^{\infty}\left[s_j(B)\right]^p\right)^{\frac{1}{p}}.$$

So,

$$\sum_{j=1}^{\infty}\left[s_j(A)+s_j(B)\right]^p \leq \left[\left(\sum_{j=1}^{\infty}\left[s_j(A)\right]^p\right)^{\frac{1}{p}} + \left(\sum_{j=1}^{\infty}\left[s_j(B)\right]^p\right)^{\frac{1}{p}}\right]^p.$$

Therefore,

$$\sum_{j=1}^{\infty}\left[s_j(A+B)\right]^p \leq 2\left[\left(\sum_{j=1}^{\infty}\left[s_j(A)\right]^p\right)^{\frac{1}{p}} + \left(\sum_{j=1}^{\infty}\left[s_j(B)\right]^p\right)^{\frac{1}{p}}\right]^p < \infty.$$

Thus, $A + B \in \mathscr{C}_p(X)$. Q.E.D.

Proposition 4.4.2 *$\mathscr{C}_p(X)$ $(p > 0)$ is a two-sided ideal of $\mathscr{L}(X)$, i.e., if $A \in \mathscr{C}_p(X)$ and $B \in \mathscr{L}(X)$, then the operators AB and BA are in $\mathscr{C}_p(X)$.* ◇

Proof Let $A \in \mathscr{C}_p(X)$ and $B \in \mathscr{L}(X)$. By Proposition 4.1.4, we have

$$s_j(BA) \leq \|B\|s_j(A) \ (j = 1, 2, \ldots),$$

and

$$s_j(AB) \leq \|B\|s_j(A) \ (j = 1, 2, \ldots).$$

Hence,

$$\sum_{j=1}^{\infty}\left[s_j(BA)\right]^p \leq \|B\|^p\sum_{j=1}^{\infty}\left[s_j(A)\right]^p < \infty,$$

and

$$\sum_{j=1}^{\infty}\left[s_j(AB)\right]^p \leq \|B\|^p\sum_{j=1}^{\infty}\left[s_j(A)\right]^p < \infty.$$

Thus, the operators AB and BA are in $\mathscr{C}_p(X)$. Q.E.D.

Definition 4.4.2 An operator A is said with resolvent belong to the Carleman-class $\mathscr{C}_p$ $(p > 0)$ if the resolvent set is non-empty and there is $\lambda_0 \in \rho(A)$ such that $(\lambda_0 - A)^{-1}$ belong to the Carleman-class $\mathscr{C}_p(X)$. ◇

4.5 Fredholm Determinant

Definition 4.5.1 Let K be an operator of trace class on a Hilbert space X, and $(\mu_n)_n$ be the sequence of non-zero eigenvalues of K, counted according to their algebraic multiplicities. Then, the Fredholm determinant $d(\lambda)$ associated to K is given by

$$d(\lambda) = \prod_{n=1}^{\infty}(1 - \lambda\mu_n).$$

For an operator K belonging to the Carleman-class $\mathscr{C}_p$ for some $p > 1$, it is then necessary to go back to the definition of $d(\lambda)$ since the product is no longer always convergent. One sets

$$d(\lambda) = \prod_{k=1}^{\infty}(1 - \lambda\mu_k)e^{\lambda\mu_k + \cdots + \frac{1}{m-1}(\lambda\mu_k)^{m-1}},$$

where m is an integer not less than 1 such that $1 < p \leq m$. ◇

We recall that the Fredholm determinant of A^{-1} is the function defined by

$$d(\xi) = \prod_{k=1}^{\infty}\left(1 - \frac{\xi}{\xi_k}\right)^{\nu_k},$$

where the numbers ξ_k, $k = 1, 2, \ldots$ are the unrepeated eigenvalues of A and the integers ν_k are the algebraic multiplicities of ξ_k. Let us also recall that the geometric multiplicity of ξ_k is the dimension of the eigenspace $N(A - \xi_k)$ as opposed to the algebraic multiplicity which is the dimension ν_k of the generalized eigenspace

$$H_k = \bigcup_{m\geq 1} N(A - \xi_k)^m.$$

4.6 Notes and Remarks

- Isolated singular points of the resolvent naturally exhibit in the case of the eigenvalues of an operator with compact resolvent or non-zero eigenvalues of a compact operator. In both cases, the singularity is a pole of order $n_0(\lambda_0)$ (the index of the eigenvalue λ_0).
- Let K be a compact operator acting in a separable Hilbert space X. It is well known that the resolvent,

$$R_\lambda = (I - \lambda K)^{-1},$$

of the operator K is a meromorphic operator function of the parameter λ. Let $\Delta(\lambda)$ be an entire function of λ whose zeros coincide, with due regard to multiplicity, with the poles of R_λ. Then, the operator R_λ can be represented in the form

$$R_\lambda = \frac{D(\lambda)}{\Delta(\lambda)},$$

where $D(\lambda)$ is now the entire operator function

$$D(\lambda) = \sum_{n=0}^{\infty} \lambda^n Q^{(n)},$$

and $Q^{(n)}$ are bounded operators.

(i) Inequality (4.1.8) shows that, on a separable Hilbert space, every compact operator is a uniform limit of a sequence of finite-rank operators.
(ii) Theorem 4.1.9 can be considered as an equivalent definition of the singular values of a compact operator.
(iii) $\mathscr{C}_p(X) := \mathscr{C}_{p,p}(X) \subset \mathscr{C}^{(0)}_{p,\infty}(X)$ (the inclusion being strict).
(iv) $\mathscr{C}_p(X) = \mathscr{C}_{p,p}(X)$. If $p \geq 1$, then $N_{p,p}(\cdot)$ becomes a norm on $\mathscr{C}_p(X)$. In fact, there are on $\mathscr{C}_{p,q}(X)$ norms that induce its quasi-normed topologies for $1 < p < \infty$ and $1 \leq q \leq \infty$ but the only case where $N_{p,q}(\cdot)$ is a norm is that of Carleman-classes of order greater than or equal to 1.
(v) From [1, Theorem 26, Chap. XI, Sect. 9], we can show that there exist $M > 0$ and $L > 0$ such that

$$\|d(\lambda)(I - \lambda K)^{-1}\| \leq M e^{L|\lambda|^p} \text{ for any } \lambda \in \mathbb{C},$$

where $d(\cdot)$ is given in (4.4.1).
(vi) As $K = A^{-1}$, we have the following estimates:

(a) $\|d(\lambda)[I + \lambda(A - \lambda)^{-1}]\| \leq M e^{L|\lambda|^p}$ for any $\lambda \in \mathbb{C}$, and
(b) $\|(A - \lambda)^{-1}\| \leq \dfrac{1}{|\lambda|}\left(1 + \dfrac{M e^{L|\lambda|^p}}{d(\lambda)}\right)$ for any $\lambda \in \mathbb{C}$.

(vii) If there is $\lambda_0 \in \rho(A)$ such that $(\lambda_0 - A)^{-1}$ belong to the Carleman-class $\mathscr{C}_p$, then for all $\lambda \in \rho(A)$, $(\lambda - A)^{-1}$ belong to the Carleman-class $\mathscr{C}_p$ with $p \geq 1$. In fact, based on the resolvent identity (Proposition 1.2.2) or [3], we have

$$(\lambda_0 - A)^{-1} - (\lambda - A)^{-1} = (\lambda - \lambda_0)(\lambda - A)^{-1}(\lambda_0 - A)^{-1}. \tag{4.6.1}$$

By applying Proposition 4.4.2, we conclude that the operator

$$(\lambda - \lambda_0)(\lambda - A)^{-1}(\lambda_0 - A)^{-1}$$

belongs to the Carleman-class $\mathscr{C}_p$. Hence, by using Eq. (4.6.1) and Proposition 4.4.1, for $p \geq 1$, we have $(\lambda - A)^{-1}$ which belong to the Carleman-class $\mathscr{C}_p$.

References

1. N. Dunford, J.T. Schwartz, *Linear Operators* (Part III, Wiley-Interscience, New-York, 1971)
2. I.C. Gohberg, M.G. Krein, *Introduction to the Theory of Linear Non-self-Adjoint Operators in Hilbert Space* (Amer. Math. Soci, Providence, 1969)
3. A. Jeribi, Spectral theory and applications of linear operators and block operator matrices. (Springer, Cham, 2015). p. 599. (ISBN: 978-3-319-17565-2/hbk; 978-3-319-17566-9/ebook). xvi
4. A. Jeribi, *Denseness, Bases and Frames in Banach Spaces and Applications* (de Gruyter, Berlin, 2018)
5. V. B. Lidskii, Summability of series in the principal vectors of non-self-adjoint operators. Am. Math. Soc. Transl. (2), 40 (1964)
6. A. Pietsch, *Eigenvalues and s-Numbers, Cambridge studies in Advanced Mathematics*, vol. 13 (Cambridge University Press, 1987)
7. A.A. Shkalikov, On the basis property of root vectors of a perturbed self-adjoint operator. Proc. Steklov Inst. Math. **269**, 284–298 (2010)
8. S. Yakubov, *Completeness of Root Functions of Regular Differential Operators* (Longman, Harlow, 1994)

Chapter 5
The Evolutionary Problem

In this chapter, we recall some facts related to the evolutionary problem such as Hille-Yosida theorem, differentiability of the semigroup. We also give some properties of fractional operators and expansion of solution on generalized eigenvectors of operators in Hilbert space. In this chapter some elementary properties of semigroups are given.

5.1 Semigroups

5.1.1 Basic Elementary Properties of Semigroups

Some elementary properties of semigroups are given in the following. Let $T(t)$ be a strongly continuous semigroup on a Banach space X. We show that $\|T(t)\|$ is bounded on some neighborhood of the origin, that is, there exist $\delta > 0$ and $M > 1$ depending on δ such that

$$\|T(t)\| \leq M$$

for all $t \in [0, \delta]$. If this does not hold, then there exists a sequence $(t_n)_n$, $t_n \to 0^+$ such that

$$\|T(t_n)\| \geq n.$$

Hence, by the uniform boundedness theorem (Theorem 2.6.1), there exists one z such that $\{\|T(t_n)z\|\}$ is unbounded; but this contradicts the strong continuity at the origin. If we set $t = m\delta + \tau$ with $0 \leq \tau \leq \delta$, then

A. Jeribi, *Perturbation Theory for Linear Operators*,
https://doi.org/10.1007/978-981-16-2528-2_5

$$\begin{aligned}\|T(t)\| &\leq \|T(\delta)\|^m \|T(\tau)\| \\ &\leq M^{1+m} \\ &\leq MM^{\frac{t}{\delta}} \\ &= Me^{tw},\end{aligned} \tag{5.1.1}$$

where $w = \delta^{-1} \log M$. So, $\|T(t)\|$ is bounded on every finite subinterval of $[0, \infty)$.

We claim that $T(t)$ is strongly continuous for all $t \in [0, \infty)$. In fact, let $t > 0$ and $s \geq 0$. By using Eq. (5.1.1), we have

$$\begin{aligned}\|T(t+s)z - T(t)z\| &\leq \|T(t)\|\|T(s)z - z\| \\ &\leq Me^{wt}\|T(s)z - z\|.\end{aligned}$$

Hence, we may conclude that

$$\lim_{s \to 0^+} \|T(t+s)z - T(t)z\| = 0.$$

Moreover, for $t > 0$ and $s \geq 0$ sufficiently small, we have

$$\|T(t-s)z - T(t)z\| \leq \|T(t-s)\|\|z - T(s)z\|.$$

Thus,

$$\lim_{s \to 0^-} \|T(t+s)z - T(t)z\| = 0,$$

and $T(t)z$ is continuous. This prove the claim.

Let $z \in X$ and $\varepsilon > 0$. By the strong continuity of the semigroup $T(t)$, we can choose a $\tau > 0$ such that

$$\|T(s)z - z\| \leq \varepsilon$$

for all $s \in [0, \tau]$. For $t \in [0, \tau]$, we have

$$\begin{aligned}\left\| \frac{1}{t} \int_0^t T(s)z\mathrm{d}s - z \right\| &= \left\| \frac{1}{t} \int_0^t [T(s)z - z]\mathrm{d}s \right\| \\ &\leq \frac{1}{t} \int_0^t \|T(s)z - z\|\mathrm{d}s \\ &\leq \frac{1}{t} \int_0^t \varepsilon \mathrm{d}s = \varepsilon.\end{aligned}$$

Hence,

$$\frac{1}{t}\int_0^t T(s)z\,ds \to z \text{ as } t \to 0^+. \tag{5.1.2}$$

Consider the constant w_0 is, therefore, called the growth bound of the semigroup defined by

$$w_0 = \inf_{t>0}\left(\frac{1}{t}\log(\|T(t)\|)\right).$$

Let $t_0 > 0$ be a fixed number and

$$M = \sup_{t\in[0,t_0]} \|T(t)\|.$$

Then, for every $t \geq t_0$, there exists $n \in \mathbb{N}$ such that

$$nt_0 \leq t < (n+1)t_0.$$

Hence,

$$\begin{aligned}\frac{\log\|T(t)\|}{t} &= \frac{\log\|T^n(t_0)T(t-nt_0)\|}{t}\\ &\leq \frac{n\log\|T(t_0)\|}{t} + \frac{\log M}{t}\\ &= \frac{\log\|T(t_0)\|}{t_0}\frac{nt_0}{t} + \frac{\log M}{t}.\end{aligned}$$

The latter is smaller than or equal to $\frac{\log\|T(t_0)\|}{t_0} + \frac{\log M}{t}$ (respectively, $\frac{\log\|T(t_0)\|}{t_0}\frac{t-t_0}{t} + \frac{\log M}{t}$) if $\log\|T(t_0)\|$ is positive (respectively, negative). Thus,

$$\limsup_{t\to\infty} \frac{\log\|T(t)\|}{t} \leq \frac{\log\|T(t_0)\|}{t_0} < \infty,$$

and since t_0 is arbitrary, we have the following

$$\begin{aligned}\limsup_{t\to\infty}\frac{\log\|T(t)\|}{t} &\leq \inf_{t>0}\frac{\log\|T(t)\|}{t}\\ &\leq \liminf_{t\to\infty}\frac{\log\|T(t)\|}{t}.\end{aligned}$$

So,

$$w_0 = \inf_{t>0}\frac{\log\|T(t)\|}{t} = \lim_{t\to\infty}\frac{\log\|T(t)\|}{t} < \infty. \tag{5.1.3}$$

Let $w > w_0$, then there exists a t_0 such that

$$\frac{\log \|T(t)\|}{t} < w$$

for $t \geq t_0$, that is,

$$\|T(t)\| \leq e^{wt}$$

for $t \geq t_0$. But

$$\|T(t)\| \leq M_0$$

for $0 \leq t \leq t_0$, and so with

$$M_w = \begin{cases} M_0 & \text{if } w > 0 \\ e^{-wt_0} M_0 & \text{if } w < 0. \end{cases}$$

We obtain that for all $w > w_0$, there exists a constant M_w such that for all $t \geq 0$,

$$\|T(t)\| \leq M_w e^{wt}. \tag{5.1.4}$$

Let Ω be a closed subset of $\mathbb{R}$, and $(\Omega, \mathscr{B}, \mathrm{d}t)$ be the measure space with measurable subsets $\mathscr{B}$ and the Lebesgue measure $\mathrm{d}t$.

Lemma 5.1.1 (Curtain and Zwart [3, Lemma A.5.3]) *Let X be a separable Hilbert space, and let $f_1, f_2 : \Omega \longrightarrow X$ be two weakly measurable functions. The complex-valued function $\langle f_1(t), f_2(t)\rangle$ defined by the inner product of these functions is a measurable function.* ◇

For functions from $\mathbb{R}$ to a separable Hilbert space X, there is a simple criterion to test whether a function is Lebesgue integrable.

Lemma 5.1.2 (Curtain and Zwart [3, Lemma A.5.5]) *Let X be a separable Hilbert space and $f(t) : \Omega \longrightarrow X$. Then,*

$$\int_E f(t)\mathrm{d}t$$

is well defined as a Lebesgue integral for $E \in \mathscr{B}$ if, and only if, the function $\langle z, f(t)\rangle$ is measurable for every $z \in X$ and

$$\int_E \|f(t)\|\mathrm{d}t < \infty.$$ ◇

Next, let's consider

$$\int_0^\tau T(\tau - s)F(s)\mathrm{d}s,$$

where $T(t)$ is a C_0-semigroup on a separable Hilbert space X, $F(\cdot) \in \mathscr{L}(U, X)$, U is a Hilbert space, F is weakly measurable, and $\|F\| \in L_1(0, \tau)$. Since $T^*(t)$ is also a C_0-semigroup, $T^*(t)z$ is continuous and so strongly measurable. Furthermore, by definition, we have $F(s)u$ is weakly measurable. Hence, Lemma 5.1.1 shows that

$$\langle z, T(\tau - s)F(s)u\rangle = \langle T^*(\tau - s)z, F(s)u\rangle$$

is measurable in s for all $z \in X$, $u \in U$. So, from Lemma 5.1.2, we get for each $u \in U$,

$$\int_0^\tau T(\tau - s)F(s)u\mathrm{d}s$$

is a well defined Lebesgue integral. However,

$$\int_0^\tau T(\tau - s)F(s)\mathrm{d}s$$

need not be a well defined Lebesgue integral, since the integrand will not be uniformly measurable in general. Considering that

$$\int_0^\tau T(\tau - s)Bu(s)\mathrm{d}s,$$

where $T(t)$ is a C_0-semigroup on a separable Hilbert space X, $B \in \mathscr{L}(V, X)$, V is a separable Hilbert space and $u \in L_1([0, \tau]; V)$. Then,

$$\langle z, T(\tau - s)Bu(s)\rangle$$

is measurable in s for all $z \in X$ and

$$\int_0^\tau \|T(\tau - s)Bu(s)\|\mathrm{d}s \leq M_w e^{w\tau}\|B\| \int_0^\tau \|u(s)\|\mathrm{d}s < \infty.$$

So, by Lemma 5.1.2, the integral is well defined either as a Pettis or as a Lebesgue integral.

Lemma 5.1.3 (Curtain and Zwart [3, Lemma A.3.36]) *If $(x_n)_n$ is a weakly convergent sequence in a normed linear space with a weak limit x, then $(x_n)_n$ is uniformly bounded in norm and*

$$\|x\| \leq \liminf \|x_n\| < \infty. \qquad \diamond$$

Lemma 5.1.4 (Curtain and Zwart [3, Lemma A.4.8]) *For a closed linear operator A, on a Banach space X, we have the resolvent operator is a holomorphic function on $\rho(A)$, and the derivative is given by*

$$\frac{d}{d\lambda}(\lambda - A)^{-1} = -(\lambda - A)^{-2}. \qquad \diamond$$

5.1.2 The Infinitesimal Generator of a Continuous Semigroup

Let $T(t)$ be a strongly continuous semigroup on a Hilbert space X with infinitesimal generator A. Let $s > 0$ and consider

$$\begin{aligned}\frac{T(t+s)z_0 - T(t)z_0}{s} &= T(t)\frac{(T(s) - I)z_0}{s} \\ &= \frac{T(s) - I}{s}T(t)z_0.\end{aligned}$$

If $z_0 \in \mathscr{D}(A)$, the middle limit exists as $s \to 0^+$, and hence, the other limits also exist. In particular, $T(t)z_0 \in \mathscr{D}(A)$ and the strong right derivative of $T(t)z_0$ equals

$$AT(t)z_0 = T(t)Az_0.$$

For $t > 0$ and s sufficiently small, we have

$$\frac{T(t-s)z_0 - T(t)z_0}{-s} = T(t-s)\frac{(T(s) - I)z_0}{s}.$$

Hence, the strong left derivative exists and equals $T(t)Az_0$. So,

(i) for $z_0 \in \mathscr{D}(A)$, $T(t)z_0 \in \mathscr{D}(A)$ for all $t \geq 0$, and
(ii) $\dfrac{d}{dt}(T(t)z_0) = AT(t)z_0 = T(t)Az_0$ for $z_0 \in \mathscr{D}(A)$, $t > 0$.

By induction we can prove

$$\frac{d^n}{dt^n}(T(t)z_0) = A^n T(t)z_0 = T(t)A^n z_0$$

for $z_0 \in \mathscr{D}(A^n)$, $t > 0$. Take any $z^* \in X$ and $z_0 \in \mathscr{D}(A)$. Then,

$$\langle z^*, T(t)z_0 - z_0\rangle = \int_0^t \frac{d}{du}\langle z^*, T(u)z_0\rangle du,$$

and hence,

$$\langle z^*, T(t)z_0 - z_0\rangle = \int_0^t \langle z^*, T(u)Az_0\rangle \mathrm{d}u \quad \text{for } z_0 \in \mathscr{D}(A)$$
$$= \left\langle z^*, \int_0^t T(u)Az_0 \mathrm{d}u \right\rangle.$$

z^* was arbitrary, which proves

$$T(t)z_0 - z_0 = \int_0^t T(s)Az_0 \mathrm{d}s$$

for $z_0 \in \mathscr{D}(A)$. Now, we claim that $\mathscr{D}(A)$ is dense in X. Considering the following, for any $z \in X$

$$\frac{T(s) - I}{s} \int_0^t T(u)z\mathrm{d}u = \frac{1}{s} \int_0^t T(s+u)z\mathrm{d}u - \frac{1}{s} \int_0^t T(u)z\mathrm{d}u.$$

Such integrals are well defined since $T(t)$ is strongly continuous (see Lemma 5.1.2). Letting $\rho = s + u$ in the second integral, we get

$$\frac{T(s) - I}{s} \int_0^t T(u)z\mathrm{d}u = \frac{1}{s} \int_s^{t+s} T(\rho)z\mathrm{d}\rho - \frac{1}{s} \int_0^t T(u)z\mathrm{d}u$$
$$= \frac{1}{s} \left[\int_t^{t+s} T(\rho)z\mathrm{d}\rho + \int_s^t T(\rho)z\mathrm{d}\rho - \int_s^t T(u)z\mathrm{d}u - \int_0^s T(u)z\mathrm{d}u \right].$$

Thus,

$$\frac{T(s) - I}{s} \int_0^t T(u)z\mathrm{d}u = \frac{1}{s} \left[\int_0^s (T(t+u) - T(u))z\mathrm{d}u \right]$$
$$= \frac{1}{s} \int_0^s T(u)(T(t) - I)z\mathrm{d}u.$$

Now, as $s \to 0^+$, the right-hand side tends to $(T(t) - I)z$ (see (5.1.2)). Hence,

$$\int_0^t T(u)z\mathrm{d}u \in \mathscr{D}(A)$$

and

$$A\int_0^t T(u)z\mathrm{d}u = (T(t) - I)z.$$

Furthermore,

$$\frac{1}{t}\int_0^t T(u)z\mathrm{d}u \to z \text{ as } t \to 0^+,$$

and hence, for any $z \in X$, there exists a sequence in $\mathscr{D}(A)$ that tends to z. This shows that

$$\overline{\mathscr{D}(A)} = X$$

and prove the claim. So,

$$\int_0^t T(s)z\mathrm{d}s \in \mathscr{D}(A)$$

and

$$A\int_0^t T(s)z\mathrm{d}s = T(t)z - z$$

for all $z \in X$, and $\mathscr{D}(A)$ is dense in X.

It is easy to see that A is a closed linear operator. In fact, let $(z_n)_n$ be a sequence in $\mathscr{D}(A)$ converging to z such that Az_n converges to y. Then,

$$\|T(s)Az_n - T(s)y\| \le Me^{ws}\|Az_n - y\|$$

and so,

$$T(s)Az_n \to T(s)y$$

uniformly on $[0, t]$. Now, since $z_n \in \mathscr{D}(A)$, we have that

$$T(t)z_n - z_n = \int_0^t T(s)Az_n\mathrm{d}s.$$

Using the Lebesgue dominated convergence (Theorem 1.5.1), we can see that

$$T(t)z - z = \int_0^t T(s)y\mathrm{d}s.$$

Thus,

$$\lim_{t\to 0^+}\frac{T(t)z - z}{t} = \lim_{t\to 0^+}\frac{1}{t}\int_0^t T(s)y\mathrm{d}s = y.$$

Hence, $z \in \mathscr{D}(A)$ and

$$Az = y,$$

which proves that A is closed.

Let $C_0^\infty(\mathbb{R}_{\geq 0})$ be the class of all real-valued functions on $\mathbb{R}_{\geq 0}$ having continuous derivatives of all orders and having compact support contained in the open right half-line $(0, \infty)$. If $\psi \in C_0^\infty(\mathbb{R}_{\geq 0})$ so does $\psi^{(r)}$, the rth derivative of ψ, and $\psi(u)T(u)z$ is a continuous vector-valued function from $\mathbb{R}_{\geq 0}$ into X. Let Z_0 be the set defined by

$$Z_0 := \left\{ g = \int_0^\infty \psi(u)T(u)z\mathrm{d}u,\ z \in X \text{ and } \psi \in C_0^\infty(\mathbb{R}_{\geq 0}) \right\}, \tag{5.1.5}$$

where $T(t)$ is a strongly continuous semigroup on a Hilbert space X with infinitesimal generator A. These are well defined by Lemma 5.1.2. We shall show that $Z_0 \subset \mathscr{D}(A^r)$ for $r \geq 1$ and that Z_0 is dense in X. Indeed, for sufficiently small s, we have

$$\begin{aligned}\frac{T(s) - I}{s}g &= \frac{1}{s}\int_0^\infty \psi(u)[T(u+s)z - T(u)z]\mathrm{d}u\\ &= \frac{1}{s}\int_s^\infty [\psi(u-s) - \psi(u)]T(u)z\mathrm{d}u - \frac{1}{s}\int_0^s \psi(u)T(u)z\mathrm{d}u.\end{aligned}$$

But

$$\frac{\psi(u-s) - \psi(u)}{s} \to \dot{\psi}(u) \text{ as } s \to 0^+,$$

uniformly with respect to u, and the last expression is zero for sufficiently small s, since the support ψ is contained in $(0, \infty)$. Thus, $g \in \mathscr{D}(A)$ and

$$Ag = -\int_0^\infty \dot{\psi}(u)T(u)z\mathrm{d}u.$$

Repeating this argument, we see that $g \in \mathscr{D}(A^r)$ for all $r > 0$, and

$$A^r g = (-1)^r \int_0^\infty \psi^{(r)}(u)T(u)z\mathrm{d}u$$

which shows that

$$Z_0 \subset \bigcap_{r=1}^{\infty} \mathscr{D}(A^r).$$

Suppose now that the closure of Z_0 is not X. Then, there must exist a $z_0 \in X$ such that

$$\langle z_0, g \rangle = 0$$

for all $g \in Z_0$ and $\|z_0\| = 1$. Thus,

$$\left\langle z_0, \int_0^{\infty} \psi(u)T(u)z du \right\rangle = \int_0^{\infty} \psi(u)\langle z_0, T(u)z \rangle du = 0$$

for all $\psi \in C_0^{\infty}(\mathbb{R}_{\geq 0})$ and $z \in X$. But $\langle z_0, T(u)z_0 \rangle$ is continuous with $\|z_0\| = 1$. Hence, there exists a $\psi \in C_0^{\infty}(\mathbb{R}_{\geq 0})$ such that

$$\int_0^{\infty} \psi(u)\langle z_0, T(u)z_0 \rangle du \neq 0.$$

This is a contradiction, and so $\overline{Z_0} = X$. Hence, $\bigcap_{n=1}^{\infty} \mathscr{D}(A^n)$ is dense in X.

5.1.3 Hille-Yosida Theorem

Lemma 5.1.5 *Let $T(t)$ be a C_0-semigroup with infinitesimal generator A and with growth bound w_0. If $Re(\lambda) > w > w_0$, then $\lambda \in \rho(A)$, and for all $z \in X$, we have*

$$(\lambda - A)^{-1} z = \int_0^{\infty} e^{-\lambda t} T(t) z dt$$

and

$$\|(\lambda - A)^{-1}\| \leq \frac{M}{Re\lambda - w}. \qquad \diamond$$

Proof Let

$$R_{\lambda} z = \int_0^{\infty} e^{-\lambda t} T(t) z dt,$$

$z \in X$, $Re\lambda > w$. This operator is well defined by Lemma 5.1.2 since by (5.1.4), we have

$$\|e^{-\lambda t}T(t)z\| \leq Me^{(w-Re\lambda)t}\|z\|,$$

and $\langle z_1, T(t)z\rangle$ is a continuous function. Furthermore, we have

$$\begin{aligned}\|R_\lambda\| &\leq M\int_0^\infty e^{-(Re\lambda-w)t}\mathrm{d}t\\ &= \frac{M}{Re\lambda - w},\end{aligned}$$

and so, R_λ is bounded. We shall now show that $R_\lambda z \in \mathscr{D}(A)$ and

$$(\lambda - A)R_\lambda z = z$$

for all $z \in X$. In fact,

$$\begin{aligned}\frac{T(s)-I}{s}R_\lambda z &= \frac{1}{s}\int_0^\infty e^{-\lambda t}[T(s+t)-T(t)]z\mathrm{d}t\\ &= \frac{1}{s}\left[\int_s^\infty e^{-\lambda(u-s)}T(u)z\mathrm{d}u - \int_0^\infty e^{-\lambda t}T(t)z\mathrm{d}t\right]\\ &= \frac{1}{s}\left[e^{\lambda s}\int_0^\infty e^{-\lambda u}T(u)z\mathrm{d}u - e^{\lambda s}\int_0^s e^{-\lambda u}T(u)z\mathrm{d}u - \int_0^\infty e^{-\lambda t}T(t)z\mathrm{d}t\right]\\ &= \frac{e^{\lambda s}-1}{s}\int_0^\infty e^{-\lambda t}T(t)z\mathrm{d}t - \frac{e^{\lambda s}}{s}\int_0^s e^{-\lambda t}T(t)z\mathrm{d}t.\end{aligned}$$

Thus,

$$\begin{aligned}AR_\lambda z &= \lim_{s\to 0^+}\frac{T(s)-I}{s}R_\lambda z\\ &= \lambda R_\lambda z - z\end{aligned}$$

for all $z \in X$. Moreover,

$$\begin{aligned}R_\lambda Az &= \int_0^\infty e^{-\lambda t}T(t)Az\mathrm{d}t\\ &= A\int_0^\infty e^{-\lambda t}T(t)z\mathrm{d}t\\ &= AR_\lambda z,\end{aligned}$$

$z \in \mathscr{D}(A)$. Hence,

$$R_\lambda(\lambda - A)z = z,$$

$z \in \mathscr{D}(A)$ and

$$(\lambda - A)R_\lambda z = z,$$

$z \in X$. Thus,

$$(\lambda - A)^{-1} = R_\lambda. \qquad \text{Q.E.D.}$$

Lemma 5.1.6 *Let $T(t)$ be a C_0-semigroup with infinitesimal generator A and with growth bound w_0. If $Re(\lambda) > w > w_0$, then $\lambda \in \rho(A)$, and for all $z \in X$, we have*

$$\lim_{\alpha \to \infty} \alpha(\alpha - A)^{-1} z = z$$

for all $z \in X$, where α is constrained to be real. ◇

Proof Since the domain of A is dense in X, we can always find an $x \in \mathscr{D}(A)$ such that

$$\|x - z\| \le \varepsilon,$$

for any given $\varepsilon > 0$. Choose $\alpha_0 \in \mathbb{R}$ such that

$$\|(\alpha - A)^{-1}\| \le \frac{\varepsilon}{\|Ax\|}$$

and

$$\left| \frac{\alpha}{\alpha - w} \right| \le 2$$

for all real $\alpha > \alpha_0$. We know that this is possible by Lemma 5.1.5. Calculating $\|\alpha(\alpha - A)^{-1}z - z\|$ for $\alpha > \alpha_0$ gives

$$\begin{aligned} \|\alpha(\alpha - A)^{-1}z - z\| &= \|\alpha(\alpha - A)^{-1}z - \alpha(\alpha - A)^{-1}x + \alpha(\alpha - A)^{-1}x - x + x - z\| \\ &\le \|\alpha(\alpha - A)^{-1}(z - x)\| + \|(\alpha - A + A)(\alpha - A)^{-1}x - x\| + \|x - z\| \\ &\le \left| \frac{\alpha M}{\alpha - w} \right| \|z - x\| + \|(\alpha - A)^{-1}Ax\| + \|x - z\| \\ &\le (2M + 2)\varepsilon. \end{aligned}$$

This holds for every $\varepsilon > 0$. Thus,

$$\lim_{\alpha \to \infty} \alpha(\alpha - A)^{-1} z = z,$$

where α is real. Q.E.D.

We recall the Hille-Yosida theorem (see Curtain and Zwart [3]).

Theorem 5.1.1 (Hille-Yosida theorem) *A necessary and sufficient condition for a closed, densely defined, and linear operator A on a Hilbert space X to be the infinitesimal generator of a C_0-semigroup is that there exist real numbers M, w, such that for all real $\alpha > w$, $\alpha \in \rho(A)$, the resolvent set of A, and*

$$\|(\alpha - A)^{-r}\| \leq \frac{M}{(\alpha - w)^r}$$

for all $r \geq 1$. In this case

$$\|T(t)\| \leq Me^{wt}. \qquad \diamond$$

Proof Necessity: Using the same notation as in Lemma 5.1.5, we have that $\alpha \in \rho(A)$ if $\alpha > w > w_0$, and

$$(\alpha - A)^{-1}z = \int_0^\infty e^{-\alpha t}T(t)z\mathrm{d}t$$

for $z \in X$ and $\alpha > w > w_0$. We want to differentiate this with respect to α. Differentiating the function $e^{-\alpha t}T(t)z\chi_{[0,\infty)}(t)$, with respect to α, gives $-te^{-\alpha t}T(t)z\chi_{[0,\infty)}(t)$, where $\chi_{[0,\infty)}(.)$ denotes the characteristic functions of the intervals $[0, \infty)$. Furthermore, we have

$$\| - te^{-\alpha t}T(t)z\chi_{[0,\infty)}(t)\| \leq te^{-wt}M_1e^{w_1 t}\chi_{[0,\infty)}(t)\|z\|,$$

where w_1 satisfies $w > w_1 > w_0$ (see (5.1.3)). We can differentiate $(\alpha - A)^{-1}z$, obtaining

$$\frac{d(\alpha - A)^{-1}z}{d\alpha} = \int_0^\infty (-t)e^{-\alpha t}T(t)z\mathrm{d}t.$$

Similarly, one can show that

$$\frac{d^{r-1}(\alpha - A)^{-1}z}{d\alpha^{r-1}} = \int_0^\infty (-t)^{r-1}e^{-\alpha t}T(t)z\mathrm{d}t$$

for $r \geq 1$. Thus,

$$\begin{aligned}\left\|\frac{d^{r-1}(\alpha - A)^{-1}}{d\alpha^{r-1}}\right\| &\leq M\int_0^\infty t^{r-1}e^{-(\alpha - w)t}\mathrm{d}t \\ &= M(r-1)!(\alpha - w)^{-r}.\end{aligned}$$

However, the resolvent is holomorphic on $\rho(A)$, and so from the resolvent equation (see also Lemma 5.1.4), we have

$$\frac{d^{r-1}(\alpha - A)^{-1}}{d\alpha^{r-1}} = (-1)^{r-1}(r-1)!(\alpha - A)^{-r}$$

for $\alpha \in \rho(A)$. Hence, for real $\alpha > w > w_0$, we have

$$\|(\alpha - A)^{-r}\| \leq \frac{M}{(\alpha - w)^r}.$$

Sufficiency: Set $A_\alpha = \alpha^2(\alpha - A)^{-1} - \alpha I$, $\alpha > w$. Then, $A_\alpha \in \mathscr{L}(X)$, and we may construct the C_0-semigroup given by

$$T^\alpha(t) = e^{tA_\alpha} = e^{-\alpha t} \sum_{n=0}^{\infty} \frac{(\alpha^2 t)^n}{n!}(\alpha - A)^{-n}.$$

We shall show that the strong limit of $T^\alpha(t)$ exists as $\alpha \to \infty$, and it is the desired semigroup, $T(t)$. First, we show that

$$\|A_\alpha z - Az\| \to 0$$

as $\alpha \to \infty$ for $z \in \mathscr{D}(A)$. From Lemma 5.1.6, we have that

$$\alpha(\alpha - A)^{-1} z \to z$$

as $\alpha \to \infty$ for all $z \in X$. Now,

$$A_\alpha z = \alpha(\alpha - A)^{-1} A z,$$

so that $A_\alpha z \to Az$ as $\alpha \to \infty$ for $z \in \mathscr{D}(A)$. Note that

$$\begin{aligned} \|T^\alpha(t)\| &\leq e^{-\alpha t} \sum_{n=0}^{\infty} \frac{(\alpha^2 t)^n}{n!} \frac{M}{(\alpha - w)^n} \\ &= M e^{-\alpha t} e^{(\frac{\alpha^2}{\alpha - w})t} \\ &= M e^{(\frac{\alpha w}{\alpha - w})t}. \end{aligned} \tag{5.1.6}$$

So, $T^\alpha(t)$ is uniformly bounded on compact time intervals for sufficiently large α. Now,

$$(\alpha - A)^{-1}(\mu - A)^{-1} = (\mu - A)^{-1}(\alpha - A)^{-1}.$$

Thus,

$$A_\alpha A_\mu = A_\mu A_\alpha$$

and

$$A_\alpha T^\mu(t) = T^\mu(t) A_\alpha.$$

So, for $z \in \mathscr{D}(A)$, the following holds

$$\begin{aligned} T^{\alpha}(t)z - T^{\mu}(t)z &= \int_0^t \frac{d}{ds}\left(T^{\mu}(t-s)T^{\alpha}(s)z\right) ds \\ &= \int_0^t T^{\mu}(t-s)(A_{\alpha} - A_{\mu})T^{\alpha}(s)z ds \\ &= \int_0^t T^{\mu}(t-s)T^{\alpha}(s)(A_{\alpha} - A_{\mu})z ds. \end{aligned}$$

For α larger than $2|w|$, from (5.1.6), we have that

$$\begin{aligned} \|T^{\alpha}(t)\| &\leq Me^{(\frac{\alpha w}{\alpha - w})t} \\ &\leq Me^{2|w|t}. \end{aligned}$$

Thus, for α and μ larger than $2|w|$, it follows that

$$\begin{aligned} \|T^{\alpha}(t)z - T^{\mu}(t)z\| &\leq \int_0^t Me^{2|w|(t-s)}Me^{2|w|s}\|(A_{\alpha} - A_{\mu})z\| ds \\ &= M^2 te^{2|w|t}\|(A_{\alpha} - A_{\mu})z\|. \end{aligned}$$

But,

$$\|(A_{\alpha} - A_{\mu})z\| \to 0$$

as $\alpha, \mu \to \infty$, since

$$A_{\alpha}z \to Az$$

as $\alpha \to \infty$. Thus, $T^{\alpha}(t)z$ is a Cauchy sequence and so, it converges to what is called $T(t)z$. Using the uniform boundedness of $T^{\alpha}(t)z$ and the fact that $\mathscr{D}(A)$ is dense in X, we may extend this convergence to every $z \in X$. Again, using the uniform boundedness of $T^{\alpha}(t)$ and the equation above, notice that $T^{\alpha}(t)z$ converges to $T(t)z$ uniformly on compact time intervals. Furthermore, from both (5.1.6) and Lemma 5.1.3, we conclude that

$$\begin{aligned} \|T(t)z\| &\leq \liminf_{\alpha \to \infty} \|T^{\alpha}(t)z\| \\ &\leq \liminf_{\alpha \to \infty} Me^{(\frac{\alpha w}{\alpha - w})t}\|z\| \\ &= Me^{wt}\|z\|. \end{aligned}$$

Now, we only have to show that $T(t)$ is a strongly continuous semigroup with the infinitesimal generator A. For all $z \in X$, we have

$$\begin{aligned} T(t+s)z &= \lim_{\alpha\to\infty} T^\alpha(t+s)z \\ &= \lim_{\alpha\to\infty} T^\alpha(t)T^\alpha(s)z \\ &= T(t)T(s)z. \end{aligned}$$

In addition, $T(0) = I$, and the strong continuity is a consequence of the uniform convergence on compact intervals. The following holds

$$\|T^\alpha(t)A_\alpha z - T(t)Az\| \leq \|T^\alpha(t)\|\|A_\alpha z - Az\| + \|T^\alpha(t)Az - T(t)Az\|$$

and hence, $T^\alpha(t)A_\alpha z$ converges strongly to $T(t)Az$ as $\alpha \to \infty$, uniformly on compact intervals for $z \in \mathscr{D}(A)$. Thus, we may apply the Lebesgue Dominated convergence theorem (Theorem 1.5.1) to

$$T^\alpha(t)z - z = \int_0^t T^\alpha(s)A_\alpha z \mathrm{d}s$$

to obtain

$$T(t)z - z = \int_0^t T(s)Az\mathrm{d}s$$

for $z \in \mathscr{D}(A)$. So, the infinitesimal generator $\widetilde{A}$ of $T(t)$ is an extension of A, since

$$\widetilde{A}z = \lim_{t\to 0^+} \frac{T(t)z - z}{t} = Az$$

for $z \in \mathscr{D}(A)$. Now, if $\alpha > w$, then

$$(\alpha - A)(\mathscr{D}(A)) = X,$$

and by Sect. 5.1.2, it follows that

$$(\alpha - \widetilde{A})(\mathscr{D}(\widetilde{A})) = X.$$

Since $A(\mathscr{D}(A)) = \widetilde{A}(\mathscr{D}(A))$, then

$$(\alpha - \widetilde{A})(\mathscr{D}(A)) = (\alpha - \widetilde{A})(\mathscr{D}(\widetilde{A})).$$

Thus,

$$\mathscr{D}(A) = \mathscr{D}(\widetilde{A}),$$

and this completes the proof of the theorem. Q.E.D.

5.1.4 The Differentiability of the Semigroup

Let $\{T(t)\}_{t\geq 0}$ be a C_0-semigroup on a Banach space X and A be its generator. Assume that A has a discrete spectrum, that is,

$$\sigma(A) = \sigma_p(A) = \{\lambda_n;\, n \in \mathbb{N}\}.$$

For each $\lambda_n \in \sigma(A)$, we note that $E(\lambda_n, A)$ as its Riesz projection on X. We define the $\{T(t)\}$-invariant spectral subspace of X by

$$Sp(A) := \overline{span\left\{\sum_{j=0}^{m} E(\lambda_j, A)x \text{ such that } x \in X;\ \forall m \in \mathbb{N}\right\}},$$

and another $\{T(t)\}$-invariant spectral subspace by

$$M_\infty := \{x \in X \text{ such that } E(\lambda, A)x = 0, \forall\, \lambda \in \sigma(A)\}\,.$$

Clearly,

$$Sp(A) \bigcap M_\infty = \{0\}$$

and

$$\overline{Sp(A) + M_\infty} \subseteq X.$$

We define also the operators

$$D_n := (A - \lambda_n)E(\lambda_n, A)$$

and

$$D_n^0 := E(\lambda_n, A),$$

where $\lambda_n \in \sigma(A)$. Then, for each $n \in \mathbb{N}$, D_n is a bounded linear operator with the property that

$$D_n^k := (A - \lambda_n)^k E(\lambda_n, A)$$

and

$$D_n^{m_a(\lambda_n)} = 0,$$

where $m_a(\lambda_n)$ is the algebraic multiplicity of λ_n.
Suppose that A satisfies the following conditions:

(A1) there exist positive constants M_1, ρ_1 and ρ_3 such that

$$\sum_{k=0}^{m_a(\lambda_n)} \frac{t^k \|D_n^k\|}{k!} \leq M_1 e^{-\rho_1 Re\lambda_n} e^{\rho_3 t}, \quad \forall n \in \mathbb{N}, t \geq 0.$$

(A2) there exists a $\tau_0 > 0$ such that the series

$$\sum_{n=1}^{\infty} e^{Re\lambda_n \tau_0}$$

converges.

If the conditions $(A1)$ and $(A2)$ are fulfilled, then for each $\lambda_n \in \sigma(A)$ and any $x \in X$, we have

$$\begin{aligned} T(t)E(\lambda_n, A)x &= e^{\lambda_n t} \sum_{k=0}^{m_a(\lambda_n)} \frac{t^k}{k!}(A - \lambda_n)^k E(\lambda_n, A)x \\ &= e^{\lambda_n t} \sum_{k=0}^{m_a(\lambda_n)} \frac{t^k}{k!} D_n^k x. \end{aligned}$$

Since

$$\left\| \sum_{k=0}^{m_a(\lambda_n)} \frac{t^k}{k!} D_n^k x \right\| \leq \sum_{k=0}^{m_a(\lambda_n)} \frac{t^k \|D_n^k\|}{k!} \|x\|,$$

condition $(A1)$ ensures that, there exist positive constants M_1, ρ_1 and ρ_3 such that

$$\|T(t)E(\lambda_n, A)\| \leq M_1 e^{Re(\lambda_n)(t-\rho_1)} e^{\rho_3 t},$$

for all $n \in \mathbb{N}$ and all $t \geq 0$. Then, as $t \geq \tau_0 + \rho_1$, it holds that

$$\sum_{n=1}^{\infty} \|T(t)E(\lambda_n, A)\| \leq M_1 e^{\rho_3 t} \sum_{n=1}^{\infty} e^{Re(\lambda_n)(t-\rho_1)}.$$

Since X is a Banach space, the series

$$\sum_{n=1}^{\infty} T(t)E(\lambda_n, A)$$

converges in the operator norm. Now, we define the operator $T_1(t)$ for each $t \geq \tau_0 + \rho_1$ by

$$\begin{cases} T_1(t) : X \longrightarrow Sp(A) \\ \qquad u \longrightarrow T_1(t)u = \sum_{n=1}^{\infty} T(t)E(\lambda_n, A)u = \sum_{n=1}^{\infty} E(\lambda_n, A)T(t)u. \end{cases} \tag{5.1.7}$$

Lemma 5.1.7 *Let $T(t)$ be a C_0-semigroup on a Banach space X and A be its generator. Suppose that A satisfies the hypotheses $(A1)$ and $(A2)$. Then, $T_1(t)$, defined in (5.1.7), is compact for $t \geq \tau_0 + \rho_1$ and $T_1(t)x \in Sp(A)$ for any $x \in X$.* ◇

Proof For each n, $T(t)E(\lambda_n, A)$ is compact, so

$$\sum_{n=1}^{\infty} T(t)E(\lambda_n, A)$$

is also compact. So, $T_1(t)$ is compact for $t \geq \tau_0 + \rho_1$ and $T_1(t)x \in Sp(A)$. This completes the proof. Q.E.D.

Lemma 5.1.8 *Let $t \geq \tau_0 + \rho_1$ and $x \in X$. Then,*

$$\lim_{h \to 0} \|T_1(t+h)x - T_1(t)x\| = 0.$$ ◇

Proof For each $\lambda_n \in \sigma(A)$, $x \in X$, $t \geq \tau_0 + \rho_1$ and $s \geq 0$, we get

$$\begin{aligned} T(s)T_1(t)x &= \sum_{n=1}^{\infty} T(s)T(t)E(\lambda_n, A)x \\ &= \sum_{n=1}^{\infty} T(t)E(\lambda_n, A)T(s)x. \end{aligned}$$

Thus,

$$\begin{aligned} T(s)T_1(t)x &= T_1(t)T(s)x \\ &= \sum_{n=1}^{\infty} T(t+s)E(\lambda_n, A)x \\ &= T_1(t+s)x. \end{aligned}$$

So, for $t+h > \tau_0 + \rho_1$ and $t > \tau_0 + \rho_1$, we get

$$\|T_1(t+h)x - T_1(t)x\| = \|[T(t+h-(\tau_0+\rho_1)) - T(t-(\tau_0+\rho_1))]T_1(\tau_0+\rho_1)x\|.$$

Hence,

$$\lim_{h\to 0} \|T_1(t+h)x - T_1(t)x\| = 0. \qquad \text{Q.E.D.}$$

Now, we define $T_2(t)$ for $t \in [\tau_0 + \rho_1, \infty)$ by

$$\begin{cases} T_2(t) : X \longrightarrow M_\infty \\ \qquad\quad u \longrightarrow T_2(t)u = T(t)u - T_1(t)u. \end{cases} \tag{5.1.8}$$

We assume that the spectrum of A satisfies

(A3) there exist constants $M_2 > 0$ and $\rho_2 > 0$ such that

$$|Im\lambda_n| \le M_2 e^{-\rho_2 Re\lambda_n}.$$

Theorem 5.1.2 (Xu and Yung [16]) *Let $T(t)$ be a C_0-semigroup on a Banach space X and A be its generator. Suppose that A satisfies the hypotheses $(A1)$ and $(A2)$. Then, the two family of operators parameterized on $[\tau_0 + \rho_1, \infty)$, $T_1(t)$, defined in (5.1.7), and $T_2(t)$, defined in (5.1.8), satisfy*

(i) $T_1(t)$ and $T_2(t)$ are strongly continuous,
(ii) $T_j(t)T(s) = T(s)T_j(t) = T_j(t+s)$, for $t \ge \tau_0 + \rho_1$, $s \ge 0$, $j = 1, 2$, and
(iii) $T(t)$ has the following decomposition $T(t) = T_1(t) + T_2(t), \quad t \ge \tau_0 + \rho_1$.

In addition, if the spectrum of A satisfies the condition $(A3)$, then for each $x \in X$, $T_1(t)x$ is differentiable in $(\tau_0 + \rho_1 + \rho_2, \infty)$. ◇

Proof Let $t \ge \tau_0 + \rho_1$. By using Lemma 5.1.8, we have

$$\lim_{h\to 0} \|T_1(t+h)x - T_1(t)x\| = 0.$$

Now, clearly, $\{T_2(t)\}_{t\ge\tau_0+\rho_1}$ are bounded linear operators on X and strongly continuous with respect to t. Furthermore,

$$\begin{aligned} T_2(t)T(s) &= T(t)T(s) - T_1(t)T(s) \\ &= T(t+s) - T_1(t+s) \\ &= T_2(t+s), \end{aligned}$$

$t \ge \tau_0 + \rho_1$, $s \ge 0$, and

$$E(\lambda_n, A)T_2(t) = E(\lambda_n, A)T(t) - E(\lambda_n, A)T_1(t) = 0,$$

where $\lambda_n \in \sigma(A)$. So,

$$T_2(t)x \in M_\infty.$$

We now further suppose that $(A3)$ holds. For $x \in X, t \geq \tau_0 + \rho_1 + \rho_2$, we have

$$\begin{aligned}\left\|\sum_{n=1}^{\infty} \lambda_n e^{\lambda_n t} \sum_{k=0}^{m_a(\lambda_n)} \frac{t^k}{k!} D_n^k x\right\| &\leq \sum_{n=1}^{\infty} |\lambda_n| e^{Re\lambda_n t} \sum_{k=0}^{m_a(\lambda_n)} \frac{t^k \|D_n^k\|}{k!} \|x\| \\ &\leq M_1 e^{\rho_3 t} \sum_{n=1}^{\infty} |\lambda_n| e^{Re\lambda_n (t-\rho_1)} \|x\| \\ &\leq M_1 \left(\frac{1}{\rho_2} + M_2\right) e^{\rho_3 t} \sum_{n=1}^{\infty} e^{Re\lambda_n (t-\rho_1-\rho_2)} \|x\|.\end{aligned}$$

So, the series

$$\sum_{n=1}^{\infty} AT(t)E(\lambda_n, A)x = \sum_{n=1}^{\infty} \left[\lambda_n e^{\lambda_n t} \sum_{k=0}^{m_a(\lambda_n)} \frac{t^k}{k!} D_n^k + e^{\lambda_n t} \sum_{k=0}^{m_a(\lambda_n)-1} \frac{t^k}{k!} D_n^{k+1}\right] x$$

converges absolutely for $t \geq \tau_0 + \rho_1 + \rho_2$. Since A is a closed linear operator, it holds that

$$A \sum_{n=1}^{\infty} T(t)E(\lambda_n, A)x = \sum_{n=1}^{\infty} AT(t)E(\lambda_n, A)x.$$

Therefore, for $t > \tau_0 + \rho_1 + \rho_2$, $T_1(t)x$ is differentiable and

$$\begin{aligned}\frac{dT_1(t)x}{\mathrm{d}t} &= \sum_{n=1}^{\infty} AT(t)E(\lambda_n, A)x \\ &= AT_1(t)x.\end{aligned}$$

The proof is then complete. Q.E.D.

Corollary 5.1.1 *Let $T(t)$ be a C_0-semigroup on a Banach space X and A be its generator. Suppose that A satisfies the conditions $(A1)$-$(A3)$ and the generalized eigenvectors of A are complete in X, then $T(t)$ is a differentiable semigroup for $t > \max\{\tau_0 + \rho_1 + \rho_2, \tau_0 + \rho_1 + h\}$.* ◇

Proof We can assume without loss of generality that $T(t)$ is exponential stable. If the system of generalized eigenvectors of A is complete in X, i.e., $Sp(A) = X$, then the desired result follows immediately by using Theorem 5.1.2. Q.E.D.

Corollary 5.1.2 *Let $T(t)$ be a C_0-semigroup on a Banach space X and A be its generator. Suppose that A satisfies the conditions $(A1)$-$(A3)$. In addition, if the restriction of the resolvent of A to M_∞ is an entire function with values in X of finite exponential type h, then $T(t)$ is a differentiable semigroup for $t > \max\{\tau_0 + \rho_1 + \rho_2, \tau_0 + \rho_1 + h\}$.* ◇

Proof From Theorem 5.1.2, we have a decomposition

$$T(t)x = T_1(t)x + T_2(t)x,$$

$t \geq \tau_0 + \rho_1$, with $T_1(t)x \in Sp(A)$ and $T_2(t)x \in M_\infty$. If $M_\infty = \{0\}$, then

$$T_2(t) = 0,$$

so the desired result follows. If $M_\infty \neq \{0\}$, for each $x \in M_\infty$, we have $T(t)x \in M_\infty$, and $R(\lambda, A)x$ is an entire function of finite exponential type h, which is independent of x. From the theory of semigroups of linear bounded operators, we have

$$R(\lambda, A)z = \int_0^\infty e^{-\lambda t} T(t)z \mathrm{d}t,$$

$z \in X$, $Re\lambda \geq 0 > w_0(A)$. So,

$$R(\lambda, A)x = \int_0^\infty e^{-\lambda t} T(t)x \mathrm{d}t,$$

$x \in M_\infty$, $Re\lambda \geq 0 > w_0(A)$. Therefore, for each $f \in X^*$, $\langle R(\lambda, A)x, f\rangle$ is in $H^2(\mathbb{C}_+)$. Note that $\langle R(\lambda, A)x, f\rangle$ is an entire function of exponential type at most h. The Wiener-Paley theorem says that there is a function $T_{x,f}(t) \in L_2[0, h]$ such that

$$\langle R(\lambda, A)x, f\rangle = \int_0^h e^{-\lambda t} T_{x,f}(t) \mathrm{d}t$$

for all $\lambda \in \mathbb{C}$. The uniqueness theorem on the Fourier transform ensures that

$$T_{x,f}(t) = \langle T(t)x, f\rangle.$$

Therefore,

$$T(t)x = 0,$$

for all $x \in M_\infty$, $t > h$. Now, for any $x \in X$, we have

$$T(\tau_0 + \rho_1)x = T_1(\tau_0 + \rho_1)x + T_2(\tau_0 + \rho_1)x,$$

and for $t > \tau_1 := \max\{\tau_0 + \rho_1 + \rho_2, \tau_0 + \rho_1 + h\}$, $0 < \varepsilon < t - \tau_1$,

$$\begin{aligned} T(t)x &= T(t-h-\varepsilon-\tau_0-\rho_1)T(h+\varepsilon)T(\tau_0+\rho_1)x \\ &= T(t-h-\varepsilon)T_1(h+\varepsilon)x \\ &= T_1(t)x. \end{aligned}$$

This, together with Theorem 5.1.2, asserts that $T(t)x$ is differentiable for $t \in (\tau_1, \infty)$. The proof is then complete. Q.E.D.

5.2 Fractional Operators

A linear operator A in a Banach space X is said to be of type (ω, M) if

(i) A is densely defined and closed, and

(ii) the resolvent set of $-A$ contains the sector $|\arg\lambda| < \pi - \omega$ of the complex plane and $\lambda(\lambda + A)^{-1}$ is uniformly bounded in each smaller sector $|\arg\lambda| < \pi - \omega - \varepsilon$, $\varepsilon > 0$, with

$$\|\lambda(\lambda + A)^{-1}\| \leq M$$

for $\lambda > 0$.

Let X be a Hilbert space. An operator $-A$ is said to be dissipative if

$$Re\langle Au, u\rangle \geq 0$$

for every $u \in \mathscr{D}(A)$, and $-A$ is maximal dissipative if it has no proper dissipative extension. It is known that a closed and maximal dissipative operator is densely defined that $-A$ is closed and maximal dissipative if, and only if, the operator $-A^*$ is, and also if, and only if, the operator $-A$ is the infinitesimal generator of a contraction semigroup $\{\exp(-tA)\}_{0<t<\infty}$ that is,

$$\|\exp(-tA)\| \leq 1.$$

We shall say that A is accretive if $-A$ is dissipative.

5.2.1 Dunford Integral

5.2.1.1 Fractional Power

If X is a Hilbert space, A is of type $(\pi/2, 1)$ if, and only if, A is closed and maximal accretive. If A is of type (ω, M), the fractional power A^α, $0 < \alpha < 1$, can be defined, indirectly, by

$$(\lambda + A^\alpha)^{-1} = \frac{\sin \pi\alpha}{\pi} \int_0^\infty \frac{\mu^\alpha}{\lambda^2 + 2\lambda\mu^\alpha \cos \pi\alpha + \mu^{2\alpha}} (\mu + A)^{-1} \mathrm{d}\mu, \tag{5.2.1}$$

which is valid at least for $\lambda > 0$. A^α is an operator of type $(\alpha\omega, M)$. If $\alpha\omega < \pi/2$, $-A^\alpha$ is the infinitesimal generator of an analytic semigroup, that is, the semigroup

$$T_{t,\alpha} = \exp(-tA^\alpha)$$

has an analytic continuation to the sector

$$|\arg t| < \frac{\pi}{2} - \alpha\omega,$$

$T_{t,\alpha}$ and $t\mathrm{d}T_{t,\alpha}/\mathrm{d}t$ being uniformly bounded for

$$|\arg t| < \frac{\pi}{2} - \alpha\omega - \varepsilon,$$

$\varepsilon > 0$. If we make the additional assumption that A^{-1} is bounded, (5.2.1) is equivalent to the Dunford integral

$$(\lambda + A^\alpha)^{-1} = -\frac{1}{2\pi i} \int_C (\lambda + z^\alpha)^{-1} (z - A)^{-1} dz, \tag{5.2.2}$$

where the integration path C runs in the resolvent set of A from $\infty e^{-i\theta}$ to $\infty e^{i\theta}$, $\omega < \theta \leq \pi$, avoiding the negative real axis and 0. In this case (5.2.1) and (5.2.2) are also valid for $\lambda = 0$. We now prove several lemmas that are not stated in [7]. In what follows, it is always assumed that, A is of type $(\frac{\pi}{2}, M)$ and that $0 < \alpha < 1$. We note that $A + \varepsilon$ is also of type $(\frac{\pi}{2}, M)$ if $\varepsilon > 0$; this is a direct consequence of the assumption.

Let $\varepsilon > 0$. Since $(I + \varepsilon A)^{-1}$ is bounded, (5.2.1) is valid for $\lambda = 0$ if A is replaced by $I + \varepsilon A$. Since

$$\|(\mu + 1 + \varepsilon A)^{-1}\| \leq M(\mu + 1)^{-1},$$

it follows that

$$\|(I + \varepsilon A)^{-\alpha}\| \leq \frac{\sin \pi\alpha}{\pi} \int_0^\infty \mu^{-\alpha} M(\mu + 1)^{-1} \mathrm{d}\mu = M.$$

So,

$$\|(I + \varepsilon A)^{-\alpha}\| \leq M.$$

Note that $I + \varepsilon A$ is also of type $(\frac{\pi}{2}, M)$ so that $(I + \varepsilon A)^\alpha$ exists.

For brevity, we write $A_\varepsilon = A + \varepsilon$. We get

$$A_\varepsilon^{-\alpha} A_\varepsilon^{-(1-\alpha)} = A_\varepsilon^{-(1-\alpha)} A_\varepsilon^{-\alpha} = A_\varepsilon^{-1}.$$

It is easy to see that A_ε^{-1}, $A_\varepsilon^{-\alpha}$ etc. are bounded for $\varepsilon > 0$. Hence, follows that

$$A_\varepsilon^{-(1-\alpha)} A_\varepsilon \subset A_\varepsilon^{\alpha}.$$

Now, (5.2.1) is true for $\lambda = 0$ if A is replaced by A_ε. Replacing α by $1-\alpha$ and applying both members of (5.2.1) to $A_\varepsilon u$, where $u \in \mathscr{D}(A)$, we thus obtain

$$A_\varepsilon^{\alpha} u = \frac{\sin \pi\alpha}{\pi} \int_0^\infty \mu^{\alpha-1} (\mu + A_\varepsilon)^{-1} A_\varepsilon u \mathrm{d}\mu, \quad \varepsilon > 0, \quad u \in \mathscr{D}(A). \tag{5.2.3}$$

Subtracting from (5.2.3) a similar expression with ε replaced by η, $0 < \eta < \varepsilon$, we have

$$A_\varepsilon^{\alpha} u - A_\eta^{\alpha} u = \frac{\sin \pi\alpha}{\pi} \left[\int_0^\delta \mu^{\alpha-1} (\mu + A_\varepsilon)^{-1} A_\varepsilon u \mathrm{d}\mu - \int_0^\delta \mu^{\alpha-1} (\mu + A_\eta)^{-1} A_\eta u \mathrm{d}\mu \right.$$
$$\left. + (\varepsilon - \eta) \int_\delta^\infty \mu^{\alpha} (\mu + A_\varepsilon)^{-1} (\mu + A_\eta)^{-1} u \mathrm{d}\mu \right].$$

Noting that

$$\|(\mu + A_\varepsilon)^{-1} A_\varepsilon u\| \le (1 + M)\|u\|$$

and

$$\|(\mu + A_\varepsilon)^{-1}\| \le M\mu^{-1},$$

this gives

$$\|A_\varepsilon^{\alpha} u - A_\eta^{\alpha} u\| \le \frac{\sin \pi\alpha}{\pi} \left[2(1 + M) \int_0^\delta \mu^{\alpha-1} \mathrm{d}\mu + M^2 (\varepsilon - \eta) \int_\delta^\infty \mu^{\alpha-2} \mathrm{d}\mu \right] \|u\|$$
$$= \frac{\sin \pi\alpha}{\pi} \left[2(1 + M)\alpha^{-1}\delta^{\alpha} + M^2(\varepsilon - \eta)(1 - \alpha)^{-1}\delta^{\alpha-1} \right] \|u\|.$$

Taking $\delta = (\varepsilon - \eta)M^2/2(1 + M)$, we obtain

$$\|A_\varepsilon^{\alpha} u - A_\eta^{\alpha} u\| \le c(\varepsilon - \eta)^{\alpha} \|u\|,$$

$0 < \eta < \varepsilon$, $u \in \mathscr{D}(A)$;

$$c = 2^{1-\alpha} M^{2\alpha} (1 + M)^{1-\alpha} \sin \pi\alpha / \pi\alpha(1 - \alpha)$$

depends only on α and M. This shows that

$$\lim_{\varepsilon \to 0} A_\varepsilon^\alpha u = Bu$$

exists for $u \in \mathscr{D}(A)$, and that

$$\|A_\varepsilon^\alpha u - Bu\| \le c\varepsilon^\alpha \|u\|, \quad \varepsilon > 0, \quad u \in \mathscr{D}(A). \tag{5.2.4}$$

Now, $\mathscr{D}(A)$ is a core of A_ε^α; by this, we mean that the closure of the restriction of A_ε^α to $\mathscr{D}(A)$ is A_ε^α itself. To see this, it suffices to note that, for any $v \in \mathscr{D}(A_\varepsilon^\alpha)$, the sequence

$$v_n = (1 + n^{-1} A_\varepsilon)^{-1} v \in \mathscr{D}(A_\varepsilon) = \mathscr{D}(A)$$

has the property that $v_n \to v$ and

$$A_\varepsilon^\alpha v_n = (1 + n^{-1} A_\varepsilon)^{-1} A_\varepsilon^\alpha v \to A_\varepsilon^\alpha v, \ n \to \infty.$$

Here, we used the relation

$$A_\varepsilon^\alpha (1 + n^{-1} A_\varepsilon)^{-1} v = (1 + n^{-1} A_\varepsilon)^{-1} A_\varepsilon^\alpha v.$$

This follows from the fact that

$$(1 + n^{-1} A_\varepsilon)^{-1} A_\varepsilon^{-\alpha} w = A_\varepsilon^{-\alpha} (1 + n^{-1} A_\varepsilon)^{-1} w$$

with $w = A_\varepsilon^\alpha v$. The fact that $\mathscr{D}(A)$ is a core of A_ε^α, combined with (5.2.4), shows that the closure B^{**} of B exists (B was defined only on $\mathscr{D}(A)$) and has the same domain with A_ε^α. At the same time the inequality (5.2.4) is extended to all $u \in \mathscr{D}(B^{**})$, with B replaced by B^{**}. Thus,

$$\mathscr{D}(A_\varepsilon^\alpha) = \mathscr{D}(B^{**})$$

is independent of ε and $A_\varepsilon^\alpha u \to B^{**} u$ uniformly for $\|u\| \le 1$, $u \in \mathscr{D}(B^{**})$. From this it follows easily that $(B^{**} + \lambda)^{-1}$ exists as a bounded operator with domain X for $\lambda > 0$ and that

$$(A_\varepsilon^\alpha + \lambda)^{-1} \to (B^{**} + \lambda)^{-1}$$

for $\varepsilon \to 0$ in the uniform operator topology. Since, however, we know that

$$(A_\varepsilon^\alpha + \lambda)^{-1} \to (A^\alpha + \lambda)^{-1}$$

(see [7]), it follows that

$$B^{**} = A^\alpha.$$

So, $\mathscr{D}[(A+\varepsilon)^\alpha] = \mathscr{D}[A^\alpha]$ and

$$\|(A+\varepsilon)^\alpha u - A^\alpha u\| \le c\varepsilon^\alpha \|u\|, \quad u \in \mathscr{D}(A^\alpha),$$

where the constant c depends only on α and M. For $A^\alpha = B^{**}$ is the closure of B and B was the restriction of $B^{**} = A^\alpha$ to $\mathscr{D}(A)$, we have $\mathscr{D}(A)$ is a core of A^α. Since

$$\|A(A+\lambda)^{-1}\| \le 1 + M,$$

the integral is absolutely convergent near $\lambda = 0$. If $u \in \mathscr{D}(A)$,

$$\begin{aligned}\|A(A+\lambda)^{-1}u\| &= \|(A+\lambda)^{-1}Au\| \\ &\le M\lambda^{-1}\|Au\|\end{aligned}$$

so that the integral is absolutely convergent also for $\lambda \to \infty$. Furthermore, we know that

$$A^\alpha u = \frac{\sin \pi\alpha}{\pi} \int_0^\infty \lambda^{\alpha-1} A(\lambda + A)^{-1} u \mathrm{d}\lambda, \tag{5.2.5}$$

is valid if A is replaced by $A_\varepsilon = A + \varepsilon$, see (5.2.3). Thus, (5.2.5) follows from (5.2.3) by going to the limit $\varepsilon \to 0$, and that

$$A_\varepsilon(\lambda + A_\varepsilon)^{-1}u \to A(\lambda + A)^{-1}u \text{ (dominated convergence).}$$

So, for each $u \in \mathscr{D}(A)$, we have (5.2.5) the integral being absolutely convergent.

5.2.1.2 Estimate of Fractional Power

Let K be a compact operator on a Hilbert space X. If the resolvent of K satisfies the estimate

$$\|(K-\lambda)^{-1}\| \le \frac{M}{|\lambda|}, \tag{5.2.6}$$

for λ lying in the sector

$$S = \left\{\lambda \in \mathbb{C} \text{ such that } \pi - \frac{\delta}{2} \le arg(\lambda) \le \pi + \frac{\delta}{2}\right\}, \tag{5.2.7}$$

where $\delta > 0$. Then, we can define, for $\alpha \in]0, 1[$, the power K^α by the formula

$$K^\alpha = \frac{\sin(\alpha\pi)}{\pi} \int_0^\infty \lambda^\alpha \left[(K+\lambda)^{-1} - \frac{1}{\lambda}\right] \mathrm{d}\lambda. \tag{5.2.8}$$

Theorem 5.2.1 (Kato [8]) *Let T be a maximal accretive operator acting in a Hilbert space X. Then, there exists a unique maximal accretive square root $\sqrt{T}$ of T such that*

$$(\sqrt{T})^2 = T$$

and $\sqrt{T}$ has the following additional properties

(i) $\sqrt{T}$ is m-sectorial with the numerical range contained in the sector,

$$\left\{ z \in \mathbb{C} \text{ such that } |arg(z)| \leq \frac{\pi}{4} \right\},$$

(ii) $\mathscr{D}(T)$ is a core of $\sqrt{T}$, and
(iii) T has a compact resolvent if, and only if, $\sqrt{T}$ has. ◇

Let K_j, $j = 1, 2$ be two compact operators on a Hilbert space X such that their resolvents $(K_j - \lambda)^{-1}$ satisfy the estimates (5.2.6) for λ lying in the sector S (see (5.2.7)). Under the assumption (5.2.6), we can define, for $\alpha \in]0, 1[$, the power K_j^α, $j = 1, 2$ by the formula (5.2.8). Thus, considering the integral representation of $K_1^\alpha - K_2^\alpha$, we get

$$\begin{aligned} K_1^\alpha - K_2^\alpha &= \frac{\sin(\alpha\pi)}{\pi} \int_0^\infty \lambda^\alpha \left[(K_1 + \lambda)^{-1} - (K_2 + \lambda)^{-1} \right] \mathrm{d}\lambda \\ &= \frac{\sin(\alpha\pi)}{\pi} \left[\int_0^\rho \lambda^\alpha \left[(K_1 + \lambda)^{-1} - (K_2 + \lambda)^{-1} \right] \mathrm{d}\lambda + \right. \\ &\qquad \left. \int_\rho^\infty \lambda^\alpha \left[(K_1 + \lambda)^{-1} - (K_2 + \lambda)^{-1} \right] \mathrm{d}\lambda \right], \end{aligned}$$

where $\rho > 0$ will be chosen appropriately later. Using Eq. (5.2.6), we have

$$\left\| \int_0^\rho \lambda^\alpha \left[(K_1 + \lambda)^{-1} - (K_2 + \lambda)^{-1} \right] \mathrm{d}\lambda \right\| \leq 2M \int_0^\rho \lambda^{\alpha-1} \mathrm{d}\lambda,$$

and

$$\left\| \int_\rho^\infty \lambda^\alpha \left[(K_1 + \lambda)^{-1} - (K_2 + \lambda)^{-1} \right] \mathrm{d}\lambda \right\| \leq 2M^2 \int_\rho^\infty \lambda^{\alpha-2} \|K_1 - K_2\| \mathrm{d}\lambda.$$

That is, we get

$$\|K_1^\alpha - K_2^\alpha\| \le M\frac{\sin(\alpha\pi)}{\pi}\left[2\frac{\rho^\alpha}{\alpha} + M\|K_1 - K_2\|\frac{\rho^{\alpha-1}}{1-\alpha}\right].$$

By choosing $\rho = \frac{M}{2}\|K_1 - K_2\|$, we deduce that for $\alpha \in (0, 1)$, we have the estimate

$$\|K_1^\alpha - K_2^\alpha\| \le M\frac{\sin(\alpha\pi)}{\pi}\left[\frac{2}{\alpha} + \frac{M}{1-\alpha}\right]\|K_1 - K_2\|^\alpha. \tag{5.2.9}$$

5.2.2 *Fractional of Carleman-Class Operators*

Let K be a compact operator such that its resolvent satisfies the estimate (5.2.6). We will consider that the operator K which satisfies the condition

$$|arg\langle K\varphi, \varphi\rangle| \le \frac{\theta}{2}$$

for every $\varphi \in X$. Then, it is fairly simple to verify that

$$|arg\langle K^\alpha\varphi, \varphi\rangle| \le \frac{\alpha\theta}{2}$$

for every $\varphi \in X$ and $0 < \alpha < 1$. If

$$\lambda_0 = |\lambda_0|e^{i\beta}, \quad \left(\beta < \frac{\theta}{2}\right)$$

is a normal eigenvalue of the operator K, then

$$\lambda_0^\alpha = |\lambda_0|^\alpha e^{i\alpha\beta}$$

is a normal eigenvalue of the operator K^α and the corresponding root subspaces $\mathscr{L}_K(\lambda_0)$ and $\mathscr{L}_{K^\alpha}(\lambda_0)$ coincide. Henceforth, we shall denote by $\mathfrak{B}_n$ $(n = 0, 1, 2, \cdots)$ the set of all finite-dimensional operators of dimension $\le n$. By using Theorem 4.1.9 (theorem of Allahverdiev), we infer that

$$s_{n+1}(K) = \min_{F\in\mathfrak{B}_n}\|K - F\| = \|K - K_n\|,$$

where

$$K_n = \sum_{j=1}^{n} s_j(K)\langle\cdot, \varphi_j\rangle\Psi_j$$

is the nth partial sum of the Schmidt expansion (see (4.1.3)) of the operator K and $\{\varphi_j\}$ and $\{\Psi_j\}$ are certain orthonormal systems. In the following, we put

- $G = \frac{1}{2}(K + K^*)$ and
$$G_n = \sum_{j=1}^{n} s_j(G)\langle \cdot, \varphi_j\rangle \Psi_j,$$
where G_n is the nth partial sum of the Schmidt expansion of the operator G.
- $H = \frac{1}{2i}(K + K^*)$ and
$$H_n = \sum_{j=1}^{n} s_j(H)\langle \cdot, \varphi_j\rangle \Psi_j,$$
where H_n is the nth partial sum of the Schmidt expansion of the operator H.
- $K_n = G_n + iH_n$.

As K_n^α is a finite-dimensional operator of dimension $\leq 2n$, then

$$s_{2n+1}(K) \leq \|K^\alpha - K_n^\alpha\|$$

and, from both Eqs. (5.2.9) and (4.1.8), we obtain

$$\begin{aligned} s_{2n+1}(K) &\leq C_\alpha \|K - K_n\|^\alpha \\ &\leq C_\alpha(\|G - G_n\| + \|H - H_n\|)^\alpha \\ &\leq C_\alpha(s_{n+1}(G) + s_{n+1}(H))^\alpha \\ &\leq C_\alpha \left(s_{n+1}\left(\frac{1}{2}(K + K^*)\right) + s_{n+1}\left(\frac{1}{2i}(K + K^*)\right)\right)^\alpha . \end{aligned}$$

Now, by using the inequality of Ky Fan (Proposition 4.1.3), we can deduce that

$$s_{2n+1}(K) \leq C_\alpha(2s_{n+1}(K) + 2s_{n+1}(K^*))^\alpha.$$

Since

$$s_{n+1}(K) = s_{n+1}(K^*),$$

we have

$$s_{2n+1}(K) \leq 4^\alpha C_\alpha[s_{n+1}(K)]^\alpha,$$

and consequently

$$s_{2n}(K^\alpha) \leq C[s_n(K)]^\alpha,$$

where $C = 4^\alpha C_\alpha$. Therefore,

$$[s_{2n}(K^\alpha)]^{\frac{p}{\alpha}} \leq [Cs_n^\alpha(K)]^{\frac{p}{\alpha}} = C^{\frac{p}{\alpha}}[s_n(K)]^p.$$

Hence,

$$\sum_{n=1}^{\infty}[s_{2n}(K^{\alpha})]^{\frac{p}{\alpha}} \le C^{\frac{p}{\alpha}} \sum_{n=1}^{\infty}[s_n(K)]^p,$$

which implies that, if K belongs to the Carleman-class $\mathscr{C}_p$, then it holds that its fractional powers K^{α} belong to the Carleman-class $\mathscr{C}_{\frac{p}{\alpha}}$ for every $0 < \alpha < 1$.

Corollary 5.2.1 (Macaev and Palant [12]) *Let K be a compact operator acting on a Hilbert space X. We suppose that*

(i) $|arg\langle K\Phi, \Phi\rangle| \le \frac{\pi}{2}$ *for any* $\Phi \in X$*, and*
(ii) $K \in \mathscr{C}_p$.

Then, $\sqrt{K} \in C_{2p}$. ◇

The most complete sharp results for β-subordinate and fully subordinate perturbations can be found in the book of G. Krein [11] and [6]. Among these results, we recall the following:

Lemma 5.2.1 (Moment inequality) *Let G be a closed linear operator on X and assume that for all $s > 0$, there is $M > 0$, such that*

$$\|(G+s)^{-1}\| \le \frac{M}{1+s}. \tag{5.2.10}$$

Then, for $\alpha < \beta < \gamma$ and $\varphi \in \mathscr{D}(G^{\gamma})$, we have

$$\|G^{\beta}\varphi\| \le c(\alpha,\beta,\gamma)\|G^{\gamma}\varphi\|^{\frac{\beta-\alpha}{\gamma-\alpha}}\|G^{\alpha}\varphi\|^{\frac{\gamma-\beta}{\gamma-\alpha}}.$$

In particular, if $\alpha = 0 < \beta < \gamma = 1$, we obtain

$$\|G^{\beta}\varphi\| \le c(\beta)\|G\varphi\|^{\beta}\|\varphi\|^{1-\beta}.$$

Hence, if S is subordinate to G^{β}, then S is β-subordinate to G. ◇

Lemma 5.2.2 (Krein [11]) *Let G and S be two closed operators and suppose that Eq.* (5.2.10) *is satisfied. Then,*

(i) if S is β-subordinate to G, then S is $\frac{\beta}{\gamma}$-subordinate to G^{γ}, for all $0 < \beta < \gamma < 1$, and
(ii) if S is β-subordinate to G, then it is fully subordinate to it. ◇

5.3 Expansions on Generalized Eigenvectors of Operators in Hilbert Space

Let A be a closed linear operator with a dense domain $\mathscr{D}(A)$ in a Hilbert space X. We shall consider a solution u of the operational differential equation

$$u'(t) = -Au(t), \quad t > 0. \tag{5.3.1}$$

5.3.1 *Hypotheses*

Suppose that A satisfies the following conditions:

(A4) there exists $\delta > 0$ such that the resolvent set $\rho(A)$ of A includes the double sector

$$S = \left\{\lambda \in \mathbb{C} \text{ such that } \left|arg\lambda \pm \frac{\pi}{2}\right| \leq \delta\right\}$$

and that, there exists an integer $m \geq -1$ such that

$$\|(A - \xi)^{-1}\| \leq C(1 + |\xi|)^m$$

for every $\xi \in S$.

We assume that the operator A satisfies the following:

(A5) $A^{-1} \in \mathscr{C}_p$.

5.3.2 *Basic Properties*

Since A has a compact resolvent, its spectrum reduces to a sequence of eigenvalues with finite (algebraic) multiplicities which may lay on both sides of S. Since A is invertible, $\rho(A)$ includes a neighborhood of the origin. This allows one to define two curves C^+ and C^- entirely contained in the half planes $Re\xi > 0$ and $Re\xi < 0$, respectively, that coincide with the boundary of S far from the origin and such that all the eigenvalues of A with a positive (respectively, negative) real part should be on the right side of C^+ (respectively, on the left side of C^-) according to Fig. 5.1.

Since the spectrum of A is discrete, it is possible to define a sequence of closed curves C^j $j = 1, 2, \cdots$, located on the left side of C^- and completely contained in $\rho(A)$, and such that all the eigenvalues of $-A$ with a negative real part belonging to the union of the domains bounded by the curves C^j according to Fig. 5.1.

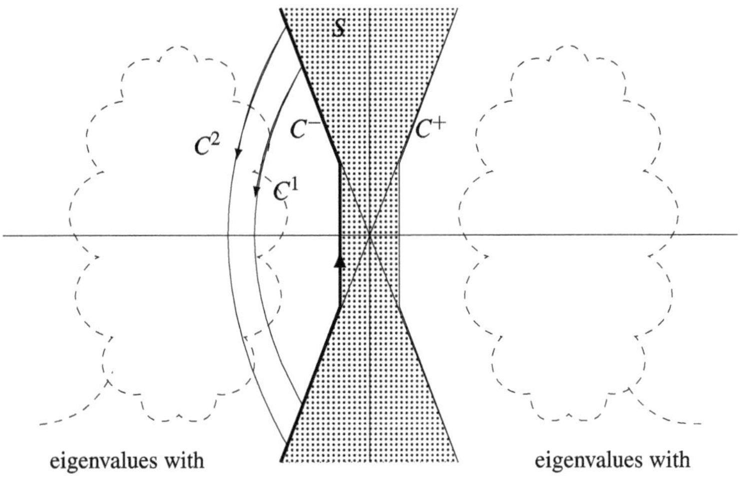

Fig. 5.1 Location of the spectrum for the operator A

Definition 5.3.1 (i) We shall say that u is a strong solution if $u \in L_2(\mathbb{R}_{\geq 0}; \mathscr{D}(A))$, where $\mathscr{D}(A)$ is equipped with the graph norm and if u fulfills (5.3.1) almost everywhere in $\mathbb{R}_{\geq 0}$ (consequently one has $u' \in L_2(\mathbb{R}_{\geq 0}; X)$).

(ii) We shall say that u is a weak solution if $u \in L_2(\mathbb{R}_{\geq 0}; X)$ and if for every $v \in \mathscr{D}(A^*)$, u fulfills

$$\langle u'(t), v\rangle = -\langle u(t), A^* v\rangle \tag{5.3.2}$$

almost everywhere in $\mathbb{R}_{\geq 0}$. ◇

We shall denote by $\mathscr{H}$ the dual space of $\mathscr{D}(A^*)$ equipped with the graph norm. The duality is defined by the scalar product of X and will be denoted in the same manner. Since A^* has dense domain, X is naturally imbedded in $\mathscr{H}$. In addition A may be continued into an invertible operator $\mathscr{A}$ from X onto $\mathscr{H}$ whose adjoint is A^* (viewed as a continuous linear operator from $\mathscr{D}(A^*)$ into X). Finally, for every $\xi \in S$, the operator $\mathscr{A} - \xi$ is invertible with inverse

$$(\mathscr{A} - \xi)^{-1} = A(A - \xi)^{-1}\mathscr{A}^{-1},$$

which is the continuation of $(A - \xi)^{-1}$ to $\mathscr{H}$. Let u be a weak solution of (5.3.1). For every $v \in \mathscr{D}(A^*)$, the function

$$\varphi_v(t) = \langle u(t), v\rangle$$

is square integrable together with its first derivative

$$\varphi_v'(t) = -\langle u(t), A^* v\rangle.$$

It follows that $\varphi_v(0)$ is well defined and that

$$|\varphi_v(0)|^2 \le C \int_0^1 (|\varphi_v(t)|^2 + |\varphi_v'(t)|^2)dt$$
$$\le C'(\|A^*v\|^2 + \|v\|^2).$$

It is, therefore, possible to define $u(0)$ by setting

$$\langle u(0), v\rangle = \varphi_v(0).$$

So, $u(0)$ is well defined as an element of X. One can rewrite (5.3.2) as

$$\langle u'(t), v\rangle = -\langle \mathscr{A}u(t), v\rangle$$

for every $v \in \mathscr{D}(A^*)$. Consequently, one has $u \in L_2(\mathbb{R}_{\ge 0}; X), u' \in L_2(\mathbb{R}_{\ge 0}; \mathscr{H})$ and

$$u'(t) = -\mathscr{A}u(t)$$

in $\mathscr{H}$ almost everywhere for $t > 0$. In particular, u (after possible modification on a null set) is continuous up to 0 with values in $\mathscr{H}$ and $u(0)$ is also such that

$$u(0) = \lim_{t\to 0^+} u(t)$$

in $\mathscr{H}$.

5.3.3 *Representation of the Solutions*

Let u be a weak solution of (5.3.1) and assume that A fulfills $(A4)$. Then, one has

$$u(t) = \frac{1}{2i\pi}\int_{C^-} e^{t\xi}(\xi + \mathscr{A})^{-1}u(0)\mathrm{d}\xi \tag{5.3.3}$$

for every $t > 0$. In fact, let U denotes the continuation of u by 0 for $t < 0$, one has

$$U' + \mathscr{A}U = u(0) \otimes \delta,$$

in the sense of tempered distributions with values in $\mathscr{H}$. By Fourier transform one gets

$$(i\tau + \mathscr{A})\mathscr{F}U = u(0).$$

Hence,

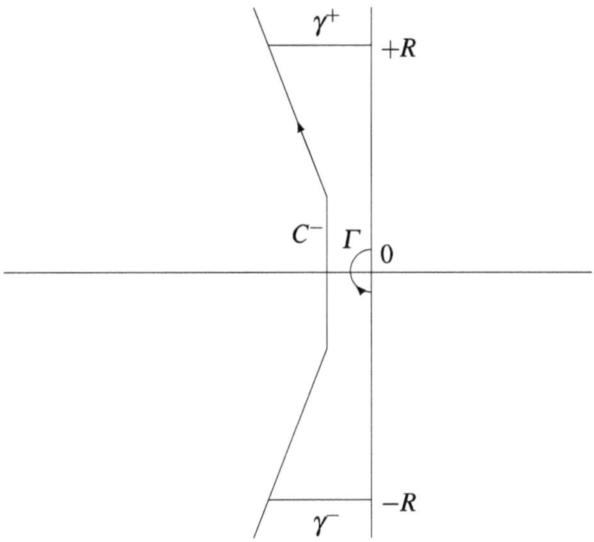

Fig. 5.2 The integration domain Γ

$$U = \mathscr{F}^{-1}(i\tau + \mathscr{A})^{-1}u(0).$$

However, since one knows that $u \in L_2(\mathbb{R}_{\geq 0}; X)$ one also has

$$u(t) = \frac{1}{2\pi}\int\limits_{\mathbb{R}} e^{it\tau}(i\tau + \mathscr{A})^{-1}u(0)d\tau$$

for almost every $t > 0$. The change of variable $\xi = i\tau$ provides the identity

$$u(t) = \frac{1}{2i\pi}\int\limits_{i\mathbb{R}} e^{t\xi}(\xi + \mathscr{A})^{-1}u(0)\mathrm{d}\xi$$

for almost every $t > 0$. One can replace integration on $i\mathbb{R}$ by integration on a curve Γ completely contained in $\rho(A)$, which coincides with $i\mathbb{R}$ away from 0 which is bypassed by the left, according to Fig. 5.2.
Finally, one has

$$\langle u(t), v\rangle = \frac{1}{2i\pi}\int\limits_{\Gamma} e^{t\xi}\langle u(0), (\overline{\xi} + A^*)^{-1}v\rangle \mathrm{d}\xi \tag{5.3.4}$$

for almost every $t > 0$ and all $v \in \mathscr{D}(A^*)$. In order to proceed, we shall assume, in addition, that $v \in \mathscr{D}((A^*)^{k+1})$ with an integer $k \geq 0$ which will be chosen later, and we shall use the elementary identity

$$(A^* + \xi)^{-1} v = (-1)^k \xi^{-k} (A^* + \xi)^{-1} (A^*)^k v + \sum_{1 \le j \le k} (-1)^{j-1} \xi^{-j} (A^*)^{j-1} v. \tag{5.3.5}$$

It follows from (5.3.5) that the function in the integral (5.3.4), that we will denote temporarily by $\varphi(\xi)$, has the growth

$$|\varphi(\xi)| \le O(|\xi|^{-1}),$$

when $k = m + 1$. Deforming the integration path, one concludes immediately that

$$\begin{aligned}
\langle u(t), v \rangle &= \int_{\Gamma} \varphi(\xi) d\xi \\
&= \lim_{R \to \infty} \int_{\Gamma \bigcap \{|Im\xi| \le R\}} \varphi(\xi) d\xi \\
&= \lim_{R \to \infty} \left[\int_{C^- \bigcap \{|Im\xi| \le R\}} \varphi(\xi) d\xi + \int_{\gamma^+} \varphi(\xi) d\xi + \int_{\gamma^-} \varphi(\xi) d\xi \right] \\
&= \lim_{R \to \infty} \left[\int_{C^- \bigcap \{|Im\xi| \le R\}} \varphi(\xi) d\xi \right] \\
&= \int_{C^-} \varphi(\xi) d\xi,
\end{aligned}$$

where $\gamma^{\pm}$ denote the segments

$$\{\xi = \pm i R + x;\; -R\tan(\delta) \le x \le 0\},$$

respectively, according to Fig. 5.2. Actually the integrals on $\gamma^{\pm}$ are estimated as follows:

$$\begin{aligned}
\left| \int_{\gamma^{\pm}} \varphi(\xi) d\xi \right| &= O\left(\int_0^{R\tan(\delta)} \frac{e^{-tx}}{\sqrt{x^2 + R^2}} dx \right) \\
&= O\left(\int_0^{\tan(\delta)} \frac{e^{-tRy}}{\sqrt{1 + y^2}} dy \right).
\end{aligned}$$

They tend to zero as $R \to \infty$. To sum up, we have shown that

$$\langle u(t), v\rangle = \frac{1}{2i\pi}\int\limits_{C^-} e^{t\xi}\langle u(0), (\overline{\xi} + A^*)^{-1}v\rangle \mathrm{d}\xi$$

for every $v \in \mathscr{D}((A^*)^{m+2})$. However, since $\mathscr{D}(A^*)$ is dense in X and A^* is invertible, one sees that $\mathscr{D}((A^*)^{m+2})$ is also dense in $\mathscr{D}(A^*)$. The identity (5.3.3) in $\mathscr{H}$ follows for almost every $t > 0$ since the integrand has exponential decay in X along C^-. We conclude that the two sides of (5.3.3) are continuous with values in $\mathscr{H}$. Consequently, the identity (5.3.3) holds for every $t > 0$ and u is continuous with values in X.

5.3.4 *The Simple Case of an Operator with Nuclear Resolvent*

Theorem 5.3.1 (Geymonat and Grisvard [4]) *We assume that A fulfills (A4) and (A5) for some $p < 1$. Then, one can find a sequence of closed curves C^j such that for every weak solution of (5.3.1), one has*

$$u_j(t) \to u(t),$$

in X uniformly on every half-line $[t_0, \infty[$ with $t_0 > 0$, where u_j is defined by

$$u_j(t) = \frac{1}{2i\pi}\int\limits_{C^j} e^{t\xi}(\xi + \mathscr{A})^{-1}u(0)\mathrm{d}\xi. \qquad \diamond$$

Proof The function $d(\cdot)$ is an entire function is of "exponential type p"; i.e.,

$$d(\xi) = O(e^{|\xi|^{p+\varepsilon}})$$

for every $\varepsilon > 0$. By using the minimum principle (see Theorem 2.3.1), there exists a sequence of radii $r_j \to \infty$ such that for $|\xi| = r_j$ one has

$$|d(\xi)| \geq e^{-|\xi|^{p+\varepsilon}} \tag{5.3.6}$$

for every $\varepsilon > 0$. Moreover, the Carleman-Hadamard inequality for A^{-1} gives

$$\|d(\xi)(I - \xi A^{-1})^{-1}\| \leq e^{K|\xi|^p},$$

for some $K \geq 0$, or equivalent

$$\|d(\xi)A(A - \xi)^{-1}\| \leq e^{K|\xi|^p}.$$

From this, follows

$$\|d(\xi)(\mathscr{A} - \xi)^{-1}\|_{\mathscr{H}\to X} \leq \|\mathscr{A}^{-1}\|_{\mathscr{H}\to X}e^{K|\xi|^p}, \tag{5.3.7}$$

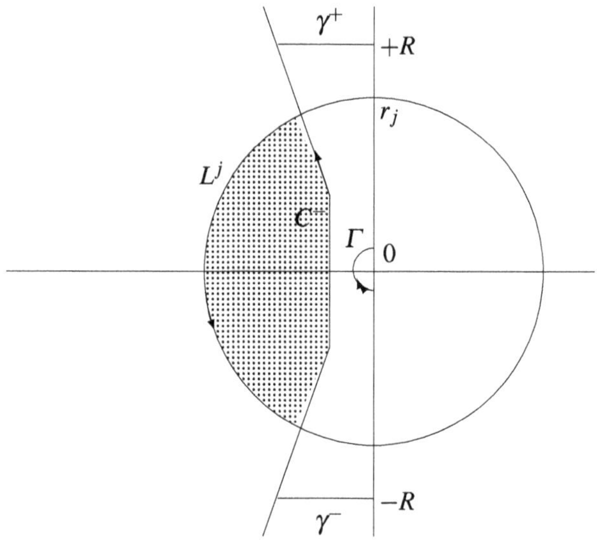

Fig. 5.3 The domains C^j and L^j

where $\|\cdot\|_{\mathscr{H}\to X}$ denotes the norm in the space of continuous linear operators from $\mathscr{H}$ into X. Combining both (5.3.6) and (5.3.7) one gets the following estimate

$$\|(\mathscr{A}-\xi)^{-1}\|_{\mathscr{H}\to X} \leq \|\mathscr{A}^{-1}\|_{\mathscr{H}\to X} e^{K|\xi|^p+|\xi|^{p+\varepsilon}} \tag{5.3.8}$$

on the circles of radius r_j. To conclude, we shall define now the curve C^j as following: it is the curve obtained by joining the part of C^- which is included in the disk of radius r_j with the arc of the circle (denoted L^j) with radius r_j defined by

$$\alpha = \frac{\pi}{2} + \delta \leq arg\xi \leq \frac{3\pi}{2} - \delta = \beta,$$

according to Fig. 5.3, where the domain bounded by C^j has been dotted.
It is clear that to conclude, it will suffice to show that

$$\frac{1}{2i\pi}\int_{L^j} e^{t\xi}(\xi+\mathscr{A})^{-1}u(0)\mathrm{d}\xi \to 0$$

uniformly for $t > t_0$. This is evident since by (5.3.8) one has

$$\|e^{t\xi}(\xi+\mathscr{A})^{-1}u(0)\|_X \leq e^{-t_0 r_j|\cos(\theta)|+Kr_j^p+r_j^{p+\varepsilon}}\|\mathscr{A}^{-1}\|_{\mathscr{H}\to X}\|u(0)\|_{\mathscr{H}}$$

for

$$\frac{\pi}{2} + \delta \leq \theta = arg\xi \leq \frac{3\pi}{2} - \delta.$$

Consequently,

$$\left\| \frac{1}{2i\pi} \int_{L^j} e^{t\xi}(\xi + \mathscr{A})^{-1} u(0) \mathrm{d}\xi \right\|_X \leq$$

$$\frac{1}{2\pi} \int_{\alpha}^{\beta} e^{-t_0 r_j |\cos(\theta)| + K r_j^p + r_j^{p+\varepsilon}} \|\mathscr{A}^{-1}\|_{\mathscr{H} \to X} \|u(0)\|_{\mathscr{H}} r_j d\theta.$$

Since $p < 1$, then this last integral tends to 0 as $j \to \infty$ and one may also choose ε such that $p + \varepsilon < 1$. Q.E.D.

Now, let's explain the function u_j. We will denote by H_k the generalized eigenspace of ξ_k; i.e., the finite-dimensional subspace (with dimension ν_k) generated by the kernels of the powers of $A - \xi_k$. Following A. E. Taylor [15], this is also the image of the projector

$$P_k = \frac{1}{2i\pi} \int_{\gamma^k} (\xi - A)^{-1} \mathrm{d}\xi,$$

where γ^k is a small curve around ξ_k directly oriented and containing no other eigenvalue of A. Obviously, this projector admits a continuation as a continuous linear operator from $\mathscr{H}$ onto H_k, which may be written as

$$\mathscr{P}_k = \frac{1}{2i\pi} \int_{\gamma^k} (\xi - \mathscr{A})^{-1} \mathrm{d}\xi.$$

The restriction A_k of A to H_k has only the eigenvalue ξ_k. It can be written in an appropriate basis as a matrix made up of a finite number of Jordan blocks with only ξ_k on the diagonals. If one denotes by D^j the domain bounded by C^j (i.e., the dotted domain in Fig. 5.3) one sees, with the help of Dunford's operational calculus, that

$$u_j(t) = \sum w_k(t),$$

where the summation is over those ξ_k such that $-\xi_k \in D^j$ and, where

$$w_k(t) = \frac{1}{2i\pi} \int_{\gamma^k} e^{-t\xi} (\xi - \mathscr{A})^{-1} \mathscr{P}_k u(0) \mathrm{d}\xi.$$

The function $w_k(t)$ is solution of

$$w_k \in L_2(\mathbb{R}_{\geq 0}; H_k),$$

$$w_k'(t) = -A_k w_k(t),$$

and

$$w_k(0) = \mathscr{P}_k u(0).$$

It is well known that such a function may be written as an exponential polynomial since it is the solution of a finite-dimensional differential system. One has

$$w_k(t) = e^{-t\xi_k} p_k(t),$$

where p_k is a polynomial of a degree less than or equal to $\nu_k - 1$ with values in H_k. Summing up we have proved.

Theorem 5.3.2 (Geymonat and Grisvard [4]) *Assume that A fulfills $(A4)$ and $(A5)$ for some $p < 1$. Then, one can find a sequence of radii $r_j \to \infty$ such that for every weak solution of (5.3.1), the expansion*

$$u(t) = \sum_j \left\{ \sum_{Re\xi_k>0,\ r_j<|\xi_k|<r_{j+1}} e^{-t\xi_k} p_k(t) \right\} \tag{5.3.9}$$

holds, where p_k is a polynomial of degree less than or equal to $\nu_k - 1$ with values in H_k and, where the series in j converges in X uniformly on every half-line $[t_0; \infty[$ with $t_0 > 0$. ◇

When all the eigenvalues ξ_k of the operator A are semi-simple, meaning

$$H_k = N(A - \xi_k),$$

we can simplify the calculations which give the terms of the expansion (5.3.9). In fact, the restriction of $\mathscr{A}$ to H_k is reduced to the multiplication by ξ_k. Hence,

$$\begin{aligned} w_k(t) &= \frac{1}{2i\pi} \int_{\gamma^k} e^{t\xi} (\xi + \xi_k)^{-1} \mathscr{P}_k u(0) \mathrm{d}\xi \\ &= e^{-t\xi_k} \mathscr{P}_k u(0) \end{aligned}$$

with

$$\mathscr{P}_k u(0) = \sum_l \frac{\langle u(0), \psi_{k,l} \rangle}{\langle \varphi_{k,l}, \psi_{k,l} \rangle} \varphi_{k,l},$$

where the sum is over l such that $1 \le l \le \nu_k$ (the dimension of H_k), $\varphi_{k,l}$ and $\psi_{k,l}$ are, respectively, a basis of $N(A - \xi_k)$ and $N(A^* - \overline{\xi_k})$. Furthermore

$$\langle \varphi_{k,i}, \psi_{k,j} \rangle = 0$$

for $i \neq j$. Since $\psi_{k,l} \in \mathscr{D}(A^*)$, then we can observe that the scalar product $\langle u(0), \psi_{k,l}\rangle$ is meaningful for arbitrary $u(0)$ in $\mathscr{H}$. The expansion (5.3.9) then, reads

$$u(t) = \sum_j \left\{ \sum_{Re\xi_k>0,\ r_j<|\xi_k|<r_{j+1}} e^{-t\xi_k} \sum_l \frac{\langle u(0), \psi_{k,l}\rangle}{\langle \varphi_{k,l}, \psi_{k,l}\rangle} \varphi_{k,l} \right\}.$$

5.3.5 The Limit Case of an Operator with an Almost Nuclear Resolvent

We called that A has an almost nuclear resolvent if $(A5)$ holds for every $p > 1$. It is important to say that the interesting case is when the resolvent of A is actually not nuclear. Since the product

$$\prod_{k=1}^{\infty} \left(1 - \frac{\xi}{\xi_k}\right)^{\nu_k}$$

is no longer always convergent, it is then necessary to go back to the definition of $d(\xi)$. One sets

$$d(\xi) = \prod_{k=1}^{\infty} \left[\left(1 - \frac{\xi}{\xi_k}\right) e^{\frac{\xi}{\xi_k}}\right]^{\nu_k}.$$

This is an entire function of exponential type p for every $p > 1$. Consequently,

$$d(\xi) = O(e^{|\xi|^p}).$$

Furthermore, by using the Carleman-Hadamard's inequality, we have

$$\|d(\xi)(I - \xi A^{-1})^{-1}\| \leq e^{K|\xi|^p},$$

from which Eq. (5.3.7) follows again for every $p > 1$. We will declare a new version of Theorem 5.3.1 in which we shall add an assumption on the exponential growth (in the strict sense) of $d(\xi)$ (Fig. 5.4).

One assumes that A fulfills $(A4)$ and $(A5)$ for every $p > 1$. In addition, one assumes that there exists c such that

$$d(\xi) = O(e^{c|\xi|}) \tag{5.3.10}$$

and that there exist a sequence of real numbers $r_j \to \infty$ and numbers $a > 0$ and $b \geq 0$ such that

$$|d(\xi)| \geq ae^{-br_j} \tag{5.3.11}$$

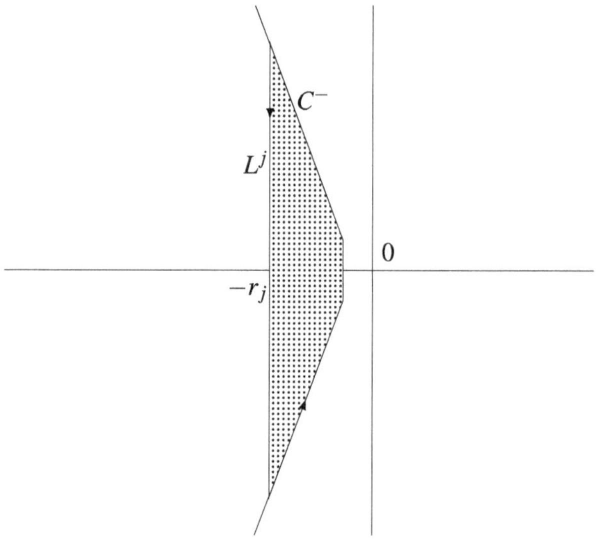

Fig. 5.4 Resolvent set of the almost nuclear operator

for $Re\xi = -r_j$ (see Fig. 5.4). We define the curve C^j as follows: it is the closed curve obtained by joining the part of C^- included in the half-plane defined by $Re\xi \geq -r_j$ and the line segment (denoted L^j) whose abcissa is $-r_j$ defined by

$$|Im\xi| \leq \frac{r_j}{\tan\delta}.$$

First, let us clarify the behavior of

$$\|d(\xi)(\mathscr{A} + \xi)^{-1}\|_{\mathscr{H}\to X}$$

on the left of C^-. On the one hand, Eq. (5.3.7) implies that this function has exponential growth of type p for every $p > 1$ on the left of C^-. On the other hand, it follows from both $(A4)$ and (5.3.10) that it grows exponentially in the usual sense on C^-. Since the sector on the left of C^- is strictly convex, then this allows one to apply Phragmén-Lindelöf theorem. Hence, the existence of c' such that

$$\|d(\xi)(\mathscr{A} + \xi)^{-1}\|_{\mathscr{H}\to X} = O(e^{c'|\xi|}) \tag{5.3.12}$$

on the left of C^- follows. It is enough to show that

$$\frac{1}{2i\pi}\int_{L^j} e^{t\xi}(\xi + \mathscr{A})^{-1}u(0)\mathrm{d}\xi \to 0$$

uniformly for $t > t_0$. By using both (5.3.10) and (5.3.12) the integrand admits on L^j the following estimate

$$\begin{aligned} e^{t\xi}(\xi + \mathscr{A})^{-1}u(0)\| &= O(e^{-t_0 r_j + c'|\xi| + b r_j}) \\ &= O(e^{(-t_0 + b + \frac{c'}{\sin\delta}) r_j}). \end{aligned}$$

We conclude by assuming

$$t_0 > b + \frac{c'}{\sin\delta}.$$

So, there exists $t_0 > 0$ such that for every weak solution of (5.3.1), one has

$$u_j(t) \to u(t)$$

in X uniformly in $[t_0, \infty[$.

5.4 Notes and Remarks

- The condition $(A1)$ is a condition on the action of the operator A on each root subspace. If we take $t = 0$, then the condition $(A1)$ is just

$$\|E(\lambda_n, A)\| \leq M_1 e^{-\rho_1 Re\lambda_n}.$$

So, the condition $(A1)$ includes the case that

$$\sup_n \|E(\lambda_n, A)\| = \infty.$$

It also requires that the norm $\|E(\lambda_n, A)\|$ grows not faster than $e^{-\rho_1 Re\lambda_n}$ as $Re\lambda_n \to -\infty$.
- The condition $(A2)$ is a requirement on the spectral distribution of the operator A.
- The condition $(A3)$ is a requirement on the spectral distribution of the operator A. The condition $(A3)$ is also a spectral condition for the differentiable semigroup. It is equivalent to the condition

$$|\lambda_n| \leq M_2 e^{-\rho_2 Re\lambda_n},$$

for all $n \in \mathbb{N}$.
- Since the spectrum of A is not, in general, included in one and only one of the half planes $Re\xi > 0$ or $Re\xi < 0$, then it is worth observing that neither the operator A nor $-A$ is, in general, the infinitesimal generator of a distribution semigroup. The assumption $(A4)$ is actually an ellipticity assumption since one assumes the invert-

ibility of the "symbol" $(i\tau + A)^{-1}$ of the differential operator $D_t + A$ obtained by Fourier transform.

- The fractional powers of the operator A^α, for $0 \le \alpha \le 1$, can be defined in a natural way if both the operators A and A^α are closed and maximal accretive. Such fractional powers have been defined, in Banach spaces, for a more general class of linear operators by the authors Balakrishnan [1, 2], Glushko and Krein [5], Jeribi [6], Kato [7], Krasnosel'skii and Pustylnik [9], Krasnosel'skii and Sobolevskii [10], Sobolevskii [13], Solomiak [14], and Yosida [17].

(i) The simplest example of an operator, in Hilbert space for which condition (5.2.10) is satisfied, is a definite positive selfadjoint operator.

(ii) When u is a strong solution, we have shown that

$$u(t) = \frac{1}{2i\pi} \int_{C^-} e^{t\xi} (\xi + A)^{-1} u(0) \mathrm{d}\xi,$$

since in that case, we have $u(0) \in X$.

(iii) Under the same assumptions one has the expansion similar to (5.3.9):

$$u(t) = \sum_j \left\{ \sum_{r_j < Re\xi_k < r_{j+1}} e^{-t\xi_k} p_k(t) \right\}, \tag{5.4.1}$$

where p_k is a polynomial whose degree is at most $\nu_k - 1$ with values in H_k and, where the series in j converges in X uniformly for $t \ge t_0$.

References

1. A.V. Balakrishnan, An operator calculus for infinitesimal generators of semigroups. Trans. Am. Math. Soc. **91**, 330–353 (1959)
2. A.V. Balakrishnan, Fractional powers of closed operators and the semigroups generated by them. Pac. J. Math. **10**, 419–437 (1960)
3. R.F. Curtain, H.J. Zwart, *An Introduction To Infinite Dimentional Linear Systems Theory* (Springer, New-York, 1995)
4. G. Geymonat, P. Grisvard, Expansions on generalized eigenvectors of operators arising in the theory of elasticity. Differ. Inte. Equ. **4**(3), 459–481 (1991)
5. V.P. Glushko, S.G. Krein, Fractional powers of differential operators and the imbedding theorems. Doklady Akad. Nauk **122**, 963–966 (Russian) (1958)
6. A. Jeribi, *Denseness, Bases and Frames in Banach Spaces and Applications* (de Gruyter, Berlin, 2018)
7. T. Kato, Note on fractional powers of linear operators. Proc. Jpn. Acad. **36**, 94–96 (1960)
8. T. Kato, *Perturbation Theory for Linear Operators* (Springer, Berlin, 1980)
9. M.A. Krasnosel'skii, E.I. Pustylnik, The use of fractional powers of operators in studying Fourier series of eigenfunctions of differential operators. Doklady Akad. Nauk. **122**, 978–981 (Russian) (1958)

10. M.A. Krasnosel'skii, P.E. Sobolevskii, Fractional power of operators defined on Banach spaces. Doklady Akad. Nauk **129**, 499–502 (Russian) (1959)
11. S.G. Krein, Linear differential equations in Banach space. Am. Math. Soc. (1971)
12. V.I. Macaev, J.A. Palant, On the powers of a bounded dissipative operator, (Russian) Ukrain. Mat. Z. **14**, 329–337 (1962)
13. P. E. Sobolevskii, On equations with operators forming an acute angle. Doklady Akad. Nauk, **116**, 754–757 (Russian) (1957)
14. M.Z. Solomiak, The application of the semigroup theory to the study of differential equations in Banach spaces. Doklady Akad. Nauk **122**, 766–769 (Russian) (1958)
15. A.E. Taylor, *Introduction to Functional Analysis* (Wiley, New-York, 1958)
16. G.Q. Xu, S.P. Yung, Properties of a class of $\mathscr{C}_0$-semigroups on a Banach spaces and their applications. J. Math. Anal. Appl. **328**, 245–256 (2007)
17. K. Yosida, Fractional powers of infinitesimal generators and the analyticity of the semigroups generated by them. Proc. Jpn. Acad. **36**, 86–89 (1960)

Chapter 6
Completeness Criteria of the Space of Generalized Eigenvectors of Non-Self-Adjoint Operators

In this chapter, we recall some Keldysh results that are not only specific for Hilbert spaces, but they have been formulated for operators in Banach spaces. We also discuss some theorems on denseness of the generalized eigenvectors of a compact operator or an operator with compact resolvent and different conditions assuring the completeness of the system of root subspaces. In the selection of the material, we present as fully as the various existing results. Indeed, we give some completeness criteria of the system of generalized eigenvectors of a compact operator or an operator with compact resolvent on a Hilbert space. The following example shows that this completeness is not verified in general. In fact, let A be the operator defined by

$$\begin{cases} A : \mathscr{D}(A) \subset H^1(0,\pi) \longrightarrow L_2(0,\pi) \\ \qquad \varphi \longrightarrow A\varphi = \dfrac{d\varphi}{dx} \\ \mathscr{D}(A) = \{\varphi \in H^1(0,\pi) \text{ such that } \varphi(0) = 0\} \end{cases} \tag{6.0.1}$$

with an inverse K, given by

$$K\varphi(x) = \int_0^x \varphi(t)\mathrm{d}t. \tag{6.0.2}$$

It is easy to see that

$$\sigma(K) = \sigma_c(K) = \{0\}, \quad \sigma(A) = \emptyset$$

and

$$s_n(K) = \frac{1}{n - \frac{1}{2}}. \tag{6.0.3}$$

A. Jeribi, *Perturbation Theory for Linear Operators*,
https://doi.org/10.1007/978-981-16-2528-2_6

This excludes the existence of eigenvectors and, a fortiori, the existence of a total system for such vectors.

6.1 Keldysh Results

6.1.1 In Hilbert Space

In order to prove Keldysh theorem, we need the following two theorems:

Theorem 6.1.1 (Gohberg and Krein [17, Lemma 7.1, p. 252]) *Let K be a compact operator on a Hilbert space X and let H be a normal and injective compact operator on X such that all its eigenvalues, except perhaps a finite number, are outside the sector*

$$\{z \in \mathbb{C} \text{ such that } \theta_1 < arg z < \theta_2\}, \ (0 \leq \theta_1 < \theta_2 \leq 2\pi).$$

Then, for all $\varepsilon > 0$, we have

$$\lim_{|\xi| \to \infty} \|K(I - \xi H)^{-1}\| = 0$$

uniformly on the sector

$$\{z \text{ such that } \theta_1 + \varepsilon \leq arg z \leq \theta_2 - \varepsilon\};$$

i.e., for all $\varepsilon > 0$ and for all $\delta > 0$, there is $r > 0$ such that for $|\xi| \geq r$, we have

$$\|K(I - \xi H)^{-1}\| < \delta$$

for all ξ satisfying

$$\theta_1 + \varepsilon \leq arg \xi \leq \theta_2 - \varepsilon. \quad \diamond$$

Definition 6.1.1 A sequence of vectors $\{f_1, f_2, f_3, \ldots\}$ in a Hilbert space X is said to be complete if the zero vector alone is perpendicular to every f_n. ◇

We will recall the Liouville theorem.

Theorem 6.1.2 *Let $f(z)$ be an analytic function on the complex plane. If $f(z)$ is bounded, then $f(z)$ is constant.* ◇

The objective of this part is to obtain the necessary conditions on an operator in order to ensure that the density of the space generated by the generalized eigenvectors. For this, let A be a closed and densely defined linear operator on a Hilbert space X. We suppose that the operator A is self-adjoint and belongs to Carleman-class $\mathscr{C}_p$, $p > 0$. Let $B = A(I + K)$, where K is a compact operator on X and suppose that the operator B vanishes only at zero. Then, the operator $I + K$ vanishes only at zero, and consequently is invertible since K is compact. Since B is one-to-one, then

$$A = B(I + K)^{-1}$$

is one-to-one. Moreover, the operator

$$\begin{aligned} T &:= I - (I + K)^{-1} \\ &= (I + K - I)(I + K)^{-1} \\ &= K(I + K)^{-1} \end{aligned}$$

is compact. By applying Theorem 6.1.1 to both the self-adjoint operator A and to the compact operator T, we get, for all $\varepsilon > 0$, there is $r > 0$ such that in the region

$$F_\varepsilon = \{z \in \mathbb{C} \text{ such that } \varepsilon \leq |argz| \leq \pi - \varepsilon\},$$

for $|\xi| \geq r$, the operator $T(I - \xi A)^{-1}$ satisfies the following estimate

$$\|T(I - \xi A)^{-1}\| < 1.$$

Moreover, we have

$$\begin{aligned} I - \xi B &= I - \xi A(I + K) \\ &= [(I + K)^{-1} - \xi A](I + K) \\ &= [(-I + (I + K)^{-1}) + I - \xi A](I + K) \\ &= (I - T - \xi A)(I + K) \\ &= (I - T(I - \xi A)^{-1})(I - \xi A)(I + K). \end{aligned}$$

Therefore, on the part

$$F_{\varepsilon,r} := F_\varepsilon \bigcap \{\xi \text{ such that } |\xi| \geq r\},$$

the operator $I - T(I - \xi A)^{-1}$ is invertible with inverse

$$\left(I - T(I - \xi A)^{-1}\right)^{-1} = \sum_{n=0}^{\infty} \left(I - T(I - \xi A)^{-1}\right)^n.$$

However, we have

$$\|(I - \xi A)^{-1}\| \leq \frac{1}{\sin \varepsilon}.$$

Hence,

$$\|(I - \xi B)^{-1}\| \leq \frac{\|(I + K)^{-1}\|}{(1 - q)\sin \varepsilon}. \tag{6.1.1}$$

Let $\mathscr{F}$ be the adherence of the subspace spanned by the generalized eigenvectors of the operator A. In what follows, we will show the equalities

$$\mathscr{F} = X.$$

If this is not the case, i.e., $\mathscr{F} \neq X$, then by using Proposition 2.1.2, we have

$$B_1 := QBQ$$

is a Volterra operator, where $Q = I - P$ and P denotes the orthogonal projection which projects X on $\mathscr{F}$. By applying Proposition 2.1.3 to both the bounded operator $\widetilde{B} = I - \xi B$ and to the orthogonal projection P, we deduce that $Q\widetilde{B}Q + P$ is invertible and

$$(Q\widetilde{B}Q + P)^{-1} = Q\widetilde{B}^{-1}Q + P.$$

Moreover,

$$\begin{aligned} Q\widetilde{B}Q + P &= Q(I - \xi B)Q + P \\ &= Q(Q - \xi BQ) + P \\ &= Q - \xi QBQ + P \\ &= I - \xi B_1. \end{aligned}$$

Hence,

$$(I - \xi B_1)^{-1} = Q(I - \xi B)^{-1}Q + P, \ (\xi \in F_\varepsilon).$$

Thus, the inequality (6.1.1) implies that $(I - \xi B_1)^{-1}$ is bounded on F_ε and $B \in \mathscr{C}_p(X)$. So,

$$B_1 = QBQ \in \mathscr{C}_p(X).$$

Hence,

$$s_n(B_1) = o(n^{-\frac{1}{p}}).$$

Since B is a Volterra operator, then according to Theorem 4.2.2, we have

$$\log \|(I - \xi B_1)^{-1}\| = o(|\xi|^{\frac{1}{p}}) \quad (|\xi| \to \infty). \tag{6.1.2}$$

So, for $\varepsilon > 0$, there is p_0 such that, for all $p \geq p_0$, we have

$$\frac{\log \|(I - \xi B_1)^{-1}\|}{|\xi|^{\frac{1}{p}}} < \varepsilon.$$

Hence,

$$\log \|(I - \xi B_1)^{-1}\| < \varepsilon |\xi|^{\frac{1}{p}},$$

which implies that

$$\|(I - \xi B_1)^{-1}\| < e^{\varepsilon|\xi|^{\frac{1}{p}}}.$$

We choose $\varepsilon < \frac{\pi}{p}$. Since the operator function is bounded on both the straight areas

$$|arg\xi| \leq \varepsilon$$

and

$$|\pi - arg\xi| \leq \varepsilon.$$

Hence, by taking into account equality (6.1.2), Theorem 2.4.2 implies that $(I - \xi B_1)^{-1}$ is bounded inside these sectors. Thus, the operator function $(I - \xi B_1)^{-1}$ is bounded on the whole complex plane. Hence, according to Liouville's theorem (see Theorem 6.1.2), we have $(I - \xi B_1)^{-1}$ is constant. So,

$$(I - \xi B_1)^{-1} = I.$$

Consequently,

$$B_1 = QBQ = 0.$$

Now,

$$\begin{aligned} QBQ &= (I - P)B(I - P) \\ &= (I - P)B - (I - P)BP \\ &= B - PB - BP + PBP \\ &= B - PB \\ &= (I - P)B \\ &= QB. \end{aligned}$$

Thus,

$$B^*Q = 0.$$

Hence,

$$B^*Q(X) = 0,$$

which contradicts the fact that $B^* = (I + K^*)A$ is one-to-one. So, we have the following theorem.

Theorem 6.1.3 (Gohberg and Krein [17, Theorem 8.1, p. 257]) *Let A be a closed and densely defined linear operator on a Hilbert space X. We suppose that the operator A is self-adjoint and belongs to Carleman-class $\mathscr{C}_p$, $p > 0$. Let $B = A(I + K)$, where K is a compact operator on X. If the operator B vanishes only at zero, then the system of its root vectors is complete on X.* ◇

Corollary 6.1.1 *If A and K satisfy the conditions of Theorem 6.1.3, then the system of the root vectors of $B = (I + K)A$ is complete in X.* ◇

Proof In fact, $N(I + K) = \{0\}$. Therefore, and according to Fredholm theory, $I + K$ is an invertible operator. Now, the result follows from Theorem 6.1.3. Q.E.D.

Some applications of Theorem 6.1.3 are given in [4–16, 18, 20]
Let us consider a Hilbert space X. Markus's theorem is formulated as

Theorem 6.1.4 (Markus [22, Theorem 4.3]) *Let A be a normal operator whose resolvent belongs to the Carleman-class $\mathscr{C}_p$ ($p > 0$), and whose spectrum lies on a finite number of rays* $\arg \lambda = \alpha_k$ $(k = 1, \ldots, n)$. *If B is A-compact, then the operator*

$$G = A + B$$

has compact resolvent and the system of its root vectors is complete in X. ◇

6.1.2 In Banach Space

Now, let us recall some definitions and preliminary results.

Definition 6.1.2 Let A be a linear compact operator on a Banach space X. The nth approximation number $a_n(A)$ of A is defined as

$$a_n(A) := \inf \{\|A - T\| \text{ such that } rk(T) \leq n\},$$

where $rk(T)$ designates the rank of T. ◇

Definition 6.1.3 Let A be a linear compact operator on a Banach space X. A is said to belong to the class $\mathscr{C}_p(X)$, $0 < p < \infty$ if the series

$$\sum_{n=1}^{\infty} (a_n(A))^p$$

converges, where $a_n(A)$, with $n = 1, 2, \ldots$, are the approximation numbers of A.◇

Definition 6.1.4 Let X be a Banach space and A be a linear operator on X. A is called quasiconservative if its spectrum is purely imaginary and there exists a constant $M \geq 0$ such that

$$\|(\lambda - A)^{-1}\| \leq \frac{M}{|Re\lambda|},$$

$Re\lambda \neq 0$, for all $\lambda \in \rho(A)$. ◇

Definition 6.1.5 Let X be a Banach space and A be a linear operator on X. A is said to have a quasiconservative resolvent if the resolvent set of A, $\rho(A)$, is not empty and there exists $\lambda_0 \in \rho(A)$ such that $(\lambda_0 - A)^{-1}$ is quasiconservative. ◇

In a generalization of Keldysh's theorem to operators on a Banach space self-adjointness has a suitable replacement in the property of being quasiconservative.

Theorem 6.1.5 (Lyubich [21, Theorem, p. 219]) *Let X be a Banach space and A be a quasiconservative linear operator on X. We suppose that the operator A belongs* *to the class $\mathscr{C}_p(X)$, $0 < p < \infty$. Let $B = A(I + K)$, where K is a compact operator in X. If the operator B vanishes only at zero, then the system of its root vectors is complete in X.* ◇

6.2 Denseness of the Generalized Eigenvectors of a Compact Operator or an Operator with Compact Resolvent

The results of this section, given by M.T. Aimar, A. Intissar, J.M. Paoli, may be found in [1, 3, 23].

6.2.1 Subspace Attached to an Operator with Compact Resolvent

Let X be a Hilbert space and A be a closed linear operator with compact resolvent on X. Using Proposition 2.2.2, we have

$$\sigma(A) = \sigma_p(A).$$

Let $\sigma(A) = \{\lambda_k,\ k \in \mathbb{N}\}$ (with $|\lambda_{k+1}| \geq |\lambda_k|$).Thus,

$$\sigma(A^*) = \{\overline{\lambda_k},\ k \in \mathbb{N}\}.$$

Consider $N_k := N[(\lambda_k - A)^{n_0(\lambda_k)}]$ and $F_k := R[(\lambda_k - A)^{n_0(\lambda_k)}]$, where $n_0(\lambda_k)$ is the index of the eigenvalue λ_k of A. Then, we get

$$N_k^* = N[(\overline{\lambda_k} - A^*)^{n_0(\lambda_k)}]$$

and

$$F_k^* = R[(\overline{\lambda_k} - A^*)^{n_0(\lambda_k)}].$$

Hence, $N_k = F_k^{*\perp}$ and $F_k = N_k^{*\perp}$ (see Proposition 2.2.3). Consider the following sets

$$M_\infty := \bigcap_{k=0}^{\infty} F_k,$$

$$P := \oplus_{k=0}^{\infty} N_k,$$

$$M_\infty^* := \bigcap_{k=0}^{\infty} F_k^*$$

and

$$P^* := \oplus_{k=0}^{\infty} N_k^*.$$

P is the vector subspace spanned by the generalized eigenvectors of A. It is never closed (in infinite dimensional). M_∞ is closed as a closed intersection. Now, let us establish some stability properties of F_k and N_k by A or by its resolvent.

Lemma 6.2.1 *Let $\lambda \in \mathbb{C}$ and $k \in \mathbb{N}$. Then,*

(i) $(\lambda - A)(N_k) \subset N_k$,
(ii) $(\lambda - A)(\mathcal{D}(A) \bigcap F_k) \subset F_k$, *and*
(iii) furthermore, if $\lambda \in \rho(A)$, then
$$(\lambda - A)^{-1}(F_k) \subset \mathcal{D}(A) \bigcap F_k. \qquad \diamond$$

Proof (i), (ii), and (iii) follow from the fact that for all $\lambda \in \mathbb{C}$, $\lambda - A$ commutes with $(\lambda_k - A)^{n_0(\lambda_k)}$ and, for all $\lambda \in \rho(A)$, $(\lambda - A)^{-1}$ commutes with $(\lambda_k - A)^{n_0(\lambda_k)}$, for all $\lambda_k \in \sigma(A)$. Q.E.D.

Let $\lambda_k \in \sigma(A)$. By using Lemma 6.2.1 (ii), we infer that

$$(\lambda_k - A)\left(\mathcal{D}(A) \bigcap F_k\right) \subset F_k.$$

Since

$$F_k \bigcap N_k = \{0\},$$

it is easy to see that

$$(\lambda_k - A)_{|\mathcal{D}(A) \bigcap F_k}$$

is one-to-one. So, we have the following:

Lemma 6.2.2 *Let $\lambda_k \in \sigma(A)$. Then, the operator $\lambda_k - A$ maps $\mathcal{D}(A) \bigcap F_k$ into F_k is one-to-one.* $\diamond$

Let $y \in F_k$. Since $F_k \subset R(\lambda_k - A)$, then there is $x \in \mathcal{D}(A)$ such that

$$(\lambda_k - A)x = y.$$

Moreover,

$$X = N_k \oplus F_k$$

and so,

$$x = x_1 + x_2$$

with $x_1 \in N_k$ and $x_2 \in F_k$. By using (i) and (ii) of Lemma 6.2.1, we deduce that $(\lambda_k - A)x_1 \in N_k$ and $(\lambda_k - A)x_2 \in F_k$. Hence,

$$y = (\lambda_k - A)x_1 + (\lambda_k - A)x_2 \in F_k.$$

Thus,

$$y = (\lambda_k - A)x_2.$$

Furthermore, for all $y \in F_k$, there is $x \in \mathscr{D}(A) \bigcap F_k$ such that

$$(\lambda_k - A)x = y$$

and so, $(\lambda_k - A)_{|\mathscr{D}(A) \bigcap F_k}$ is onto. This proves the following:

Lemma 6.2.3 *Let $\lambda_k \in \sigma(A)$. Then, the operator $\lambda_k - A$ maps $\mathscr{D}(A) \bigcap F_k$ into F_k is onto.* ◇

Lemma 6.2.4 *Let λ_k and $\lambda_{k'}$ in $\sigma(A)$ with $k \neq k'$. Then, the operator $\lambda_k - A$ maps $\mathscr{D}(A) \bigcap F_k \bigcap F_{k'}$ is one-to-one onto $F_k \bigcap F_{k'}$.* ◇

Proof We notice that

$$(\lambda_k - A)\left(\mathscr{D}(A) \bigcap F_k \bigcap F_{k'}\right) \subset F_k \bigcap F_{k'}$$

and

$$(\lambda_k - A)_{|\mathscr{D}(A) \bigcap F_k \bigcap F_{k'}}$$

is one-to-one. Let $y \in F_k \bigcap F_{k'}$. By using both Lemmas 6.2.2 and 6.2.3, we infer that there exists a unique $x \in \mathscr{D}(A) \bigcap F_k$ such that

$$(\lambda_k - A)x = y.$$

Now, it suffices to prove that $x \in F_{k'}$. For this purpose, let us write $x = x_1 + x_2$, where $x_1 \in N_{k'}$ and $x_2 \in F_{k'}$. So,

$$y = (\lambda_k - A)x_1 + (\lambda_k - A)x_2$$

with $(\lambda_k - A)x_1 \in N_{k'}$ and $(\lambda_k - A)x_2 \in F_{k'}$. Hence, $y \in F_{k'}$ and

$$y = (\lambda_k - A)x_2.$$

The one-to-one of $(\lambda_k - A)_{|\mathscr{D}(A) \bigcap F_k \bigcap F_{k'}}$ implies that

$$x = x_2 \in F_{k'}.$$

This completes the proof. Q.E.D.

Theorem 6.2.1 *Let $\lambda \in \mathbb{C}$. The restriction*

$$(\lambda - A)_{|\mathscr{D}(A) \bigcap M_\infty} : \mathscr{D}(A) \bigcap M_\infty \longrightarrow M_\infty$$

is one-to-one onto M_∞. ◇

Proof Let $\lambda \in \mathbb{C}$ and let $k \in \mathbb{N}$. By using Lemma 6.2.1(ii), we get

$$(\lambda - A)\left(\mathscr{D}(A)\bigcap F_k\right) \subset F_k.$$

Hence,

$$(\lambda - A)\left(\mathscr{D}(A)\bigcap M_\infty\right) \subset M_\infty.$$

So, M_∞ is invariant under all $\lambda - A$. In order to prove that

$$(\lambda - A)_{|\mathscr{D}(A)\bigcap M_\infty} : \mathscr{D}(A)\bigcap M_\infty \longrightarrow M_\infty$$

is one-to-one, we discuss two cases.

<u>1st *case*</u> If $\lambda \in \rho(A)$, then the restriction

$$(\lambda - A)_{|\mathscr{D}(A)\bigcap M_\infty} : \mathscr{D}(A)\bigcap M_\infty \longrightarrow M_\infty$$

is one-to-one.

<u>2nd *case*</u> If $\lambda = \lambda_k \in \sigma(A)$,

$$(\lambda - A)_{|\mathscr{D}(A)\bigcap M_\infty}$$

is a restriction of

$$(\lambda_k - A)_{|\mathscr{D}(A)\bigcap F_k} : \mathscr{D}(A)\bigcap F_k \longrightarrow F_k$$

and therefore, it is a restriction of one-to-one mapping, so is one-to-one.
We claim that

$$(\lambda - A)_{|\mathscr{D}(A)\bigcap M_\infty}$$

is onto. In fact, let $y \in M_\infty$. We will discuss two cases. First, if $\lambda \in \rho(A)$, then we have

$$y = (\lambda - A)(\lambda - A)^{-1}y.$$

By using Lemma 6.2.1(iii), we have for all $k \in \mathbb{N}$,

$$(\lambda - A)^{-1}y \in \mathscr{D}(A)\bigcap F_k.$$

Hence, $(\lambda - A)^{-1}y \in \mathscr{D}(A)\bigcap M_\infty$ and $(\lambda_k - A)_{|\mathscr{D}(A)\bigcap M_\infty}$ is onto. Second, if $\lambda = \lambda_k \in \sigma(A)$, then by using both Lemmas 6.2.2 and 6.2.3, there is a unique $x \in \mathscr{D}(A)\bigcap F_k$ such that

$$y = (\lambda_k - A)x.$$

Let $k' \neq k$, and by using Lemma 6.2.4, there is a unique $x' \in \mathscr{D}(A)\bigcap F_k \bigcap F_{k'}$ such that

$$y = (\lambda_k - A)x',$$

and the injectivity, gives

$$x = x'.$$

So, for all $k' \in \mathbb{N}$, $x \in F_{k'}$. Hence, $x \in \mathscr{D}(A) \bigcap M_\infty$. This proves the claim and completes the proof of the theorem. Q.E.D.

According to Theorem 6.2.1, we have $\sigma(A_{|M_\infty}) = \emptyset$. This implies that for all $\lambda \in \rho(A)$, $(\lambda - A_{|M_\infty})^{-1}$ is a quasi-nilpotent operator, i.e.,

$$\sigma((\lambda - A_{|M_\infty})^{-1}) = \{0\}. \tag{6.2.1}$$

Corollary 6.2.1 *$M_\infty = \{0\}$ or M_∞ has an infinite dimensional.* ◇

Proof Suppose that $M_\infty \neq \{0\}$. If M_∞ has a finite dimension, then by using Theorem 6.2.1, M_∞ is one-to-one onto $\mathscr{D}(A) \bigcap M_\infty$. Hence,

$$\mathscr{D}(A) \bigcap M_\infty = M_\infty$$

and so, $M_\infty \subset \mathscr{D}(A)$. By using the closed graph theorem, the operator $A_{|M_\infty}$ is bounded and

$$\sigma(A_{|M_\infty}) = \emptyset.$$

These two statements are contradictory according to Liouville's theorem (see Theorem 6.1.2). Q.E.D.

Lemma 6.2.5 *Let $A_{|M_\infty} : \mathscr{D}(A) \bigcap M_\infty \longrightarrow M_\infty$. Then, for all $\lambda \in \rho(A)$,*

$$(\lambda - A_{|M_\infty})^{-1} = (\lambda - A)^{-1}_{|M_\infty}.$$

In particular, $A_{|M_\infty}$ has a compact resolvent. ◇

Proof Let $\lambda \in \rho(A)$. By using Theorem 6.2.1, we deduce that

$$(\lambda - A)^{-1}(M_\infty) = \mathscr{D}(A) \bigcap M_\infty \subset M_\infty$$

and

$$(\lambda - A)\left(\mathscr{D}(A) \bigcap M_\infty\right) \subset M_\infty.$$

By restriction, we have

$$(\lambda - A)_{|M_\infty}(\lambda - A)^{-1}_{|M_\infty} = id_{M_\infty}$$

and

$$(\lambda - A)^{-1}_{|M_\infty}(\lambda - A)_{|M_\infty} = id_{\mathscr{D}(A)\bigcap M_\infty}.$$

Since the restriction of a compact operator on a Hilbert space is compact, then $A_{|M_\infty}$ has a compact resolvent. Q.E.D.

Proposition 6.2.1 *The Hilbert space X has the orthogonal direct sum decompositions*

$$X = M_\infty \oplus \overline{P^*} = M^*_\infty \oplus \overline{P}. \qquad \diamond$$

Proof From both $F^*_k = N_k^\perp$ and $F_k = N_k^{*\perp}$, we may deduce that

$$\overline{P^\perp} = P^\perp = \bigcap_{k=0}^{\infty} F^*_k = M^*_\infty.$$

Hence,

$$X = \bigcap_{k=0}^{\infty} F^*_k \oplus \overline{P}.$$

The statements

$$X = M_\infty \oplus \overline{P^*}$$

can be proved similarly. Q.E.D.

Now, we may give a useful property of M_∞ which characterizes it in terms of an extension of the meromorphic function

$$\begin{cases} \rho(A) \longrightarrow X \\ \quad \lambda \longrightarrow (\lambda - A)^{-1}\varphi, \end{cases}$$

where $\varphi \in X$. Let $\lambda_0 \in \mathbb{C}$. We will discuss two cases. First, if $\lambda_0 \in \rho(A)$, then the function

$$\lambda \longrightarrow (\lambda - A)^{-1}\varphi$$

is defined and analytic in a neighborhood of λ_0, for all φ in X. Second, if $\lambda_0 \in \sigma(A)$, then there is $\varepsilon > 0$ such that, for $\lambda \in \mathbb{D}(\lambda_0, \varepsilon)\backslash\{\lambda_0\}$, we have

$$(\lambda - A)^{-1}\varphi = \sum_{n=0}^{+\infty}(\lambda - \lambda_0)^n A_n\varphi + \sum_{n=1}^{n_0(\lambda)}(\lambda - \lambda_0)^n A_{-n}\varphi,$$

where $\mathbb{D}(\lambda_0, \varepsilon)$ is the open disc with center λ_0 and radius ε in $\mathbb{C}$ and, the

$$(A_n)_{-n_0(\lambda)\le n<\infty}$$

are given in Theorem 4.2.1. Let $\varphi \in M_\infty$, then

$$\varphi \in R[(\lambda_0 - A)^{n_0(\lambda)}] = N(A_{-1}).$$

Thus,

$$A_{-1}\varphi = 0.$$

By using the fact that

$$A_{n-1} = (A - \lambda_0)^{-n} A_{-1}$$

for $n \leq -1$, we have $A_n\varphi = 0$ for all $n \leq -1$. So, for $\varphi \in R[(\lambda_0 - A)^{n_0(\lambda)}]$, the singular part of $(\lambda - A)^{-1}\varphi$ at λ_0 vanishes. Hence, $(\lambda - A)^{-1}\varphi$ extends to an analytic function in λ_0. From the definition of

$$M_\infty = \bigcap_{\lambda_k \in \sigma(A)} F_k,$$

if $\varphi \in M_\infty$, then the function $(\lambda - A)^{-1}\varphi$ extends to an entire function on the complex plane, where $F_k = R[(\lambda_k - A)^{n_0(\lambda_k)}]$.

Inversely, if $\varphi \notin M_\infty$, then there is $\lambda_k \in \sigma(A)$ such that $\varphi \notin F_k$. The Laurent development of $(\lambda - A)^{-1}\varphi$ in the neighborhood of λ_k shows that this function is not bounded in norm in the neighborhood of λ_k and therefore, can't be extended. So,

$$M_\infty = \{\varphi \in X \text{ such that } \lambda \longrightarrow (\lambda - A)^{-1}\varphi \text{ extends in an entire function on } \mathbb{C}\}.$$

Let $\varphi \in M_\infty$ and $\psi \in X$. For $\lambda_0 \in \rho(A)$, the analyticity of the function

$$f(\lambda) = \langle (\lambda - A)^{-1}\varphi, \psi \rangle$$

in the neighborhood of λ_0 is deduced from the analyticity of the resolvent of the operator A. For $\lambda_0 \in \sigma(A)$ of index $n_0(\lambda_0)$, there exists $\alpha_0 > 0$ such that

$$\{\lambda \in \mathbb{C} \text{ such that } |\lambda - \lambda_0| < \alpha_0\} \subset \rho(A)$$

and, we have the Laurent development

$$f(\lambda) = \langle (\lambda - A)^{-1}\varphi, \psi \rangle = \sum_{k=0}^{\infty} (\lambda - \lambda_0)^k \langle A_k\varphi, \psi \rangle + \sum_{k=1}^{n_0(\lambda_0)} \frac{\langle B_k\varphi, \psi \rangle}{(\lambda - \lambda_0)^k},$$

where B_1 is the projection on $N(\lambda_0 - A)$ along $R(\lambda_0 - A)$ and, where other coefficients of the Laurent series check

$$(\lambda_0 - A)A_0 = I - B_1, \tag{6.2.2}$$

$$(-1)^k(\lambda_0 - A)^k A_k = A_0 \text{ for } k \geq 0, \tag{6.2.3}$$

and

$$(-1)^k(\lambda_0 - A)^k B_1 = B_{k+1} \text{ for all } 0 \leq k \leq n_0(\lambda_0) - 1.$$

Since $N(B_1) = R(\lambda_0 - A)$, we deduce that, if $\varphi \in M_\infty$, then φ is in $R(\lambda_0 - A)$ and so, in $N(B_1)$. Hence, $B_1\varphi = 0$. Thus,

$$B_{k+1}\varphi = 0$$

for all $0 \leq k \leq n_0(\lambda_0) - 1$. So, f is analytic at any point in the neighborhood of λ_0. It remains to show that f is analytic at the point λ_0. For this, we will show that

$$f(\lambda_0) = \langle A_0\varphi, \psi\rangle.$$

Using Eq. (6.2.2), we get $A_0\varphi \in \mathscr{D}(A)$ and, using Eq. (6.2.3), we have $A_k\varphi \in \mathscr{D}(A^k)$ and $A_0\varphi \in R[(\lambda_0 - A)^k]$. Consequently, $A_0\varphi \in R(\lambda_0 - A)$ and so,

$$A_0\varphi \in \mathscr{D}(A) \bigcap R(\lambda_0 - A).$$

Moreover, since $\varphi \in M_\infty \subset N(B_1)$, then

$$(\lambda_0 - A)A_0\varphi = (I - B_1)\varphi = \varphi$$

and

$$(\lambda_0 - A)[(\lambda_0 - A)_{|M_\infty}]^{-1}\varphi = \varphi$$

with

$$[(\lambda_0 - A)_{|M_\infty}]^{-1}\varphi \in \mathscr{D}(A) \bigcap R(\lambda_0 - A).$$

Hence,

$$A_0\varphi = [(\lambda_0 - A)_{|M_\infty}]^{-1}\varphi.$$

Thus,

$$\langle A_0\varphi, \psi\rangle = \langle[(\lambda_0 - A)_{|M_\infty}]^{-1}\varphi, \psi\rangle = f(\lambda_0).$$

So, the application

$$\lambda \longrightarrow f(\lambda) = \langle(\lambda - A)^{-1}\varphi, \psi\rangle,$$

$\varphi \in M_\infty$ and $\psi \in X$ is analytic throughout the complex field.

6.2.2 Completeness Criteria of the System of Generalized Eigenvectors of an Operator with Compact Resolvent

Let $\varphi \in M_\infty$. We denote by f_φ the entire function which extends

$$\lambda \longrightarrow (\lambda - A)^{-1}\varphi.$$

For $\psi \in X$, let us consider

$$\begin{cases} F_{\varphi,\psi} : \mathbb{C} \longrightarrow X \\ \qquad\quad \lambda \longrightarrow \langle f_\varphi(\lambda), \psi\rangle. \end{cases} \tag{6.2.4}$$

f_φ is an entire function with values in X. We will first consider the case where $\lambda \in \rho(A)$. Let $\xi_0 \in \rho(A)$ such that $(\xi_0 - A)^{-1} \in \mathscr{C}^{(0)}_{p,\infty}(X)$. We have $I + (\lambda - \xi_0)(\xi_0 - A)^{-1}$ is invertible and

$$(\lambda - A)^{-1}\varphi = (\xi_0 - A)^{-1}[I + (\lambda - \xi_0)(\xi_0 - A)^{-1}]^{-1}\varphi.$$

Hence,

$$\|(\lambda - A)^{-1}\varphi\| \leq \|(\xi_0 - A)^{-1}\|\|[I + (\lambda - \xi_0)(\xi_0 - A)^{-1}]^{-1}\varphi\|.$$

Moreover, since $(\xi_0 - A)^{-1}_{|M_\infty}$ is quasi-nilpotent (see (6.2.1)), we have

$$I + (\lambda - \xi_0)(\xi_0 - A)^{-1}_{|M_\infty} : M_\infty \longrightarrow M_\infty$$

is one-to-one onto M_∞. So, for all $y \in M_\infty$, there is $x \in M_\infty$ such that

$$(I + (\lambda - \xi_0)(\xi_0 - A)^{-1}_{|M_\infty})x = (I + (\lambda - \xi_0)(\xi_0 - A)^{-1})x = y.$$

Hence,

$$x = (I + (\lambda - \xi_0)(\xi_0 - A)^{-1})^{-1}y = (I + (\lambda - \xi_0)(\xi_0 - A)^{-1}_{|M_\infty})^{-1}y.$$

So, we have

$$\|(\lambda - A)^{-1}\varphi\| \leq \|(\xi_0 - A)^{-1}\|\|[I + (\lambda - \xi_0)(\xi_0 - A)^{-1}_{|M_\infty}]^{-1}\varphi\|.$$

Since $(\xi_0 - A)^{-1}_{|M_\infty}$ is quasi-nilpotent and, by using (4.1.13), we have

$$s_n((\xi_0 - A)^{-1}_{|M_\infty}) \leq s_n((\xi_0 - A)^{-1}).$$

Hence,

$$s_n((\xi_0 - A)^{-1}) = o(n^{-\frac{1}{p}}).$$

By using Macaev's theorem, we have

$$\log(\|[I + (\lambda - \xi_0)(\xi_0 - A)^{-1}_{|M_\infty}]^{-1}\|) = o(|\lambda - \xi_0|^p) = o(|\lambda|^p)$$

as $|\lambda| \to \infty$. Hence, for all $\varepsilon > 0$, there is $B > 0$ such that

$$\|[I + (\lambda - \xi_0)(\xi_0 - A)^{-1}_{|M_\infty}]^{-1}\| \leq Be^{\varepsilon|\lambda|^p},$$

for all $\lambda \in \mathbb{C}$. We can choose $A \leq \|(\xi_0 - A)^{-1}\|\|\varphi\| B$ in order to obtain

$$\|(\lambda - A)^{-1}\varphi\| \leq Ae^{\varepsilon|\lambda|^p}$$

for all $\lambda \in \rho(A)$. Now, since $\rho(A)$ is dense in $\mathbb{C}$, then for all $\varepsilon > 0$, there is $A > 0$ such that

$$\|f_\varphi(\lambda)\| \leq Ae^{\varepsilon|\lambda|^p}, \tag{6.2.5}$$

for all $\lambda \in \mathbb{C}$.

Lemma 6.2.6 *Let A be a closed linear operator with a dense domain $\mathscr{D}(A)$ on a Hilbert space X. We suppose that*

(i) there exists $\lambda \in \rho(A)$ such that $(\lambda - A)^{-1}$ is a compact operator,
(ii) there exists $\xi_0 \in \rho(A)$ such that $(\xi_0 - A)^{-1} \in \mathscr{C}^{(0)}_{p,\infty}(X)$, for some $p > 0$, and
(iii) there are $\gamma_1, \gamma_2, \ldots, \gamma_m$, m half-lines from the origin of the complex plane such that
(a) the angle between any two adjacent half-lines is either less than or equal to $\frac{\pi}{p}$ (which implies that $m \geq [2p] + 1$, where $[2p]$ designates the entire part of $2p$, except for around $2p$ strictly positive or we can have $m = 2p$).
(b) $\|(\lambda - A)^{-1}\|$ is bounded for $|\lambda|$ sufficiently large on each of these half-lines i.e., there is $M > 0$ and $R > 0$ such that

$$\|(\lambda - A)^{-1}\| \leq M,$$

for $|\lambda| \geq R$ and $\lambda \in \bigcup_{1 \leq i \leq m} \gamma_i$.

Then, $F_{\varphi,\psi}$, given in (6.2.4), is bounded on $\mathbb{C}$. ◇

Proof Let $\varphi \in M_\infty$. For $\psi \in X$, let us consider $F_{\varphi,\psi}$ given in (6.2.4). By using (6.2.5), and for all $\varepsilon > 0$, there is $A > 0$ such that

$$|F_{\varphi,\psi}(\lambda)| \leq Ae^{\varepsilon|\lambda|^p}$$

for all $\lambda \in \mathbb{C}$. For $(\varphi, \psi) \in M_\infty \times X$, the function $F_{\varphi,\psi}$ is bounded on each of the m rays $\gamma_1, \gamma_2, \ldots, \gamma_m$. In fact, for $|\lambda|$ sufficiently large, i.e., $|\lambda| > R$, we have

$$|F_{\varphi,\psi}(\lambda)| \leq \|(\lambda - A)^{-1}\|\|\varphi\|\|\psi\|$$

on $\gamma_1, \gamma_2, \ldots, \gamma_m$, and on the closed disk $\overline{\mathbb{D}}(0, R)$, $|F_{\varphi,\psi}(\lambda)|$ is bounded by continuity. Hence, by using Phragmén-Lindelöf theorem (Theorem 2.4.1), $F_{\varphi,\psi}$ is bounded on all angular sectors limited by $\gamma_1, \gamma_2, \ldots, \gamma_m$, i.e., $F_{\varphi,\psi}$ is bounded on $\mathbb{C}$. Q.E.D.

Lemma 6.2.7 *Under the hypotheses of Lemma 6.2.6, we have $M_\infty = \{0\}$.* ◇

Proof Let $\varphi \in M_\infty$. For $\psi \in X$, let us consider $F_{\varphi,\psi}$ given in (6.2.4). Using Lemma 6.2.6, we have $F_{\varphi,\psi}$ is bounded on $\mathbb{C}$. Moreover, since $F_{\varphi,\psi}$ is an entire function, the Liouville theorem (see Theorem 6.1.2) implies that $F_{\varphi,\psi}$ is a constant function. So,

$$\frac{\mathrm{d}}{\mathrm{d}\lambda} F_{\varphi,\psi}(\lambda) = 0.$$

Since

$$\frac{\mathrm{d}}{\mathrm{d}\lambda} F_{\varphi,\psi}(\lambda) = \frac{\mathrm{d}}{\mathrm{d}\lambda}\langle f_\varphi(\lambda), \psi\rangle = \left\langle \frac{\mathrm{d}}{\mathrm{d}\lambda} f_\varphi(\lambda), \psi \right\rangle,$$

we have

$$\left\langle \frac{\mathrm{d}}{\mathrm{d}\lambda} f_\varphi(\lambda), \psi \right\rangle = 0$$

for all $\psi \in X$. Hence,

$$\frac{\mathrm{d}}{\mathrm{d}\lambda} f_\varphi(\lambda) = 0$$

and so, f_φ is a constant function. Thus, for all $\varphi \in M_\infty$, the function $\rho(A) \longrightarrow X$, $\lambda \longrightarrow (\lambda - A)^{-1}\varphi$ is constant. Since A has a compact resolvent, there is $\lambda \in \rho(A)$ such that $\lambda \neq 0$ and $2\lambda \in \rho(A)$. So,

$$(2\lambda - A)^{-1}\varphi = (\lambda - A)^{-1}\varphi = \psi.$$

Hence,

$$\varphi = (2\lambda - A)\psi = (\lambda - A)\psi.$$

Thus, $\psi = 0$. Therefore, for all $\varphi \in M_\infty$, $f_\varphi = 0$. For $\lambda \in \rho(A)$, we have

$$f_\varphi(\lambda) = (\lambda - A)^{-1}\varphi = 0$$

and so, $\varphi = 0$. Thus, $M_\infty = \{0\}$. Q.E.D.

The following result, given by Aimar et al. [1, 3], can be formulated as follows:

Theorem 6.2.2 *Under the hypotheses of Lemma 6.2.6, the system of generalized eigenvectors of A is complete in X.* ◇

Proof Using Lemma 6.2.7, we have $M_\infty = \{0\}$. By using Notes and Remarks 6.4 (v), we have

$$M_\infty^* = \{0\}$$

and

$$\overline{P'} = M_\infty^{*\perp} = X,$$

where P' is the subspace spanned by the generalized eigenvectors of A. This completes the proof of theorem. Q.E.D.

Corollary 6.2.2 *Let $A \in \mathscr{C}(X)$ with compact resolvent. We suppose that*

(i) there is $\xi_0 \in \rho(A)$ such that $(\xi_0 - A)^{-1} \in \mathscr{C}_{p,\infty}(X)$, i.e.,

$$s_n((\xi_0 - A)^{-1}) = O(n^{-\frac{1}{p}}) \text{ as } n \to \infty,$$

(ii) there are $\gamma_1, \gamma_2, \ldots, \gamma_m$, m half-lines from the origin of the complex plane such that the angle between any two adjacent half-lines is strictly less than $\frac{\pi}{p}$ and $\|(\lambda - A)^{-1}\|$ is bounded, for $|\lambda|$ sufficiently large, on $\gamma_1, \gamma_2, \ldots, \gamma_m$.

Then, the system of the generalized eigenvectors of A is complete in X. ◇

Proof It is sufficient to notice that all the angles between γ_i are increased by $\frac{\pi}{p'}$ with $p' > p$ and

$$(\xi_0 - A)^{-1} \in \mathscr{C}_{p,\infty}(X) \subset \mathscr{C}_{p'}(X) \subset \mathscr{C}_{p,\infty}^{(0)}(X).$$

The result follows from Theorem 6.2.2. Q.E.D.

Corollary 6.2.3 *Let $A \in \mathscr{C}(X)$ with compact resolvent. We suppose that*

(i) there is $\xi_0 \in \rho(A)$ such that $(\xi_0 - A)^{-1} \in \mathscr{C}_p(X)$, and

(ii) there are m rays $\gamma_1, \gamma_2, \ldots, \gamma_m$ centered at the origin divided on the whole complex plan such that the angle between any two adjacent half-lines is strictly less than $\frac{\pi}{p}$ and

$$\|(\lambda - A)^{-1}\| = O\left(\frac{1}{|\lambda|}\right)$$

on each $\gamma_1, \gamma_2, \ldots, \gamma_m$.

Then, the system of the generalized eigenvectors of A is complete in X. ◇

Let A be a closed and linear operator with a dense domain $\mathscr{D}(A)$ on a Hilbert space X. We assume that

(i) there exists $\lambda \in \rho(A)$, such that $(\lambda - A)^{-1}$ is a compact operator,
(ii) there exists $\xi_0 \in \rho(A)$, such that $(\xi_0 - A)^{-1} \in \mathscr{C}_{p,\infty}^{(0)}(X)$ for some $p > 0$,
(iii) for $|\lambda|$ sufficiently large, the resolvent $(\lambda - A)^{-1}$ is bounded on each of the m rays $\gamma_1, \gamma_2, \ldots, \gamma_m$ centered at the origin divided on the whole complex plan, such that the angle between any two adjacent half-lines is strictly less than $\frac{\pi}{p}$ ($m \geq [2p] + 1$, where $[2p]$ designates the entire part of $2p$), and

(iv) there is a sequence $\xi_n \in \rho(A)$ such that $(\xi_n - A)^{-1}$ converges weakly to 0, when $|\xi_n|$ tends to infinity.

By using both Macaev theorem (see Theorem 4.2.2) and hypothesis (iii), the function

$$f(\lambda) = \langle (\lambda - A)^{-1}\varphi, \psi \rangle,$$

verifies the assumptions of the principle of Phragmén-Lindelöf theorem (see Theorem 2.4.1), where $\lambda \in \mathbb{C}$, $\varphi \in M_\infty$ and $\psi \in X$. Hence, f is bounded on all sectors limited by the half-lines $\gamma_1, \gamma_2, \ldots, \gamma_m$ and, then is also bounded on the whole $\mathbb{C}$. So, f is an entire function which is bounded on $\mathbb{C}$ and therefore constant. Moreover, by using the hypothesis (iv), we deduce that f is identically nul. Thus, for all $\varphi \in M_\infty$ and $\psi \in X$,

$$\langle (\lambda - A)^{-1}\varphi, \psi \rangle = 0.$$

Hence,

$$(\lambda - A)^{-1}\varphi = 0.$$

Since $(\lambda - A_{|M_\infty})^{-1}$ is one-to-one, we deduce that $M_\infty = \{0\}$. So, the system of the generalized eigenvectors of A is complete in X.
Using the above reasoning, we have the following:

Theorem 6.2.3 *Let A be a closed and linear operator with a dense domain $\mathscr{D}(A)$ on a Hilbert space X. We suppose that*

(i) there exists $\lambda \in \rho(A)$ such that $(\lambda - A)^{-1}$ is a compact operator,
(ii) there exists $\xi_0 \in \rho(A)$ such that $(\xi_0 - A)^{-1} \in \mathscr{C}_{p,\infty}(X)$, for some $p \geq 1$, and
(iii) the numerical range of A, $W(A)$, contained in angular sector with apex the origin and strictly lower opening at $\frac{\pi}{p}$.

Then, the system of generalized eigenvectors of A is complete in X. ◇

6.2.3 A Density Result of the Space Generated by the Generalized Eigenvectors of a Compact Operator

6.2.3.1 Completeness of Generalized Eigenvectors

Let X be a Hilbert space and $K \in \mathscr{C}^{(0)}_{p,\infty}(X)$ for some $p > 0$. Assume that the complex plane is divided by m half-lines $\gamma_1, \gamma_2, \ldots, \gamma_m$ centered at the origin such that

(i) the angle formed by two adjacent half-lines is less than or equal to $\frac{\pi}{p}$, and
(ii) on each of these half-lines, there is $N \in \mathbb{N}$, such that

$$\|(\lambda - K)^{-1}\| = O(|\lambda|^{-N}) \tag{6.2.6}$$

when λ tends to 0.

Let P' be the space generated by the generalized eigenvectors associated with the non-zero eigenvalues of K. We claim that

$$K^N(X) \subset \overline{P'}. \tag{6.2.7}$$

It suffices to prove that

$$\overline{P'}^{\perp} \subset K^N(X)^{\perp}.$$

Note that

$$\lambda K^{*(N-1)}(\lambda - K^*)^{-1} - K^{*(N-1)} = K^{*N}(\lambda - K^*)^{-1}$$

which gives us, by induction on N,

$$K^{*N}(\lambda - K^*)^{-1} = \lambda^N(\lambda - K^*)^{-1} - \lambda^{N-1} - \lambda^{N-2}K^* - \cdots - K^{*(N-1)}, \tag{6.2.8}$$

for $\lambda \in \rho(K^*)$. By using Sect. 4.2.2, we deduce that the function $\lambda \longrightarrow \lambda^N(\lambda - K^*)^{-1}\varphi$ extends to an entire function for all $\varphi \in (P')^{\perp}$. So, applying Eq. (6.2.8), we conclude that

$$\lambda \longrightarrow K^{*N}(\lambda - K^*)^{-1}\varphi$$

extends to an entire function on $\mathbb{C}$ for all $\varphi \in (P')^{\perp}$. Moreover,

$$\|K^{*N}(\lambda - K^*)^{-1}\| \leq \|K^*\|^N \|(\lambda - K^*)^{-1}\|$$

and $\|(\lambda - K^*)^{-1}\|$ is bounded on $\{z \in \mathbb{C}$ such that $|z| \geq R\}$ for $R > \|K^*\|$. Hence, the function extending $K^{*N}(\lambda - K^*)^{-1}\varphi$ for $\varphi \in P'^{\perp}$ is bounded on $\mathbb{C}$ (since it is continuous and bounded on $\{z \in \mathbb{C}$ such that $|z| \leq R\}$). The Liouville's theorem (see Theorem 6.1.2) implies that it is constant and, noticing that for K^* is bounded

$$\lim_{\lambda \to \infty} \|(\lambda - K^*)^{-1}\| = 0,$$

it is null. So, for all $\varphi \in P'^{\perp}$, and for all $\lambda \in \rho(K^*)$, we have

$$K^{*N}(\lambda - K^*)^{-1}\varphi = 0.$$

Since both K^{*N} and $(\lambda - K^*)^{-1}$ commute, we have

$$(\lambda - K^*)^{-1}K^{*N}\varphi = 0$$

and $(\lambda - K^*)^{-1}$ is one-to-one. Thus, for all $\varphi \in P'^{\perp}$,

$$K^{*N}\varphi = 0.$$

Hence, for all $\varphi \in P'^{\perp}$, and for all $\psi \in X$,

$$\langle K^{*N}\varphi, \psi\rangle = \langle \varphi, K^N\psi\rangle = 0.$$

So, the fact that $\varphi \in P'^{\perp}$ implies that $\varphi \in (K^N(X))^{\perp}$. So,

$$K^N(X) \subset \overline{P'}.$$

This prove the claim. Now, if we change the condition (6.2.6) by

$$\|(\lambda - K)^{-1}\| = O(|\lambda|^{-1})$$

when λ tends to 0. Then, the system of generalized eigenvectors of K is complete in X. In fact, by using (6.2.7), we have

$$K(X) \subset \overline{P'},$$

where P' is the space generated by the generalized eigenvectors associated with the non-zero eigenvalues of K. Hence,

$$\overline{K(X)} \subset \overline{P'}.$$

Moreover, Therem 1.2.5 gives

$$X = N(K) + \overline{K(X)} \subset N(K) + \overline{P'}.$$

Hence, the system of generalized eigenvectors of K is complete in X. Furthermore, if we change the condition (6.2.6) by

$$\|(I - \lambda K)^{-1}\| = O(1)$$

when $|\lambda|$ tends to ∞. Then, the system of generalized eigenvectors of K is complete in X.

6.2.3.2 Hypotheses on the Resolvent

Let A be a closed and linear operator with a dense domain $\mathscr{D}(A)$ on a Hilbert space X. Suppose that

(i) there exists $\lambda \in \rho(A)$ such that $(\lambda - A)^{-1}$ is a compact operator,
(ii) there exists $\xi_0 \in \rho(A)$ such that $(\xi_0 - A)^{-1} \in \mathscr{C}^{(0)}_{p,\infty}(X)$ for some $p > 0$, and
(iii) the complex plane is divided by m half-lines $\gamma_1, \gamma_2, \ldots, \gamma_m$ centered at the origin such that the angle formed by two adjacent half-lines is less than or equal to $\frac{\pi}{p}$, and on each of these half-lines, one has

$$\|(\lambda - A)^{-1}\| = O(|\lambda|^N)$$

for $|\lambda| \to \infty$ ($N \in \mathbb{N}$).

Then, the system of generalized eigenvectors of A is complete in X. In fact, without loss of generality, we assume that $\xi_0 = 0 \in \rho(A)$. Let $\mu = \frac{1}{\lambda}$ and $K = A^{-1}$. We have

$$(\mu - K)^{-1} = \lambda - \lambda^2(\lambda - A)^{-1} = \frac{1}{\mu} - \frac{1}{\mu^2}\left(\frac{1}{\mu} - A\right)^{-1}.$$

For $\mu \to 0$, on each of these m half-lines, one has

$$\left\|\left(\frac{1}{\mu} - A\right)^{-1}\right\| = O\left(\frac{1}{|\mu|^N}\right)$$

and

$$\left\|\frac{1}{\mu^2}\left(\frac{1}{\mu} - A\right)^{-1}\right\| = O\left(\frac{1}{|\mu|^{N+2}}\right).$$

So,

$$\|(\mu - K)^{-1}\| = O\left(\frac{1}{|\mu|^{N+2}}\right)$$

on the half-right with range by $z \longrightarrow \frac{1}{z}$ in the neighborhood of O. According to (6.2.7), the generalized eigenvectors associated with the non-zero eigenvalues of K generate a subspace P' which satisfies

$$K^{N+2}(X) \subset \overline{P'}.$$

Since $K(X) = \mathcal{D}(A)$ is dense in X and, by induction, we have

$$\overline{K^N(X)} \supset K(\overline{K^{N-1}(X)}) \supset X.$$

Thus,

$$\overline{K^{N+2}(X)} = X = \overline{P'}$$

and P' can be seen as the space generated by the generalized eigenvectors of A.

Theorem 6.2.4 (Macaev) *Let X be a Hilbert space and $K \in \mathcal{C}_{p,\infty}^{(0)}(X)$ for some $p > 0$. Suppose that the numerical range of K, $W(K)$, contained in angular sector with apex the origin and strictly lower opening at $\frac{\pi}{p}$. Then, the system of generalized eigenvectors of K is complete in X.* ◇

Proof Let

$$S := \{z \in \mathbb{C} \text{ such that } \theta_1 \leq \arg(z) \leq \theta_2\}$$

be an angular sector with origin O. For $\lambda \notin S$, we have

$$\text{dist}(\lambda, S) = |\lambda| \sin(\varepsilon),$$

where

$$\varepsilon = \min(|\arg(\lambda) - \theta_1|, |\arg(\lambda) - \theta_2|).$$

Hence,

$$\text{dist}(\lambda, W(K)) \geq |\lambda| \sin(\varepsilon),$$

$$\|(\lambda - K)^{-1}\| \leq \frac{1}{|\lambda| \sin(\varepsilon)},$$

and this inequality is verified on all straight with origin O exterior to S. Hence, the system of generalized eigenvectors of K is complete in X. Q.E.D.

Lemma 6.2.8 (Xu and Wang [25]) *Let X be a separable Hilbert space and A be the generator of a C_0-semigroup. Suppose that $R(\lambda, A)$ is a Riesz operator and, there exist a real number ρ and $[\rho]+1$ rays $\mu_0, \mu_1, \mu_2, \ldots, \mu_{[\rho]}$ such that*

$$\|R(\lambda, A)\| = O(e^{|\rho|^{\rho}}),$$

$$arg\mu_0 = \frac{\pi}{2} < arg\mu_1 < arg\mu_2 < \cdots < arg\mu_{[\rho]} = \frac{3\pi}{2}$$

and

$$arg\mu_{k+1} - arg\mu_k < \frac{\pi}{2([\rho]+1)}.$$

If $R(\lambda, A)$ is bounded on each μ_j, $j = 1, 2, 3, \ldots, [\rho]-1$, then the system of generalized eigenvectors of A is complete in X. ◇

6.3 Completeness of the System of Root Subspaces

6.3.1 Riesz Projection

Let A be an operator on a Banach space X with a compact isolated part $\sigma \subset \sigma(A)$ of the spectrum. Let Γ be a simply closed, positively oriented integration contour with σ in its interior and $\sigma(A)\backslash\sigma$ in its exterior. Then,

$$P = \frac{i}{2\pi} \int_{\Gamma} (A - z)^{-1} \mathrm{d}z$$

defines a projection such that $R(P)$ and $N(P)$ are A-invariant, $R(P) \subset \mathscr{D}(A)$, $\sigma(A_{|R(P)}) = \sigma$, and $\sigma(A_{|N(P)}) = \sigma(A)\backslash\sigma$. P does not depend on the particular choice of Γ.

6.3.2 Root Subspaces

Let A be an operator with a compact resolvent on a Hilbert space X and let M be the subspace generated by all root subspaces of A, i.e., the set of all finite linear combinations of root vectors of A. Let P be the Riesz projection of A^* corresponding to an eigenvalue $\lambda \in \sigma(A^*)$. It is easy to see $\lambda \in \sigma(A^*)$ if, and only if, $\overline{\lambda} \in \sigma(A)$. Let P be the Riesz projection of A^* corresponding to λ, then P^* is the Riesz projection of A corresponding to $\overline{\lambda}$. Since $R(P^*) \subset M$, then

$$M^{\perp} \subset R(P^*)^{\perp} = N(P). \tag{6.3.1}$$

Furthermore, let $v \in M$ and $z \in \rho(A^*)$, then both Av and $(A - \overline{z})^{-1}v$ are in M and,

$$u \in M^{\perp} \bigcap \mathscr{D}(A^*) \Rightarrow \langle A^*u, v\rangle = \langle u, Av\rangle = 0,$$
$$u \in M^{\perp} \Rightarrow \langle (A^* - z)^{-1}u, v\rangle = \langle u, (A - \overline{z})^{-1}v\rangle = 0.$$

Thus, $M^{\perp}$ is A^*- and $(A^* - z)^{-1}$-invariant for every $z \in \rho(A^*)$ and this implies that

$$\rho(A^*) \subset \rho(A^*_{|M^{\perp}}).$$

Since A has a compact resolvent, the same holds for A^* and $A^*_{|M^{\perp}}$. Hence, if $\lambda \in \sigma(A^*_{|M^{\perp}})$, then λ is an eigenvalue of $A^*_{|M^{\perp}}$, i.e.,

$$A^*u = \lambda u$$

for some $u \in M^{\perp}\backslash\{0\}$. In particular, λ is an eigenvalue of A^* and, $u \in R(P)$, where P is the Riesz projection of A^* corresponding to λ. Furthermore, (6.3.1) implies $u \in M^{\perp} \subset N(P)$. Thus, $u = 0$, which is a contradiction. Consequently,

$$\sigma(A^*_{|M^{\perp}}) = \emptyset.$$

So,

$$\rho(A^*_{|M^{\perp}}) = \mathbb{C}. \tag{6.3.2}$$

6.3.3 System of Subspaces

Definition 6.3.1 A system of subspaces in a Hilbert space X is called complete if the subspace generated by the system is dense in X. ◇

Let A be an operator with a compact resolvent on a Hilbert space X with a scalar product $\langle \cdot, \cdot \rangle$. We suppose that there exists a sequence of bounded regions $(U_k)_{k=0}^{\infty}$ such that

$$\mathbb{C} = \bigcup_{k \in \mathbb{N}} U_k,$$

$\partial U_k \subset \rho(A)$ for all k, and there is a constant $C \geq 0$ such that

$$\|(A - z)^{-1}\| \leq C$$

for $z \in \partial U_k$, $k \in \mathbb{N}$. Consider M, the subspace generated by all root subspaces of the operator A. Consider the holomorphic function defined by

$$f(z) = \langle (A^*_{|M^\perp} - z)^{-1} u, v \rangle,$$

for all $u, v \in M^\perp$. From (6.3.2), we have that its domain of definition is $\mathbb{C}$. Since

$$\|(A^*_{|M^\perp} - z)^{-1}\| \leq \|(A^* - z)^{-1}\| = \|(A - \overline{z})^{-1}\| \text{ for } \overline{z} \in \rho(A),$$

then

$$|f(z)| \leq C\|u\|\|v\|$$

holds for $\overline{z} \in \partial U_k$. From the maximum principle, we infer that

$$|f(z)| \leq C\|u\|\|v\|$$

for every $z \in \mathbb{C}$. In view of Liouville's theorem (see Theorem 6.1.2), the function f is constant. Since u and v are arbitrary, then the mapping

$$z \longrightarrow (A^*_{|M^\perp} - z)^{-1}$$

is also constant. Let $u \in M^\perp$, then

$$(A^*_{|M^\perp})^{-1} u = (A^*_{|M^\perp} - I)^{-1} u.$$

Thus,

$$(A^*_{|M^\perp} - I)(A^*_{|M^\perp})^{-1} u = u.$$

So,

$$(A^*_{|M^\perp})^{-1} u = 0.$$

Consequently, $u = 0$ and $M^{\perp} = \{0\}$, i.e., $M \subset X$ is dense. So, the system of root subspaces of the operator A is complete.

Corollary 6.3.1 (Wyss [24, Corollary 2.4]) *Let A be an operator with a compact resolvent on a Hilbert space X. Suppose that almost all eigenvalues of A lie in a finite number of pairwise disjoint sectors*

$$\Omega_j = \{z \in \mathbb{C} \text{ such that } |argz - \theta_j| < \psi_j\}$$

with $0 < \psi_j \leq \frac{\pi}{4}$, $j = 1, \ldots, n$. If there are constants C, $r_0 \geq 0$ such that

$$\|(A - z)^{-1}\| \leq C$$

for $z \notin \Omega_1 \bigcup \cdots \bigcup \Omega_n$, $|z| \geq r_0$ and, for each sector Ω_j, there is a sequence $(x_k)_k$ with $x_k \to \infty$ and

$$\|(A - z)^{-1}\| \leq C$$

for $z \in \Omega_j$, $Re(e^{-i\theta_j} z) = x_k$, $k \in \mathbb{N}$, then the system of root subspaces of A is complete in X. ◇

6.4 Notes and Remarks

- The theory of non-self-adjoint operators in Hilbert space is an important branch of functional analysis that attracts the attention of mathematicians, physicists, and engineers. In contrast with the theory of self-adjoint operators, no spectral decomposition nor even any theorem hold in the abstract theory of non-self-adjoint operators.
- Based on the study of the resolvent of non-self-adjoint operators, some progress was made in 1951 by M. V. Keldysh to begin the development of a new and profound technique in the theory of non-self-adjoint operators. In fact, M. V. Keldysh established theorems of the completeness of eigenvectors and associated vectors and also theorems concerning the asymptotic properties of the eigenvalues for a wide class of polynomial bundles of non-self-adjoint operators. These theorems made it possible to obtain important results in boundary value problems for partial differential equations. They also led to strong new results for ordinary differential equations. Moreover, on the basis of the study of the resolvent, it has been possible to obtain tests for the existence of a complete set of invariant subspaces of such operators.
- In his book [21], Yu. I. Lyubich, N. K. Nikol'skij, I. Tweddle gave an appreciable development as a generalization of Keldysh's theorem to operators on a Banach space. They investigated the completeness of quasiconservative operators instead of self-adjoint ones.

- In recent times, a remarkable development has taken place in the theory of compact operators or operators with compact resolvent in order to check the denseness of the generalized eigenvectors. In this context, we can cite the contributions of Aimar et al. [1–4], Jeribi [19], Paoli [23], and Xu and Wang [25].

(i) We will emphasize the fact that, if X is a Hilbert space, then the approximation numbers (a-numbers) of a linear compact operator A coincide with the singular numbers (s-numbers) (see [17, p. 26]) of A (see [17, Theorem 2.1, p. 28] or [21, p. 216]).

(ii) $\sigma(A_{|\mathscr{D}(A)\bigcap M_\infty}) = \emptyset$.

(iii) Proposition 6.2.1 gives a necessary and sufficient condition for the existence of a total system of generalized eigenvectors. In fact, $\overline{P} = X$ if, and only if, $M_\infty^* = \{0\}$.

(iv) The system of generalized eigenvectors of A is dense in X if, and only if, $M_\infty^* = \{0\}$.

(v) We notice that, if an operator A satisfies the hypotheses of Lemma 6.2.6, it is also true for his adjoint. In fact, since $\rho(A^*) = \overline{\rho(A)}$ and $(\overline{\xi_0} - A^*)^{-1} = ((\xi_0 - A)^{-1})^*$, then

$$(\xi_0 - A)^{-1} \in \mathscr{C}_{p,\infty}^{(0)}(X)$$

implies that

$$(\overline{\xi_0} - A^*)^{-1} \in \mathscr{C}_{p,\infty}^{(0)}(X)$$

(since $s_n(K) = s_n(K^*)$ for all $K \in \mathscr{L}(X)$). Moreover, the fact that $\|(\lambda - A)^{-1}\|$ is bounded for $|\lambda|$ sufficiently large on each of the m rays $\gamma_1, \gamma_2, \ldots, \gamma_m$ implies that $\|(\lambda - A^*)^{-1}\|$ is bounded for $|\lambda|$ sufficiently large on each of the m rays $\overline{\gamma_1}, \overline{\gamma_2}, \ldots, \overline{\gamma_m}$, where

$$\overline{\gamma_i} = \{\overline{z} \text{ such that } z \in \gamma_i\}.$$

The symmetry $z \longrightarrow \overline{z}$ leaves invariant the angles (in absolute values) between all γ_i. Hence, A satisfies the hypotheses of Lemma 6.2.6 if, and only if, A^* satisfies the hypotheses of Lemma 6.2.6.

(vi) We can not extend the conclusion of Theorem 6.2.2 if the resolvent is only in $\mathscr{C}_{p,\infty}(X)$. In fact, let A be the operator defined in (6.0.1) and $K := A^{-1}$ be the operator defined in (6.0.2). Using Eq. (6.0.3), we have

$$K \in \mathscr{C}_{1,\infty}(L_2(0, \pi)).$$

Secondly, by clarifying the resolvent $(\lambda - A)^{-1}$, we see that it is bounded across original half-line O located in the half-plane $Re(\lambda) \leq 0$. It was therefore right, taking the half-lines of imaginary pure positive and negative, that there is γ_1 and γ_2 with $(\gamma_1, \gamma_2) = \pi$ and $(\lambda - A)^{-1}$ bounded on γ_1 and γ_2, but no completeness of generalized eigenvectors.

(vii) If an operator A satisfies the hypotheses of Theorem 6.2.3, it is also true for its conjugates A^* since $\rho(A^*) = \overline{\rho(A)}$ and $(\overline{\lambda} - A^*)^{-1} = [(\lambda - A)^{-1}]^*$ for all $\lambda \in \rho(A)$. In fact, it is easy to check the assumptions (i) and (iii) of Theorem 6.2.3. As to (iv) of Theorem 6.2.3, it suffices to note that the weak convergence to 0 of $(\xi_n - A^*)^{-1}$ when $|\xi_n|$ tends to infinity is equivalent to $(\overline{\xi_n} - A)^{-1}$. The only point slightly less trivial is the verification of the hypothesis (ii) of Theorem 6.2.3 resulting from the important property $s_n(K) = s_n(K^*)$ (see Proposition 4.1.1).

References

1. M.T. Aimar, A. Intissar, J.M. Paoli, Densité des vecteurs propres généralisés d'une classe d'opérateurs compacts non auto-adjoints et applications. Commun. Math. Phys. **156**, 169–177 (1993)
2. M.T. Aimar, A. Intissar, J.M. Paoli, Quelques nouvelles propriétés de régularité de l'opérateur de Gribov. Commun. Math. Phys. **172**(3), 461–466 (1995)
3. M.T. Aimar, A. Intissar, J.M. Paoli, Critères de complétude des vecteurs propres généralisés d'une classe d'opérateurs non auto-adjoints compacts ou à résolvante compacte et applications. Publ. Res. Inst. Math. Sci. **32**(2), 191–205 (1996)
4. M. T. Aimar, A. Intissar, A. Jeribi, On an unconditional basis of generalized eigenvectors of the non-self-adjoint Gribov operator in Bargmann space, J. Math. Anal. Appl. **231**, 588–602 (1999)
5. N. Ben Ali, A. Jeribi, On the Riesz basis of a family of analytic operators in the sens of Kato and application to the problem of radiation of a vibrating structure in a light fluid, J. Math. Anal. Appl. **320**(1), 78–94 (2006)
6. S. Charfi, A. Damergi, A. Jeribi, On a Riesz basis of finite-dimensional invariant subspaces and application to Gribov operator in Bargmann space. Line. Mult. Alge. **61**(11), 1577–1591 (2013)
7. S. Charfi, A. Intissar, A. Jeribi, Expansion of solution in terms of generalized eigenvectors for a rectilinear transport equation. Transport Theory Statist. Phys. **38**, 1–16 (2009)
8. S. Charfi, A. Jeribi, I. Walha, Riesz basis property of families of nonharmonic exponential and application to a problem of a radiation of a vibrating structure in a light fluid. Numerical Functional Analysis and Optimization, **32**(4), 370–382 (2011)
9. H. Ellouz, I. Feki, A. Jeribi, On a Riesz basis of exponentials related to the eigenvalues of an analytic operator and application to a non-self-adjoint problem deduced from a perturbation method, J. Math. Phys. **54**, 112101 (2013)
10. H. Ellouz, I. Feki, A. Jeribi, On the asymptotic behavior of the eigenvalues of an analytic operator in the sense of Kato and applications, Serdica J. Math. **45**(1), 55–88 (2019)
11. H. Ellouz, I. Feki, A. Jeribi, On a Riesz basis of exponentials related to a family of analytic operators and application. J. Pseudo-Differ. Oper. Appl. **10**(4), 999–1014 (2019)
12. H. Ellouz, I. Feki, A. Jeribi, Non-orthogonal Fusion Frames of an Analytic Operator and Application to a One-Dimensional Wave Control System. Mediterr. J. Math. **16**(2), 16:52 (2019)
13. H. Ellouz, I. Feki, A. Jeribi, On the basis property of root vectors related to a non-self-adjoint analytic operator and applications, (2021)
14. I. Feki, A. Jeribi, R. Sfaxi, On an unconditionl basis of generalized eigenvectors of an analytic operator and application to a problem of radiation of a vibrating structure in a light fluid, J. Math. Anal. Appl. **375**, 261–269 (2011)
15. I. Feki, A. Jeribi, R. Sfaxi, On a Schauder basis related to the eigenvectors of a family of non-self-adjoint analytic operators and applications, Anal. Math. Phys. **3**(4), 311–331 (2013)

16. I. Feki, A. Jeribi, R. Sfaxi, On a Riesz basis of eigenvectors of a non-self-adjoint analytic operator and applications. Line. Mult. Alge. **62**(8), 1049–1068 (2014)
17. I.C. Gohberg, M.G. Krein, *Introduction to the Theory of Linear Non-self-adjoint Operators in Hilbert Space* (Amer. Math. Soci, Providence, 1969)
18. A. Intissar, A. Jeribi, I. Walha, Riesz basis of exponential family for a hyperbolic system. J. Appl. Anal. **25**(1), 13–23 (2019)
19. A. Jeribi, *Denseness, Bases and Frames in Banach Spaces and Applications* (de Gruyter, Berlin, 2018)
20. A. Jeribi, A. Intissar, On an Riesz basis of generalized eigenvectors of the nonself-adjoint problem deduced from a perturbation method for sound radiation by a vibrating plate in a light fluid. J. Math. Anal. Appl. **292** 1–16 (2004)
21. YuI Lyubich, N.K. Nikol'skij, I. Tweddle, *Functional Analysis I: Linear Functional Analysis, Encyclopaedia of Mathematical Sciences*, vol. 19 (Springer, Berlin, 1992)
22. A.S. Markus, *Introduction to the Spectral Theory of Polynomial Operator Pencils*, vol. 71 (Amer. Math. Soci., Providence, RI, 1988)
23. J.M. Paoli, *Conditions suffisantes de densité du sous-espace engendré par les vecteurs propres généralisés d'un opérateur compact ou à résolvante compacte sur un espace de Hilbert et applications aux oprateurs de Gribov* (Thèse de l'université de Corse, France, 1993)
24. C. Wyss, Riesz bases for p-subordinate perturbations of normal operator. J. Funct. Anal. **258**, 208–240 (2010)
25. G.Q. Xu, S.H. Wang, The completeness of systems of generalized eigenfunctions of generators of Riesz semigroups (Chinese). Acta Math. Sinica **39**(2), 263–267 (1996)

Chapter 7
Bases on Hilbert and Banach Spaces

In this chapter, considerable attention has been devoted to bases on Hilbert and Banach spaces. We derive their basic properties, one of the most important of which is the fact that the coefficient functionals associated with a basis are automatically continuous. The Hamel, Schauder, orthonormal, and Riesz basis are studied. We provide some examples of compact, finite rank and Hilbert-Schmidt operators and we continue the development of basis theory by examining Riesz bases acting in a separable Hilbert space and we recall the $\mathscr{L}$-basis in $L_2[0, T]$. For this, we begin this chapter with some notions on the bases of a vector space, which are the bases that we are familiar with from linear algebra.

7.1 Some Notions on the Bases of a Vector Space

7.1.1 On a Schauder Basis

Let l^p $(1 \leq p < \infty)$ be the Banach space, of all infinite sequences of scalars $b = \{b_n\}_{n=1}^{\infty}$ such that

$$\|b\|_p = \left(\sum_{n=1}^{\infty} |b_n|^p\right)^{\frac{1}{p}} < \infty.$$

The vector operations are coordinatewise. In each of these spaces, the "natural basis" $\{e_1, e_2, e_3, \ldots\}$, where $e_n = (0, 0, \ldots, 0, 1, 0, \ldots)$, and the 1 appears in the nth position, is easily seen to be a Schauder basis. If $b = \{b_n\}$ is in l^p, then the obvious expansion

A. Jeribi, *Perturbation Theory for Linear Operators*,
https://doi.org/10.1007/978-981-16-2528-2_7

$$b = \sum_{n=1}^{\infty} b_n e_n$$

is valid. It follows, for example, that since l^∞ is not separable, it cannot possess a basis.

Let f be a continuous function on the interval [0, 1]. Then, the nth Bernstein polynomial for f is

$$B_n(x) = \sum_{k=0}^{n} C_n^k f\left(\frac{k}{n}\right) x^k (1-x)^{n-k}, \quad n = 1, 2, 3, \ldots.$$

By using Akhiezer [1], we get

$$\sup_{x \in [0,1]} |f(x) - B_n(x)| \to 0$$

as $n \to \infty$. Hence,

$$f(x) = \lim_{n \to \infty} B_n(x)$$

is uniformly on [0, 1]. Let $C[a, b]$ be the Banach space of all continuous functions on the closed finite interval $[a, b]$, together with the norm

$$\|f\| = \sup_{x \in [a,b]} |f(x)|.$$

The Weierstrass approximation theorem asserts that the polynomials are dense in $C[a, b]$, i.e., if f is continuous on $[a, b]$, then for every positive number ε, there is a polynomial P such that the inequality

$$|f(x) - P(x)| < \varepsilon$$

holds throughout the interval $[a, b]$. For a given continuous function, a sequence of approximating polynomials can even be given explicitly. The most elegant representation is due to Bernstein. Since every polynomial can be uniformly approximated on a closed interval by a polynomial with rational coefficients, the preceding remarks show that the space $C[a, b]$ is separable; in fact, it has a basis.

Theorem 7.1.1 (Schauder) *The space $C[a, b]$ possesses a basis.* ◇

Proposition 7.1.1 *Every Banach space X with a Schauder basis $(x_n)_{n=1}^{\infty}$ is separable.* ◇

Definition 7.1.1 (*Approximation property*) A Banach space is called to have the approximation property if each linear compact operator is a limit of finite-rank operators. The converse is always true. ◇

Definition 7.1.2 A set of vectors $\{\varphi_n\}_{n=1}^{\infty}$ on a Banach space X is said to be ω-linearly independent if the equality

$$\sum_{n=1}^{\infty} c_n \varphi_n = 0$$

is possible only for $c_n = 0$ $(n = 1, 2, 3, \ldots)$. ◇

Definition 7.1.3 A sequence $(e_n)_{n=1}^{\infty}$ is a basis for a separable Hilbert space X if

(i) it is complete, i.e., $\langle u, e_n \rangle = 0 \ \forall n \in \mathbb{N}^* \Longrightarrow u = 0$,
(ii) the e_n are independent, i.e., if

$$\sum_{n=1}^{\infty} c_n e_n = 0$$

for some sequence $(c_n)_{n=1}^{\infty} \in l^2$, then $c_n = 0$ for all $n \in \mathbb{N}^*$. ◇

Definition 7.1.4 A sequence $(x_n)_{n=1}^{\infty}$ in a Banach space X is said minimal if

$$x_m \notin \overline{\text{span}}\{x_n\}_{n \neq m}$$

for every m. ◇

7.1.2 The Coefficient Functionals

If $\{x_1, x_2, x_3, \ldots\}$ is a basis for a Banach space X, then every vector x, in the space, has a unique series expansion of the form

$$x = \sum_{n=1}^{\infty} c_n x_n.$$

It is clear that each coefficient c_n, is a linear function of x. If we denote this linear function by f_n, then

$$c_n = f_n(x),$$

and we may write

$$x = \sum_{n=1}^{\infty} f_n(x) x_n.$$

The functionals f_n $(n = 1, 2, 3, \ldots)$, thus defined, are called the coefficient functionals associated with the basis $\{x_n\}_{n=1}^{\infty}$. They play an essential role in many parts of

Banach space theory; perhaps the most important statement about them is that they are continuous. Consider the vector space Y consisting of those sequences of scalars $\{c_n\}_{n=1}^{\infty}$ for which the series

$$\sum_{n=1}^{\infty} c_n x_n$$

is convergent in X. If $\{c_n\}_{n=1}^{\infty} \in Y$, then the number

$$\|\{c_n\}\| = \sup_n \left\| \sum_{i=1}^{n} c_i x_i \right\|$$

satisfies all the properties of a norm. It is not difficult to see that Y, endowed with this norm, is a Banach space. The verification that Y is a Banach space is typical of standard completeness arguments, and since it poses no special difficulties, it is left as an exercise for the reader (see [15]). Let T be the operator defined by

$$\begin{cases} T : Y \longrightarrow X \\ \quad \{c_n\} \longrightarrow \displaystyle\sum_{n=1}^{\infty} c_n x_n, \end{cases} \tag{7.1.1}$$

T is obviously linear. Since $\{x_n\}_{n=1}^{\infty}$ is a basis, it is also one-to-one and onto. Since

$$\left\| \sum_{n=1}^{\infty} c_n x_n \right\| \le \sup_n \left\| \sum_{i=1}^{n} c_i x_i \right\|,$$

it follows that T is continuous, and the open mapping theorem then guarantees that T^{-1} is also continuous. Let

$$x = \sum_{n=1}^{\infty} c_n x_n$$

be an arbitrary element of X. Then,

$$\begin{aligned} |c_n| &= \frac{\|c_n x_n\|}{\|x_n\|} \\ &\le \frac{\left\| \sum_{i=1}^{n} c_i x_i \right\| + \left\| \sum_{i=1}^{n-1} c_i x_i \right\|}{\|x_n\|} \\ &\le 2 \frac{\sup_n \left\| \sum_{i=1}^{n} c_i x_i \right\|}{\|x_n\|} \\ &= 2 \frac{\|T^{-1} x\|}{\|x_n\|} \\ &\le 2 \frac{\|T^{-1}\| \|x\|}{\|x_n\|}. \end{aligned} \tag{7.1.2}$$

Lemma 7.1.1 *Let $\{x_n\}_{n=1}^{\infty}$ be a basis for a Banach space X. Consider the vector space Y consisting of those sequences of scalars $\{c_n\}_{n=1}^{\infty}$ for which the series*

$$\sum_{n=1}^{\infty} c_n x_n$$

is convergent in X. Then, Y is a Banach space which isomorphic to X. ◇

Proof If $\{c_n\}_{n=1}^{\infty} \in Y$, then the number

$$\|\{c_n\}\| = \sup_n \left\| \sum_{i=1}^{n} c_i x_i \right\|$$

satisfies all the properties of a norm. The space Y, endowed with this norm, is a Banach space. Moreover, the continuous mapping $T : Y \longrightarrow X$, defined in (7.1.1), is one-to-one and onto. This proves that X and Y are isomorphic, which completes the proof of lemma. Q.E.D.

Let $\{x_n\}_{n=1}^{\infty}$ be a basis for a Banach space X and Y be the vector space consisting of those sequences of scalars $\{c_n\}_{n=1}^{\infty}$ for which the series

$$\sum_{n=1}^{\infty} c_n x_n$$

is convergent in X. By Lemma 7.1.1, we have Y is a Banach space and, both X and Y are isomorphic. Suppose now that

$$x = \sum_{n=1}^{\infty} c_n x_n$$

is a fixed but arbitrary element of X. If $\{f_n\}_{n=1}^{\infty}$ is the associated sequence of coefficient functionals of $\{x_n\}_{n=1}^{\infty}$, then by using Eq. (7.1.2), we have

$$\begin{aligned} |f_n(x)| &= |c_n| \\ &\leq 2\frac{\|T^{-1}\|}{\|x_n\|}\|x\| \end{aligned}$$

for every n, where T is the operator defined in (7.1.1). This proves that each f_n is continuous i.e., $f_n \in X^*$ (the dual of X), and that

$$\|f_n\| \leq 2\frac{\|T^{-1}\|}{\|x_n\|}.$$

Choosing $M = 2\|T^{-1}\|$, we have

$$\|x_n\| \|f_n\| \leq M$$

for every n. So,

$$\begin{aligned} 1 &= f_n(x_n) \\ &\leq \|f_n\| \|x_n\| \\ &\leq M \ (n = 1, 2, 3, \ldots). \end{aligned}$$

7.2 Orthonormal Bases in Hilbert Space

In a separable Hilbert space X, a distinguished role is played by those Schauder bases that are orlhonormal (the basis vectors are mutually perpendicular and each has a unit length). An equivalent characterization of such bases is that they are complete orthonormal sequences. It follows readily from this characterization that every separable Hilbert space has an orthonormal basis. The most important property of an orthonormal basis is the simplicity of all basis expansions. If $\{e_1, e_2, e_3, \ldots\}$ is an orthonormal basis for a Hilbert space X, then for every element $f \in X$, we have the Fourier expansion

$$f = \sum_{n=1}^{\infty} \langle f, e_n \rangle e_n.$$

The inner product $\langle f, e_n \rangle$ is called the nth Fourier coefficient of f (relative to $\{e_n\}$). Applying the Pythagorean formula to this series, the result is Parseval's identity

$$\|f\|^2 = \sum_{n=1}^{\infty} |\langle f, e_n \rangle|^2.$$

The validity of Parseval's identity for every vector in the space is both necessary and sufficient for an orthonormal sequence to be a basis cf. [17].

In l^2, the "natural basis" is orthonormal.

We can prove that each of the systems

$$e_n(t) = \sqrt{2} \sin(nt), \ n \geq 1, \ \text{ or } \ e_n(t) = \sqrt{2} \cos(nt), \ n \geq 0,$$

constitutes an orthonormal basis for $L_2(0, \pi)$.

In $L_2[-\pi, \pi]$, with the inner product

$$\langle f, g \rangle = \frac{1}{2\pi} \int_{-\pi}^{\pi} f(t)\overline{g(t)} dt,$$

the complex trigonometric system $\{e^{int}\}_{-\infty}^{+\infty}$ constitutes an orthonormal basis. The system is orthonormal is obvious. The following theorem proves that it is complete.

Theorem 7.2.1 (Young [34]) *The trigonometric system* $\{e^{int}\}_{-\infty}^{+\infty}$ *is complete in* $L_2[-\pi, \pi]$. ◇

Example 7.1 Let $S : l^2 \longrightarrow l^2$ be the shift operator, defined by

$$\begin{cases} Se_0 = 0, \\ Se_{k+1} = e_k, \end{cases}$$

where $(e_k)_{k=0}^{\infty}$ is the standard orthonormal basis of l^2. Then,

$$e_k \in \mathscr{L}_S(0)$$

for every k and

$$0 \in \sigma_p(S),$$

Hence $(e_k)_{k=0}^{\infty}$ is an orthonormal basis of root vectors.

Example 7.2 The Hardy space H^2 consists of all analytic functions f in the open unit disk $\mathbb{D}(0, 1)$ of the complex plane whose Taylor coefficients are square-summable, i.e.,

$$f(z) = \sum_{n=0}^{\infty} c_n z^n$$

with

$$\sum_{n=0}^{\infty} |c_n|^2 < \infty.$$

The inner product in H^2 of two functions f and g is, by definition,

$$\langle f, g \rangle = \sum_{n=0}^{\infty} a_n \overline{b_n},$$

where

$$f(z) = \sum_{n=0}^{\infty} a_n z^n$$

and

$$g(z) = \sum_{n=0}^{\infty} b_n z^n.$$

Let $e_n(z) = z^n$ for $z \in \mathbb{D}(0, 1)$ $(n = 0, 1, 2, \ldots)$. Then, $\{e_n\}_{n=0}^{\infty}$ form an orthonormal basis for the Hardy space H^2.

Theorem 7.2.2 (The Hilbert-Schmidt theorem) *Let A be a self-adjoint compact operator on a separable Hilbert space X. Then, there is an orthonormal basis $\{\phi_n\}_1^\infty$ of X such that*

$$A\phi_n = \lambda_n \phi_n \text{ and } \lim_{n\to\infty} \lambda_n = 0. \qquad \diamond$$

Proof For each eigenvalue of A choose an orthonormal basis for the set of eigenvectors corresponding to the eigenvalue. The collection of all these vectors, $\{\phi_n\}_{n=1}^\infty$, is an orthonormal set since eigenvectors corresponding to distinct eigenvalues are orthogonal. Let $\mathscr{M}$ be the closure of the span of $\{\phi_n\}_{n=1}^\infty$. Since A is self-adjoint and $A : \mathscr{M} \longrightarrow \mathscr{M}$, $A : \mathscr{M}^\perp \longrightarrow \mathscr{M}^\perp$. Let $\widetilde{A}$ be the restriction of A to $\mathscr{M}^\perp$. Then, $\widetilde{A}$ is self-adjoint and compact since A is. By the Riesz Schauder theorem (see Theorem 2.2.2), if any $\lambda \neq 0$ is in $\sigma(\widetilde{A})$, it is an eigenvalue of $\widetilde{A}$, and thus of A. Therefore, the spectral radius of $\widetilde{A}$ is zero since the eigenvectors of A are in $\mathscr{M}$. Because $\widetilde{A}$ is self-adjoint, it is the zero operator on $\mathscr{M}^\perp$ by Theorem 1.1.1. Thus, $\mathscr{M}^\perp = \{0\}$ since if $\varphi \in \mathscr{M}^\perp$, then

$$A\varphi = 0$$

which implies that $\varphi \in \mathscr{M}$. Therefore,

$$\mathscr{M} = X.$$

The fact that $\lambda_n \to 0$ is a consequence of the first part of the Riesz Schauder theorem (see Theorem 2.2.2(ii)) which says that each non-zero eigenvalue has finite multiplicity and the only possible limit point of the λ_n is zero. Q.E.D.

In the following, we give a canonical form or representation of Schmidt of a compact operator. In fact, let K be a compact operator on a separable Hilbert space X. Then, it follows from Theorem 1.6.1(ii) that K^* is compact. Hence, so is K^*K. Furthermore, K^*K is self-adjoint. Hence, by using the Hilbert-Schmidt theorem (Theorem 7.2.2), there is an orthonormal set $\{\psi_n\}_1^N$ of X such that

$$K^*K\psi_n = \lambda_n \psi_n$$

with $\lambda_n \neq 0$, and so is K^*K the zero operator on the subspace orthogonal to $\{\psi_n\}_1^N$. Since K^*K is positive, it is easy to see that each $\lambda_n > 0$ for all n. We consider $\mu_n = \sqrt{\lambda_n}$ and

$$\phi_n = \frac{1}{\mu_n} K\psi_n.$$

Let $f \in X$. Since $\{\psi_n\}_1^N$ is an orthonormal set of X, then

$$f = \sum_{n=1}^{N} \langle f, \psi_n \rangle \psi_n.$$

Hence,

$$\begin{aligned} Kf &= K\left(\sum_{n=1}^{N}\langle f, \psi_n\rangle \psi_n\right) \\ &= \sum_{n=1}^{N}\langle f, \psi_n\rangle K\psi_n \\ &= \sum_{n=1}^{N}\mu_n\langle f, \psi_n\rangle \phi_n. \end{aligned}$$

Since

$$\lim_{n\to\infty}\lambda_n = 0,$$

then

$$\lim_{n\to\infty}\mu_n = \lim_{n\to\infty}\sqrt{\lambda_n} = 0.$$

Now, it is easy to see that the family $\{\phi_n\}_1^N$ is orthonormal. Indeed, we have

$$\begin{aligned} \langle \phi_n, \phi_k\rangle &= \left\langle \frac{1}{\mu_n}K\psi_n, \frac{1}{\mu_k}K\psi_k\right\rangle \\ &= \frac{1}{\mu_n\mu_k}\langle K\psi_n, K\psi_k\rangle \\ &= \frac{1}{\mu_n\mu_k}\langle K^*K\psi_n, \psi_k\rangle \\ &= \frac{1}{\mu_n\mu_k}\langle \lambda_n\psi_n, \psi_k\rangle \\ &= \frac{\mu_n}{\mu_k}\langle \psi_n, \psi_k\rangle \\ &= \delta_{nk}, \end{aligned}$$

where δ_{nk} is the symbol of Kroneker, i.e.,

$$\delta_{nk} = \begin{cases} 0 & \text{if } n \neq k \\ 1 & \text{if } n = k. \end{cases}$$

So, there exist (not necessarily complete) two orthonormal sets $\{\psi_n\}_{n=1}^N$ and $\{\phi_n\}_{n=1}^N$ of vector of X and positive real numbers $(\mu_n)_{n=1}^N$ with $\lim_{n\to\infty}\mu_n = 0$ such that

$$K = \sum_{n=1}^{N}\mu_n\langle \psi_n, \cdot\rangle \phi_n. \tag{7.2.1}$$

The sum in (7.2.1), which may be finite or infinite, converges in norm. The numbers (μ_n) are called the singular values of K.

7.3 Examples of Compact Operators

7.3.1 Finite-Rank Operator

Theorem 7.3.1 *Every finite-rank operator is compact.* ◇

Proof An operator of finite rank is bounded and it transforms a bounded set into a bounded set. However, in a finite-dimensional space, a bounded set is relatively compact (see [16]). Q.E.D.

7.3.2 Hilbert-Schmidt Operator

A second example of compact operators is given by the family of Hilbert-Schmidt operators that are defined only on a separable Hilbert space. To enter, we need the following result:

Proposition 7.3.1 *Let X be a separable Hilbert space and let $\{e_n\}_{n=1}^{\infty}$ be an orthonormal basis of X. Then, for all bounded operator A on X, the number (finite or infinite)*

$$\|A\|_2^2 = \sum_{n=1}^{\infty} \|Ae_n\|^2$$

does not depend on the orthonormal basis considered. ◇

Proof Let A be a bounded operator and $\{\widetilde{e_n}\}_1^{\infty}$ another orthonormal basis of X. We have

$$\begin{aligned}\sum_{n=1}^{\infty} \|A\widetilde{e_n}\|^2 &= \sum_{n=1}^{\infty}\sum_{j=1}^{\infty} |\langle A\widetilde{e_n}, e_j\rangle|^2 \\ &= \sum_{n=1}^{\infty}\sum_{j=1}^{\infty} |\langle \widetilde{e_n}, A^* e_j\rangle|^2.\end{aligned}$$

For $j, n \in \mathbb{N}^*$, let

$$u_{j,n} = |\langle \widetilde{e_n}, A^* e_j\rangle|^2.$$

We have $u_{j,n} \geq 0$ for all $j, n \geq 1$. Hence, according to Fubini theorem, we have

$$\begin{aligned}\sum_{n=1}^{\infty} \|A\widetilde{e_n}\|^2 &= \sum_{n=1}^{\infty}\sum_{j=1}^{\infty} u_{j,n} \\ &= \sum_{j=1}^{\infty}\sum_{n=1}^{\infty} u_{j,n} \\ &= \sum_{j=1}^{\infty}\sum_{n=1}^{\infty} |\langle \widetilde{e_n}, A^* e_j\rangle|^2 \\ &= \sum_{j=1}^{\infty} \|A^* e_j\|^2.\end{aligned}$$

In particular, taking the same basis, we have

$$\sum_{n=1}^{\infty} \|A^* e_n\|^2 = \sum_{n=1}^{\infty} \|Ae_n\|^2.$$

Thus,

$$\sum_{n=1}^{\infty} \|A\widetilde{e_n}\|^2 = \sum_{n=1}^{\infty} \|Ae_n\|^2.$$

Q.E.D.

Definition 7.3.1 A Hilbert-Schmidt operator on a separable Hilbert space X is a bounded operator A on X for which

$$\|A\|_2^2 := \sum_{n=1}^{\infty} \|Ae_n\|^2 < \infty,$$

where $\{e_n\}_1^\infty$ is an orthonormal basis of X. ◇

Example 7.3 The Volterra operator V on $L_2(0, 1)$, defined by

$$(Vf)(t) = \int_0^t f(\tau)\mathrm{d}\tau, \;\; 0 < t < 1,$$

is a Hilbert-Schmidt operator.

Theorem 7.3.2 *Let X be a separable Hilbert space. Then, all the Hilbert-Schmidt operator on X is compact.* ◇

Proof Let $\{e_n\}_1^\infty$ be an orthonormal basis of X and A be a Hilbert-Schmidt operator on X. Let $N \geq 1$, we denote by P_N the orthogonal projection on the subspace spanned by $\{e_1, e_2, \ldots e_N\}$. The operator $A_N = AP_N$ is a finite rank and is equal to N and, we have $A_N e_j = Ae_j$ if $j \leq N$ and $A_N e_j = 0$ if $j > N + 1$. So,

$$(A - A_N)e_j = \begin{cases} 0 & \text{if } j \leq N \\ Ae_j & \text{if } j \geq N + 1. \end{cases}$$

As a result of inequality (7.8.1),

$$\begin{aligned} \|A - A_N\|^2 &\leq \|A - A_N\|_2^2 \\ &= \sum_{n=N+1}^{\infty} \|Ae_n\|^2. \end{aligned}$$

The second member of the last inequality tends to 0 when N tends to infinity, since it is the remains of a convergent series. The linear operator A is the limit of a sequence of finite-rank operators on X. Hence, A is compact. Q.E.D.

7.4 Equivalent Bases

7.4.1 Image of a Basis Under a Topological Isomorphism

Let $T : X \longrightarrow Y$ be a topological isomorphism, where X and Y are two Banach spaces. If y is any element of Y, then $T^{-1}y \in X$. If $\{x_n\}_{n=1}^{\infty}$ is a basis for X, then there are unique scalars (c_n) such that

$$T^{-1}y = \sum c_n x_n.$$

Since T is continuous, this implies that

$$y = T(T^{-1}y) = \sum c_n T x_n.$$

Suppose

$$y = \sum b_n T x_n$$

is another representation of y. Then, since T^{-1} is continuous, we have

$$T^{-1}y = \sum b_n x_n$$

and hence,

$$b_n = c_n$$

for each n since $\{x_n\}_{n=1}^{\infty}$ is a basis for X. Thus, $\{Tx_n\}_{n=1}^{\infty}$ is a basis for Y.

7.4.2 Definitions

Definition 7.4.1 Two Schauder bases, $\{b_n\}_{n=1}^{\infty}$ in X and $\{c_n\}_{n=1}^{\infty}$ in Y are said to be equivalent if, there exist two positive constants c and C such that for every integer N and all sequences $\{\alpha_n\}_{n=1}^{\infty}$ of scalars, we have

$$c \left\| \sum_{n=1}^{N} \alpha_n b_n \right\|_X \leq \left\| \sum_{n=1}^{N} \alpha_n c_n \right\|_Y \leq C \left\| \sum_{n=1}^{N} \alpha_n b_n \right\|_X . \qquad \diamond$$

The simplest and perhaps the most obvious way of constructing new bases from old is through an isomorphism of the underlying space. Thus, if $\{v_n\}_{n=1}^{\infty}$ is a fixed but arbitrary basis for a Banach space X and if the bounded invertible operator T transforms $\{v_n\}_{n=1}^{\infty}$ into $\{w_n\}_{n=1}^{\infty}$, that is, if $Tv_n = w_n$, for $n = 1, 2, 3, \ldots$, then $\{w_n\}_{n=1}^{\infty}$ is also a basis for X. In fact, it is easy to see that $\{v_n\}_{n=1}^{\infty}$ and $\{w_n\}_{n=1}^{\infty}$ are equivalent in the following sense.

Definition 7.4.2 Let X be a Banach space. Two bases $\{v_n\}_{n=1}^{\infty}$ and $\{w_n\}_{n=1}^{\infty}$ of X are said to be equivalent if

$$\sum_{n=1}^{\infty} c_n v_n$$

is convergent if, and only if,

$$\sum_{n=1}^{\infty} c_n w_n$$

is convergent. ◇

7.4.3 *Characterization of Equivalent Bases*

The property of being linked by an isomorphism of the space completely characterizes equivalent bases. Let X be a Banach space. Two bases $\{v_n\}_{n=1}^{\infty}$ and $\{w_n\}_{n=1}^{\infty}$ of X are equivalent if, and only if, there exists a bounded invertible operator $T : X \longrightarrow X$ such that

$$Tv_n = w_n, \text{ for every } n.$$

In fact, the sufficiency is trivial. Suppose that $\{v_n\}_{n=1}^{\infty}$ and $\{w_n\}_{n=1}^{\infty}$ are equivalent bases for X. If $x \in X$, with

$$x = \sum_{n=1}^{\infty} c_n v_n,$$

then the series

$$\sum_{n=1}^{\infty} c_n w_n$$

converges to an element Tx in X. The function T, thus defined, is clearly linear, one-to-one and onto, and $Tv_n = w_n$ for every n. To show that T is a bounded invertible operator, we define functions T_n by setting

$$T_n x = \sum_{i=1}^{n} c_i w_i.$$

Then,

$$Tx = \lim_{n\to\infty} T_n x$$

for every x. Since each T_n is bounded, it follows that T is bounded (by the Banach Steinhaus theorem (Theorem 2.6.1)). The open mapping theorem guarantees that T is invertible.

7.4.4 Near Bases

Theorem 7.4.1 (Paley-Wiener) *Let X be a Banach space, $\{\varphi_n\}_{n=1}^{\infty}$ be a basis for X, and assume that $\{f_n\}_{n=1}^{\infty}$ is a sequence of elements of X verifying*

$$\left\| \sum_{i=1}^{n} c_i(\varphi_i - f_i) \right\| \leq \lambda \left\| \sum_{i=1}^{n} c_i \varphi_i \right\|$$

for some constant λ, $0 \leq \lambda < 1$, and all choices of the scalars $c_1, \ldots, c_n$, $(n = 1, 2, 3, \ldots)$. Then, $\{f_n\}_{n=1}^{\infty}$ is a basis for X equivalent to $\{\varphi_n\}_{n=1}^{\infty}$. ◇

Proof It follows, by assumption, that the series

$$\sum_{n=1}^{\infty} c_n(\varphi_n - f_n)$$

is convergent whenever the series

$$\sum_{n=1}^{\infty} c_n \varphi_n$$

is convergent. Define a mapping $T : X \longrightarrow X$ by setting

$$T\left(\sum_{n=1}^{\infty} c_n \varphi_n\right) = \sum_{n=1}^{\infty} c_n(\varphi_n - f_n).$$

Evidently, T is linear and bounded, and

$$\|T\| \leq \lambda < 1.$$

Thus, the operator $I - T$ is invertible (one-to-one and onto). Since

$$(I - T)\varphi_n = f_n, \text{ for every } n,$$

the result follows. Q.E.D.

Let $\{\varphi_n\}_{n=1}^{\infty}$ be a basis for a Banach space X and let $\{f_n\}_{n=1}^{\infty}$ be the associated sequence of coefficient functionals. Let $\{\psi_n\}_{n=1}^{\infty}$ be a sequence of vectors in X for which

$$\sum_{n=1}^{\infty} \|f_n\| \|\varphi_n - \psi_n\| < 1. \tag{7.4.1}$$

Put

$$\lambda = \sum_{n=1}^{\infty} \|f_n\| \|\varphi_n - \psi_n\|.$$

If

$$x = \sum c_i \varphi_i$$

is an arbitrary finite sum, then

$$\begin{aligned}
\left\| \sum c_i (\varphi_i - \psi_i) \right\| &= \left\| \sum f_i(x)(\varphi_i - \psi_i) \right\| \\
&\leq \sum \|f_i(x)(\varphi_i - \psi_i)\| \\
&\leq \lambda \|x\| \\
&= \lambda \left\| \sum c_i \varphi_i \right\|.
\end{aligned}$$

Since $0 \leq \lambda < 1$, by using Theorem 7.4.1, we have $\{\psi_n\}_{n=1}^{\infty}$ is a basis for X equivalent to $\{\varphi_n\}_{n=1}^{\infty}$. The following theorem is now immediate.

Theorem 7.4.2 (Krein-Milman-Rutman theorem) *If $\{\varphi_n\}_{n=1}^{\infty}$ is a basis for a Banach space X, then there exist numbers $\alpha_n > 0$ with the following property: If $\{\psi_n\}_{n=1}^{\infty}$ is a sequence of vectors in X for which*

$$\|\varphi_n - \psi_n\| < \alpha_n, \ n = 1, 2, \ldots,$$

then $\{\psi_n\}_{n=1}^{\infty}$ is a basis for X equivalent to $\{\varphi_n\}_{n=1}^{\infty}$. ◇

Proof Let $\{f_n\}_{n=1}^{\infty}$ be the sequence of coefficient functionals associated with the basis $\{\varphi_n\}_{n=1}^{\infty}$. Let $\alpha_n > 0$ such that

$$\sum_{n=1}^{\infty} \alpha_n \|f_n\| < 1.$$

If $\|\varphi_n - \psi_n\| < \alpha_n$, then

$$\|\varphi_n - \psi_n\| \|f_n\| < \alpha_n \|f_n\|.$$

Hence,

$$\sum_{n=1}^{m} \|\varphi_n - \psi_n\| \|f_n\| < \sum_{n=1}^{m} \alpha_n \|f_n\|.$$

So, passing the limit when $m \to \infty$, we have

$$\sum_{n=1}^{\infty} \|\varphi_n - \psi_n\| \|f_n\| < \sum_{n=1}^{\infty} \alpha_n \|f_n\| < 1.$$

Hence, Eq. (7.4.1) is verified. So, $\{\psi_n\}_{n=1}^{\infty}$ is a basis of X equivalent for $\{\varphi_n\}_{n=1}^{\infty}$. This completes the proof. Q.E.D.

Corollary 7.4.1 *If X is a Banach space with a basis, then every dense subset of X contains a basis.* ◇

Corollary 7.4.2 *The space $C[a, b]$ has a basis consisting entirely of polynomials.* ◇

7.5 Hilbert Bases

A Hilbert basis (named after David Hilbert), [5, 6] is a generalization of Hilbert spaces of the classical notion of an orthonormal basis in linear algebra for euclidean spaces (or Hermitian in the complex case) which are finite dimensional.

Definition 7.5.1 Let X be a pre-Hilbert space and $\{e_n\}_{n=1}^{\infty}$ be a sequence of mutually orthogonal vectors of X.

(i) $\{e_n\}_{n=1}^{\infty}$ is called a Hilbert basis of X, if $\{e_n\}_{n=1}^{\infty}$ is orthonormal (i.e., elements $\{e_n\}_{n=1}^{\infty}$ with have norm 1 and pairwise orthogonal) and $\{e_n\}_{n=1}^{\infty}$ is total, i.e., the linear span of $\{e_n\}_{n=1}^{\infty}$ is dense in X.
(ii) Every vector in X can be written uniquely as an infinite linear combination of the vectors in the basis. ◇

Theorem 7.5.1 (Beck et al. [4]) *(Density of orthogonal polynomials) If w is an exponential decay function, i.e., there is $\alpha > 0$ such that*

$$\int_I e^{\alpha|x|} w(x) \mathrm{d}x < +\infty,$$

then orthogonal polynomials associated with w form a Hilbert basis of the space L_2 $(I, w\, \mathrm{d}\lambda)$. ◇

Theorem 7.5.2 ([5, 6]) *A Hilbert space X admits a countable Hilbert basis if, and only if, X is separable.* ◇

7.6 Riesz Bases

In this part, we recall the closely related concepts of the notions of Riesz bases, Riesz bases with parentheses, and Riesz bases of subspaces, see [11, 25, 26, 30, 31, 34].

7.6.1 On Riesz Bases in a Separable Hilbert Space

It is known that in Hilbert spaces, the most important bases are orthonormal bases. Second in importance are Riesz bases that are bases equivalent to some orthonormal bases. In actual fact, the simplest and perhaps the most obvious way of constructing new bases in a Hilbert space X from old is through an isomorphism acting on the space X. Thus, if $\{v_n\}_{n=1}^{\infty}$ is a fixed but arbitrary basis for X and if there exists an isomorphism J that transforms $\{v_n\}_{n=1}^{\infty}$ into $\{w_n\}_{n=1}^{\infty}$, then $\{w_n\}_{n=1}^{\infty}$ is also a basis for X. In fact, it is easy to see that $\{v_n\}_{n=1}^{\infty}$ and $\{w_n\}_{n=1}^{\infty}$ are equivalent in the following sense:

Definition 7.6.1 Let X be a separable Hilbert space. A basis $\mathscr{R} := \{\varphi_n\}_{n=1}^{\infty}$ in X is called to be Riesz basis for X, if it is equivalent to an orthonormal basis; i.e., $\varphi_n = Ge_n$ for all $n \in \mathbb{N}^*$, where $\{e_n\}_{n=1}^{\infty}$ is an orthonormal basis for X and G is a bounded invertible (one-to-one and onto) operator on X. ◇

Example 7.4 1- Let X be a separable Hilbert space and $\{e_n\}_{n=1}^{\infty}$ be an orthonormal basis for X. Then,

(i) $\{2e_1, e_2, e_3, \ldots\}$ is a Riesz basis for X with bounds 1 and 2.
(ii) $\{e_1, \frac{e_2}{2}, \frac{e_3}{3}, \ldots\}$ is a complete orthogonal sequence and it is a basis for X, but it does not possess a lower bound and hence is not a Riesz basis. 2- The sequence $\{|t|^{\alpha}e^{int}\}_{-\infty}^{+\infty}$, $0 < \alpha < \frac{1}{2}$, is a bounded basis for $L_2(0, 2\pi)$ that is not a Riesz basis.

A Riesz basis $\{\varphi_n\}_{n=1}^{\infty}$ for a Hilbert space X is necessarily bounded, that is to say,

$$0 < \inf_n \|\varphi_n\| \leq \sup_n \|\varphi_n\| < \infty.$$

In fact, if $\{\varphi_n\}_{n=1}^{\infty}$ is obtained from the orthonormal basis $\{e_n\}_{n=1}^{\infty}$ by means of the bounded invertible operator G, then for every n,

$$\frac{1}{\|G^{-1}\|} \leq \|\varphi_n\| \leq \|G\|.$$

If $\{\varphi_n\}_{n=1}^{\infty}$ is a Riesz basis, then there exists a bounded invertible operator S for which

$$Se_n = \frac{e_n}{\|\varphi_n\|}, \ (n = 1, 2, 3, \ldots).$$

Therefore, the operator GS is bounded and invertible, and

$$(GS)e_n = \frac{\varphi_n}{\|\varphi_n\|}, \ (n = 1, 2, 3, \ldots).$$

So, it follows readily from this that if $\{\varphi_n\}_{n=1}^{\infty}$ is a Riesz basis, then the sequence of unit vectors

$$\left\{\frac{\varphi_n}{\|\varphi_n\|}\right\}_{n=1}^{\infty}$$

is also a Riesz basis. We will say that a sequence of vectors forms a Riesz basis with defect N in X if they form a Riesz basis of their closed linear span and if this subspace has a defect N.
To a Riesz basis $\mathscr{R} := \{\varphi_n\}_{n=1}^{\infty}$, we associate a constant

$$r(\mathscr{R}) := \|G^{-1}\|^{-1}, \tag{7.6.1}$$

where G is the operator defined in Definition 7.6.1. This constant depends only on $\mathscr{R}$. In fact, let $\{e_n\}_{n=1}^{\infty}$ and $\{\widetilde{e_n}\}_{n=1}^{\infty}$ be two orthonormal bases, G and $\widetilde{G}$ two bounded invertible operators such that

$$\varphi_n = Ge_n = \widetilde{G}\widetilde{e_n}$$

for all $n \in \mathbb{N}^*$. Let U be the operator defined by $Ue_n = \widetilde{e_n}$ for all $n \in \mathbb{N}^*$. We note that U is bounded and unitary. Indeed, we have

$$\langle e_n, e_n \rangle = \langle \widetilde{e}_n, \widetilde{e}_n \rangle = 1.$$

Hence,

$$\langle Ue_n, Ue_n \rangle = \langle e_n, e_n \rangle.$$

This implies that,

$$\langle U^*Ue_n, e_n \rangle = \langle e_n, e_n \rangle.$$

Thus,

$$\langle (U^*U - I)e_n, e_n \rangle = 0.$$

Hence,

$$(U^*U - I)e_n = 0.$$

So,

$$U^*U - I = 0,$$

or,

$$U^*U = I.$$

Similarly, we can prove that $UU^* = I$. Therefore, U is a unitary operator. Thus,

$$G = \widetilde{G}U,$$

so,

$$\widetilde{G}^{-1} = UG^{-1}.$$

Hence,

$$\|\widetilde{G}^{-1}\| \leq \|G^{-1}\|. \tag{7.6.2}$$

Similarly, we consider the operator $\widetilde{U}$ defined by $\widetilde{U}\widetilde{e_n} = e_n$ for all $n \in \mathbb{N}^*$. We note as well that $\widetilde{U}$ is bounded and unitary, again, we have

$$\widetilde{G} = G\widetilde{U}.$$

So,

$$G^{-1} = \widetilde{U}\widetilde{G}^{-1}.$$

Hence,

$$\|G^{-1}\| \leq \|\widetilde{G}^{-1}\|. \tag{7.6.3}$$

Thus, inequality (7.6.2) and (7.6.3) imply that

$$\|G^{-1}\| = \|\widetilde{G}^{-1}\|.$$

Proposition 7.6.1 $\{\varphi_n\}_{n=1}^{\infty}$ *in a separable Hilbert space X is a Riesz basis if, and only if, it satisfies the following property that the $\langle e_n, u\rangle$ obey*

$$A\|u\|^2 \leq \sum_{n=0}^{\infty} |\langle e_n, u\rangle|^2 \leq B\|u\|^2,$$

for strictly positive constants A and B independent of u. ◇

The following equivalent characterizations of Riesz basis can be found in the R. M. Young's book [34].

Proposition 7.6.2 (Young [34, Theorem 9, p. 32]) *Let X be a separable Hilbert space. Then, the following statements below are equivalent:*

(i) The sequence $\{\varphi_n\}_{n=1}^{\infty}$ forms a Riesz basis for the space X.
(ii) The sequence $\{\varphi_n\}_{n=1}^{\infty}$ is ω-linearly independent and there are numbers $A, B > 0$ such that

$$A\|\varphi\|^2 \leq \sum_{n=1}^{\infty} |\langle \varphi, \varphi_n\rangle|^2 \leq B\|\varphi\|^2 \textit{ for all } \varphi \in X. \tag{7.6.4}$$

(iii) The sequence $\{\varphi_n\}_{n=1}^{\infty}$ *is complete in* X*, and there exist positive constants* A *and* B *such that for an arbitrary positive integer* n *and arbitrary scalars* c_1, $c_2, \ldots, c_n$*, one has*

$$A\sum_{i=1}^{n}|c_i|^2 \leq \left\|\sum_{i=1}^{n} c_i\varphi_i\right\|^2 \leq B\sum_{i=1}^{n}|c_i|^2. \qquad \diamond$$

Definition 7.6.2 The constants A and B of Eq. (7.6.4) are said lower and upper bounds of the Riesz basis $\{\varphi_n\}_{n=1}^{\infty}$. ◇

The second result of this section is formulated as follows:

Proposition 7.6.3 *Let* X *be a separable Hilbert space. If* $\{f_n\}_{n=1}^{\infty}$ *is a Riesz basis of* X *and* $(a_n)_{n=1}^{\infty}$ *is a sequence of scalars such that*

$$0 < \inf_n |a_n|$$

and

$$\sup_n |a_n| < +\infty,$$

then $\{a_n f_n\}_{n=1}^{\infty}$ *is also a Riesz basis of the space* X. ◇

Proof Let $\{c_n\}_{n=1}^{\infty}$ be a sequence of scalars such that

$$\sum_{n=1}^{\infty} c_n a_n f_n = 0.$$

Since $\{f_n\}_{n=1}^{\infty}$ is w-linearly independent, then

$$c_n a_n = 0$$

for all $n \in \mathbb{N}^*$. Furthermore, we get

$$0 < \inf_n |a_n|.$$

Thus, $a_n \neq 0$ for all $n \in \mathbb{N}^*$. Consequently,

$$c_n = 0$$

for each $n \in \mathbb{N}^*$. So, $\{a_n f_n\}_{n=1}^{\infty}$ is w-linearly independent. Now, let $\varphi \in X$, we get

$$\alpha(\varphi) \leq \sum_{n=1}^{\infty} |\langle\varphi, a_n f_n\rangle|^2 = \sum_{n=1}^{\infty} |a_n|^2 |\langle\varphi, f_n\rangle|^2 \leq \beta(\varphi), \tag{7.6.5}$$

where

$$\alpha(\varphi) = \left(\inf_n |a_n|\right)^2 \sum_{n=1}^{\infty} |\langle \varphi, f_n \rangle|^2$$

and

$$\beta(\varphi) = \left(\sup_n |a_n|\right)^2 \sum_{n=1}^{\infty} |\langle \varphi, f_n \rangle|^2.$$

The fact that $\{f_n\}_{n=1}^{\infty}$ is a Riesz basis in X, then due to Proposition 7.6.2, there are A, $B > 0$ such that

$$A\|\varphi\|^2 \leq \sum_{n=1}^{\infty} |\langle \varphi, f_n \rangle|^2 \leq B\|\varphi\|^2.$$

So, Eq. (7.6.5) implies that

$$A_1\|\varphi\|^2 \leq \sum_{n=1}^{\infty} |\langle \varphi, a_n f_n \rangle|^2 \leq B_1\|\varphi\|^2,$$

where $A_1 = (\inf_n |a_n|)^2\ A$ and $B_1 = \left(\sup_n |a_n|\right)^2 B$. Q.E.D.

Example 7.5 We consider the following eigenvalue problem

$$\begin{cases} y'''(x) = \lambda y'(x), & x \in]0, 1[\\ y'(0) + y(0) = 0 \\ y'(1) = 0 \\ y''(0) = 0. \end{cases} \tag{7.6.6}$$

Then, the family of eigenfunctions related to (7.6.6) does not form a Riesz basis in the space $L_2[0, 1]$. Indeed, assume that

$$y(x) = -z(0) + \int_0^x z(s)\mathrm{d}s. \tag{7.6.7}$$

Then, substituting it into the system (7.6.6), we obtain

$$\begin{cases} z''(x) = \lambda z(x), & x \in]0, 1[\\ z'(0) = 0 \\ z(1) = 0. \end{cases} \tag{7.6.8}$$

Hence, by using an elementary calculations, we reveal that the eigenvalues $\{\lambda_n\}_{n=0}^{\infty}$ of the system (7.6.8) have the form

$$\lambda_n = -\left(\frac{\pi}{2}(2n+1)\right)^2, \quad n = 0, 1, 2, \ldots$$

and the corresponding eigenfunctions are given by

$$z_n(x) = \cos\left(\frac{\pi}{2}(2n+1)x\right), \quad n = 0, 1, 2, \ldots.$$

In view of Eq. (7.6.7), we obtain that the eigenfunctions associated to (7.6.6) are

$$y_n(x) = -1 + \frac{2}{\pi(2n+1)}\sin\left(\frac{\pi}{2}(2n+1)x\right).$$

Now, we show that the family $\{y_n\}_{n=0}^{\infty}$ is dense in the space $L_2[0, 1]$. To this interest, let $f \in L_2[0, 1]$ such that $\langle f, y_n\rangle = 0$, for all $n \in \mathbb{N}$. So, we get

$$\begin{aligned}\langle f, y_n\rangle &= \int_0^1 f(x)\left(-1 + \frac{2}{\pi(2n+1)}\sin\left(\frac{\pi}{2}(2n+1)x\right)\right)\mathrm{d}x\\ &= -\int_0^1 f(x)\mathrm{d}x + \int_0^1 f(x)\frac{2}{\pi(2n+1)}\sin\left(\frac{\pi}{2}(2n+1)x\right)\mathrm{d}x\\ &= 0.\end{aligned}$$

This implies that

$$\int_0^1 f(x)\mathrm{d}x = \frac{2}{\pi(2n+1)}\int_0^1 f(x)\sin\left(\frac{\pi}{2}(2n+1)x\right)\mathrm{d}x = a_n.$$

Since $a_n \to 0, n \to \infty$, we obtain

$$\int_0^1 f(x)\mathrm{d}x = 0 \text{ and } a_n = 0 \text{ pour tout } n \in \mathbb{N}.$$

As $\{\sin\left(\frac{\pi}{2}(2n+1)x\right)\}_{n=0}^{\infty}$ is dense in $L_2[0, 1]$, then $f = 0$. Consequently, the family $\{y_n\}_{n=0}^{\infty}$ is dense in the space $L_2[0, 1]$. Furthermore, since it is not minimal, then the family of eigenfunctions $\{y_n\}_{n=0}^{\infty}$ is not a Riesz basis. In fact, let $g \neq 0 \in L_2[0, 1]$ such that $\langle g, y_n\rangle = 0$, for all $n \in \mathbb{N}^*$. By using the same reasoning as above, we get

$$\int_0^1 g(x)\mathrm{d}x = \frac{2}{\pi(2n+1)}\int_0^1 g(x)\sin\left(\frac{\pi}{2}(2n+1)x\right)\mathrm{d}x = 0, \ n \in \mathbb{N}^*.$$

Thus, g is orthogonal to 1. Let us consider E_1 the subspace spanned by the vectors $\{\sin\left(\frac{\pi}{2}(2n+1)x\right)\}_{n\neq 0}$ and E_2 the subspace spanned by the vector $\{\sin\left(\frac{\pi}{2}x\right)\}$. It is easy to see that E_1 is orthogonal to E_2. Moreover,

$$\left\langle 1, \sin\left(\frac{\pi}{2}x\right)\right\rangle = \frac{2}{\pi} \neq 0.$$

So, the family $\{y_n\}_{n=0}^{\infty}$ is not minimal in $L_2[0, 1]$.

Example 7.6 Let's consider the eigenvalue problem (7.6.6) in the space $H^1([0, 1])$ equipped with the inner product

$$\langle f, g\rangle = f(0)\overline{g(0)} + \int_0^1 f'(x)\overline{g'(x)}\mathrm{d}x.$$

Then, the family of eigenfunctions related to Eq. (7.6.6) and given by

$$y_n(x) = -1 + \frac{2}{\pi(2n+1)}\sin\left(\frac{\pi}{2}(2n+1)x\right),$$

is minimal in $H^1([0, 1])$ and does not form a Riesz basis in $H^1([0, 1])$. First, we begin by showing that it is dense in $H^1([0, 1])$. So, let $f \in H^1([0, 1])$ such that

$$\langle f, y_n\rangle = 0,$$

for all $n \in \mathbb{N}$. Hence, we get

$$\begin{aligned}
\langle f, y_n\rangle &= f(0)\overline{y_n(0)} + \int_0^1 f'(x)\overline{y_n'(x)}\mathrm{d}x \\
&= -f(0) + \int_0^1 f'(x)\cos\left(\frac{\pi}{2}(2n+1)x\right)\mathrm{d}x \\
&= 0.
\end{aligned}$$

This implies that

$$\int_0^1 f'(x)\cos\left(\frac{\pi}{2}(2n+1)x\right)\mathrm{d}x = f(0) = 0.$$

As $\{\cos\left(\frac{\pi}{2}(2n+1)x\right)\}_{n=0}^{\infty}$ is dense in $H^1([0, 1])$, then $f' = 0$ and therefore $f = 0$. So, $\{y_n\}_{n=0}^{\infty}$ is dense in $H^1([0, 1])$. Further, it is minimal. Indeed, there exists a family $\{v_n(x)\}_{n=0}^{\infty}$ given by

$$v_n(x) = \frac{4}{\pi(2n+1)} \sin\left(\frac{\pi}{2}(2n+1)x\right)$$

such that

$$\langle y_n, v_k \rangle = \delta_{nk}.$$

Now, it remains to show that the family $\{y_n\}_{n=0}^{\infty}$ is not a Riesz basis. Indeed, if it is a Riesz basis then

$$\langle f, y_n \rangle \to 0, \text{ as } n \to \infty \text{ and } f \in H^1([0, 1]).$$

However, for $f = 1$ we obtain

$$\langle 1, y_n \rangle = -1.$$

Consequently, $\{y_n\}_{n=0}^{\infty}$ is not a Riesz basis in $H^1([0, 1])$.

Example 7.7 Let the differential operator in $L_2[0, 1]$ be defined by

$$\begin{cases} u''(x) + \lambda^2 u(x) = 0, & x \in]0, 1[\\ u(0) = 0 \\ u'(0) - u(1) = 0. \end{cases}$$

Then, its eigenvalues are

$$\lambda_n \approx an + ib\log(n), \ a, b \in \mathbb{R}$$

and its root vectors u_n corresponding to λ_n do not form a basis in $L_2[0, 1]$. Indeed, it is easy to check that

(i) $|Im(\lambda_n)| \to \infty, \quad n \to \infty$
(ii) $\{\lambda_n\}_n$ are with finite multiplicity
(iii) $Im(\lambda_n) = o(Re(\lambda_n)), \quad n \to \infty.$

Consequently, the family of root vectors $\{u_n\}_{n=1}^{\infty}$ does not form a basis in $L_2[0, 1]$.

7.6.2 *Riesz Basis of Jordan Chains*

Let $G(X \longrightarrow X)$ be an operator with compact resolvent. Let λ_k, $k \in \mathbb{N}$, be the eigenvalues of G. Since G has compact resolvent, we have $|\lambda_k| \to \infty$. Let $c \geq 0$, $0 \leq p < 1$, and $\mu_k \in \mathbb{C}\backslash\{0\}$, $k \in \mathbb{N}$, such that

$$|\mu_k - \lambda_k| \leq c|\lambda_k|^p \text{ for almost all } k.$$

From

$$
\begin{aligned}
|\lambda_k| &\le |\mu_k| + |\mu_k - \lambda_k| \\
&\le |\mu_k| + c|\lambda_k|^p
\end{aligned}
$$

we obtain

$$
\left(1 - \frac{c}{|\lambda_k|^{1-p}}\right)|\lambda_k| \le |\mu_k|
$$

for almost all k. Therefore,

$$
\frac{|\lambda_k|}{2} \le |\mu_k|
$$

for almost all k and $|\mu_k| \to \infty$. We also have

$$
\begin{aligned}
|\mu_k| &\le |\lambda_k| + c|\lambda_k|^p \\
&= \left(1 + \frac{c}{|\lambda_k|^{1-p}}\right)|\lambda_k|,
\end{aligned}
$$

which implies

$$
|\mu_k| \le 2|\lambda_k|
$$

for almost all k. Now, suppose that $(x_j)_j$ is a Riesz basis of Jordan chains of G and let J be an isomorphism such that $(Jx_j)_j$ is an orthonormal basis. Then, $(Jx_j)_j$ consists of Jordan chains of JGJ^{-1}, and we may thus assume that $J = I$ and

$$
G = \sum_{k=0}^{\infty}(\lambda_k + N_k)P_k,
$$

where P_k are orthogonal projections onto $\mathscr{L}_G(\lambda_k)$,

$$
N_k : R(P_k) \longrightarrow R(P_k)
$$

are nilpotent operators, and for every k, there is an orthonormal basis of $R(P_k)$ such that the matrix representing N_k in this basis is block diagonal with blocks of the form

$$
\begin{pmatrix}
0 & 1 & & \\
 & \ddots & \ddots & \\
 & & \ddots & 1 \\
 & & & 0
\end{pmatrix}.
$$

We decompose G as

$$
G = G_0 + S_1 + S_2
$$

with

$$G_0 = \sum_{k=0}^{\infty} \mu_k P_k,$$

$$S_1 = \sum_{k=0}^{\infty} (\lambda_k - \mu_k) P_k,$$

and

$$S_2 = \sum_{k=0}^{\infty} N_k P_k.$$

Then, G_0 is a normal operator with compact resolvent, spectrum $\{\mu_k,\ k \in \mathbb{N}\}$, and

$$\mathscr{L}_G(\lambda_k) = R(P_k) \subset \mathscr{L}_{G_0}(\mu_k).$$

According to (2.1.5) and (2.1.6), we have $u \in \mathscr{D}(G)$ if, and only if,

$$\sum_{k=0}^{\infty} \|(\lambda_k + N_k) P_k u\|^2 < \infty$$

and analogous characterizations hold for the domains of G_0 and S_1. We have

$$\begin{aligned}
\|(\lambda_k + N_k) P_k u\| &\leq (|\lambda_k| + 1)\|P_k u\| \\
&\leq (|\mu_k| + c|\lambda_k|^p + 1)\|P_k u\| \\
&\leq (|\mu_k| + 2^p c|\mu_k|^p + 1)\|P_k u\| \\
&\leq (2 + 2^p c)|\mu_k|\|P_k u\|
\end{aligned}$$

for almost all k, hence $\mathscr{D}(G_0) \subset \mathscr{D}(G)$. Using Lemma 1.3.5, we also have

$$\begin{aligned}
\|(\lambda_k + N_k) P_k u\| &\geq \frac{1}{2}|\lambda_k|\|P_k u\| \\
&\geq \frac{1}{4}|\mu_k|\|P_k u\|
\end{aligned}$$

for almost all k. This implies $\mathscr{D}(G) \subset \mathscr{D}(G_0)$ and hence

$$\mathscr{D}(G) = \mathscr{D}(G_0).$$

Since $\mu_k \neq 0$, we have

$$|\lambda_k - \mu_k| \leq C|\mu_k|^p$$

for all $k \in \mathbb{N}$ with some appropriate constant C. The estimate

$$
\begin{aligned}
\left\| \sum_{k=0}^{\infty} (\lambda_k - \mu_k) P_k u \right\|^2 &= \sum_{k=0}^{\infty} |\lambda_k - \mu_k|^2 \| P_k u \|^2 \\
&\leq \sum_{k=0}^{\infty} C^2 |\mu_k|^{2p} \| P_k u \|^2 \\
&\leq C^2 \left(\sum_{k=0}^{\infty} |\mu_k|^2 \| P_k u \|^2 \right)^p \left(\sum_{k=0}^{\infty} \| P_k u \|^2 \right)^{1-p}
\end{aligned}
$$

implies that $\mathscr{D}(G_0) \subset \mathscr{D}(S_1)$ and that S_1 is p-subordinate to G_0. Since S_2 is bounded and $0 \in \rho(G_0)$, S_2 is also p-subordinate to G_0. Thus, we have proved the following.

Lemma 7.6.1 (Wyss [30, Lemma 3.4.12]) *Let $G(X \longrightarrow X)$ be an operator with compact resolvent and has a Riesz basis of Jordan chains. Let λ_k, $k \in \mathbb{N}$, be the eigenvalues of G, $c \geq 0$, $0 \leq p < 1$, and $\mu_k \in \mathbb{C}\backslash\{0\}$, $k \in \mathbb{N}$, such that*

$$
|\mu_k - \lambda_k| \leq c|\lambda_k|^p \textit{ for almost all } k.
$$

Then, there is an isomorphism $J : X \longrightarrow X$, a normal operator $G_0(X \longrightarrow X)$ with compact resolvent, and an operator $S_0(X \longrightarrow X)$ p-subordinate to G_0 such that

$$
J\mathscr{D}(G) = \mathscr{D}(G_0),
$$

$$
JGJ^{-1} = G_0 + S_0,
$$

and

$$
\sigma(G_0) = \{\mu_k,\ k \in \mathbb{N}\}.
$$

In addition, J maps the Riesz basis of Jordan chains of G onto an orthonormal basis of eigenvectors of G_0 such that $x \in \mathscr{L}_G(\lambda_k)$ implies $Jx \in \mathscr{L}_{G_0}(\mu_k)$. ◇

7.6.3 Basis Property of the Exponential Family

We recall the following theorem on the basis property of the exponential family $\{e^{i\lambda_n t}\}$.

Theorem 7.6.1 (Pavlov [21]) *Let $\Lambda := \{\lambda_n\}_1^\infty$ be a countable set of the complex numbers. The family $\{e^{i\lambda_n t}\}_1^\infty$ forms a Riesz basis in $L_2(0, T)$ if, and only if, the following conditions are satisfied:*

(i) the sequence $\{\lambda_n\}_1^\infty$ lies in a strip parallel to the real axis:

$$
\sup_n |Im\lambda_n| < \infty;
$$

(ii) $\{\lambda_n\}_1^\infty$ *is separated in the sense that*

$$\delta(\Lambda) := \inf_{m \neq n} |\lambda_n - \lambda_m| > 0;$$

(iii) *the generating function of the family* $\{e^{i\lambda_n t}\}_1^\infty$ *on the interval* $(0, T)$ *satisfies the Muckenhoupt condition*

$$\sup_{I \in \mathscr{J}} \left\{ \frac{1}{|I|^2} \int_I |f(x+ih)|^2 \mathrm{d}x \int_I |f(x+ih)|^{-2} \mathrm{d}x \right\} < \infty$$

for some $h \in \mathbb{R}$*, where* $\mathscr{J}$ *is the set of all intervals of the real axis.* ◇

7.6.4 Hilbert-Schmidt Operators

Let $\{\varphi_n\}_{n=1}^\infty$ be a Riesz basis on X. There is an bounded invertible operator G and an orthonormal basis $\{e_n\}_{n=1}^\infty$ of X such that

$$\varphi_n = Ge_n \text{ for all } n.$$

Consider the operators

$$Su = \sum_{n=1}^\infty \langle \varphi_n, u \rangle \phi_n, \tag{7.6.9}$$

$$Vu = \sum_{n=1}^\infty \langle \varphi_n, u \rangle \varphi_n, \tag{7.6.10}$$

and

$$Wu = \sum_{n=1}^\infty \langle \phi_n, u \rangle \varphi_n, \tag{7.6.11}$$

where $\{\phi_n\}_{n=1}^\infty$ is a sequence of vectors of X. The space $N(S)$ is trivial if, and only if, the ϕ_n are linearly independent since any sequence $(c_n)_{n=1}^\infty \in l^2$ is the set of "Fourier coefficients" of some vector $u \in X$. The space $N(W)$ is trivial if, and only if, the ϕ_n are a basis since any vector orthogonal to all the ϕ_n lies in $N(W)$, and the reverse inclusion follows from the linear independence of the vectors φ_n. Moreover, V is self-adjoint. Since the $(\varphi_n)_{n=1}^\infty$ form a Riesz basis,

$$\sigma(V) \subset [A, B],$$

where A and B are given in (7.6.4). Since $A > 0$, V^{-1} exists and, we have

$$\begin{aligned}
\|V^{-1}(S-V)u\| &= \frac{1}{A}\left\|\sum_{n=1}^{\infty}\langle\varphi_n, u\rangle(\phi_n-\varphi_n)\right\| \\
&\leq \frac{1}{A}\left(\sum_{n=1}^{\infty}|\langle\varphi_n, u\rangle|\right)^{\frac{1}{2}}\left(\sum_{n=1}^{+\infty}\|\phi_n-\varphi_n\|^2\right)^{\frac{1}{2}} \\
&< \frac{B^{\frac{1}{2}}}{A}\left(\sum_{n=1}^{+\infty}\|\phi_n-\varphi_n\|^2\right)^{\frac{1}{2}}\|u\| \\
&< \|u\|
\end{aligned} \tag{7.6.12}$$

and similarly,

$$\begin{aligned}
\|V^{-1}(W-V)u\| &\leq \frac{1}{A}\left\|\sum_{n=1}^{\infty}\langle\phi_n-\varphi_n, u\rangle\varphi_n\right\| \\
&< \|u\|.
\end{aligned} \tag{7.6.13}$$

Let us define the operator T by

$$\begin{cases} T: X \longrightarrow X \\ \quad f \longrightarrow \displaystyle\sum_{n=1}^{\infty}\langle f, e_n\rangle(\varphi_n-\phi_n). \end{cases} \tag{7.6.14}$$

Theorem 7.6.2 *Let $\{\varphi_n\}_{n=1}^{\infty}$ be a Riesz basis on X, $\{e_n\}_{n=1}^{\infty}$ be an orthonormal basis for X such that*

$$\varphi_n = Ge_n \text{ for all } n,$$

where G is a bounded invertible operator and $\{\phi_n\}_{n=1}^{\infty}$ is a sequence of vectors of X w-linearly independent. Let T be the operator defined in (7.6.14). Then, T is a Hilbert-Schmidt operator. ◇

Proof Let $f \in X$, we have

$$\begin{aligned}
\|Tf\| &= \left\|\sum_{n=1}^{\infty}\langle f, e_n\rangle(\varphi_n-\phi_n)\right\| \\
&\leq \sum_{n=1}^{\infty}|\langle f, e_n\rangle|\|\varphi_n-\phi_n\|.
\end{aligned}$$

By the Hölder inequality (Lemme 1.3.2), we have

$$\sum_{n=1}^{\infty} |\langle f, e_n \rangle| \, \|\varphi_n - \phi_n\| \leq \left(\sum_{n=1}^{\infty} |\langle f, e_n \rangle|^2 \right)^{\frac{1}{2}} \left(\sum_{n=1}^{\infty} \|\varphi_n - \phi_n\|^2 \right)^{\frac{1}{2}}.$$

So,

$$\|Tf\| \leq \|f\| \left(\sum_{n=1}^{\infty} \|\varphi_n - \phi_n\|^2 \right)^{\frac{1}{2}}.$$

Hence,

$$\|T\| \leq \left(\sum_{n=1}^{\infty} \|\varphi_n - \phi_n\|^2 \right)^{\frac{1}{2}}.$$

Thus, T is a bounded linear operator. Moreover, since

$$\begin{aligned} Te_n &= \sum_{k=1}^{\infty} \langle e_n, e_k \rangle (\varphi_k - \phi_k) \\ &= \sum_{k=1}^{\infty} \delta_{nk} (\varphi_k - \phi_k) \\ &= \varphi_n - \phi_n, \end{aligned}$$

then

$$\sum_{n=1}^{\infty} \|Te_n\|^2 = \sum_{n=1}^{\infty} \|\varphi_n - \phi_n\|^2 < \infty.$$

So, T is a Hilbert-Schmidt operator. This completes the proof. Q.E.D.

7.6.5 Perturbation of Riesz Bases in a Separable Hilbert Space

Let $\mathscr{R} := \{\varphi_n\}_{n=1}^{\infty}$ be a Riesz basis on X. Then, there is a bounded invertible operator G and an orthonormal basis $\{e_n\}_{n=1}^{\infty}$ of X such that

$$\varphi_n = Ge_n$$

for each n. Consider $\{\phi_n\}_{n=1}^{\infty}$ a sequence of vectors of X w-linearly independent such that

$$\sum_{n=1}^{\infty} \|\phi_n - \varphi_n\|^2 < \infty,$$

and T be the operator defined in (7.6.14). By Theorem 7.6.2, we have T is a Hilbert-Schmidt operator. Hence, T is compact. Consider the linear operator $\widetilde{G} = G - T$. It is easy to see that $\widetilde{G}$ is bounded. Moreover, $\widetilde{G}$ is injective. In fact, let $x \in N(\widetilde{G})$. So,

$$\widetilde{G}x = 0.$$

Or,

$$\begin{aligned}\widetilde{G}x &= Gx - Tx \\ &= G\left(\sum_{n=1}^{\infty}\langle x, e_n\rangle e_n\right) - \sum_{n=1}^{\infty}\langle x, e_n\rangle(\varphi_n - \phi_n) \\ &= \sum_{n=1}^{\infty}\langle x, e_n\rangle\varphi_n - \sum_{n=1}^{\infty}\langle x, e_n\rangle(\varphi_n - \phi_n) \\ &= \sum_{n=1}^{\infty}\langle x, e_n\rangle\phi_n.\end{aligned}$$

The fact that $\{\phi_n\}_{n=1}^{\infty}$ is w-linearly independent, then

$$\langle x, e_n\rangle = 0 \text{ for all } n.$$

So, $x = 0$. Moreover, we have

$$\widetilde{G} = G - T = G(I - G^{-1}T).$$

Hence,

$$I - G^{-1}T = G^{-1}\widetilde{G}$$

is also injective. Since $G^{-1}T$ is compact, then by using Corollary 2.2.1, we have $I - G^{-1}T$ is invertible. Hence, together with the fact that G is invertible, leads to

$$\widetilde{G} = G - T = G(I - G^{-1}T)$$

is invertible. By using the equality $\widetilde{G}e_n$ as

$$\widetilde{G}e_n = \sum_{k=1}^{\infty}\langle e_n, e_k\rangle\phi_k = \phi_n,$$

we get that the sequence of vectors $\{\phi_n\}_{n=1}^{\infty}$ is equivalently to the orthonormal basis $\{e_n\}_{n=1}^{\infty}$. So, $\{\phi_n\}_{n=1}^{\infty}$ is also a Riesz basis on X. Now, let us now recall a result due to J. Lutgen [19, Theorem 1.3]:

Theorem 7.6.3 (Lutgen [19, Theorem 1.3]) *Let X be a separable Hilbert space. If $\mathscr{R} := \{\varphi_n\}_{n=1}^{\infty}$ is a Riesz basis for X and if $\{\phi_n\}_{N+1}^{\infty} (N \geq 0)$ is a sequence of vectors such that*

$$\sum_{n=N+1}^{\infty} \|\phi_n - \varphi_n\|^2 < r(\mathscr{R})^2,$$

then $\{\varphi_n\}_1^N \bigcup \{\phi_n\}_{N+1}^\infty$ *is also a Riesz basis for* X. *In particular, we call that* $\{\Phi_j\}_{j=N+1}^\infty$ *is a Riesz basis of defect* N. ◇

By the same methods we can also prove the following continuity statement for $r(\mathscr{R})$.

Lemma 7.6.2 *Let* $\mathscr{R} := \{\varphi_j\}_{j=1}^\infty$ *be a Riesz basis for a separable Hilbert space* X, $\{\psi_j\}_{j=1}^\infty \subset X$ *be any sequence of normalized vectors, and* $\delta > 0$ *be given, and define*

$$\Phi_j := \varphi_j - \varepsilon_j \psi_j,$$

$\varepsilon_j \in \mathbb{C}$ $(j \in \mathbb{N}^*)$. *If* $\|\{\varepsilon_j\}_{j=1}^\infty\|_{l^2(\mathbb{N})}$ *is sufficiently small, then* $\widetilde{\mathscr{R}} := \{\Phi_j\}_{j=1}^\infty$ *is also a Riesz basis for* X *and*

$$r(\mathscr{R})^2 - \delta < r(\widetilde{\mathscr{R}})^2 < r(\mathscr{R})^2 + \delta. \qquad \diamond$$

Theorem 7.6.4 (Schueller [24, Lemma 1.1]) *Let* X *be a separable Hilbert space and* $(\varphi_n)_{n=1}^\infty$ *be a Riesz basis of* X. *Let* $(\phi_n)_{n=1}^\infty$ *be a collection of vectors in* X *satisfying*

$$\sum_{n=1}^{+\infty} \|\phi_n - \varphi_n\|^2 < \frac{A^2}{B},$$

where A *and* B *are given in (7.6.4). Then, the vectors* $(\phi_n)_{n=1}^\infty$ *form a Riesz basis of the space* X. ◇

Proof Consider the operators S, V and W defined in (7.6.9), (7.6.10) and (7.6.11), respectively. V is self-adjoint. Since the $(\varphi_n)_{n=1}^\infty$ form a Riesz basis,

$$\sigma(V) \subset [A, B].$$

Since $A > 0$, V^{-1} exists and

$$\|V^{-1}\| < \frac{1}{A}.$$

It suffices to show that

$$\|V^{-1}(S - V)\| < 1$$

and

$$\|V^{-1}(W - V)\| < 1.$$

Using Eqs. (7.6.12) and (7.6.13), we have

$$\|V^{-1}(S - V)u\| < \|u\|$$

and

$$\|V^{-1}(W - V)u\| < \|u\|.$$

Thus, W^{-1} and S^{-1} both exist and are bounded. So, that $(\phi_n)_{n=1}^{\infty}$ is a Riesz basis of X. Q.E.D.

7.6.6 Riesz Basis of Operator-Valued Functions

Let Ω be an open connected subset of $\mathbb{C}$ and X be a separable Hilbert space. The investigations are concerned with some spectral properties of operator-valued functions of the form

$$\mathscr{A}(\lambda) := A - \lambda + \mathscr{Q}(\lambda),$$

where A is always an operator in X with compact resolvent, $\mathscr{Q}(\lambda)$ is an A-compact linear operator in X for each $\lambda \in \Omega$, i.e., $\mathscr{Q}(\lambda)(A-\mu)^{-1}$ is compact in X for any $\lambda \in \Omega$, $\mu \in \rho(A)$, and the compact-valued operator function

$$\lambda \longrightarrow \mathscr{Q}(\lambda)(A-\mu)^{-1}$$

is analytic for some (and hence, by the resolvent identity, for any) $\mu \in \rho(A)$. Note that, in the same way, it follows that

$$\lambda \longrightarrow \mathscr{Q}(\lambda)(A-\lambda)^{-1}$$

is analytic on $\rho(A)\bigcap\Omega$. Due to the relative compactness, $\mathscr{A}(\lambda)$ is a closed operator with

$$\mathscr{D}(\mathscr{A}(\lambda)) = \mathscr{D}(A)$$

for every λ. Suppose the eigenvalues $\{\lambda_j\}_{j=1}^{\infty}$ of A (listed in order of increasing magnitude) are all simple, and let $\{\overline{\mathbb{D}}(\lambda_j, \rho_j)\}_{j=1}^{\infty}$ denote the closure of any sequence of open disks of $\mathbb{C}$ with radii ρ_j, centered at the eigenvalues λ_j, and such that

$$\overline{\mathbb{D}}(\lambda_j, \rho_j)\bigcap\sigma(A) = \{\lambda_j\}.$$

In case $\overline{\mathbb{D}}(\lambda_j, \rho_j) \subset \Omega$, define the constants

$$m_j := \max_{\lambda \in \partial\overline{\mathbb{D}}(\lambda_j, \rho_j)} \|\mathscr{Q}(\lambda)(A-\lambda)^{-1}\|,$$

and

$$b_j := \rho_j \max_{\lambda \in \partial\overline{\mathbb{D}}(\lambda_j, \rho_j)} \|(A-\lambda)^{-1}\|.$$

Note that if A is normal and if $\mathrm{dist}(z, \sigma(A)) = |z-\lambda_j|$ for all $z \in \partial\overline{\mathbb{D}}(\lambda_j, \rho_j)$, then

$$b_j = 1.$$

Theorem 7.6.5 (Lutgen [19, Theorem 2.1]) *If a sequence* $\mathscr{R} := \{\varphi_j\}_{j=1}^{\infty}$ *of eigenvectors of* A *corresponding to the eigenvalues* $\{\lambda_j\}_{j=1}^{\infty}$ *forms a Riesz basis for* X *and if there is an integer* $N \geq 0$ *such that* $\overline{\mathbb{D}}(\lambda_j, \rho_j) \subset \Omega$, $m_j < 1$ *for all* $j \geq N+1$ *and*

$$\sum_{j=N+1}^{\infty} b_j^2 \left(\frac{m_j}{1-m_j} \right)^2 < r(\mathscr{R})^2,$$

then $\Sigma(\mathscr{A}) \bigcap \Omega$ *is discrete, for each* $j \geq N+1$ *the spectrum of* $\mathscr{A}$ *in* $\overline{\mathbb{D}}(\lambda_j, \rho_j)$ *consists of one simple eigenvalue* Λ_j, *and there exist corresponding eigenvectors* ψ_j *of* $\mathscr{A}$ *such that* $\{\varphi_j\}_{j=1}^{N} \bigcup \{\psi_j\}_{j=N+1}^{\infty}$ *is also a Riesz basis for* X. ◇

Without loss of generality, we may assume $0 \in \rho(A)$. The proof of the theorem uses the theory of Fredholm-valued analytic operator-functions summarized above. In order to carry this out technically, we need to replace the operator-functions $\mathscr{A}$ and $A - \lambda$ by the analytic Fredholm-valued operator-functions $V, W : \Omega \longrightarrow \mathscr{L}(X)$ defined by

$$V(\lambda) = I - \lambda A^{-1} + \mathscr{Q}(\lambda) A^{-1},$$

$$W(\lambda) = I - \lambda A^{-1}$$

so that

$$V(\lambda) = (I + \mathscr{Q}(\lambda)(A - \lambda)^{-1}) W(\lambda)$$

for $\lambda \in \rho(A) \bigcap \Omega$. Regarding these we make only three observations:

(i) $\Sigma(\mathscr{A}) \subset \Sigma(V)$,
(ii) $\Sigma_p(\mathscr{A}) = \Sigma_p(V)$ and the corresponding geometric and the algebraic multiplicities are equal,
(iii) $\Sigma_p(W) = \Sigma(W) = \sigma(A)$ and eigenspaces are identical.

The first two points follow from

$$\mathscr{A}(\lambda) A^{-1} = V(\lambda).$$

If $\{x_j\}_{j=0}^{m-1}$ is a chain of eigen- and associated vectors of V for λ_0, then

$$\{A^{-1} x_j\}_{j=0}^{m-1} \subset \mathscr{D}(A)$$

is a chain of eigen- and associated vectors of $\mathscr{A}$ for λ_0 in the strong sense. Conversely, if $y_0, \ldots, y_{m-1} \in \mathscr{D}(A)$ is a chain of eigen- and associated vectors of $\mathscr{A}$, then we can choose $x_0, \ldots, x_{m-1}$ such that

$$y_j = A^{-1} x_j.$$

Hence, $x_0, \ldots, x_{m-1}$ is a chain of eigen- and associated vectors of V. Last, since eigenvectors of W are in $\mathscr{D}(A)$ and

$$A^{-1}(A-\lambda) = W(\lambda) \text{ on } \mathscr{D}(A),$$

the point spectra and eigenspaces are identical. For $\lambda \in \rho(A)$, $A(A-\lambda)^{-1}$ is the bounded inverse of

$$W(\lambda) = (A-\lambda)A^{-1}.$$

Hence,

$$\Sigma(W) \subset \sigma(A).$$

The reverse inclusion is a special case of the first point.

Proof of Theorem 7.6.5 For $\lambda \in \rho(A) \bigcap \partial\overline{\mathbb{D}}(\lambda_j, \rho_j) \bigcap \Omega$ both factors on the right of

$$V(\lambda) = \left[I + \mathscr{Q}(\lambda)(A-\lambda)^{-1}\right] W(\lambda)$$

are boundedly invertible and the discreteness of $\Sigma(\mathscr{A}) \bigcap \Omega$ follows from Theorem 2.5.1 and the comments above. Since for $j \geq N+1$, W is normal with respect to $\overline{\mathbb{D}}(\lambda_j, \rho_j)$, the operator Rouché's theorem, and the estimate

$$\begin{aligned}\|(V(\lambda) - W(\lambda))W(\lambda)^{-1}\| &= \|\mathscr{Q}(\lambda)(A-\lambda)^{-1}\| \\ &\leq m_j \\ &< 1\end{aligned}$$

for $\lambda \in \partial\overline{\mathbb{D}}(\lambda_j, \rho_j)$ imply

$$\begin{aligned}\sum_{\lambda \in \overline{\mathbb{D}}(\lambda_j, \rho_j)} m(\lambda; V) &= \sum_{\lambda \in \overline{\mathbb{D}}(\lambda_j, \rho_j)} m(\lambda; W) \\ &= \sum_{\lambda \in \overline{\mathbb{D}}(\lambda_j, \rho_j)} m_a(\lambda) \\ &= 1,\end{aligned}$$

i.e., $\Sigma(\mathscr{A}) \bigcap \overline{\mathbb{D}}(\lambda_j, \rho_j)$ consists of precisely one simple eigenvalue Λ_j. Let ψ_j be an eigenvector of $\mathscr{A}$ corresponding to Λ_j. Then,

$$\widetilde{\psi}_j := A\psi_j$$

is an eigenvector of V for the eigenvalue Λ_j. Notice also that

$$\widetilde{\varphi}_j := -A^{-1}\varphi_j = -\varphi_j/\lambda_j$$

is an eigenvector of A for the eigenvalue λ_j, and we may take the φ_j to be normalized. We can choose a connected open set U_j such that

$$\overline{\mathbb{D}}(\lambda_j, \rho_j) \subset U_j \subset \Omega$$

and such that V and W are boundedly invertible on $U_j\backslash\{\Lambda_j\}$ and $U_j\backslash\{\lambda_j\}$, respectively. As special cases of the expansion theorem for resolvents of analytic Fredholm-valued operator-functions in [12, Sect. 7], we have the expansions

$$W(\lambda)^{-1}x = \frac{1}{\lambda - \lambda_j}\langle x, \widetilde{\varphi}_j^*\rangle \widetilde{\varphi}_j + H_1(\lambda)x \quad (\lambda \in U_j\backslash\{\lambda_j\}),$$

and

$$V(\lambda)^{-1}x = \frac{1}{\lambda - \Lambda_j}\langle x, \widetilde{\psi}_j^*\rangle \widetilde{\psi}_j + H_2(\lambda)x \quad (\lambda \in U_j\backslash\{\Lambda_j\}),$$

where $H_1, H_2 : U_j \longrightarrow \mathscr{L}(X)$ are analytic, $\widetilde{\varphi}_j^*$ is an eigenvector of the adjoint A^* for the eigenvalue $\overline{\lambda}_j$, and $\widetilde{\psi}_j^*$ is an eigenvector of

$$V^*(\lambda) := [V(\overline{\lambda})]^*$$

for $\overline{\Lambda}_j$. Furthermore, $\widetilde{\varphi}_j$ and $\widetilde{\varphi}_j^*$ satisfy the biorthogonal relation

$$\langle -A^{-1}\widetilde{\varphi}_j, \widetilde{\varphi}_j^*\rangle = \langle W'(\lambda_j)\widetilde{\varphi}_j, \widetilde{\varphi}_j^*\rangle = 1,$$

and this implies

$$\begin{aligned} \langle \widetilde{\varphi}_j, \widetilde{\varphi}_j^*\rangle &= \langle A^{-1}\widetilde{\varphi}_j, A^*\widetilde{\varphi}_j^*\rangle \\ &= \lambda_j\langle A^{-1}\widetilde{\varphi}_j, \widetilde{\varphi}_j^*\rangle \\ &= -\lambda_j \neq 0. \end{aligned}$$

With Cauchy's theorem and an integral formulas, it follows that

$$\begin{aligned} \frac{1}{2\pi i}\int\limits_{\partial\overline{\mathbb{D}}(\lambda_j,\rho_j)} \left(V(\lambda)^{-1} - W(\lambda)^{-1}\right)\widetilde{\varphi}_j \mathrm{d}\lambda &= \langle \widetilde{\varphi}_j, \widetilde{\psi}_j^*\rangle \widetilde{\psi}_j + \lambda_j\widetilde{\varphi}_j \\ &= \langle \widetilde{\varphi}_j, \widetilde{\psi}_j^*\rangle \widetilde{\psi}_j - \varphi_j \\ &= A\left(\langle \widetilde{\varphi}_j, \widetilde{\psi}_j^*\rangle \psi_j + \widetilde{\varphi}_j\right) \\ &= A\frac{1}{\lambda_j}\left(\lambda_j\langle \widetilde{\varphi}_j, \widetilde{\psi}_j^*\rangle \psi_j - \varphi_j\right). \end{aligned}$$

Hence,

$$\begin{aligned}\Phi_j - \Psi_j &= \varphi_j - \lambda_j \langle \widetilde{\varphi}_j, \widetilde{\psi}_j^* \rangle \psi_j \\ &= \frac{1}{2\pi i} \int\limits_{\partial \overline{\mathbb{D}}(\lambda_j, \rho_j)} A^{-1} \left(V(\lambda)^{-1} - W(\lambda)^{-1} \right) \varphi_j \mathrm{d}\lambda, \qquad (7.6.15)\end{aligned}$$

where

$$\Phi_j := \varphi_j - \varepsilon_j \psi_j$$

and

$$\Psi_j := (\lambda_j \langle \widetilde{\varphi}_j, \widetilde{\psi}_j^* \rangle - \varepsilon_j) \psi_j.$$

Here, the complex numbers ε_j are taken to be zero whenever

$$\langle \widetilde{\varphi}_j, \widetilde{\psi}_j^* \rangle \neq 0$$

and otherwise $\varepsilon_j \neq 0$ so that the Ψ_j are all non-zero and form a "rescaled" sequence of eigenvectors of $\mathscr{A}$. Further, we choose the l^2-norm of $\{\varepsilon_j\}_{j=1}^\infty$ sufficiently small that $\{\Phi_j\}_{j=1}^\infty$ is also a Riesz basis for X and

$$r(\mathscr{R})^2 - \delta < r(\{\Phi_j\}_{j=1}^\infty)^2 < r(\mathscr{R})^2 + \delta,$$

where

$$\delta := r(\mathscr{R})^2 - \sum_{j=N+1}^{\infty} b_j^2 \left(\frac{m_j}{1 - m_j} \right)^2 > 0.$$

This choice is possible according to Lemma 7.6.2. We now attempt to show that $\{\Psi_j\}_{j=N+1}^\infty$ is a perturbation of $\{\Phi_j\}_{j=N+1}^\infty$ in the sense of Theorem 7.6.3 by estimating the integral above. The integrand in (7.6.15) can be rewritten for $\lambda \in \partial \overline{\mathbb{D}}(\lambda_j, \rho_j)$ as

$$\begin{aligned} A^{-1} \left(V(\lambda)^{-1} - W(\lambda)^{-1} \right) &= -A^{-1} V(\lambda)^{-1} \left(V(\lambda) - W(\lambda) \right) W(\lambda)^{-1} \varphi_j \\ &= -A^{-1} V(\lambda)^{-1} \mathscr{Q}(\lambda) A^{-1} W(\lambda)^{-1} \varphi_j \\ &= -A^{-1} V(\lambda)^{-1} \mathscr{Q}(\lambda) (A - \lambda)^{-1} \varphi_j \\ &= -A^{-1} W(\lambda)^{-1} \left(I + \mathscr{Q}(\lambda)(A - \lambda)^{-1} \right)^{-1} \mathscr{Q}(\lambda)(A - \lambda)^{-1} \varphi_j \\ &= -(A - \lambda)^{-1} \left(I + \mathscr{Q}(\lambda)(A - \lambda)^{-1} \right)^{-1} \mathscr{Q}(\lambda)(A - \lambda)^{-1} \varphi_j. \end{aligned}$$

Because $m_j < 1$, we have the standard estimate

$$\begin{aligned} \|(I + Q(\lambda)(A - \lambda)^{-1})^{-1}\| &\leq (1 - \|Q(\lambda)(A - \lambda)^{-1}\|)^{-1} \\ &\leq (1 - m_j)^{-1} \end{aligned}$$

for all $\lambda \in \partial \overline{\mathbb{D}}(\lambda_j, \rho_j)$ $(j \geq N + 1)$. Thus,

$$\|\Phi_j - \Psi_j\| \le b_j \frac{m_j}{1 - m_j}$$

for $j \geq N + 1$, and

$$\begin{aligned}\sum_{j=N+1}^{\infty} \|\Phi_j - \Psi_j\|^2 &\le \sum_{j=N+1}^{\infty} b_j^2 \left(\frac{m_j}{1 - m_j}\right)^2 \\ &= r(\mathscr{R})^2 - \delta \\ &< r\left(\{\Phi_j\}_{j=1}^{\infty}\right)^2.\end{aligned}$$

A consequence from Theorem 7.6.3 that

$$\{\Phi_j\}_{j=1}^N \bigcup \{\Psi_j\}_{j=N+1}^{\infty}$$

is a Riesz basis for X. Moreover, we may replace $\Phi_1, \ldots, \Phi_N$ by $\varphi_1, \ldots, \varphi_N$ without altering this property provided, it suffices to choose $\varepsilon_1, \ldots, \varepsilon_N$ small enough. Q.E.D.

7.6.7 *Riesz Basis of Subspaces*

Definition 7.6.3 Let X be a separable Hilbert space and Λ be an arbitrary index set. We denote the subspace generated by a family $(V_\lambda)_{\lambda \in \Lambda}$ of subspaces $V_\lambda \subset X$ by

$$\sum_{\lambda \in \Lambda} V_\lambda := \left\{ x_{\lambda_1} + \cdots + x_{\lambda_n} \text{ such that } x_{\lambda_j} \in V_{\lambda_j}, \text{ and } \lambda_j \in \Lambda, \text{ where } n \in \mathbb{N} \right\}.$$

The family $(V_\lambda)_{\lambda \in \Lambda}$ is called to be complete if

$$\sum_{\lambda \in \Lambda} V_\lambda$$

is dense in X. ◇

Definition 7.6.4 Let X be a separable Hilbert space. A sequence of closed subspaces $\{V_n\}_{n=0}^{\infty}$ is called a Riesz basis of subspaces of X if, there exists an isomorphism J of X, such that $\{J V_n\}_{n=0}^{\infty}$ is a complete system of pairwise orthogonal subspaces. ◇

Definition 7.6.5 Let X be a Banach space. A series

$$\sum_{n=1}^{\infty} x_n$$

is unconditionally convergent if, for every sequence $\{\varepsilon_n\}_{n=1}^{\infty}$, with $\varepsilon_n \in \{-1, 1\}$, the series

$$\sum_{n=1}^{\infty} \varepsilon_n x_n$$

converges. ◇

Definition 7.6.6 A sequence $\{v_k\}_{k=0}^{\infty}$ in a Hilbert space X is called a Riesz basis with parentheses if, there exists a Riesz basis of closed subspaces $\{V_k\}_{k=0}^{\infty}$ of X and a subsequence $(n_k)_k$ of $\mathbb{N}$ with $n_0 = 0$ such that the family $(v_{n_k}, \dots, v_{n_{k+1}-1})$ is a basis of V_k. In this case, every $x \in X$ has a unique representation

$$x = \sum_{k=0}^{\infty} \Big(\sum_{j=n_k}^{n_{k+1}-1} \alpha_j v_j \Big), \quad \alpha_j \in \mathbb{C},$$

where the series over k converges unconditionally. ◇

In the following we give a characterization for Riesz bases of subspaces, (see for example Gohberg and Krein [11] and Wyss [30] for the proof).

Proposition 7.6.4 *For a sequence $(V_k)_{k=0}^{\infty}$ of closed subspaces of X, the following assertions are equivalent:*

(i) *$(V_k)_{k=0}^{\infty}$ is a Riesz basis of subspaces for X.*
(ii) *The sequence $(V_k)_{k=0}^{\infty}$ is complete and there is $c \geq 1$ satisfying*

$$c^{-1} \sum_{k \in F} \|x_k\|^2 \leq \Big\| \sum_{k \in F} x_k \Big\|^2 \leq c \sum_{k \in F} \|x_k\|^2 \tag{7.6.16}$$

for all finite subsets $F \subset \mathbb{N}$ and $x_k \in V_k$.
(iii) *Every element $x \in X$ has a unique representation in the form*

$$x = \sum_{k=0}^{\infty} x_k$$

with $x_k \in V_k$, where the convergence of the series is unconditional. ◇

Consider $(W_k)_{k \in \Lambda}$ a Riesz basis of subspaces of X with constant c_0 and $(V_{kj})_{j \in J_k}$ a Riesz bases of subspaces of W_k for all $k \in \Lambda$ with common constant c_1. Using the fact that $(W_k)_{k \in \Lambda}$ is complete in X and $(V_{kj})_{j \in J_k}$ is complete in W_k for every $k \in \Lambda$, then the family $(V_{kj})_{k \in \Lambda,\ j \in J_k}$ is complete in X. Consider now $F \subset \Lambda$ as finite, $F_k \subset J_k$ as finite for each $k \in F$, and $x_{kj} \in V_{kj}$. Then, in view of (7.6.16), we get

$$\begin{aligned}\left\| \sum_{k\in F,\ j\in F_k} x_{kj} \right\|^2 &\le c_0 \sum_{k\in F} \left\| \sum_{j\in F_k} x_{kj} \right\|^2 \\ &\le c_0 \sum_{k\in F} c_1 \sum_{j\in F_k} \|x_{kj}\|^2 \\ &= c_0 c_1 \sum_{k\in F,\ j\in F_k} \|x_{kj}\|^2. \end{aligned}$$

Similarly, we can see

$$\left\| \sum_{k\in F,\ j\in F_k} x_{kj} \right\|^2 \ge c_0^{-1} c_1^{-1} \sum_{k\in F,\ j\in F_k} \|x_{kj}\|^2.$$

Thus, we obtain the following.

Lemma 7.6.3 (Wyss [31, Lemma 3.3]) *Let $(W_k)_{k\in\Lambda}$ be a Riesz basis of subspaces of X with constant c_0. Let $(V_{kj})_{j\in J_k}$ be Riesz bases of subspaces of W_k for all $k \in \Lambda$ with common constant c_1. Then, the family $(V_{kj})_{k\in\Lambda,\ j\in J_k}$ is a Riesz basis of subspaces of X with constant c_0c_1.* ◇

Let X be a Hilbert space and let $(P_k)_{k=0}^{\infty}$ be a sequence of projections in X satisfying

$$P_j P_k = 0$$

for $j \neq k$. Suppose that the family $(R(P_k))_{k=0}^{\infty}$ is complete in X and that

$$\sum_{k=0}^{\infty} |\langle P_k x, y\rangle| \le C\|x\|\|y\| \quad \text{for all } x,\ y \in X \tag{7.6.17}$$

with some constant $C \geq 0$. From

$$\begin{aligned}\left| \left\langle \sum_{k=0}^{n} P_k x, y \right\rangle \right| &\le \sum_{k=0}^{n} |\langle P_k x, y\rangle| \\ &\le C\|x\|\|y\|\end{aligned}$$

we conclude that

$$\left\| \sum_{k=0}^{n} P_k \right\| \le C$$

for all $n \in \mathbb{N}$. This assertion remains valid after an arbitrary rearrangement of the sequence $(P_k)_{k=0}^{\infty}$ since (7.6.17) still holds for the rearranged sequence. An application of Eqs. (1.3.2), (2.1.2), and Proposition 7.6.4, shows that

$$(R(P_k))_{k=0}^{\infty} \tag{7.6.18}$$

is a Riesz basis of subspaces of X with constant $c = 4C^2$.

Theorem 7.6.6 (Markus [20, Lemma 6.1]) *Let $\{P_k\}_{k=1}^{\infty}$ be a sequence of disjoint projections (i.e., $P_j P_k = \delta_{jk} P_k$). If the sequence of subspaces*

$$R_k = R(P_k)$$

($k \in \mathbb{N}^*$) i s complete in X, then it is a basis for X if, and only if,

$$\sup_n \left\| \sum_{k=1}^{n} P_k \right\| < \infty. \qquad \diamond$$

Begining with related concepts of finitely spectral Riesz basis, see [32] for more details.

Definition 7.6.7 Let G be a closed operator on a separable Hilbert space X. We call that a Riesz basis of subspaces $\{V_k\}_{k=0}^{\infty}$ of X is finitely spectral for G if each V_k is finite dimensional, G-invariant, $V_k \subset \mathscr{D}(G)$, the sets $\sigma(G_{|V_k})$ are pairwise disjoint, and the space

$$\sum_{k=0}^{\infty} V_k$$

is a *core* for the operator G. $\diamond$

Let G be closed with non-empty resolvent set and $\dim \mathscr{L}_G(\lambda) < \infty$ for all $\lambda \in \sigma_p(G)$, where $\mathscr{L}_G(\lambda)$ is given in (1.10.1). Then, for the assertions

(i) G has a Riesz basis of Jordan chains,
(ii) the root subspaces $\mathscr{L}_G(\lambda)$ of G form a Riesz basis, and
(iii) G has a finitely spectral Riesz basis of subspaces,

we have $(i) \Rightarrow (ii) \Rightarrow (iii)$. In fact, $(ii) \Longrightarrow (iii)$ is trivial. For $(i) \Longrightarrow (ii)$, consider for each eigenvalue $\lambda \in \sigma_p(G)$ the subspace V_λ generated by all Jordan chains from the basis which correspond to λ. Then, $(V_\lambda)_{\lambda\in\sigma_p(G)}$ is a Riesz basis of subspaces and $V_\lambda = \mathscr{L}_G(\lambda)$.

7.7 $\mathscr{L}$-Basis in $L_2(0, T)$

Let A be a densely defined and closed linear operator in a Banach space X. Consider the evolutionary equation in X

$$\begin{cases} \dfrac{d\psi}{dt}(t) = A\psi(t), \quad t > 0 \\ \psi(0) = \psi_0. \end{cases} \tag{7.7.1}$$

Let A be a generator of a C_0-semigroup $T(t)$ on X. Then, the solution of (7.7.1) can be written into

$$\psi(t) = T(t)\psi_0.$$

In particular, if the spectrum of A, $\sigma(A)$, is of the form

$$\sigma(A) = \sigma_1(A)\bigcup\sigma_2(A),$$

where

$$\sigma_2(A) = \{\lambda_k\}_{k=1}^r$$

is consisted of isolated eigenvalues of A with finite multiplicity, then when $r < \infty$, the solution of (7.7.1) has the following form

$$\psi(t) = \sum_{k=1}^{r} e^{\lambda_k t} P_k(t, \psi_0) + R(t, \psi_0), \tag{7.7.2}$$

where $P_k(t, \psi_0)$ denotes the X-valued polynomials in t. In the most important case when $r = \infty$, it is natural to ask whether the statement (7.7.2) still remains true. When a set

$$\Lambda := \{\lambda_n,\ n \in \mathbb{N}\}$$

is separated, its elements are allowed to repeat themselves. In this case, we will denote by m_n the number of times repeated and said the multiplicity of the element $\lambda_n \in \Lambda$. To every $\lambda_n \in \Lambda$, we associate it with an exponential family given by

$$\mathscr{E}^{(n)}(\Lambda) := \{e^{\lambda_n t}, te^{\lambda_n t}, \ldots, t^{m_n-1}e^{\lambda_n t}\},\ n \geq 1, \tag{7.7.3}$$

where m_n is the multiplicity of $\lambda_n \in \Lambda$.

Definition 7.7.1 We called that the family of exponentials, $\{\mathscr{E}^{(n)}(\Lambda)\}_{n=1}^{\infty}$, is an $\mathscr{L}$-basis in $L_2[0, T]$ if it is a Riesz basis in the subspace

$$U = \overline{\operatorname{span}\left\{\sum_{n=1}^{N}\sum_{j=0}^{m_n-1} \alpha_{n,j} t^j e^{\lambda_n t},\ \ \forall N \geq 1\right\}}. \tag{7.7.4}$$

◇

Denote by $\sharp G$ the number of elements in the set G (or the sequence G), taking the multiplicity into account in this case, and let's pose

$$n^+(r) := \sup_{x\in\mathbb{R}} \sharp\left\{Im\Lambda\bigcap[x, x+r)\right\},$$

$$n^-(r) := \inf_{x\in\mathbb{R}} \sharp \left\{ Im\Lambda \bigcap [x, x+r) \right\}.$$

Define the upper $\mathscr{D}^+(\Lambda)$ and the lower $\mathscr{D}^-(\Lambda)$ uniform densities of Λ by, respectively,

$$\mathscr{D}^+(\Lambda) := \lim_{r\to\infty} \frac{n^+(r)}{r} \tag{7.7.5}$$

and

$$\mathscr{D}^-(\Lambda) := \lim_{r\to\infty} \frac{n^-(r)}{r}.$$

Note that due to the subadditivity of $n^+(r)$ and the superadditivity of $n^-(r)$, both limits exist.

Theorem 7.7.1 (Avdonin and Moran [2, Theorem 3]) *Let $\Lambda = \{\lambda_n\}_{n=1}^\infty$ be a sequence with property*

$$\sup_{n\geq 1} |Re\lambda_n| < \infty.$$

If Λ is separated as a set and

$$\sup_n m_n < \infty,$$

then the following statements are valid.

(i) For any $T < 2\pi\mathscr{D}^-(\Lambda)$, there exists a subfamily $\mathscr{E}_0$ of $\{\mathscr{E}^{(n)}(\Lambda)\}_{n\geq 1}$ that forms a Riesz basis in $L_2(0, T)$, and the family

$$\{\mathscr{E}^{(n)}(\Lambda)\}_{n\geq 1} \backslash \mathscr{E}_0$$

has infinite elements.

(ii) For any $T > 2\pi\mathscr{D}^+(\Lambda)$, the family

$$\{\mathscr{E}^{(n)}(\Lambda)\}_{n\geq 1}$$

forms an $\mathscr{L}$-basis in $L_2(0, T)$, and it can be extended to a family $\mathscr{E}_1$ that forms a Riesz basis in this space. The family $\mathscr{E}_1 \backslash \{\mathscr{E}^{(n)}(\Lambda)\}_{n\geq 1}$ also contains infinite elements. ◇

7.8 Notes and Remarks

- In 1905, G. Hamel [14] proved in a particular case, via a general method, the existence of a basis for a vector space of infinite dimension. Accurately, using the axiom of choice, he demonstrated the existence of a basis of $\mathbb{R}$, seen as a vector space over the rational. This is the first such demonstration of the existence of such a basis for a vector space of infinite dimension. However, when dealing with normed spaces there are many good reasons why one doesn't want to restrict itself to just Hamel basis and needs to create infinite series. For this reason, some bases were discussed in literature such as the Haar basis which was given in 1909 [13]. Moreover, in 1910, Faber [10] discussed a basis for continuous functions on an interval, sometimes called a Faber Schauder system. In 1927, Schauder [22, 23] described the most important basis called Schauder basis.

- On the other hand, it is well known that every Hilbert space has the approximation property which is not the case of Banach spaces and we find that Enflo [9] published the first counterexample in 1973 about this case. Later, many other counterexamples were found and we can cite the space of bounded operators on l^2 which does not have the approximation property [28].

- Since every Banach space having a Schauder basis is separable, the question (Banach's basis problem) whether the converse is true, i.e., whether every infinite-dimensional separable Banach space has a Schauder basis. This question is known as the classical basis problem for Banach spaces [8, 9, 18, 27, 34, 35]. It was posed by Stefan Banach in his book, "Théorie des opérations linéaires" (see [3]) and it remained open for forty years despite the efforts of many good mathematicians, until P. Enflo found a counterexample in 1973 [9]. In fact, he showed how to construct a separable infinite-dimensional Banach space lacking the approximation property. Therefore, every Banach space with a basis has the approximation property. In fact, a distinguished role is played by those Schauder bases that are orthonormal bases in a separable Hilbert space X. An equivalent characterization of such bases is that they are complete orthonormal sequences which means that every separable Hilbert space has an orthonormal basis. More accurately, a Hilbert basis (named after David Hilbert) [5, 6] is a generalization of Hilbert spaces of the classical notion of orthonormal basis in linear algebra for euclidean spaces (or Hermitian in the complex case), which are finite dimensional. Hence, the most important property of an orthonormal basis is the simplicity of all basis expansions. Second, in importance are Riesz bases that are bases equivalent to some orthonormal bases. Thus, if $\{v_n\}_{n=1}^{\infty}$ is a fixed but arbitrary basis for X and if there exists an isomorphism J that transforms $\{v_n\}_{n=1}^{\infty}$ into $\{w_n\}_{n=1}^{\infty}$, then $\{w_n\}_{n=1}^{\infty}$ is also a basis for X.

- Considerable attention has also been devoted by B. S. Pavlov [21] to investigate theorems on the Riesz basis of exponentials in order to give necessary and sufficient conditions assuring the existence of such Riesz basis. On the other hand, the notion Riesz basis of finite-dimensional invariant subspaces was introduced by Wyss in [31] and is equivalent to the existence of a Riesz basis with parentheses of Jordan chains with the additional property that each Jordan chain is completely contained inside some parentheses. Riesz bases of this kind are frequently used in the literature, for example, by Jeribi [17], Markus [20], and Tretter [29].

- Every vector space has a Hamel basis that is finitely linearly independent and whose finite linear span is the entire space. However, when we deal with normed spaces there are many causes why we do not want to restrict ourselves to just finite linear combinations. In normed space, we can use infinite series since we have the notion of convergence.
- In a Hilbert space, the most important bases are orthonormal bases. Second in importance are Riesz bases.
- It should be noted here that a bounded basis for a Hilbert space is not necessary a Riesz basis.
- Every Hilbert space has the approximation property. There are, however, Banach spaces which do not; Per Enflo published the first counterexample in a 1973 article. Later, many other counterexamples were found. The space of bounded operators on l^2 does not have the approximation property [28]. In the Proposition 7.1.1, every Banach space having a Schauder basis is separable. The question (Banach's basis problem) whether the converse is true, i.e., whether every infinite-dimensional separable Banach space has a Schauder basis. This question is known as the classical basis problem for Banach spaces [8, 9, 34, 35]. It was posed by Stefan Banach in his book, "Théorie des opérations linéaires" (see [3]). It remained open for forty years despite the efforts of many good mathematicians, until Per Enflo found a counterexample in 1973 [9]. In fact, P. Enflo showed how to construct a separable infinite dimensional Banach space lacking the approximation property. Therefore, it is of some interest that every Banach space with a basis has the approximation property.

(i) Schauder bases of a vector space of finite dimension are its bases in the algebraic sense.
(ii) An algebraic basis of a Banach space of infinite dimension is never countable, it is a consequence of the theorem of Baire, so is not a Schauder basis.
(iii) The number $\|A\|_2$ is called the norm of Hilbert-Schmidt of A.
(iv) Any Hilbert-Schmidt operator A is bounded, and satisfies the following inequality

$$\|A\|^2 \leq \|A\|_2^2. \tag{7.8.1}$$

In fact, for all $x \in X$, we have

$$\begin{aligned}\|Ax\|^2 &= \sum_{n=1}^{\infty} |\langle Ax, e_n\rangle|^2 \\ &= \sum_{n=1}^{\infty} |\langle x, A^* e_n\rangle|^2 \\ &\le \sum_{n=1}^{\infty} \|x\|^2 \|A^* e_n\|^2 \\ &\le \|x\|^2 \sum_{n=1}^{\infty} \|A^* e_n\|^2 \\ &\le \|A\|_2^2 \|x\|^2.\end{aligned}$$

(v) Using Zorn's lemma and the Gram Schmidt process, one can show that every Hilbert space admits a Hilbert basis.

(vi) [6, 7] Legendre polynomials:

$$\left\{ p_n(x) := \left(\frac{\mathrm{d}}{\mathrm{d}x}\right)^n ((x^2 - 1)^n);\ n \in \mathbb{N} \right\}$$

form a Hilbert basis of $L_2(]-1, 1[,\ \mathrm{d}x)$.

(vii) [6, 7] Laguerre polynomials:

$$\left\{ p_n(x) := \frac{1}{n!} e^x \left(\frac{\mathrm{d}}{\mathrm{d}x}\right)^n (e^{-x} x^n);\ n \in \mathbb{N} \right\}$$

form a Hilbert basis of $L_2([0, +\infty[, e^{-x}\, \mathrm{d}x)$.

(viii) [6, 7] Hermite polynomials:

$$\left\{ p_n(x) := e^{x^2} \left(\frac{\mathrm{d}}{\mathrm{d}x}\right)^n (e^{-x^2});\ n \in \mathbb{N} \right\}$$

form a Hilbert basis of $L_2(\mathbb{R}, e^{-x^2}\, \mathrm{d}x)$.

(ix) It is seen that if the generating function f of $\{e^{i\lambda_n t}\}_1^\infty$ is of sine type, conditions (i) and (iii) of Theorem 7.6.1 are always satisfied.

(x) Let X be a Hilbert space and A be an operator on X. Since each finite-dimensional A-invariant subspace admits a basis consisting of Jordan chains, then a consequence immediate that a Riesz basis of finite-dimensional A-invariant subspaces is equivalent to a Riesz basis with parentheses of Jordan chains such that every Jordan chain lies inside some parenthesis. In view of Lemma 7.6.3, a Riesz basis of finite-dimensional invariant subspaces where almost all subspaces are one dimensional is equivalent to a Riesz basis of eigenvectors and finitely many Jordan chains.

(xi) In the situation of the Proposition 7.6.4, assertion (iii) is equivalent to the existence of a Riesz basis with parentheses of Jordan chains and with the additional property that Jordan chains corresponding to the same eigenvalue lie inside the same parentheses.

References

1. N.I. Akhiezer, *Theory of approximation* (Ungar, New-York, 1956)
2. S. Avdonin, W. Moran, Ingham-type inequalities and Riesz bases of divided differences. Int. J. Appl. Math. Comput. Sci. **11**(4), 803–820 (2001)
3. S. Banach, *Théorie des opérations linéaires* (Warszawa, Monografje Matematyczne, 1932)
4. V. Beck, J. Malick and G. Peyré, Objectif agrégation, diteur(s) : H et K, Amazon (2005)
5. H. Brezis, Analyse fonctionnelle théorie et applications, Ed. Masson, (1983)
6. J.B. Conway, *A course in functional analysis*, 2nd edn. (Springer, New-York, 1990)
7. V.R. Courant, R. Hilbert, *Methods of mathematical physics* (Interscience, 1962)
8. A. M. Davie, The Banach approximation problem, J. Appr. Theo., Collection of articles dedicated to G. G. Lorentz on the occasion of his sixty-fifth birthday, IV. 13 : 392-394 (1975)
9. P. Enflo, A counterexample to the approximation problem in Banach spaces. Acta Math. **130**(3–4), 309–317 (1973)
10. G. Faber, Uber die orthogonalfunktionen des herrn haar. Deutsche Math. Ver (in German) **19**, 104–112 (1910)
11. I.C. Gohberg, M.G. Krein, *Introduction to the theory of linear non-self-adjoint operators in Hilbert space* (Amer. Math. Soci, Providence, 1969)
12. I.C. Gohberg, E.I. Sigal, An operator generalization of the logarithmic residue theorem and the theorem of Rouché. Math. USSR-Sb. **13**(4), 603–625 (1971)
13. A. Haar, Zur theorie der orthogonalen Funktionensysteme. Mathematische Annalen **69**(3), 331–371 (1910)
14. G. Hamel, Eine basis aller zahlen und die unstetigen losungen der funktionalgleichung $f(x+y) = f(x) + f(y)$, Math. Ann. **60**(3), 459–462 (1905)
15. C. Heil, *A basis theory primer* (Birkhauser, New-York, 1998)
16. A. Jeribi, *Spectral theory and applications of linear operators and block operator matrices* (Springer-Verlag, New-York, 2015)
17. A. Jeribi, *Denseness, bases and frames in Banach spaces and applications* (de Gruyter, Berlin, 2018)
18. M. G. Krein, D. P. Milman and M. A. Rutman, On one property of basis in a Banach space, Notes of Math. Soc., Kharkov, 16, 106-110 (1940)
19. J. Lutgen, A note on Riesz bases of eigenvectors of certain holomorphic operator-functions. J. Math. Anal. Appl. **255**, 358–373 (2001)
20. A. S. Markus, Introduction to the spectral theory of polynomial operator pencils, Amer. Math. Soci., Providence, RI, 71 (1988)
21. B.S. Pavlov, Basicity of an exponential systems and Muckenhoupt's condition. Soviet Math. Dokl. **20**, 655–659 (1979)
22. J. Schauder, Zur theorie stetiger abbildungen in funktionalraumen. Math. Z. **26**, 47–65 (1927)
23. J. Schauder, Eine eigenschaft des haarschen orthogonalsystems. Math. Z. **28**, 317–320 (1928)
24. A. Schueller, Uniqueness for near-constant data in fourth-order inverse eigenvalue problems. J. Math. Anal. Appl. **258**, 658–670 (2001)
25. I. Singer, *Bases in Banach spaces I* (Springer-Verlag, 1970)
26. I. Singer, *Bases in Banach spaces II* (Springer-Verlag, 1981)
27. Y. Sonntag, Topologie et analyse fonctionnelle, Cours de licence avec 240 exercices et 30 problmes corrigs, Edition ellipses, (1998)

28. A. Szankowski, $\mathscr{L}(H)$ does not have the approximation property. Acta Math. **147**, 89–108 (1981)
29. C. Tretter, Spectral problems for systems of differential equations $y' + A_0 y = \alpha A_1 y$ with α-polynomial boundary conditions. Math. Nachr. **214**, 129–172 (2000)
30. C. Wyss, Perturbation theory for Hamiltonian operator matrices and Riccati equations, PhD thesis, University of Bern, (2008)
31. C. Wyss, Riesz bases for p-subordinate perturbations of normal operator. J. Func. Anal. **258**, 208–240 (2010)
32. C. Wyss, Hamiltonians with Riesz bases of generalised eigenvectors and Riccati Equations, arXiv:1005.5336v1
33. , (2010)
34. R.M. Young, *An introduction to nonharmonic Fourier series* (Academic press, London, 1980)
35. V. Zeh, A Counterexample to Banach's basis problem. Praktikum mit Bachelorarbeit **101–297**, (2010)

Chapter 8
On a Riesz Basis of Finite-Dimensional Invariant Subspaces

This chapter is devoted to study the Riesz basis of finite-dimensional invariant subspaces for a class of unbounded perturbations of unbounded normal operators, we study the change of the spectrum and we establish the existence of a Riesz basis of finite-dimensional invariant subspaces under an additional a priori assumption on the spectrum of the perturbed operator. The results are applied to two classes of block operator matrices. Moreover, we discuss problems on the expansion of a semigroup and a criterion for being a Riesz basis. Under different conditions on the Riesz projection, the expansion of a semigroup and existence of Riesz basis are studied. The main goal of this chapter is to study the existence of a Riesz basis with parentheses of generalized eigenvectors for a class of normal operators with compact resolvent. In fact, for a class of non-normal perturbations of normal operators we establish different conditions in terms of the spectrum which imply the existence of such Riesz bases. Finally, a Riesz basis associated to a block operator matrix is given.

8.1 Location of the Spectrum

Consider G and S two operators on a Banach space X with $\mathscr{D}(G) \subset \mathscr{D}(S)$ and $T = G + S$. Let $0 < \varepsilon < 1$ and $z \in \rho(G)$ satisfying

$$||S(G - z)^{-1}|| \leq \varepsilon. \tag{8.1.1}$$

The inequalities (8.1.1) implies that $I + S(G - z)^{-1}$ is invertible with

$$(I + S(G - z)^{-1})^{-1} = \sum_{k=0}^{\infty}(-S(G - z)^{-1})^k$$

A. Jeribi, *Perturbation Theory for Linear Operators*,
https://doi.org/10.1007/978-981-16-2528-2_8

and

$$||(I + S(G - z)^{-1})^{-1}|| \leq \frac{1}{1 - ||S(G - z)^{-1}||} \leq \frac{1}{1 - \varepsilon}.$$

As $T - z = (I + S(G - z)^{-1})(G - z)$, then $z \in \rho(T)$ and

$$||(T - z)^{-1}|| \leq ||(I + S(G - z)^{-1})^{-1}|| \, ||(G - z)^{-1}|| \leq \frac{1}{1 - \varepsilon} ||(G - z)^{-1}||.$$

Using that

$$(T - z)^{-1} = (G - z)^{-1}(I + S(G - z)^{-1})^{-1},$$

we conclude that

$$||S(T - z)^{-1}|| \leq \frac{\varepsilon}{1 - \varepsilon}.$$

Let $\Gamma \subset \rho(G)$ be a simply closed, positively oriented integration contour and (8.1.1) holds for all $z \in \Gamma$. Consider the operators

$$T_r = G + rS$$

for $r \in [0, 1]$. We have the power series expansion

$$(I + rS(G - z)^{-1})^{-1} = \sum_{k=0}^{\infty} r^k (-S(G - z)^{-1})^k,$$

$r \in [0, 1]$, which converges uniformly in $z \in \Gamma$. Consequently, $\Gamma \subset \rho(T_r)$, and

$$(T_r - z)^{-1} = (G - z)^{-1}(I + rS(G - z)^{-1})^{-1}$$

is continuous in r uniformly for $z \in \Gamma$. Hence, the Riesz projections P_r of T_r associated with Γ also depend continuously on r. Now, if

$$\|P_r - P_s\| < 1,$$

then there are isomorphisms $R(P_r) \cong R(P_s)$, $N(P_r) \cong N(P_s)$, see [16, I.4.6]. Since r ranges over a compact interval, then $\Gamma \subset \rho(T)$ and for the Riesz projections Q and P of G and T associated with Γ, there are isomorphisms $R(Q) \cong R(P)$, and $N(Q) \cong N(P)$. Now, we give the following lemma.

Lemma 8.1.1 *Let X be a Hilbert space and G be a normal operator on X satisfying*

$$\sigma(G)\bigcap\Omega(2\varphi_-,2\varphi_+)\subset\mathbb{R}_{\geq 0}$$

with $-\pi\leq\varphi_-<0<\varphi_+\leq\pi$, where the sector $\Omega(\cdot,\cdot)$ is defined in (1.13.2). Let S be an operator G-bounded with order p ($p\in[0,1[$), and $T=G+S$. Then, for all $\alpha>b$, $b/\alpha<\varepsilon<1$ and $0<\psi<\min\{-\varphi_-,\varphi_+,\pi/2\}$, there is $r_0>0$ such that the sets

$$\rho_1=\left\{z\in\overline{\Omega(\varphi_-,\varphi_+)}\text{ such that }|z|\geq r_0, z\notin\Omega(\psi)\right\},$$
$$\rho_2=\left\{z=x+iy\in\overline{\Omega(\psi)}\text{ such that }|z|\geq r_0, |y|\geq\alpha x^p\right\},$$

and

$$\rho_3=\left\{z=x+iy\in\overline{\Omega(\psi)}\text{ such that }|z|\geq r_0, |y|\leq\alpha x^p\leq dist(z,\sigma(G))\right\}$$

satisfy $\rho_1\bigcup\rho_2\bigcup\rho_3\subset\rho(T)$ and for $z\in\rho_1\bigcup\rho_2\bigcup\rho_3$, we get

$$||S(G-z)^{-1}||\leq\varepsilon,$$

$$||(T-z)^{-1}||\leq\frac{(1-\varepsilon)^{-1}}{dist(z,\sigma(G))},$$

and

$$||S(T-z)^{-1}||\leq\frac{\varepsilon}{1-\varepsilon}.$$

Furthermore, there is a constant $M>0$ such that (Fig. 8.1)

$$||(T-z)^{-1}||\leq M,$$

for all $z\in\rho_1\bigcup\rho_2\bigcup\rho_3$. ◇

Proof Let $z\in\rho_1\bigcup\rho_2\bigcup\rho_3$. By using the spectral theorem for normal operators (see Eq. (1.13.1)), and writing $d=\text{dist}(z,\sigma(G))$, we obtain

$$||(G-z)^{-1}||=\sup_{\lambda\in\sigma(G)}\frac{1}{|\lambda-z|}=\frac{1}{d}$$

and

$$\begin{aligned}||G(G-z)^{-1}||&=||I+z(G-z)^{-1}||\\&\leq 1+\frac{|z|}{d}.\end{aligned}$$

Since S is G-bounded with order p, by using above the following inequality holds

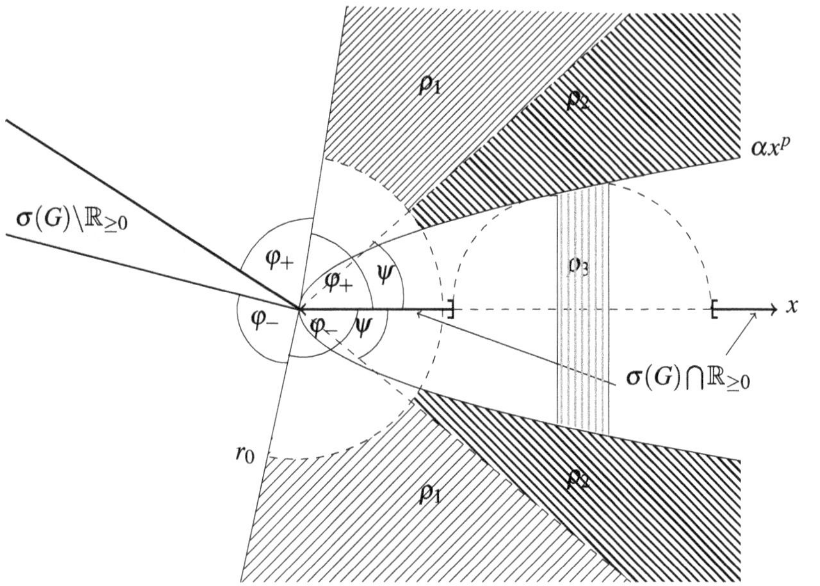

Fig. 8.1 The situation of Lemma 8.1.1

$$
\begin{aligned}
||S(G-z)^{-1}u|| &\leq b||G(G-z)^{-1}u||^p||(G-z)^{-1}u||^{1-p} + a||(G-z)^{-1}u|| \\
&\leq \left[b\left(1+\frac{|z|}{d}\right)^p \frac{1}{d^{1-p}} + \frac{a}{d}\right]||u||
\end{aligned}
$$

for each element $u \in X$. In order to get the desired estimates, it suffices to show that

$$
C = b\left(1+\frac{|z|}{d}\right)^p \frac{1}{d^{1-p}} + \frac{a}{d} \leq \varepsilon.
$$

Note that if $p = 0$, then we may suppose without loss of generality that $a = 0$. For $z = x + iy$, the analysis of the geometry of the situation leads to

$$
\varphi_- \leq arg(z) \leq -\pi/2 \text{ or } \pi/2 \leq arg(z) \leq \varphi_+ \Rightarrow d \geq |z|, \tag{8.1.2}
$$
$$
\max\{\varphi_-, -\pi/2\} \leq arg(z) \leq \min\{\varphi_+, \pi/2\} \Rightarrow d \geq |y|. \tag{8.1.3}
$$

Further, we obtain

$$
\psi \leq |arg(z)| \leq \pi/2 \Rightarrow |y| \geq |z| \sin\psi, \tag{8.1.4}
$$
$$
|arg(z)| \leq \psi \Rightarrow x \geq |z| \cos\psi. \tag{8.1.5}
$$

In what follows, three cases will be discussed:
First case $z \in \rho_1$: If Eq. (8.1.2) is satisfied, together with $d \geq |z|$, for r_0 large enough, then

$$C \leq 2^p b|z|^{p-1} + \frac{a}{|z|} \leq \varepsilon.$$

If $\psi \leq |arg(z)| \leq \pi/2$, then in virtue of both (8.1.3) and (8.1.4) implies $d \geq |z|\sin\psi$. Thus, we conclude, for r_0 large enough, that

$$C \leq b\Big(1 + \frac{1}{\sin\psi}\Big)^p \frac{1}{(|z|\sin\psi)^{1-p}} + \frac{a}{|z|\sin\psi} \leq \varepsilon.$$

Second case $z \in \rho_2$: In view of both Eqs. (8.1.3), (8.1.5) and the following inequality $|y| \geq \alpha x^p$, it is easy to see that

$$d \geq \alpha x^p.$$

Moreover, if $p = 0$, then $d \geq \alpha$ and

$$C = b/d \leq b/\alpha \leq \varepsilon.$$

On the one hand, if $p > 0$, the Minkowski inequality yields

$$\begin{aligned}\Big(1 + \frac{|z|}{d}\Big)^p &\leq \Big(1 + \frac{x + |y|}{d}\Big)^p \\ &\leq 1 + \frac{x^p + |y|^p}{d^p} \\ &\leq 1 + \frac{\alpha^{-1}d + d^p}{d^p} \\ &= 2 + \frac{1}{\alpha}d^{1-p}.\end{aligned}$$

On the other hand, taking account of the fact that $d \geq \alpha x^p$ and Eq. (8.1.5), we obtain

$$d \geq \alpha(|z|\cos\psi)^p,$$

together with the estimate $b/\alpha < \varepsilon$, the last inequality implies the following

$$C \leq 2bd^{p-1} + ad^{-1} + b/\alpha \leq \varepsilon$$

for r_0 sufficiently large.

Third case $z \in \rho_3$: By using the definition of ρ_3, we get $d \geq \alpha x^p$. Hence, the argument used in the second case adapts easily.

To complete the proof of the lemma, we mention that if $z \in \rho_1$, then

$$d \geq |z| \geq r_0 > 0$$

or

$$d \geq |z|\sin\psi \geq r_0 \sin\psi > 0.$$

Furthermore, if $z \in \rho_2 \bigcup \rho_3$, then

$$d \geq \alpha(|z| \cos \psi)^p \geq \alpha(r_0 \cos \psi)^p > 0.$$

Thus, d^{-1} is bounded independently of z and so, we infer that $||(T - z)^{-1}||$ is uniformly bounded. Q.E.D.

The first main important result about the location of the spectrum of T, is given by the following theorem:

Theorem 8.1.1 *Let X be a Hilbert space, G be a normal operator on X whose spectrum lies on finitely many rays $e^{i\theta_j}\mathbb{R}_{\geq 0}$ with $0 \leq \theta_j < 2\pi$, $j = 1, \cdots, n$. If $T = G + S$, where S is G-bounded with order p ($p \in [0, 1[$), then for every $\alpha > b$, there exists $r_0 > 0$ such that the spectrum of T, $\sigma(T)$, satisfies the following*

$$\sigma(T) \subset B_{r_0}(0) \bigcup \bigcup_{j=1}^{n} \{e^{i\theta_j}(x + iy); x \geq 0 \text{ and } |y| \leq \alpha x^p\},$$

where $B_{r_0}(0)$ denotes the open ball with radius r_0 around 0 in $\mathbb{R}^n$ or $\mathbb{C}^n$ (see Fig. 8.2). Moreover, if G has a compact resolvent, then so has T. ◇

Proof Without loss of generality, we may assume that $\theta_1 < \theta_2 < \cdots < \theta_n$ and put $\theta_0 = \theta_n - 2\pi$, $\theta_{n+1} = \theta_0 + 2\pi$. After a rotation by θ_j, we should use the Lemma 8.1.1 to each sector $\Omega(\theta_{j-1}, \theta_{j+1})$. In other words, the lemma will be applied to the three operators $e^{-i\theta_j}G$, $e^{-i\theta_j}S$ and $e^{-i\theta_j}T$ with

$$\varphi_+ = \frac{\theta_{j+1} - \theta_j}{2},$$

$$\varphi_- = \frac{\theta_{j-1} - \theta_j}{2},$$

and some suitable ε. Thus, $z \in \sigma(T)$, $\frac{\theta_{j-1}+\theta_j}{2} \leq \arg z \leq \frac{\theta_j+\theta_{j+1}}{2}$, $|z| \geq r_0$ implies

$$z \in \{e^{i\theta_j}(x + iy) \text{ such that } x \geq 0 \text{ and } |y| \leq \alpha x^p\},$$

with some $r_0 \geq 0$ for each $j = 1, \cdots, n$. Writing the identity

$$(T - z)^{-1} = (G - z)^{-1}(I + S(G - z)^{-1})^{-1}$$

for $z \in \rho(G) \bigcap \rho(T)$, and if G has compact resolvent, then we conclude that T has compact resolvent too (Fig. 8.2). Q.E.D.

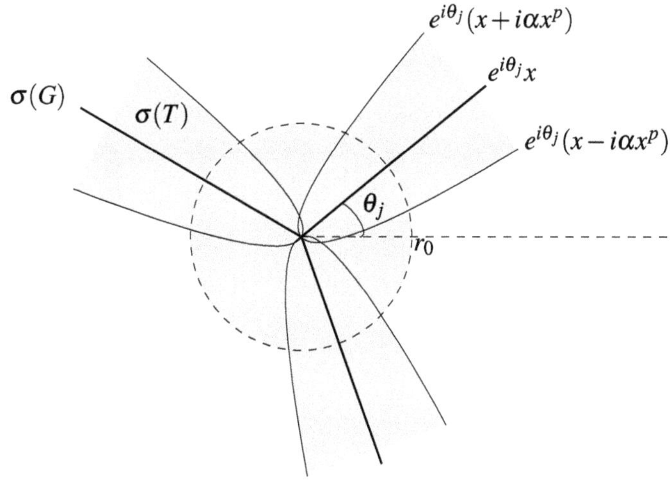

Fig. 8.2 The spectrum after G-bounded with order p perturbation

8.2 Riesz Basis

8.2.1 On a Riesz Basis of Finite-Dimensional Invariant Subspaces

The second main important result, in this section, about the existence of Riesz basis of finite-dimensional T-invariant subspaces, is given by the following theorem:

Theorem 8.2.1 *Let X be a Hilbert space, and G be a normal operator on X, having compact resolvent. Assume that the spectrum of G, $\sigma(G)$, lies on a finite number of rays from the origin and for some $p \in [0, 1[$*

$$\liminf_{r\to\infty} \frac{n(\mathbb{D}(0,r), G)}{r^{1-p}} < \infty. \tag{8.2.1}$$

If the operator $S = T - G$ is G-bounded with order p, then T admits a Riesz basis of finite-dimensional T-invariant subspaces. ◇

To prove Theorem 8.2.1 we need the following elementary results. For $\alpha > b$, let $(x_k)_{k\geq 1}$ be a sequence of positive number, and a constants c_1, c_2 such that

$$\alpha x_1^{p-1} \leq \tan\varphi,$$

$$x_n^{1-p} - x_k^{1-p} \geq c_1(n-k) \text{ for } n \geq k$$

and

$$\text{dist}(x_k, \sigma(G)) \geq c_2 x_k^p \text{ for } k \geq 1.$$

Consider γ_k the lines

$$\gamma_k := \{z = x_k + iy \in \mathbb{C} \text{ such that } |y| \leq \alpha x_k^p\},$$

and $\Gamma_{r,\pm}$ the contours

$$\Gamma_{r,\pm} := \{z = x + iy \in \mathbb{C} \text{ such that } x \geq r > 0,\ y = \pm\alpha x^p\}.$$

By using the previous resolvent estimates, we obtain an estimate for a sequence of Riesz projections associated with the parabola $\Gamma_{r,\pm}$.

Lemma 8.2.1 *Let X be a Hilbert space and G be a normal operator on X with compact resolvent. If $\sigma(G) \bigcap \Omega(2\varphi) \subset \mathbb{R}_{\geq 0}$ with $0 < \varphi \leq \pi/2$ (see the definition of $\Omega(\cdot)$ in (1.13.2)), S is G-bounded with order p ($p \in [0, 1]$), $T = G + S$ and, there is $M \geq 0$ such that $\gamma_k \subset \varrho(T)$ and*

$$||S(T - z)^{-1}|| \leq M$$

for all, $z \in \gamma_k$ $k \geq 1$, then there exist $r_0 > 0$, $k_0 \geq 1$ such that $x_{k_0} \geq r_0$ and the following holds:

- *If Γ_k with $k \geq k_0$ is the positively oriented boundary of the region enclosed by γ_k, $\Gamma_{r_0,-}$ γ_{k+1} and $\Gamma_{r_0,+}$, then $\Gamma_k \subset \rho(T)$.*
- *If P_k is the Riesz projection of T associated with Γ_k, then*

$$\sum_{k=k_0}^{\infty} |\langle P_k u, v\rangle| \leq C\|u\|\|v\| \text{ for all } u,\ v \in X \tag{8.2.2}$$

with some constant $C \geq 0$. ◇

Proof First, we claim the existence of $r_0 > 0$ is assured by Lemma 1.13.1. Second, since

$$x_n^{1-p} - x_k^{1-p} \geq c_1(n - k) \text{ for } n \geq k,$$

then $(x_k)_k$ tends monotonically to infinity and we choose k_0 such that

$$x_{k_0} \geq r_0.$$

By virtue of Lemma 8.1.1, $||S(T - z)^{-1}||$ is uniformly bounded on $\Gamma_{r_0,\pm}$. Hence, there exists a constant $M_0 \geq 0$ such that

$$\Gamma_k \subset \varrho(G) \bigcap \varrho(T)$$

and

$$||S(T-z)^{-1}|| \leq M_0$$

for all $z \in \Gamma_k$ $k \geq k_0$. Now, to establish the validity of (8.2.2), we shall verify that

$$\sum_{k=k_0}^{\infty} |\langle (P_k - Q_k)u, v\rangle| \leq C\|u\|\|v\|,$$

where Q_k is the Riesz projection of G with Γ_k satisfying

$$\begin{aligned} P_k - Q_k &= \frac{i}{2\pi}\int_{\Gamma_k} ((T-z)^{-1} - (G-z)^{-1})dz \\ &= \frac{-i}{2\pi}\int_{\Gamma_k} (T-z)^{-1}S(G-z)^{-1}dz. \end{aligned}$$

Hence, one gets

$$|\langle (P_k - Q_k)u, v\rangle| \leq \frac{1}{2\pi}\int_{\Gamma_k} \|\big((T-z)^{-1}\big)^* v\| \|S(G-z))^{-1}u\||dz|.$$

Since $(T-z)^{-1} = (G-z)^{-1}(I + S(T-z)^{-1})$, we have

$$\|\big((T-z)^{-1}\big)^* v\| \leq (1 + \|S(T-z)^{-1}\|)\|((G-z)^{-1})^* v\|$$

and, hence

$$\|((T-z)^{-1})^* v\| \leq (1 + M_0)\|\big((G-z)^{-1}\big)^* v\|.$$

Moreover, using the fact G is normal, we have

$$\|\big((G-z)^{-1}\big)^* v\| = \|(G-z)^{-1}v\|$$

and, we deduce that

$$\begin{aligned} &\sum_{k=k_0}^{\infty} |\langle (P_k - Q_k)u, v\rangle| \\ &\leq \frac{1+M_0}{2\pi}\sum_{k=k_0}^{\infty}\int_{\Gamma_k} \|S(G-z)^{-1}u\|\|(G-z)^{-1}v\||dz| \\ &\leq \frac{1+M_0}{2\pi}\bigg(\int_{\Gamma_{r_0,\pm}} + 2\sum_{k=k_0}^{\infty}\int_{\gamma_k}\bigg)\|S(G-z)^{-1}u\|\|(G-z)^{-1}v\||dz|. \quad (8.2.3) \end{aligned}$$

Concerning the first term of (8.2.3), the p-boundedness together with the Schwartz inequality allow us to write

$$\int_{\Gamma_{r_0,\pm}} \|S(G-z)^{-1}u\|\|(G-z)^{-1}v\||dz|$$

$$\leq b\int_{\Gamma_{r_0,\pm}} \|G(G-z)^{-1}u\|^p\|(G-z)^{-1}u\|^{1-p}\|(G-z)^{-1}v\||dz| \\ +a\int_{\Gamma_{r_0,\pm}} \|(G-z)^{-1}u\|\|(G-z)^{-1}v\||dz|. \quad (8.2.4)$$

Applying the Holder inequality and Lemma 1.13.1 in the first term of (8.2.4), we get

$$\int_{\Gamma_{r_0,\pm}} \|G(G-z)^{-1}u\|^p\|(G-z)^{-1}u\|^{1-p}\|(G-z)^{-1}v\||dz|$$

$$= \int_{\Gamma_{r_0,\pm}} \left(|z|^{\frac{p(p-2)}{2}}\|G(G-z)^{-1}u\|^p\right)\left(|z|^{-\frac{p(p-2)}{2}}\|(G-z)^{-1}u\|^{1-p}\|(G-z)^{-1}v\|\right)|dz|$$

$$\leq \left(\int_{\Gamma_{r_0,\pm}} |z|^{p-2}\|G(G-z)^{-1}u\|^2|dz|\right)^{p/2}$$

$$\times\left(\int_{\Gamma_{r_0,\pm}} |z|^p\|(G-z)^{-1}u\|^{\frac{2(1-p)}{2-p}}\|(G-z)^{-1}v\|^{\frac{2}{2-p}}|dz|\right)^{\frac{2-p}{2}}$$

$$\leq (C_2\|u\|^2)^{p/2}\left(\int_{\Gamma_{r_0,\pm}} |z|^{\frac{p(1-p)}{2-p}}\|(G-z)^{-1}u\|^{\frac{2(1-p)}{2-p}}|z|^{\frac{p}{2-p}}\|(G-z)^{-1}v\|^{\frac{2}{2-p}}|dz|\right)^{\frac{2-p}{2}}$$

$$\leq (C_2\|u\|^2)^{\frac{p}{2}}\left(\int_{\Gamma_{r_0,\pm}} |z|^p\|(G-z)^{-1}u\|^2|dz|\right)^{\frac{1-p}{2}}\left(\int_{\Gamma_{r_0,\pm}} |z|^p\|(G-z)^{-1}v\|^2|dz|\right)^{\frac{1}{2}}$$

$$\leq C_1^{(2-p)/2}C_2^{\frac{p}{2}}\|u\|\|v\|.$$

For the second term of (8.2.4), we can easily see that the following inequality yields

$$\int_{\Gamma_{r_0,\pm}} \|(G-z)^{-1}u\|\|(G-z)^{-1}v\||dz| \leq \int_{\Gamma_{r_0,\pm}} |z|^p\|(G-z)^{-1}u\|\|(G-z)^{-1}v\||dz|$$

for $|z| \geq r_0 >> 1$. Now, by virtue of Schwartz inequality and Lemma 1.13.1, we get

$$\int_{\Gamma_{r_0,\pm}} \|(G-z)^{-1}u\|\|(G-z)^{-1}v\||dz| \leq \Big(\int_{\Gamma_{r_0,\pm}} |z|^p\|(G-z)^{-1}u\|^2|dz|\Big)^{\frac{1}{2}} \times\Big(\int_{\Gamma_{r_0,\pm}} |z|^p\|(G-z)^{-1}v\|^2|dz|\Big)^{\frac{1}{2}} \leq C_1\|u\|\|v\|.$$

Hence,

$$\int_{\Gamma_{r_0,\pm}} \|S(G-z)^{-1}u\|\|(G-z)^{-1}v\||dz| \leq (b\,C_1^{(2-p)/2}C_2^{\frac{p}{2}} + a\,C_1)\|u\|\|v\|.$$

In the same way, using Lemma 1.13.2, we obtain

$$\sum_{k=k_0}^{\infty}\int_{\gamma_k} \|S(G-z)^{-1}u\|\|(G-z)^{-1}v\||dz| \leq (b\,C_1^{(2-p)/2}C_2' + a\,C_1')\|u\|\|v\|.$$ Q.E.D.

We denote by

$$N_+(r_1, r_2, G) := n(]r_1, r_2[, G)$$

the sum of the multiplicities of all eigenvalues of the operator G in the open interval $]r_1, r_2[$,

$$N_+(r_1, r_2, G) = \sum_{\lambda\in\sigma_p(G)\bigcap]r_1,r_2[} \dim \mathcal{L}_G(\lambda).$$

Lemma 8.2.2 *If G is normal with compact resolvent, $\sigma(G)\bigcap\Omega(2\varphi)\subset\mathbb{R}_{\geq 0}$ with $0<\varphi\leq\frac{\pi}{2}$ (see the definition of $\Omega(\cdot)$ in (1.13.2)), S is G-bounded with order p and with bound b ($p\in[0,1[$), $T=G+S$, $l>b$, $0\leq l_0<l-b$ and $\eta>0$, then there are constants C_0, C_1, $r_0>0$ such that for every $r\geq r_0$, there is a set $E_r\subset\mathbb{C}$ with the following properties:*

(i) E_r is the union of finitely many discs with radii summing up to at most ηr^p.
(ii) For each $z\in\overline{\Omega(\varphi)}\backslash E_r$ with $|Rez-r|\leq l_0r^p$, we have $z\in\varrho(T)$,

$$\|(T-z)^{-1}\| \leq \frac{C_0C_1^m}{r^p},$$

and

$$\|S(T-z)^{-1}\| \leq C_0C_1^m,$$

where $m=N_+(r-lr^p, r+lr^p, G)$. ◇

Proof First, let's take $l_1 \in]l_0, l-b[$ and α, $\widetilde{b}$ satisfying $b < \widetilde{b} < \alpha < l - l_1$. Let $r \geq r_0$ and we choose r_0 large enough. Thus, we can presume that $r - lr^p > 0$. In the next, we denote by $\lambda_1, \cdots, \lambda_n$ the eigenvalues of G in $\Delta_r =]r - lr^p, r + lr^p[$ and by $P_1, \cdots, P_n$ the orthogonal projections onto the corresponding eigenspaces. Furthermore, we define

$$K_r = \sum_{j=1}^{n} (\lambda_j - \widetilde{\lambda}_j) P_j,$$

with

$$\widetilde{\lambda}_j = \begin{cases} r - lr^p \text{ if } \lambda_j < r, \\ r + lr^p \text{ if } \lambda_j \geq r. \end{cases}$$

Then, it is easy to see that $G_r = G - K_r$ is a normal operator with $\sigma(G_r) \bigcap \Omega(2\varphi) \subset \mathbb{R}_{\geq 0}$ and $\Delta_r \subset \varrho(G_r)$. The operator K_r has rank m and satisfies

$$\|K_r\| \leq lr^p.$$

Since $\lambda_j / \widetilde{\lambda}_j \leq r/(r - lr^p)$ for all j, then it suffices to verify that

$$\|Gu\| \leq \frac{r}{r - lr^p} \|G_r u\|.$$

By observing that $1 - lr^{p-1} \to 1$ as $r \to \infty$ and $b < \widetilde{b}$, allows us to conclude the following inequality

$$\begin{aligned} \|Su\| &\leq b\|Gu\| \|u\|^{1-p} + a\|u\| \\ &\leq b\Big(\frac{1}{1 - lr^{p-1}}\Big)^p \|G_r u\|^p \|u\|^{1-p} + a\|u\| \\ &\leq \widetilde{b}\|G_r u\|^p \|u\|^{1-p} + a\|u\|, \end{aligned}$$

provided r_0 is sufficiently large. Hence, S is p-bounded to G_r with bound less than or equal to $\widetilde{b}$. Now, we want to prove the following

$$|x - r| \leq l_1 r^p \Longrightarrow]x - \alpha x^p, x + \alpha x^p[\subset \rho(G_r) \tag{8.2.5}$$

for r_0 sufficiently large. Let x such that $|x - r| \leq l_1 r^p$. Since the function $x \longrightarrow x - \alpha x^p$ is monotonically increasing for large x, then

$$x - \alpha x^p \geq r - l_1 r^p - \alpha(r - l_1 r^p)^p \geq r - l_1 r^p - \alpha r^p > r - lr^p$$

for r_0 large enough. Moreover,

$$\alpha(1+l_1r^{p-1})^p \leq l - l_1$$

holds for large r and, we find

$$x+\alpha x^p \leq r+l_1r^p+\alpha(r+l_1r^p)^p \leq r+lr^p.$$

In view of $\Delta_r \subset \rho(G_r)$, then Eq. (8.2.5) is now proved. Now, we aim to apply the Lemma 2.5.1 to the following function

$$d(z)=\det(I+K_r(G_r+S-z)^{-1}),$$

$z \in \rho(G_r+S)$, and the following sets

$$U_r=\{x+iy \text{ such that } |x-r|<l_1r^p, \text{ and } |y|<4br^p\},$$

$$F_r=\{x+iy \text{ such that } |x-r|<l_0r^p, \text{ and } |y|<3br^p\}.$$

For r_0 sufficiently large, we have the inclusion $U_r \subset \widetilde{\Omega}(\varphi)$. In view of (8.2.5), we can apply Lemma 8.1.1 to G_r+S with some $\varepsilon \in]b/\alpha, 1[$; we find $U_r \subset \rho(G_r+S)$ and, for $z \in U_r$,

$$\operatorname{dist}(z,\sigma(G_r)) \geq lr^p - l_1r^p > \alpha r^p,$$

$$\|(G_r+S-z)^{-1}\| \leq \frac{(1-\varepsilon)^{-1}}{\alpha r^p},$$

and

$$\|S(G_r+S-z)^{-1}\| \leq \frac{\varepsilon}{1-\varepsilon}.$$

So,

$$\begin{aligned}|d(z)| &\leq (1+\|K_r\|\|(G_r+S-z)^{-1}\|)^m\\ &\leq \left(1+\frac{l(1-\varepsilon)^{-1}}{\alpha}\right)^m\\ &= c_0^m\end{aligned}$$

on U_r with $c_0>0$. For $z \in \rho(T)\bigcap U_r$, we have

$$I=(I+K_r(G_r+S-z)^{-1})(I-K_r(T-z)^{-1}).$$

By applying Lemma 2.5.1 (now with $\varepsilon=2/3$) to the operator T and $z_r=r+i2br^p \in F_r$, we find $z_r \in \rho(T)$ and

$$\|(T-z_r)^{-1}\| \leq \frac{3}{2br^p}.$$

Hence,

$$
\begin{aligned}
\frac{1}{|d(z_r)|} &= |\det(I - K_r(T - z_r)^{-1})| \\
&\leq \left(1 + \frac{3l}{2b}\right)^m \\
&= c_1^m,
\end{aligned}
$$

where $c_1 > 0$. The use of Lemma 2.5.1 yields a constant $C > 0$ depending only on b, l_0, l_1 and η such that for every $r \geq r_0$, there exists a union E_r of discs with radii summing up to at most ηr^p and

$$
|d(z)| \geq c_0^{-mC} c_1^{-m(1+C)}
$$

for all $z \in F_r \backslash E_r$. For $z \in F_r \backslash E_r$, we will obtain $I + K_r(G_r + S - z)^{-1}$ is invertible with

$$
\begin{aligned}
\|(I + K_r(G_r + S - z)^{-1})^{-1}\| &\leq \frac{c_0^m}{|d(z)|} \\
&\leq (c_0 c_1)^{(1+C)m}.
\end{aligned}
$$

Consequently, $z \in \rho(T)$ with

$$
(T - z)^{-1} = (G_r + S - z)^{-1}(I + K_r(G_r + S - z))^{-1}))^{-1}
$$

and

$$
\begin{aligned}
\|(T - z)^{-1}\| &\leq \frac{(1-\varepsilon)^{-1}}{\alpha r^p}(c_0 c_1)^{(1+C)m} \\
&\leq \frac{C_0 C_1^m}{r^p},
\end{aligned}
$$

$$
\begin{aligned}
\|S(T - z)^{-1}\| &\leq \frac{\varepsilon}{1-\varepsilon}(c_0 c_1)^{(1+C)m} \\
&\leq C_0 C_1^m
\end{aligned}
$$

with appropriate constants C_0, C_1 depending on b, l, l_0, l_1, η, α, and ε only. Finally, if we consider $z = x + iy \in \overline{\Omega(\varphi)}$ with

$$
|x - r| \leq l_0 r^p
$$

and

$$
|y| \geq 3br^p,
$$

then

$$
2bx^p \leq 2b(r + l_0 r^p)^p \leq 3br^p \leq |y|
$$

holds for r_0 sufficiently large. Applying now Lemma 8.1.1, again with $\varepsilon = 2/3$, we obtain $z \in \rho(T)$ and

$$\|(T-z)^{-1}\| \le \frac{3}{|y|} \le \frac{1}{br^p} \le \frac{C_0 C_1^m}{r^p},$$

$$\|S(T-z)^{-1}\| \le 2 \le C_0 C_1^m$$

for $C_0 \ge \max(2, b^{-1})$ and $C_1 \ge 1$. Q.E.D.

Lemma 8.2.3 *If G is normal with compact resolvent, $\sigma(G) \bigcap \Omega(2\varphi) \subset \mathbb{R}_{\ge 0}$ with $0 < \varphi \le \frac{\pi}{2}$ (see the definition of $\Omega(\cdot)$ in (1.13.2)), S is G-bounded with order p and with bound b, $p \in [0, 1[$, and $T = G + S$, then for l_0, $q > 0$, there are constants C_0, C_1, $r_0 > 0$ such that for every $r \ge r_0$ the following holds:*

- *For every $z = x + iy$ with $|x - r| \le l_0 r^p$, $|y| \le 2bx^p$, there exists $q_1 \in]0, q[$ such that if $|w - z| = q_1 r^p$, then $w \in \rho(T)$, and*

$$\|(T-w)^{-1}\| \le \frac{C_0 C_1^m}{r^p},$$

where $m = N_+(r - lr^p, r + lr^p, G)$ with $l = b + 2(l_0 + q)$. ◇

Proof We apply Lemma 8.2.2 for $l = b + 2(l_0 + q)$, $l_0 + q$ replacing l_0, and $\eta = q/3$. For z as in the claim and $|w - z| \le qr^p$, we have $|arg w| \le \varphi$ (for r_0 large enough) and

$$|Re w - r| \le (l_0 + q)r^p.$$

Now, the sum of the diameters of the discs in E_r is at most

$$2\eta r^p < qr^p,$$

where E_r is given in Lemma 8.2.2. Hence, there exists $q_1 \in]0, q[$ such that $w \notin E_r$ for $|w - z| = q_1 r^p$ and the claim is proved. Q.E.D.

Lemma 8.2.4 *Let G be normal with compact resolvent, $\sigma(G) \bigcap \Omega(2\varphi) \subset \mathbb{R}_{\ge 0}$ with $0 < \varphi \le \frac{\pi}{2}$ (see the definition of $\Omega(\cdot)$ in (1.13.2)), S is G-bounded with order p and with bound b, $p \in [0, 1[$, and $T = G + S$. Suppose that there is a sequence $(r_k)_{k\ge 1}$ of positive numbers tending monotonically to infinity and some $l > b$, $m \in \mathbb{N}_{\ge 1}$ such that*

$$N_+(r_k - lr_k^p, r_k + lr_k^p, G) \le m \text{ for all } k \ge 1.$$

Then, there exists constants C, $r_0 > 0$, $\alpha > b$, and a sequence $(x_k)_{k\ge 1}$ in $\mathbb{R}_{\ge 0}$ tending monotonically to infinity such that the following holds:

(i) $z \in \overline{\Omega(\varphi)}$ *with* $Re z = x_k$ *implies* $z \in \rho(T)$,

$$\|(T-z)^{-1}\| \le C,$$

(ii) *the contours* $\Gamma_\pm$, γ_k *from (1.13.3) and (1.13.4) satisfy* $\Gamma_\pm$, *and* $\gamma_k \subset \rho(T)$, *and*
(iii) *if* P_k *is the Riesz projection of* T *associated with the region enclosed by* γ_k, Γ_-, γ_{k+1}, Γ_+, *then*

$$\sum_{k=1}^{\infty} |\langle P_k u, v\rangle| \le C\|u\|\,\|v\|$$

for all u, $v \in X$. ◇

Proof We use Lemma 8.2.2 with $l_0 = (l-b)/2$ and $\eta = l_0/2$ to $r = r_k$, $k \ge k_0$, k_0 appropriate. Since the sum of the diameters of the discs in E_r (see Lemma 8.2.2) is at most $l_0 r_k^p$ and the interval $[r_k - l_0 r_k^p, r_k + l_0 r_k^p]$ contains at most m eigenvalues of G, we can obtain an x_k such that

$$|x_k - r_k| \le l_0 r_k^p,$$

$$\operatorname{dist}(x_k, \sigma(G)) \ge \frac{l_0}{3m} r_k^p,$$

and that $z \in \overline{\Omega(\varphi)}$ with $Re z = x_k$ implies $z \in \rho(T)$,

$$\|(T-z)^{-1}\| \le \frac{C_0 C_1^m}{r_k^p},$$

$$\|S(T-z)^{-1}\| \le C_0 C_1^m.$$

Hence, $x_k/r_k \to 1$ as $k \to \infty$ and, so

$$\operatorname{dist}(x_k, \sigma(G)) \ge c_2 x_k^p$$

for $k \ge k_0$ with $c_2 > 0$ and k_0 appropriately chosen. Since $x_k \to \infty$, for every k_1, there exists $k_2 > k_1$ such that

$$x_{k_2}^{1-p} - x_{k_1}^{1-p} \ge 1.$$

Passing to an appropriate subsequence, we can thus assume that

$$x_{k+1}^{1-p} - x_k^{1-p} \ge 1$$

for all $k \in \mathbb{N}$, which yields

$$x_n^{1-p} - x_k^{1-p} \ge n - k$$

for $n > k$. Now, an application of Lemma 8.2.1 with $a = 2b$ and the sequence $(x_k)_{k \geq k_0}$, k_0 large enough, completes the proof. Q.E.D.

Proof of Theorem 8.2.1. Let $e^{i\theta_j}\mathbb{R}_{\geq 0}$ with $0 < \theta_1 < \cdots < \theta_n < 2\pi$ be the rays containing the eigenvalues of G. From Theorem 8.1.1, we know that T has a compact resolvent and that almost all of its inside sectors of the form

$$\Omega_j = \{z \in \mathbb{C} \text{ such that } |\arg z - \theta_j| < \psi_j\}$$

with $0 < \psi_j \leq \frac{\pi}{4}$, where ψ_j can be chosen such that these sectors are disjoint. Lemma 8.1.1 shows that $||(T - z)^{-1}||$ is uniformly bounded for $z \notin \theta_1 \bigcup \cdots \bigcup \theta_n$, $|z| \geq r_0$. Using the assumption (8.2.1), we can show that for each sector Ω_j, there is a sequence $(r_{jk})_{k\geq 1}$ of positive numbers tending monotonically to infinity such that

$$\sup_k N_+(r_{jk} - 2br_{jk}^p, r_{jk} + 2br_{jk}^p, e^{-i\theta_j}G) < \infty.$$

By Lemmas 8.2.1, 8.2.2 and 8.2.4, we obtain a corresponding sequence x_{jk} such that $||(T - z)^{-1}||$ is uniformly bounded for $z \in \Omega_j$,

$$Re(e^{i\theta_j}z) = x_{jk}.$$

Corollary 6.3.1 implies that the system of root subspaces of root subspaces of T is complete. Finally, the result follows from (7.6.18) and Lemma 8.2.4. Q.E.D.

Example 8.1 Consider the Laplace operator on the unit disc $\mathbb{D}(0, 1) \subset \mathbb{R}^2$ with Dirichlet boundary condition,

$$\begin{cases} G : \mathscr{D}(G) \subset L_2(\mathbb{D}(0,1)) \longrightarrow L_2(\mathbb{D}(0,1)) \\ \qquad\qquad\qquad\qquad\qquad \psi \longrightarrow G\psi = -\Delta\psi \\ \\ \mathscr{D}(T) = H^2(\mathbb{D}(0,1)) \bigcap H_0^1(\mathbb{D}(0,1)). \end{cases}$$

It is easy to see that G is positive self-adjoint with compact resolvent, and the asymptotic behavior of its spectrum is such that

$$\lim_{r\to\infty} \frac{n(\overline{\mathbb{D}}(0,r), G)}{r} = \frac{1}{4}.$$

Consider the operator $T = G + S$, where S is a bounded operator on $L_2(\mathbb{D}(0, 1))$. Then, by appliying Theorems 8.1.1 and 8.2.1 with $p = 0$, we deduce that the operator T has a compact resolvent,

$$\sigma(G) \subset \overline{\mathbb{D}}(0, \|S\|) \bigcup \{x + iy \text{ such that } x \geq 0,\ |y| \leq \|S\|\},$$

and T admits a finitely spectral l^2-decomposition.

8.2.2 A Large Gap in $\sigma(G)$ Yields a Gap in $\sigma(T)$

Let G be normal with compact resolvent, $\sigma(G) \bigcap \Omega(2\varphi) \subset \mathbb{R}_{\geq 0}$ with $0 < \varphi \leq \frac{\pi}{2}$, S p-subordinate to G with bound b, $0 \leq p < 1$, and $T = G + S$ (Fig. 8.3). Assume that there is a sequence $(r_k)_{k\geq 1}$ of non-negative numbers tending monotonically to infinity and constants $\beta \geq 0$, $\alpha > b$, $l > \beta + \alpha$ such that

$$\sigma(G) \bigcap \mathbb{R}_{\geq 0} \subset \bigcup_{k\geq 1} [r_k - \beta r_k^p, r_k + \beta r_k^p] \tag{8.2.6}$$

and

$$r_k + l r_k^p \leq r_{k+1} - l r_{k+1}^p$$

for almost all k. We set $s_k^{\pm} = r_k \pm l r_k^p$ so that

$$r_k \leq s_k^+ \leq s_{k+1}^- \leq r_{k+1}.$$

Consider $s \in [s_k^+, s_{k+1}^-]$ with $k \geq k_0$. Then,

$$\begin{aligned} s + \alpha s^p &\leq s_{k+1}^- + \alpha r_{k+1}^p \\ &= r_{k+1} - (l - \alpha) r_{k+1}^p \\ &\leq r_{k+1} - \beta r_{k+1}^p. \end{aligned}$$

Furthermore, we have

$$s - \alpha s^p \geq s_k^+ - \alpha \left(s_k^+\right)^p$$

for k_0 large enough, since the left-hand side is monotonically increasing in s for large s. In addition, the equivalent inequalities

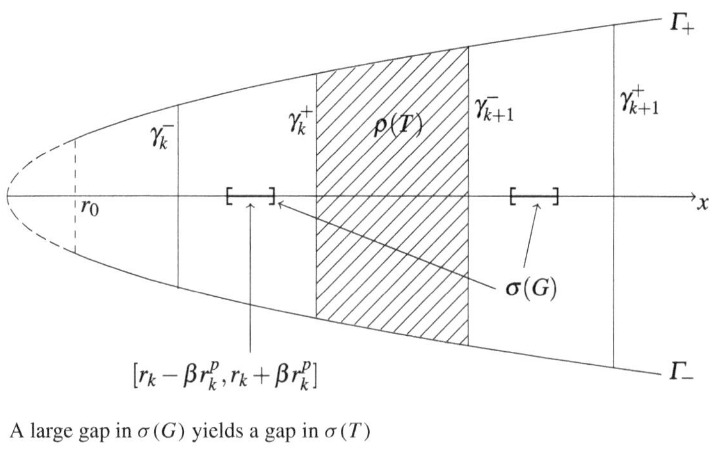

Fig. 8.3 A large gap in $\sigma(G)$ yields a gap in $\sigma(T)$

$$s_k^+ - \alpha\left(s_k^+\right)^p \ge r_k + \beta r_k^p \iff l r_k^p - \alpha(r_k + l r_k^p)^p \ge \beta r_k^p$$

hold for k_0 sufficiently large since $1 + l r_k^{p-1} \to 1$. In view of (8.2.6), we obtain for $k \ge k_0$, $s \in [s_k^+, s_{k+1}^-]$ implies

$$]s - \alpha s^p, s + \alpha s^p[\subset \rho(G).$$

With r_0 and k_0 appropriately chosen, using Lemma 8.1.1, we have the region enclosed by γ_k^+, γ_{k+1}^-, Γ^+, and Γ^- as well as the contours itself belong to $\rho(T)$ for all $k \ge k_0$. Hence, there are constants C, $r_0 > 0$, and $k_0 \ge 1$ such that the contours $\Gamma_\pm$ from (1.13.4) and

$$\gamma_k^\pm = \{x + iy \text{ such that } x = r_k \pm l r_k^p \text{ and } |y| \le \alpha x^p\}$$

with $k \ge k_0$ as well as the regions enclosed by γ_k^+, γ_{k+1}^-, Γ_+, Γ_- belong to $\rho(T)$. Furthermore, for $z \in \overline{\Omega(\varphi)}$ with $Rez = s_k^+, k \ge k_0$, $\|(T - z)^{-1}\|$ and $\|S(T - z)^{-1}\|$ are uniformly bounded. So, $z \in \overline{\Omega(\varphi)}$ with $Rez = r_k + l r_k^p$, $k \ge k_0$, implies

$$\|(T - z)^{-1}\| \le C.$$

Also,

$$\operatorname{dist}(s_k^+, G) \ge \alpha(s_k^+)^p$$

and

$$\begin{aligned} s_{k+1}^+ - s_k^+ &= r_{k+1} - r_k + l(r_{k+1}^p - r_k^p) \\ &\ge 2 l r_{k+1}^p. \end{aligned}$$

The mean value theorem then yields

$$\begin{aligned} (s_{k+1}^+)^{1-p} - (s_k^+)^{1-p} &\ge (1-p)(s_{k+1}^+)^{-p}(s_{k+1}^+ - s_k^+) \\ &\ge \frac{2l(1-p) r_{k+1}^p}{(r_{k+1} + l r_{k+1}^p)^p}, \end{aligned}$$

i.e.,

$$(s_{k+1}^+)^{1-p} - (s_k^+)^{1-p} \ge l(1-p)$$

for $k \ge k_0$, k_0 sufficiently large. By applying Lemma 8.2.1 with $x_k = s_k^+$ to get the estimate for the sum over the Riesz projections, i.e., if P_k and Q_k are the Riesz projections, respectively, of T and G associated with the region enclosed by γ_k^-, γ_k^+, Γ_+, Γ_-, then

$$\sum_{k=k_0}^{\infty} |\langle P_k u, v\rangle| \le C\|u\|\|v\|$$

for all $u, v \in X$. Using Lemma 8.1.1, we infer that $\dim R(P_k) = \dim R(Q_k)$ for $k \geq k_0$.

8.2.3 Riesz Basis

Let us replace the condition (8.2.1) by an hypotheses on the localization of the spectrum, $\sigma(G)$, of G on the rays, we obtain the following. Let X be a Hilbert space and consider the operator

$$T = G + S,$$

where G is a normal operator with compact resolvent on X, S is p-subordinate to G with bound b and $0 \leq p < 1$. Let $(r_{jk})_{k\geq 1}$ be monotonically increasing sequences of non-negative numbers satisfying the following

$$r_{jk} + l_j r_{jk}^p \leq r_{j,k+1} - l_j r_{j,k+1}^p \tag{8.2.7}$$

for almost all k with some constants $l_j > \beta_j + b$, where $\beta_j \geq 0$. Let us put

$$L_{jk} = \{e^{i\theta_j} x \text{ such that } x \geq 0 \text{ and } |x - r_{jk}| \leq \beta_j r_{jk}^p\},$$

and

$$K_{jk} = \{e^{i\theta_j}(x + iy) \text{ such that } x \geq 0,\ |x - r_{jk}| \leq (\beta_j + \alpha) r_{jk}^p, \text{ and } |y| \leq \alpha x^p\},$$

where $0 \leq \theta_1 < \cdots < \theta_n < 2\pi$, $j = 1, \cdots, n$, $k \geq 1$ and $b < \alpha < \min\{l_1 - \beta_1, \cdots, l_n - \beta_n\}$ (cf. Fig. 8.4). Assume that the spectrum, $\sigma(G)$, of the operator G lies on sequences of line segments on rays from the origin,

$$\sigma(G) \subset \bigcup_{j=1}^{n} \bigcup_{k\geq 1} L_{jk}. \tag{8.2.8}$$

According to Theorem 8.1.1 and, for every ray, Subsection 8.2.2 with α and l can be replaced, respectively, by $\widetilde{\alpha} = (\alpha + b)/2$ and $\widetilde{l}_j = \beta_j + \alpha$. Consequently, T has a compact resolvent operator and that almost every of its eigenvalues lie inside regions

$$\{e^{i\theta_j}(x + iy) \text{ such that } x \geq 0,\ |x - r_{jk}| < \widetilde{l}_j r_{jk}^p, \text{ and } |y| < \widetilde{\alpha} x^p\} \subset K_{jk}.$$

It is easy to see that the subspace corresponding to $\sigma(T) \backslash \bigcup_{j,k} K_{jk}$ together with the spectral subspaces of T corresponding to the regions K_{jk} form a Riesz basis of the space X. So,

$$n(L_{jk}, G) = n(K_{jk}, T) \tag{8.2.9}$$

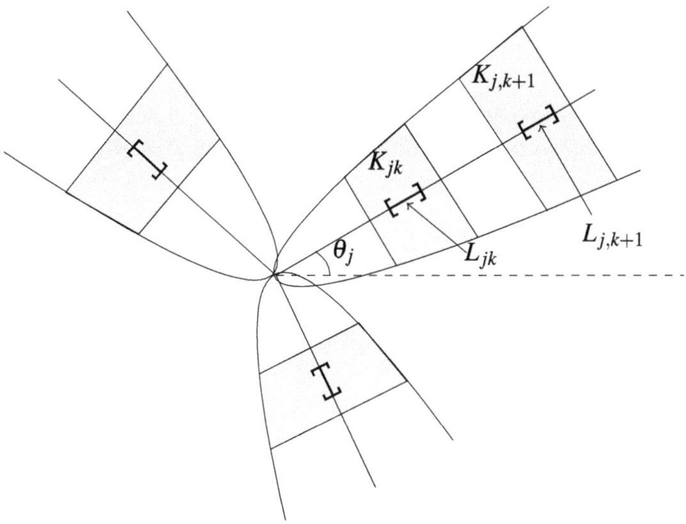

Fig. 8.4 The situation of Sect. 8.2.3

for almost every pairs (j, k). In fact, in view of Lemma 8.1.1, we get for $|z|$ large enough, $\|(T - z)^{-1}\|$ is uniformly bounded outside certain disjoint sectors Ω_j around the rays. For every ray, and according to Subsection 8.2.2, there is a sequence $(x_{jk})_{k=0}^{\infty}$ tending monotonically to infinity satisfying $\|(T - z)^{-1}\|$ is bounded for $z \in \Omega_j$,

$$Re(e^{-i\theta_j} z) = x_{jk}.$$

By using Corollary 6.3.1, we infer that the system of root subspaces of the operator T is complete. Let P_{jk} be the Riesz projection associated with K_{jk} and P_0 be the one associated with $\sigma(T)\backslash \bigcup_{j,k} K_{jk}$. Then, there is a constant $C \geq 0$ such that

$$|\langle P_0 u, v\rangle| + \sum_{j=1}^{n} \sum_{k=1}^{\infty} |\langle P_{jk} u, v\rangle| \leq C \|u\| \|v\|.$$

Hence, Eq. (7.6.18) then yields the Riesz basis property. Now, if Q_{jk} is the spectral projection of the operator G associated with L_{jk}, Sect. 8.2.2 yields

$$\dim R(Q_{jk}) = \dim R(P_{jk})$$

for almost every (j, k). Hence, (8.2.9). Moreover, if there are two constants $m, q > 0$ such that for almost every pairs (j, k) the following items

(i) $n(L_{jk}, G) \leq m$, and
(ii) if $\lambda_1, \lambda_2 \in \sigma(T) \bigcap K_{jk}$ such that $\lambda_1 \neq \lambda_2$, then $|\lambda_1 - \lambda_2| > q r_{jk}^{p}$

hold, then the root subspaces of the operator T form a Riesz basis of the space X. In fact, in order to prove that the root subspaces corresponding to the eigenvalues of the operator T in every K_{jk} form a Riesz basis of $R(P_{jk})$ with constant c independent of (j, k). We may assume, without loss of generality, that $\theta_j = 0$ and $\beta_j + \alpha + q \leq l_j$. Furthermore, we want to use the Lemma 8.2.3 with $l_0 = \beta_j + \alpha$ and the set l accordingly. By using the equation

$$r_{j,k+1} - r_{jk} \geq 2l_j r_{jk}^p,$$

it is easy to see, for r sufficiently large, that the number of elements r_{jk} in the interval $[r - lr^p, r + lr^p]$ is at most $2l/l_j$. Thus, for r sufficiently large, there exists m_0 such that

$$N_+(r - lr^p, r + lr^p, G) \leq m_0.$$

Consider $\lambda \in K_{jk}$ an eigenvalue of the operator T. By using Lemma 8.2.3, there is a constant $q_1 \in]0, q[$ such that the points w on the circle around λ with radius $q_1 r_{jk}^p$ leads to the following inequality

$$\|(T - w)^{-1}\| \leq C_0 C_1^{m_0} r_{jk}^{-p}.$$

Moreover, this circle lies inside the strip

$$|Rez - r_{jk}| \leq l_j r_{jk}^p$$

and hypothesis (ii). Consequently, λ is the only possible eigenvalue of the operator T inside that circle. Therefore, the Riesz projection P_λ for λ satisfies the following inequality

$$\begin{aligned} \|P_\lambda\| &\leq 2\pi q_1 r_{jk}^p \frac{C_0 C_1^{m_0}}{r_{jk}^p} \\ &\leq 2\pi q C_0 C_1^{m_0}. \end{aligned}$$

If $\lambda_1, \cdots, \lambda_{m_1}$ are the eigenvalues of the operator T in the sector K_{jk}, then the sum of multiplicities of all eigenvalues of T contained in K_{jk} satisfies

$$m_1 \leq n(K_{jk}, T) \leq m.$$

Hence,

$$\sum_{s=1}^{m_1} |\langle P_{\lambda_s} u, v\rangle| \leq 2\pi m q C_0 C_1^{m_0} \|u\| \|v\|.$$

According to (7.6.18), the subspaces $R(P_{\lambda_s})$, $s = 1, \cdots, m_1$, form a Riesz basis of $R(P_{jk})$ with constant c independent of k. This is true for almost every pairs (j, k). So, as an application of Lemma 7.6.3, we prove that the root subspaces of the operator T form a Riesz basis of the space X.

Remark 8.2.1 If almost every eigenvalues of G are simple and almost every line segments L_{jk} contain only one eigenvalue, then all the results of this subsection yield a Riesz basis of eigenvectors and finitely many Jordan chains for the perturbed operator T. ◇

Remark 8.2.2 If G verifies the spectral condition (8.2.8) for some $\beta_j > 0$ with the following inequalities $r_{jk} + \beta_j r_{jk}^p \leq r_{j,k+1} - \beta_j r_{j,k+1}^p$ and $N(L_{jk}, G)$ is bounded in (j, k), then

$$\sup_{r \geq 1} n(\overline{\mathbb{D}}(0, r), G) r^{p-1} < \infty.$$

A particular case, the spectral condition (8.2.1) of Theorem 8.2.1 holds. Furthermore, the first part of this subsection is applicable even if the sum of multiplicities of all eigenvalues of G contained in L_{jk}, $n(L_{jk}, G)$, is unbounded and the Eq. (8.2.1) does not hold. ◇

The inequalities (8.2.7) can be reformulated for sequences with a certain asymptotic behavior:

Lemma 8.2.5 (Wyss [27, Lemma 6.5]) *Consider the sequence of non-negative numbers given by $r_k = ck^q + d_k k^{q-1}$ with $c > 0$, $q \geq 1$ and a converging sequence $(d_k)_{k=0}^{\infty}$. Then, for l, $p \geq 0$, the relation $r_k + l r_k^p \leq r_{k+1} - l r_{k+1}^p$ holds for almost all $k \in \mathbb{N}$ if*

(i) $p < 1 - 1/q$, or
(ii) $p = 1 - 1/q$ and $l < qc^{1/q}/2$. ◇

Proof It can be shown in a straightforward way by a Taylor series expansion of r_{k+1} in k. Q.E.D.

Example 8.2 Let

$$\begin{cases} G : \mathscr{D}(G) \subset L_2(]a, b[) \longrightarrow L_2(]a, b[) \\ \qquad\qquad \psi \longrightarrow G\psi(x) = \dfrac{d^2\psi}{dx^2}(x) \\ \mathscr{D}(G) = \{\psi \in H^2(]0, 1[) \text{ such that } \psi(0) = \psi(1) = 0\}. \end{cases}$$

An easy calculation leads to G is a self-adjoint operator with compact resolvent and eigenvalues are equal to $\pi^2 k^2$, $k = 1, 2, \cdots$, which are all simple. Consider the two operators

$$\begin{cases} T : \mathscr{D}(T) \subset L_2(]0, 1[) \longrightarrow L_2(]0, 1[) \\ \qquad\qquad \psi \longrightarrow T\psi(x) = \dfrac{d^2\psi}{dx^2}(x) + \dfrac{d\psi}{dx}(x) \\ \mathscr{D}(T) = \mathscr{D}(G), \end{cases}$$

and

$$\begin{cases} S : \mathscr{D}(S) \subset L_2(]a,b[) \longrightarrow L_2(]a,b[) \\ \qquad\qquad \psi \longrightarrow G\psi(x) = \dfrac{d\psi}{dx}(x) \\ \\ \mathscr{D}(S) = \mathscr{D}(G). \end{cases}$$

By using a similar reasoning as in Example 2.3, we get

$$\|S\psi\|_{L_2(]0,1[)} \leq \|G\psi\|^{\frac{1}{2}}_{L_2(]0,1[)} \|\psi\|^{\frac{1}{2}}_{L_2(]0,1[)} \text{ for each } \psi \in \mathscr{D}(G).$$

So, S is $\frac{1}{2}$-subordinate to G with bound $b \leq 1$. Furthermore, we want to use the Sect. 8.2.3 with $p = \frac{1}{2}$, $\theta_1 = 0$, $r_{1k} = r_k = \pi^2 k^2$ and $b = 0$. The condition (8.2.7) leads to the estimate

$$r_k + l_j r_k^{\frac{1}{2}} \leq r_{k+1} - l_j r_{k+1}^{\frac{1}{2}}$$

for almost k with some $l_1 > b$ and it holds, by using Lemma 8.2.5, if and only if $l_1 < \pi$. Then, we can find l_1 such that $1 < l_1 < \pi$. Thus, (8.2.7) is satisfied and Sect. 8.2.3 is applicable and yields that for every $\alpha > 1$ almost every eigenvalues of the operator T lie inside regions

$$K_k = \{x + iy \text{ such that } |x - \pi^2 k^2| \leq \alpha\pi k, \text{ and } |y| \leq \alpha x^{\frac{1}{2}}\}.$$

Indeed, $n(K_k, T) = 1$ for almost every k. In view of Remark 8.2.1, we get T has a Riesz basis of eigenvectors and finitely many Jordan chains.

8.2.4 Sum of Multiplicities

The next proposition reverses the assertions of Theorems 8.1.1 and 8.2.1 to some extent. As a consequence, the assumptions in these theorems can be relaxed. In fact, let G be an operator on a Hilbert space X with compact resolvent and a Riesz basis of Jordan chains. Suppose that $0 \leq p < 1, \alpha \geq 0, 0 \leq \theta_j < 2\pi, j = 1, \cdots, n$, such that

(i) there exists $r_0 > 0$ with

$$\sigma(G) \subset B_{r_0}(0) \bigcup \bigcup_{j=1}^{n} \{e^{i\theta_j}(x + iy) \text{ such that } x > 0 \text{ and } |y| \leq \alpha x^p\}.$$

In order to apply Lemma 7.6.1, we need to properly choose the new eigenvalues $(\mu_l)_l$ of G_0 given the eigenvalues λ_l of G. Almost all λ_l lie inside

$$\bigcup_{j=1}^{n}\{e^{i\theta_j}(x+iy) \text{ such that } x>0 \text{ and } |y|\le \alpha x^p\}.$$

For $\lambda_l \in \{e^{i\theta_j}(x+iy)$ such that $x>0$ and $|y|\le \alpha x^p\}$, we put

$$\mu_l = e^{i\theta_j}|\lambda_l|.$$

Let $\lambda_l = e^{i\theta_j}(x+iy)$, then

$$\begin{aligned}
|\mu_l-\lambda_l| &\le \sqrt{x^2+y^2}-x+|y| \\
&= \frac{y^2}{\sqrt{x^2+y^2}+x}+|y| \\
&\le \frac{y^2}{2x}+|y| \\
&\le \frac{\alpha^2x^{2p}}{2x}+\alpha x^p \\
&= \left(\frac{\alpha^2}{2x^{1-p}}+\alpha\right)x^p \\
&\le 2\alpha|\lambda_l|^p
\end{aligned}$$

for $|\lambda_l|$ large enough. Let $\lambda_l \notin \{e^{i\theta_j}(x+iy)$ such that $x>0$ and $|y|\le\alpha x^p\}$ for every j and $\lambda_l \neq 0$, we put $\mu_l = e^{i\theta_1}|\lambda_l|$. Finally, if $\lambda_l=0$, we set $\mu_l = e^{i\theta_1}$. In particular, this choice implies $|\lambda_l| = |\mu_l|$ (if $\lambda_l \neq 0$) and $n(\mathbb{D}(0,r), G_0) = n(\mathbb{D}(0,r), G)$ for $r\ge 1$. Hence, in view of Lemma 7.6.1, there is an isomorphism $J : X \longrightarrow X$, a normal operator G_0 on X with compact resolvent, and an operator S_0 p-subordinate to G_0 such that

$$J\mathcal{D}(G) = \mathcal{D}(G_0),$$

$$JGJ^{-1} = G_0 + S_0.$$

Moreover, all eigenvalues of G_0 lie on the rays $e^{i\theta_j}\mathbb{R}_{\ge 0}$ and, we obtain

$$n(\mathbb{D}(0,r), G_0) = n(\mathbb{D}(0,r), G) \text{ for } r\ge 1.$$

Now, suppose that

(ii) almost all eigenvalues of G lie inside regions

$$K_{jk} = \{e^{i\theta_j}(x+iy) \text{ such that } r_{jk}^- \le x \le r_{jk}^+, \text{ and } |y|\le \alpha x^p\},$$

$j=1,\cdots,n$, $k\ge 1$, where $(r_{jk}^{\pm})_{k\ge 1}$ are sequences of positive numbers satisfying

$$r_{jk}^{-} \leq r_{jk}^{+} < r_{j,k+1}^{-}.$$

If $\lambda_l = e^{i\theta_j}(x+iy) \in K_{jk}$ is an eigenvalue, then we take $\mu_l = e^{i\theta_j}x$. So,

$$\begin{aligned} |\mu_l - \lambda_l| &= |y| \\ &\leq \alpha x^p \\ &\leq \alpha |\lambda_l|^p. \end{aligned}$$

If $\lambda_l \notin K_{jk}$ for every (j,k), then we take $\mu_l = e^{i\theta_1} r_{11}^{-}$. Thus,

$$n(L_{jk}, G_0) = n(K_{jk}, G)$$

whenever $(j,k) \neq (1,1)$. Thus, in view of Lemma 7.6.1, there is an isomorphism $J : X \longrightarrow X$, a normal operator G_0 on X with compact resolvent, and an operator S_0 p-subordinate to G_0 such that

$$J\mathcal{D}(G) = \mathcal{D}(G_0),$$

$$JGJ^{-1} = G_0 + S_0.$$

Moreover, all eigenvalues of G_0 lie on the line segments

$$L_{jk} = \{e^{i\theta_j}x,\ r_{jk}^{-} \leq x \leq r_{jk}^{+}\},$$

and

$$n(L_{jk}, G_0) = n(K_{jk}, G)$$

holds for almost all pairs (j,k). Finally, suppose that S is p-subordinate to G with bound b. For $u \in \mathcal{D}(G_0)$, and since $0 \in \rho(G_0)$, we obtain

$$\begin{aligned} \|JGJ^{-1}u\| &\leq \|G_0u\| + \|S_0u\| \\ &\leq \|G_0u\| + b_0\|u\|^{1-p}\|G_0u\|^p \\ &\leq (1 + b_0\|G_0^{-1}\|^{1-p})\|G_0u\|. \end{aligned}$$

Thus,

$$\begin{aligned} \|JSJ^{-1}u\| &\leq \|J\|\|SJ^{-1}u\| \\ &\leq b\|J\|\|J^{-1}u\|^{1-p}\|GJ^{-1}u\|^p \\ &\leq b\|J\|\|J^{-1}\|\|u\|^{1-p}\|JGJ^{-1}u\|^p \\ &\leq b\|J\|\|J^{-1}\|(1 + b_0\|G_0^{-1}\|^{1-p})^p\|u\|^{1-p}\|G_0u\|^p. \end{aligned}$$

Hence, if S is p-subordinate to G, then JSJ^{-1} is p-subordinate to G_0.

Example 8.3 Let g_0, g_1, and $g_2 \in L_2([0,1])$, $g_3 \in L^\infty([0,1])$. Consider the differential operator T on $L_2([0,1])$ given by

$$T = G + S,$$

with $Gu = u^{(4)}$, $Su = g_0 u + g_1 u' + g_2 u'' + g_3 u'''$,

$$\mathscr{D}(T) = \mathscr{D}(G) = \mathscr{D}(S) = \{u \in H^2[0,1] \text{ such that } V_1(u) = V_2(u) = V_3(u) = V_4(u) = 0\},$$

where the boundary conditions

$$V_j(u) = \sum_{l=0}^{3} \alpha_{jl} u^{(l)}(0) + \beta_{jl} u^{(l)}(1) = 0$$

are regular in the sense of Naimark and such that G is self-adjoint. Then, the resolvent of G is compact, and its spectrum consists of at most two sequences of eigenvalues of the form

$$\lambda_{jk} = c_j k^4 + d_{jk} k^3, \quad k \geq k_{j0}, \; j = 1, 2,$$

with $c_j \neq 0$ and converging sequences $(d_{jk})_{k \geq k_{j0}}$. Since G is self-adjoint, then c_j and $d_{jk} \in \mathbb{R}$. By using Lemma 8.2.5, we have that each sequence $(\lambda_{jk})_{k \geq k_{j0}}$ satisfies

$$|\lambda_{jk}| + a|\lambda_{jk}|^4 \leq |\lambda_{j,k+1}| - a|\lambda_{j,k+1}|^4$$

for almost k and $0 \leq a < 2c_j^{\frac{1}{4}}$. As the multiplicity of every eigenvalue is at most 4, the use of Lemma 8.2.5 and Remark 8.2.2 makes us deduce that

$$\sup_{r \geq 1} \frac{n(\mathbb{D}(0,r), G)}{r^{1-\frac{3}{4}}} < \infty.$$

For $0 \in \rho(G)$, we can now show that S is $\frac{3}{4}$-subordinate to G. For $u \in \mathscr{D}(G)$, we have

$$\begin{aligned} \|Su\|_{L_2} &= \|g_0 u + g_1 u' + g_2 u'' + g_3 u'''\|_{L_2} \\ &\leq \|g_0\|_{L_2}\|u\|_\infty + \|g_1\|_{L_2}\|u'\|_\infty + \|g_2\|_{L_2}\|u''\|_\infty + \|g_3\|_\infty \|u'''\|_{L_2}. \end{aligned}$$

Using the last equation, Sobolev and interpolation inequalities, we can find constants $b_1 \geq 0$ such that

$$\|Su\|_{L_2} \leq b_1 \|u\|_{H^3[0,1]}. \tag{8.2.10}$$

Using (8.2.10), we can find constants b_2 and $b_3 \geq 0$ such that

$$\begin{aligned}\|Su\|_{L_2} &\leq b_2\|u\|_{L_2}^{\frac{1}{4}}\|u\|_{H^4[0,1]}^{\frac{3}{4}}\\ &\leq b_3\|u\|_{L_2}^{\frac{1}{4}}(\|u\|_{L_2}+\|u^{(4)}\|_{L_2})^{\frac{3}{4}}\\ &\leq b_3(\|G^{-1}\|+1)^{\frac{3}{4}}\|u\|_{L_2}^{\frac{1}{4}}\|Gu\|_{L_2}^{\frac{3}{4}}.\end{aligned}$$

Consequently, Theorems 8.1.1 and 8.2.1 can be applied to $T = G + S$. The resolvent of T is compact, almost all eigenvalues of T lie inside regions of the form

$$\{\text{sign}(c_j)\ (x+iy) \text{ such that } x \geq 0 \text{ and } |y| \leq \alpha x^4\}$$

with some $\alpha > 0$, and T admits a finitely spectral l^2-decomposition. In particular, T has a Riesz basis of finite-dimensional T-invariant subspaces. In the case that $0 \in \sigma(G)$, we choose any $\tau \in \rho(G) \bigcap \mathbb{R}$. Then, $S + \tau$ is $\frac{3}{4}$-subordinate to $G - \tau$ and Theorems 8.1.1 and 8.2.1 can be applied to the decomposition $T = G - \tau + S + \tau$. Finally, if the boundary conditions are such that almost all eigenvalues of G are simple and the norms of the coefficient functions g_l are sufficiently small, then Sect. 8.2.3 is applicable and yields a Riesz basis of eigenvectors and finitely many Jordan chains for T.

8.2.5 *Spectral Riesz Basis of Subspaces*

Now, we state the following theorem [28, Theorem 3.7] which presents an important tool in the determining of the existence of finitely spectral Riesz basis. Let X be a Hilbert space and G be a normal operator on X, having compact resolvent. Suppose that the spectrum of the operator G lies on a finite number of rays from the origin and for some $\beta \in [0, 1[$

$$\liminf_{r\to\infty} \frac{n(\mathbb{D}(0,r), G)}{r^{1-\beta}} < \infty.$$

We assume that the operator

$$S = T - G$$

is β-subordinate to G. Then, in view of both Theorems 8.1.1 and 8.2.1, we have T has a finitely spectral Riesz basis of subspaces $\{V_k\}_{k\in\mathbb{N}}$.

8.2.6 *A Riesz Basis Associated to a Block Operator Matrix*

In applications, the time evolution of physical or mechanical system is governed by 2×2 block operator matrices. So, the spectral theory of these block matrices plays a crucial role. As a consequence of both Theorems 8.1.1 and 8.2.1, we derive the following result for a class of diagonally dominant 2×2 block operator matrix:

Theorem 8.2.2 *Let X_1 and X_2 be two Hilbert spaces and $A(X_1 \longrightarrow X_1)$ and $D(X_2 \longrightarrow X_2)$ be normal operators with compact resolvents. Suppose that the spectrum of A, $\sigma(A)$, and the spectrum of D, $\sigma(D)$, lie on finitely many rays from the origin and that for some $p \in [0, 1[$, we have*

$$\liminf_{r \longrightarrow \infty} \frac{n(\mathbb{D}(0, r), A)}{r^{1-p}} < \infty$$

and

$$\liminf_{r \longrightarrow \infty} \frac{n(\mathbb{D}(0, r), D)}{r^{1-p}} < \infty.$$

Assume that the operators $C(X_1 \longrightarrow X_2)$ and $B(X_2 \longrightarrow X_1)$ are A-bounded and D-bounded with order p, respectively. Then, the 2×2 block matrix

$$T := \begin{pmatrix} A & B \\ C & D \end{pmatrix}$$

acting on $X_1 \times X_2$ admits a Riesz basis of finite-dimensional T-invariant subspaces, and for every $\alpha > b$, there exists $r_0 > 0$ such that the spectrum of T, $\sigma(T)$, satisfy the following inclusion

$$\sigma(T) \subset B_{r_0}(0) \bigcup \bigcup_{j=1}^{n} \{e^{i\theta_j}(x + iy) \text{ such that } x \geq 0 \text{ and } |y| \leq \alpha x^p\},$$

where $0 \leq \theta_j < 2\pi$, $j = 1, \cdots, n$ and $\theta_1, \cdots, \theta_n$ are the angles of the rays on which the spectra of A and D lie. ◇

Proof Consider the following decomposition of T

$$T = G + S, \text{ where } G := \begin{pmatrix} A & 0 \\ 0 & D \end{pmatrix} \text{ and } S := \begin{pmatrix} 0 & B \\ C & 0 \end{pmatrix}.$$

In fact, we deduce that G is normal with compact resolvent,

$$\sigma(G) = \sigma(A) \bigcup \sigma(D),$$

and

$$n(\mathbb{D}(0, r), G) = n(\mathbb{D}(0, r), A) + n(\mathbb{D}(0, r), D).$$

From the fact that the operator C is A-bounded with order p, then there is strictly positive constants α and β such that

$$\|C\varphi\| \leq \alpha\|A\varphi\|^p\|\varphi\|^{1-p} + \beta\|\varphi\|, \quad \text{for every } \varphi \in \mathscr{D}(A).$$

So,

$$\|C\varphi\|^2 \leq \alpha^2\|A\varphi\|^{2p}\|\varphi\|^{2(1-p)} + \beta^2\|\varphi\|^2 + 2\alpha\beta\left(\frac{\|A\varphi\|}{\|\varphi\|}\right)^p\|\varphi\|^2.$$

Let $\gamma^2 = \beta^2 + 2\alpha\beta\left(\frac{\|A\varphi\|}{\|\varphi\|}\right)^p$. We deduce that

$$\|C\varphi\|^2 \leq \alpha^2\|A\varphi\|^{2p}\|\varphi\|^{2(1-p)} + \gamma^2\|\varphi\|^2.$$

Similarly, we also prove

$$\|B\varphi\|^2 \leq \eta^2\|D\varphi\|^{2p}\|\varphi\|^{2(1-p)} + \lambda^2\|\varphi\|^2, \quad \text{for every } \varphi \in \mathscr{D}(D).$$

Without loss of generality, we assume that there exist positive constants a and b satisfying $\alpha \leq b$, $\eta \leq b$, $\gamma \leq a$ and $\lambda \leq a$. Thus, for $u \in \mathscr{D}(A)$, $v \in \mathscr{D}(D)$, we have

$$\begin{aligned}\left\|S\begin{pmatrix}u\\v\end{pmatrix}\right\|^2 &= \|Bv\|^2 + \|Cu\|^2\\ &\leq b^2\Big(\|Dv\|^{2p}\|v\|^{2(1-p)} + \|Au\|^{2p}\|u\|^{2(1-p)}\Big) + a^2\Big(\|u\|^2 + \|v\|^2\Big).\end{aligned}$$

As a consequence of Hölder's inequality in $\mathbb{R}^2$, we obtain

$$x^p y^{1-p} + z^p t^{1-p} \leq (x+z)^p(y+t)^{1-p},$$

where x, y, z, and $t \geq 0$. We infer that

$$\left\|S\begin{pmatrix}u\\v\end{pmatrix}\right\|^2 \leq b^2\Big(\|u\|^2 + \|v\|^2\Big)^{1-p}\Big(\|Dv\|^2 + \|Au\|^2\Big)^p + a^2\Big(\|u\|^2 + \|v\|^2\Big)$$

for $u \in \mathscr{D}(A)$, $v \in \mathscr{D}(D)$. Thus, we also have

$$\|Sw\| \leq b\|w\|^{1-p}\|Gw\|^p + a\|w\|,$$

$w \in \mathscr{D}(G) = \mathscr{D}(A) \times \mathscr{D}(D)$. So, S is G-bounded with order p ($p \in [0, 1[$). To finish the proof, it is enough to apply both Theorems 8.1.1 and 8.2.1. Q.E.D.

8.2.7 Gap in the Spectrum Around the Imaginary Axis

Let X be a Hilbert space, and A be a skew-adjoint operator with compact resolvent on X. Consider the quantities $(ir_k)_{k\in\Lambda}$ for the sequence of eigenvalues of A, where $\Lambda \in \{\mathbb{Z}_{\geq 0}, \mathbb{Z}_{\leq 0}, \mathbb{Z}\}$ and $(r_k)_k$ is increasing. The operator

$$G := \begin{pmatrix} A & 0 \\ 0 & A \end{pmatrix}$$

is skew-adjoint with compact resolvent, and with spectrum

$$\sigma(G) = \{ir_k, \ k \in \Lambda\},$$

and almost every of its eigenvalues have multiplicity 2. Consider the 2×2 block operator matrix

$$T := \begin{pmatrix} A & B \\ C & A \end{pmatrix},$$

where B and $C : X \longrightarrow X$ are a bounded operators, selfadjoint and uniformly positive, satisfying $B, C \geq \gamma > 0$. We can write T in the following form

$$T = G + S,$$

where

$$S := \begin{pmatrix} 0 & B \\ C & 0 \end{pmatrix}.$$

It is easy to see that the operator S is bounded with norm $\|S\| = b := \max\{\|B\|, \|C\|\}$. Suppose that almost every eigenvalues ir_k are simple and, for some

$$l > b,$$

we have

$$r_{k+1} - r_k \geq 2l$$

for almost every $k \in \Lambda$. In view of Theorem 8.1.1, T has a compact resolvent. If z is a point outside the discs $\mathbb{D}(ir_k, b)$ with radius b around the ir_k, then

$$\text{dist}(z, \sigma(G)) > b$$

and

$$\|(G - z)^{-1}\| < b^{-1}.$$

So, $z \in \rho(T)$. It is easy to see that the operator T is J_1-skew-adjoint (i.e., $J_1 T$ is skew-adjoint), where

$$J_1 := \begin{pmatrix} 0 & -iI \\ iI & 0 \end{pmatrix}.$$

This implies that the spectrum of T, $\sigma(T)$, is symmetric with respect to the axis $i\mathbb{R}$. Furthermore, for an eigenvalue λ of T with eigenvector w and a short calculation shows

$$\begin{aligned} \gamma\|w\|^2 &\leq Re\langle J_2Tw, w\rangle \\ &\leq |Re\lambda||\langle J_2w, w\rangle| \\ &\leq |Re\lambda|\|w\|^2, \end{aligned}$$

where $\langle\cdot,\cdot\rangle$ is an inner product on $X \times X$ and J_2 is the operator defined by

$$J_2 := \begin{pmatrix} 0 & I \\ I & 0 \end{pmatrix}.$$

Thus,

$$|Re\lambda| \geq \gamma \tag{8.2.11}$$

which prove the asserted shape of the spectrum. Finally, by applying Sect. 8.2.3 with $p = \beta_1 = \beta_2 = 0$, $\theta_1 = \pi/2$, $\theta_2 = 3\pi/2$, we obtain

$$n(\mathbb{D}(ir_k, b), T) = 2$$

for almost all discs $\mathbb{D}(ir_k, b)$. So, almost all $\mathbb{D}(ir_k, b)$ contain only one skew-conjugate pair of simple eigenvalues $\lambda, -\overline{\lambda}$ with estimate (8.2.11). Thus, the spectrum of $T, \sigma(T)$, is symmetric with respect to the imaginary axis and fulfilled the following inclusion

$$\sigma(T) \subset \{z \in \mathbb{C} \text{ such that } |z - ir_k| \leq b \text{ for some } k, \text{ and } |Rez| \geq \gamma\}.$$

Since almost all root subspaces have dimension one, then the root subspaces of the operator T form a Riesz basis. This is equivalent to the existence of a Riesz basis of eigenvectors and finitely many Jordan chains.

Example 8.4 Let A be a closed linear operator with dense domain, $\mathscr{D}(X)$, on a Hilbert space X and B and C are two symmetric operators and non-negative on X. Consider the Hamiltonian 2×2 block operator matrix in the form

$$T := \begin{pmatrix} A & B \\ C & -A^* \end{pmatrix}.$$

Suppose that it has a Riesz basis of finite-dimensional T-invariant subspaces, e.g. obtained through Theorem 8.2.2. It follows immediately from this Riesz basis of subspaces that one can prove that for each subset $\sigma \subset \sigma_p(T)$,

$$U = \overline{\text{span}\{x \in \mathscr{L}_T(\lambda),\ \lambda \in \sigma\}},$$

is both T- and $(T - z)^{-1}$-invariant for any $z \in \rho(T)$. Moreover, since T is J_1-skew-symmetric, then the point spectrum of T, $\sigma_p(T)$, is symmetric with respect to the imaginary axis. If $\sigma_p(T) \bigcap i\mathbb{R} = \emptyset$ and σ contains one eigenvalue from each skew-conjugate pair in $\sigma_p(T)$, then

$$\langle J_1 x, x \rangle = 0$$

for all $x \in U$. On the other hand, it follows that U is the graph of a self-adjoint operator X, which X is a solution of the following Riccati equation

$$A^* X + XA + XBX - C = 0.$$

Additionally, if the hypotheses of Sect. 8.2.7 are satisfied, then X is even bounded, and boundedly invertible.

8.3 The Evolutionary Equation

8.3.1 C_0-Semigroup

Let A be a densely defined and closed linear operator in a Banach space X. Consider Eq. (7.7.1) the evolutionary equation in X. Let $\Lambda := \{\lambda_n,\ n \in \mathbb{N}\}$ be a complex sequence ordered in such a way that $\{Im\lambda_n,\ n \in \mathbb{N}\}$ forms a non-decreasing sequence. In what follows, we always suppose that

$$\sup_n |Re\lambda_n| < \infty.$$

Let $T(t)$ be a C_0-semigroup on Hilbert space X with infinitesimal generator A. Let $\rho_\infty(A)$ be the (maximal) component of the resolvent set, $\rho(A)$, of A that contains an interval $[r, \infty)$ and V be the closed subspace of X.

- If V is $T(t)$-invariant, then V is $(\lambda - A)^{-1}$-invariant for one λ in $\rho_\infty(A)$. Indeed, suppose $v \in V$ and $\alpha > \omega_0$, the growth bound of $T(t)$. Then, by assumption

$$(\alpha - A)^{-1} v = \int_0^\infty e^{-\alpha t} T(t) v dt \in V.$$

  Hence, $(\alpha - A)^{-1} V \subset V$.
- If V is $(\lambda - A)^{-1}$-invariant for some λ in $\rho_\infty(A)$, then V is $(\lambda - A)^{-1}$-invariant for all λ in $\rho_\infty(A)$. Indeed, we prove that assertion (ii) extends to all $\lambda \in \rho_\infty(A)$ by taking an arbitrary element x of $V^\perp$ and considering the holomorphic function

$$f(\lambda) := \langle x, (\lambda - A)^{-1} v \rangle$$

on $\rho_\infty(A)$. Using the above, we know that for $p \geq 0$,

$$(\alpha - A)^{-p-1}V = (\alpha - A)^{-p}(\alpha - A)^{-1}V \subset (\alpha - A)^{-p}V \subset \cdots \subset V.$$

Hence,

$$f^{(p)}(\alpha) = 0$$

for $p \geq 0$. Since $f(\lambda)$ is holomorphic, we infer from Theorem 2.5.4 that $f(\lambda) = 0$ on $\rho_\infty(A)$. From the fact that $x \in V^\perp$ and $z \in V$ are arbitrary, we have proved that

$$(\lambda - A)^{-1}V \subset V$$

for $\lambda \in \rho_\infty(A)$.

- If V is $(\lambda - A)^{-1}$-invariant for all λ in $\rho_\infty(A)$, then V is $T(t)$-invariant. Indeed, let α be larger than the growth bound of $T(t)$. As in the proof of the Hille-Yosida theorem (Theorem 5.1.1), we define the operator

$$A_\alpha := \alpha^2(\alpha - A)^{-1} - \alpha.$$

The operator A_α is a bounded linear operator on X. By the assumption and the definition of the operator A_α, we can see that $A_\alpha V \subset V$ for every α larger than the growth bound. From $A_\alpha \in \mathscr{L}(X)$, then

$$e^{A_\alpha t}V \subset V$$

for all $t \geq 0$. Hence, for fixed $t \geq 0$ and $v \in V$, it follows that

$$T(t)v = \lim_{\alpha\to\infty} e^{A_\alpha t}v \in V,$$

where we have used the fact that V is closed. Since $v \in V$ and $t \geq 0$ are arbitrary, we have proved that V is $T(t)$-invariant.

- If V is $T(t)$-invariant, then V is A-invariant and $T(t)_{|V}$ is a C_0-semigroup on V with infinitesimal generator A^V, where $A^V g = Ag$ for $g \in \mathscr{D}(A^V) := \mathscr{D}(A)\bigcap V$. Indeed, let v be an arbitrary element in $V\bigcap \mathscr{D}(A)$. Then, in view of Definition 2.6.2, it follows that

$$\lim_{t\to 0^+} \frac{1}{t}(T(t) - I)v = Av.$$

The left-hand side of this equation is an element of V for every $t > 0$. Hence, since V is closed, the limit is also an element of V. So, $Av \in V$. This show the first part. For the second part, it follows directly from Definition 2.6.1 that $T(t)_{|V}$ is a C_0-semigroup on V. Let A^V be the infinitesimal generator of $T(t)_{|V}$. If v is any element in V, then the limit of

$$\frac{1}{t}(T(t)_{|V} v - v)$$

exists if, and only if, the limit of

$$\frac{1}{t}(T(t)v - v)$$

exists. This implies that

$$\mathscr{D}(A^V) = \mathscr{D}(A) \bigcap V,$$

and for $v \in \mathscr{D}(A^V)$,

$$A^V v = Av.$$

8.3.2 Riesz Basis of Subspaces

Let A be the generator of a C_0-semigroup $T(t)$ on a separable Hilbert space X and satisfy the following conditions:
$(B1)$ $\sigma(A) = \sigma_1(A) \bigcup \sigma_2(A)$, where $\sigma_2(A) = \{\lambda_k\}_{k=1}^{\infty}$ consists of isolated eigenvalues of the operator A with finite multiplicity.
$(B2)$

$$\sup_{k \geq 1} m_a(\lambda_k) < \infty, \tag{8.3.1}$$

where $m_a(\lambda_k) = \dim E(\lambda_k, A)X$ and $E(\lambda_k, A)$ is the Riesz projector associated with λ_k.
Since X is a Hilbert space, then $T^*(t)$ is also a C_0-semigroup and its generator is A^*. Moreover,

$$\sigma(A^*) = \{\overline{\lambda} \in \mathbb{C} \text{ such that } \lambda \in \sigma(A)\} = \overline{\sigma_1(A)} \bigcup \overline{\sigma_2(A)}. \tag{8.3.2}$$

$(B3)$ There is a constant α such that

$$\sup\{Re\lambda \text{ such that } \lambda \in \sigma_1(A)\} \leq \alpha \leq \inf\{Re\lambda \text{ such that } \lambda \in \sigma_2(A)\} \tag{8.3.3}$$

and

$$\inf_{n \neq m} |\lambda_n - \lambda_m| > 0, \tag{8.3.4}$$

for λ_n and $\lambda_m \in \sigma_2(A)$.
Define the set

$$Sp_{\sigma_2}(A) := \left\{ \sum_{k=1}^{m} E(\lambda_k, A)\Phi, \ \forall \Phi \in X \text{ and } \forall m \in \mathbb{N} \right\}$$

and the closed subspace

$$X_2 := \overline{Sp_{\sigma_2}(A)}. \tag{8.3.5}$$

It is easy to see that X_2, given in (8.3.5), is a $T(t)$-invariant closed subspace. Hence, X_2 is A-invariant. Conditions (8.3.1), (8.3.3), and (8.3.4) imply that the set $\sigma_2(A)$ satisfies all conditions in Theorem 7.7.1. So, the family $\{\mathscr{E}^{(k)}(\sigma_2),\ k \in \mathbb{N}\}$ forms a $\mathscr{L}$-basis in $L_2[0, T]$ for $T > 2\pi\mathscr{D}^+(\sigma_2)$, where $\mathscr{E}^{(k)}(\cdot)$ (respectively, $\mathscr{D}^+(\sigma_2)$) is defined in (7.7.3) (respectively, (7.7.5)). For all $f \in X$ and $\phi \in Sp_{\sigma_2}(A)$, the function $\langle T(t)\phi, f\rangle$ is in the subspace U (the definition of U see (7.7.4)). Hence, for all $\phi \in X_2$, $\langle T(t)\phi, f\rangle \in U$ and, we have

$$\langle T(t)\phi, f\rangle = \sum_{k=1}^{\infty}\sum_{j=0}^{m_k}\frac{t^j}{j!}e^{\lambda_k t}\langle (A-\lambda_k)^j E(\lambda_k, A)\phi, f\rangle. \tag{8.3.6}$$

Let $A^{\dagger}$ and $T^{\dagger}(t)$ be the adjoint operator of A and $T(t)$ restricted to X_2, respectively. Since X_2 endowed with $\|\cdot\|_X$ is also a Hilbert space, $T^{\dagger}(t)$ is a C_0-semigroup and its generator is $A^{\dagger}$. Then, $A^{\dagger}$ has the property that each $\overline{\lambda_k}$ is an isolated eigenvalue of $A^{\dagger}$ with finite multiplicity, and we have

$$E^{\dagger}(\lambda_k, A) = E(\overline{\lambda_k}, A^{\dagger}).$$

For all $f, g \in X_2$,

$$\langle T^{\dagger}(t)g, f\rangle = \langle g, T(t)f\rangle.$$

For any $f \in Sp_{\sigma_2}(A)$,

$$f = \sum_{k=1}^{m} E(\lambda_k, A)f$$

and, it holds that

$$\begin{aligned}
\|f\|^2 = \langle f, f\rangle &= \left\langle \sum_{k=1}^{m} E(\lambda_k, A)f, f\right\rangle \\
&= \sum_{k=1}^{m}\langle E(\lambda_k, A)f, E^{\dagger}(\lambda_k, A)f\rangle. \\
&\le \left(\sum_{k=1}^{m}\|E(\lambda_k, A)f\|^2\right)^{\frac{1}{2}}\left(\sum_{k=1}^{m}\|E^{\dagger}(\lambda_k, A)f\|^2\right)^{\frac{1}{2}} \\
&\le \left(\sum_{k=1}^{m}\|E(\lambda_k, A)f\|^2\right)^{\frac{1}{2}}\left(\frac{M(e^{2wT}-1)}{2wC_1}\right)^{\frac{1}{2}}\|f\|.
\end{aligned} \tag{8.3.7}$$

Define a set in X by

$$X_1 = \mathscr{Q}(A) := \{g \in X \text{ such that } E(\lambda_k, A)g = 0 \ \ \forall k \geq 1\}. \tag{8.3.8}$$

Evidently, X_1 is a $T(t)$-invariant closed subspace, and

$$X_1 \bigcap X_2 = \{0\}.$$

Denote

$$X_2^* := \overline{\text{span}\{E^*(\lambda_k, A)X, \ \ k \geq 1\}},$$

and

$$\mathscr{Q}(A^*) := \{g \in X \text{ such that } E^*(\lambda_k, A)g = 0 \ \ \forall k \geq 1\}. \tag{8.3.9}$$

From the Riesz basis property of $\{\mathscr{E}^{(n)}(\sigma_2)\}$ in U (see below), we know that there exist positive constants C_1 and C_2 such that

$$\alpha \leq \int_0^T |\langle T(t)\phi, f\rangle|^2 dt \leq \beta,$$

where

$$\alpha = C_1 \sum_{k=1}^{\infty} \sum_{j=0}^{m_k} \left| \frac{\langle (A - \lambda_k)^j E(\lambda_k, A)\phi, f\rangle}{j!} \right|^2$$

and

$$\beta = C_2 \sum_{k=1}^{\infty} \sum_{j=0}^{m_k} \left| \frac{\langle (A - \lambda_k)^j E(\lambda_k, A)\phi, f\rangle}{j!} \right|^2.$$

Let $T(t)$ satisfy $\|T(t)\| \leq Me^{wt}$. Then, we get

$$C_1 \sum_{k=1}^{\infty} \sum_{j=0}^{m_k} \left| \frac{\langle (A - \lambda_k)^j E(\lambda_k, A)\phi, f\rangle}{j!} \right|^2 \leq M \frac{e^{2wT} - 1}{2w} \|\phi\|^2 \|f\|^2 \tag{8.3.10}$$

which implies

$$C_1 \sum_{k=1}^{\infty} |\langle E(\lambda_k, A)\phi, f\rangle|^2 \leq M \frac{e^{2wT} - 1}{2w} \|\phi\|^2 \|f\|^2.$$

Consequently,

$$C_1 \sum_{k=1}^{\infty} \|E(\lambda_k, A)\phi\|^2 \le M \frac{e^{2wT} - 1}{2w} \|\phi\|^2. \tag{8.3.11}$$

It is easy to see that $T(t)$ is a C_0-semigroup in X_2 and its generator is $A_{|X_2}$ with domain $\mathscr{D} = \mathscr{D}(A) \bigcap X_2$. Also, we can shows that each λ_k is an isolated eigenvalue of $A_{|X_2}$ with finite multiplicity. For any $f, g \in X_2$,

$$\langle T^{\dagger}(t)g, f\rangle = \langle g, T(t)f\rangle \in U$$

and, by using Eq. (8.3.6), we have

$$\langle T^{\dagger}(t)g, f\rangle = \sum_{k=1}^{\infty} \sum_{j=0}^{m_k} \frac{t^j}{j!} e^{\overline{\lambda_k} t} \langle (A^{\dagger} - \overline{\lambda_k})^j E(\overline{\lambda_k}, A^{\dagger})g, f\rangle.$$

With an entirely parallel approach, we can prove that

$$C_1 \sum_{k=1}^{\infty} \|E^{\dagger}(\lambda_k, A)g\|_{X_2}^2 \le M \frac{e^{2wT} - 1}{2w} \|g\|_{X_2}^2,$$

where $\|\cdot\|_{X_2}$ denotes the norm in X_2. For any $f \in Sp_{\sigma_2}(A)$,

$$f = \sum_{k=1}^{m} E(\lambda_k, A)f$$

and using Eq. (8.3.7), we have

$$\|f\|^2 \le \frac{M(e^{2wT} - 1)}{2wC_1} \sum_{k=1}^{m} \|E(\lambda_k, A)f\|^2.$$

Taking the limit, we have

$$\|f\|^2 \le \frac{M(e^{2wT} - 1)}{2wC_1} \sum_{k=1}^{\infty} \|E(\lambda_k, A)f\|^2 \quad \forall f \in X_2.$$

Combining it with (8.3.11) yields, for all $f \in X_2$, we have

$$\frac{2wC_1}{M(e^{2wT} - 1)} \sum_{k=1}^{\infty} \|E(\lambda_k, A)f\|^2 \le \|f\|^2 \le \frac{M(e^{2wT} - 1)}{2wC_1} \sum_{k=1}^{\infty} \|E(\lambda_k, A)f\|^2. \tag{8.3.12}$$

This means that $\{E(\lambda_k, A)X_2,\ k \ge 1\}$ is a subspace Riesz basis in X_2.

Lemma 8.3.1 *Let X be a separable Hilbert space, and let A be the generator of a C_0-semigroup $T(t)$ on X and satisfy conditions $(B1)$-$(B3)$. Then, there is a $T(t)$-invariant subspace X_1 such that $X = \overline{X_1 \oplus X_2}$.* ◇

Proof Let X_1 be the set defined in (8.3.8). It is easy to see that X_1 is a $T(t)$-invariant closed subspace, and $X_1 \bigcap X_2 = \{0\}$. In order to prove that $X = \overline{X_1 \oplus X_2}$, it remains to show that $X_1 + X_2$ is dense in X. For this, we turn to the adjoint A^* of A. By using Eq. (8.3.2), we have each $\overline{\lambda_k} \in \overline{\sigma_2(A)}$ is an isolated eigenvalue of A^* with

$$E^*(\lambda_k, A) = E(\overline{\lambda_k}, A^*).$$

It is easy to see that X_2^* and $\mathcal{Q}(A^*)$, given in (8.3.9), are $T^*(t)$-invariant closed subspace and

$$\mathcal{Q}(A^*) \bigcap X_2^* = \{0\}.$$

Furthermore, $X = \mathcal{Q}(A^*) \dot{+} X_2$ and $X = \mathcal{Q}(A) \dot{+} X_2^*$ (orthogonal sum). Now, for any $h \in X$ such that $h \perp \overline{X_1 + X_2}$, we have $h \perp X_1$ and $h \perp X_2$. Therefore, $h \in \mathcal{Q}(A^*) \bigcap X_2^*$, and hence $h = 0$. This shows that $X = \overline{X_1 \oplus X_2}$. Q.E.D.

Theorem 8.3.1 *Let X be a separable Hilbert space, and let A be the generator of a C_0-semigroup $T(t)$ on X and satisfy conditions $(B1)$-$(B3)$. Then, there exist two $T(t)$-invariant closed subspaces X_1 and X_2 with the property that $\sigma(A_{|X_1}) = \sigma_1(A)$, $\sigma(A_{|X_2}) = \sigma_2(A)$, $\{E(\lambda_k, A)X_2\}_{k=1}^{\infty}$ forms a subspace Riesz basis for X_2, and $X = \overline{X_1 \oplus X_2}$.* ◇

Proof Obviously, $\sigma(A_{|X_2}) = \sigma_2(A)$ and $\sigma(A_{|X_1}) = \sigma_1(A)$, so the result follows from above and Lemmas 8.3.1. Q.E.D.

8.3.3 *Riesz Basis*

8.3.3.1 Characterization of the Domain

Let X be a separable Hilbert space, and let A be the generator of a C_0-semigroup $T(t)$ on X and satisfy conditions $(B1)$-$(B3)$. Suppose that

$$\sup_{k\geq 1} \|E(\lambda_k, A)\|_X = M_1 < \infty.$$

Without loss of generality, we assume that $0 \in \rho(A)$. Then, for each $f \in X$, we have

$$E(\lambda_k, A)A^{-1}f = \sum_{j=1}^{m_k} \frac{(A-\lambda_k)^{j-1}E(\lambda_k, A)f}{\lambda_k^j}$$

and

$$\|E(\lambda_k, A)A^{-1}f\|^2 \leq \sum_{j=1}^{m_k} \left| \frac{\|(A-\lambda_k)^{j-1}E(\lambda_k, A)f\|}{(j-1)!} \right|^2 \sum_{j=1}^{m_k} \left| \frac{(j-1)!}{\lambda_k^j} \right|^2 .$$

Equation (8.3.10) implies that

$$C_1 \sum_{j=0}^{m_k-1} \left| \frac{\|(A-\lambda_k)^j E(\lambda_k, A)f\|}{j!} \right|^2 \leq M \frac{e^{2wT}-1}{2w} \|E(\lambda_k, A)f\|^2.$$

Equation (8.3.1) implies that

$$\sup_{k\geq 1} m_k = N < \infty,$$

and hence

$$\sum_{j=1}^{m_k} \left| \frac{(j-1)!}{\lambda_k^j} \right|^2 \leq 2\frac{(N!)^2}{|\lambda_k|^2}, \quad |\lambda_k| > 2.$$

Altogether, we have

$$\begin{aligned} \|E(\lambda_k, A)A^{-1}f\|^2 &\leq \frac{M(e^{2wT}-1)}{2wC_1} \frac{2(N!)^2}{|\lambda_k|^2} \|E(\lambda_k, A)f\|^2 \\ &\leq \frac{M(e^{2wT}-1)}{2wC_1} \frac{2(N!)^2}{|\lambda_k|^2} M_1^2 \|f\|^2. \end{aligned}$$

Since condition (8.3.4) together with

$$\sup_k |Re\lambda_k| < \infty$$

ensure that the series

$$\sum_{k=1}^{\infty} |\lambda_k|^{-2}$$

converges, so the series

$$\sum_{k=1}^{\infty} \|E(\lambda_k, A)A^{-1}f\|^2$$

also converges. From the Theorem 8.3.1, we know that

$$\sum_{k=1}^{\infty} E(\lambda_k, A)A^{-1}f$$

is an element in X_2. Setting

$$g := A^{-1}f - \sum_{k=1}^{\infty} E(\lambda_k, A)A^{-1}f,$$

we have $g \in \mathcal{Q}(A)$. So,

$$\mathcal{D}(A) \subset X_1 \oplus X_2 \subset X. \tag{8.3.13}$$

8.3.3.2 Direct Sum

Furthermore, X has the decomposition $X = X_1 \oplus X_2$ (topological direct sum) if, and only if,

$$\sup_{n\geq 1} \left\| \sum_{k=1}^{n} E(\lambda_k, A) \right\| < \infty. \tag{8.3.14}$$

In fact, suppose that X has a decomposition $X = X_1 \oplus X_2$. Define $\mathcal{P}$ be the projection from X into X_2 along X_1. Then, $\mathcal{P}$ is a bounded operator on X. For each $f \in X$, we have $\mathcal{P} f \in X_2$, and

$$\begin{aligned} \left\| \sum_{k=1}^{n} E(\lambda_k, A)\mathcal{P} f \right\| &\leq \left\| \sum_{k=1}^{n} E(\lambda_k, A) \right\|_{X_2} \|\mathcal{P} f\| \\ &\leq \left\| \sum_{k=1}^{n} E(\lambda_k, A) \right\|_{X_2} \|\mathcal{P}\| \|f\|. \end{aligned}$$

From (8.4.5), we have

$$\sup_{n\geq 1} \left\| \sum_{k=1}^{n} E(\lambda_k, A) \right\|_{X} < \infty. \tag{8.3.15}$$

Conversely, if (8.3.15) holds, then (8.3.11) and (8.3.15) imply that for any $n \in \mathbb{N}$, $f \in X$,

$$\begin{aligned} C_1 \sum_{k=1}^{n} \|E(\lambda_k, A)f\|^2 &\leq M \frac{e^{2wT} - 1}{2w} \left\| \sum_{k=1}^{n} E(\lambda_k, A)f \right\|^2 \\ &\leq M \frac{e^{2wT} - 1}{2w} \left\| \sum_{k=1}^{n} E(\lambda_k, A) \right\|_X^2 \|f\|^2. \end{aligned}$$

Therefore,

$$\sum_{k=1}^{\infty} \|E(\lambda_k, A)f\|^2 < \infty.$$

This shows that for any $f \in X$, the series

$$\sum_{k=1}^{\infty} E(\lambda_k, A)f$$

converges unconditionally in X_2. If we define

$$f_1 = f - \sum_{k=1}^{\infty} E(\lambda_k, A)f,$$

then $f_1 \in \mathcal{Q}(A) = X_1$. Thus,

$$X = X_1 \oplus X_2.$$

8.3.3.3 Riesz Basis Formed by Generalized Eigenvectors

Let X be a separable Hilbert space and assume that A, the generator a C_0-semigroup $T(t)$ on X, have compact resolvent and generate a C_0-semigroup $T(t)$ on X. Suppose that the dimensions of all eigenspaces of A are bounded uniformly and the resolvent of A can be expressed as

$$(\lambda - A)^{-1}F = \frac{G(\lambda, F)}{\Gamma(\lambda)} \tag{8.3.16}$$

with $G(\lambda, F)$ being some entire function valued in X. Assume that $\Gamma(z)$ satisfies the following properties:
(i) $\Gamma(z)$ is an analytic function on the complex plane, and the set of zeros of $\Gamma(z)$ coincides with $\sigma(A)$,
(ii) there exist constants $\varepsilon > 0$ and $N > 0$ such that as $|Rez| \geq N$, we have $|\Gamma(z)| \geq \varepsilon$, and
(iii) when $|Rez| \leq N$ and $|Imz|$ is large enough, $\Gamma(z)$ has an asymptotic expression

$$\Gamma(z) = G(z) + R(z),$$

where $R(z) = O(z^{-1})$ and $G(iz)$ is an entire function of sine type. Recall that condition (ii) says that the spectrum of A lies in the strip $|Rez| \leq N$, and condition (iii) reads that when $|z|$ is large enough, the zeros of $\Gamma(z)$ are simple and separated. Let $\{\lambda_k\}_{k=1}^{\infty}$ be an enumeration of all eigenvalues of A with

$$|Im\lambda_k| \leq |Im\lambda_{k+1}|.$$

Then, $\{\lambda_k\}_{k=1}^{\infty}$ satisfies the hypotheses $(B1)$-$(B3)$. Therefore, the family $\{E(\lambda_k, A)X_2\}_{k=1}^{\infty}$ forms a subspace Riesz basis for X_2. Note also that λ_k is of chain length one as k large enough, so we can always choose an orthonormal basis

$$\{\phi_{k,j},\ \ 1 \leq j \leq m_a(\lambda_k)\}$$

of the eigenspace of A corresponding to each λ_k. Therefore, these generalized eigenvectors form a Riesz basis for $\overline{Spe(A)}$, where
$Spe(A) :=$

$$\left\{\sum_{k=1}^{m}\sum_{j=1}^{n_k} a_{k,j}\phi_{k,j},\ a_{k,j} \in \mathbb{C},\ \phi_{k,j} \text{ are the generalized eigenvectors of } A,\ m \geq 1\right\}.$$

In particular, if the system of the generalized eigenvectors of A is complete in X, then

$$X = \overline{Spe(A)}$$

and the system

$$\{\phi_{k,j},\ \ 1 \leq j \leq m_a(\lambda_k)\}_{k=1}^{\infty}$$

forms a Riesz basis in X. Note that without the completeness assumption of the generalized eigenvectors of A, the assumptions on A and $\Gamma(z)$ alone cannot guarantee that the generalized eigenvector is a Riesz basis for X.

8.4 Notes and Remarks

In [27], C. Wyss used this notion in order to establish the existence of Riesz basis for the class of non-normal perturbations of normal operators. He gave some hypotheses on the multiplicities of the eigenvalues. In fact, he study the operator

$$T = G + S$$

acting on a Hilbert space, where G is unbounded, normal, having compact resolvent, and S is generally unbounded operator which is p-subordinate to G, i.e.,

$$\|S\varphi\| \leq b\|G\varphi\|^p\|\varphi\|^{1-p}, \text{ for every } \varphi \in \mathscr{D}(G),$$

where $p \in [0, 1]$ and b is a strictly positive constant. In this direction, some progress has been recently made for proving the existence of Riesz bases of eigenvectors associated to a non-normal perturbations of normal operators by the construction of non-trivial invariant subspaces. S. Charfi, A. Damergi and A. Jeribi considered in [5] see also [15] a non-normal operator T written as

$$T = G + S,$$

where G is a normal operator having compact resolvent and S verify the new concept of G-boundedness with order p satisfying

$$\|S\varphi\| \leq b\|G\varphi\|^p \|\varphi\|^{1-p} + a\|\varphi\|, \quad \text{for every } \varphi \in \mathscr{D}(G), \tag{8.4.1}$$

where $p \in [0, 1]$, $a > 0$ and $b > 0$.

There are many motivations to study such a generalization. Indeed, for $p = 1$, it is clear that the operator S is relatively bounded with respect to G, or simply G-bounded, i.e., $\mathscr{D}(G) \subset \mathscr{D}(S)$ and

$$\|S\varphi\| \leq b\|G\varphi\| + a\|\varphi\|, \quad \text{for every } \varphi \in \mathscr{D}(G),$$

where $a > 0$ and $b > 0$.

It is clear that the class of G-bounded perturbations of order p, given by Eq. (8.4.1), generalizes the class of p-subordinate perturbations, i.e., $\mathscr{D}(G) \subset \mathscr{D}(S)$ and

$$\|S\varphi\| \leq b\|G\varphi\|^p \|\varphi\|^{1-p}, \quad \text{for every } \varphi \in \mathscr{D}(G)$$

where $p \in [0, 1]$ and $b > 0$. If $p = 0$, the notion of subordination coincides with the boundedness of S.

It is clear that the new notion of G-boundedness with order p combines the two concepts of p-subordination and relatively boundedness.

By giving some suitable conditions on the spectrum of the operator G, the authors in [5] characterize the spectrum of the operator T. Indeed, they show that T admits a Riesz basis of generalized eigenvectors if the eigenvalue multiplicities of the operator G verify a certain asymptotic growth condition. The new situation checked in [5] is that S is not p-subordinate to G but G-bounded with order p and 0 is an eigenvalue of G. This situation will be found in Chap. 14 where the Gribov operator is treated.

- Wyss proved in [27, Theorem 6.1], that if a certain asymptotic growth condition is satisfied, then the root vectors of T form a Riesz basis property with parentheses. More precisely, C. Wyss improved some results given by Markus and Matsaev [19, 20]. Moreover, in [27, Theorem 6.2], Wyss obtained a Riesz basis with parentheses under a spectral condition of a different type without imposing such restriction on the multiplicities and he assumed that the eigenvalues of G lie on sufficiently separated line segments. Contrary to his result, many classical perturbation theorems for Riesz bases without parentheses such as Kato [16], Dunford and Schwartz [11] and Clark [6] need that almost all eigenvalues of G are simple.
  Despite the fact a Riesz basis consisting of (finite-dimensional) subspaces is equivalent to a Riesz basis with parentheses, C. Wyss [27] introduced the basis of subspaces notion to the formulation of his results. Moreover, he gave a detailed study of the change of the spectrum of a normal operator under a p-subordinate perturbation.

- The $\sigma(T_{|V_k})$ are disjoint since V_k were constructed as the ranges of Riesz projections associated with disjoint parts of $\sigma(T)$.
- [30] by Xu and Yung discussed the problems on the expansion of a semigroup and criterion for being a Riesz basis and considered the following evolutionary equation in a separable Hilbert space X

$$\begin{cases} \frac{d}{dt}\phi(t) = A\phi(t) \\ \phi(0) = \phi_0. \end{cases} \tag{8.4.2}$$

When the operator A generates a C_0-semigroup $T(t)$ on X, the solution of (8.4.2) can be expressed as

$$\phi(t) = T(t)\phi_0.$$

In particular, if

$$\sigma(A) = \sigma_1(A) \cup \sigma_2(A)$$

with $\sigma_2(A) = \{\lambda_k\}_{k=1}^r$ is consisted of isolated eigenvalues distributed in a vertical strip, then if $r < \infty$, the solution of (8.4.2) can be written as

$$\phi(t) = \sum_{k=1}^{r} e^{\lambda_k t} P_k(t, \phi_0) + R(t, \phi_0), \tag{8.4.3}$$

where $P_k(t, \phi_0)$ are X-valued polynomials in t. It is interesting to study the case when $r = \infty$, more precisely we must ask whether equality (8.4.3) still remains true. If A has a compact resolvent, then we shall ask whether the residual term $R(t, \phi_0)$ vanishes. The answer is clear in the case when A is a non-positive selfadjoint or, more generally, a skew-adjoint operator. But, the answer to this question isn't evident if A is a non-selfadjoint, or in general a non-normal operator (e.g. transport operator [18]). This situation appears in the most of applications (see the applications given at the end of this book). More precisely, when a boundary feedback control acts on a vibration system, it usually yields a non-self-adjoint operator A satisfying:

(i) A has compact resolvent and generates a C_0-semigroup,
(ii) the spectrum $\sigma(A)$ lies inside some vertical strip, and
(iii) the eigenvalues of A are separated and the algebraic multiplicity of the eigenvalues are uniformly bounded.

- Among examples that are of this type, we can cite strings [2, 3, 23], Euler-Bernoulli beams [7, 13, 14], Rayleigh beams [22], Timoshenko beams [24, 31], heat exchanger [17], etc. In order to investigate the expansion issue of (8.4.2) for these problems, [30] plan to do two things: (α) one is to show that there exists a direct sum decomposition of X: $X = X_1 + X_2$, where X_1 and X_2 are $T(t)$-invariant closed subspaces such that $\sigma(A|_{X_1}) = \sigma_1(A)$ and $\sigma(A|_{X_2}) = \sigma_2(A)$; ($\beta$) another is to show that, the non-harmonic series

$$\phi(t) = \sum_{k=1}^{\infty} e^{\lambda_k t} P_k(t, \phi_0), \tag{8.4.4}$$

converges in some sense, for each $\phi_0 \in X_2$.

For $r = \infty$, (8.4.3) holds true if these two assertions are satisfied. However, since (α) is concerned with the splitting of the spectrum for unbounded linear operators, and (β) is also concerned with establishing the basis property, some difficulties will occur. If X is separable, A has a compact resolvent and the algebraic multiplicities of the eigenvalues of A are uniformly bounded. If there is a sequence of generalized eigenvectors of A that forms an unconditional basis for X, the operator A will be said a Riesz operator and the system (8.4.2) is a Riesz system satisfying the spectrum determined growth assumption. So the stability of the system can be proved via the spectrum of A. In [30], the authors provided sufficient and necessary condition for the semigroup expansion and showed the convergence of the series (8.4.4) by establishing some property for the family of exponentials. Indeed, they proved that the solution of (8.4.2) can be expressed in terms of the generalized eigenvectors for each $\phi_0 \in X_1 \oplus X_2$. Moreover, if the Riesz projection of A satisfies

$$\sup_k \|E(\lambda_k, A)\| < \infty,$$

the result remains true for all classical solutions of (8.4.2).

(i) Section 8.2.3 yields a Riesz basis of eigenvectors and finitely many Jordan chains for the operator T if almost each eigenvalues of the operator G are simple and almost all line segments L_{jk} contain one eigenvalue only. Indeed, almost all spectral subspaces corresponding to the K_{jk} are one dimensional in this case, and the Riesz basis of subspaces is thus equivalent to the existence of a Riesz basis of eigenvectors and finitely many Jordan chains.

(ii) (8.3.12) leads to

$$\sum_{k=1}^{\infty} E(\lambda_k, A) f$$

converges unconditionally in X_2 with value f and yields that

$$\sup_{n \geq 1} \left\| \sum_{k=1}^{n} E(\lambda_k, A) \right\|_{X_2} < \infty. \tag{8.4.5}$$

(iii) From Theorem 8.3.1, it is clear that for each $\phi_0 \in X_1 \oplus X_2$, the solution of (7.7.1) can be expressed in terms of the generalized eigenvectors.

(iv) if

$$\sup_{k \geq 1} \|E(\lambda_k, A)\| < \infty,$$

it is evident from Eq. (8.3.13) that the same is true for all classical solutions of (7.7.1)

(v) The Eq. (8.3.14) is considered as a sufficient and necessary condition for the semigroup expansion. In particular, Eq. (8.3.14) yields that

$$\{E(\lambda_k, A)X,\ k \geq 1\}$$

forms a subspace Riesz basis in X if $\sigma_1(A) = \emptyset$ and $X_2 = X$.

References

1. V. Adamjan, V. Pivovarchik, C. Tretter, On a class of non-self-adjoint quadratic matrix operator pencils arising in elasticity theory. J. Oper. Theo. **47**(2), 325–341 (2002)
2. S. A. Avdonin, S. A. Ivanov, Riesz bases of exponentials and divided differences (Russian), Algebra Anal. **13**(3), 1–17 (2001) . transl. St. Petersburg Math. J. **13**(3) 339–351 (2002)
3. S. A. Avdonin, M. Tucsnark, Simultaneous controllability in sharp time for two elastic strings, ESAIM Control Optim. Calculus Variations **6**, 259–273 (electronic) (2001)
4. T.Y. Azizov, A. Dijksma, L.I. Sukhocheva, On basis properties of self-adjoint operator functions. J. Funct. Anal. **178**(2), 306–342 (2000)
5. S. Charfi, A. Damergi, A. Jeribi, On a Riesz basis of finite-dimensional invariant subspaces and application to Gribov operator in Bargmann space. Line. Mult. Alge. **61**(11), 1577–1591 (2013)
6. C. Clark, On relatively bounded perturbations of ordinary differential operators. Pac. J. Math. **25**, 59–70 (1968)
7. F. Conrad, Ö. Morgül, On the stabilization of a flexible beam with a tip mass. SIAM J. Control Optim. **36**, 1962–1986 (1998)
8. R.F. Curtain, H.J. Zwart, *An Introduction to Infinite Dimentional Linear Systems Theory* (Springer, New-York, 1995)
9. E.B. Davies, Pseudo-spectra, the harmonic oscillator and complex resonances. Proc. R. Soc. Lond. A **455**, 585–599 (1999)
10. E.B. Davies, *Linear Operators and Their Spectra* (Cambridge University Press, Cambridge, 2007)
11. N. Dunford, J. T. Schwartz, *Linear operators II*, vol. 7 (Pure Appl. Math., Interscience, New-York, (1963)
12. I.C. Gohberg, M.G. Krein, *Introduction to the Theory of Linear Non-self-adjoint Operators in Hilbert Space* (Amer. Math. Soci, Providence, 1969)
13. B.Z. Guo, On the boundary control for a hybrid system with variable coefficients. J. Opti. Theo. Appl. **114**(2), 373–395 (2002)
14. B.Z. Guo, Riesz basis property and exponential stability of controlled Euler-Bernoulli beam equations with variable coefficients. SIAM J. Control Optim. **40**, 1905–1923 (2002)
15. A. Jeribi, *Denseness, bases and frames in Banach spaces and applications* (de Gruyter, Berlin, 2018)
16. T. Kato, *Perturbation Theory for Linear Operators* (Springer, Berlin, 1980)
17. N. Kunimatsu, H. Sano, Stability analysis of heat-exchanger equations with boundary feedbacks. IMA J. Math. Control Inform. **15**, 317–330 (1998)
18. E.W. Larsen, P.F. Zweifel, On the spectrum of the linear transport operator. J. Math. Phys. **15**, 1987–1997 (1974)
19. A. S. Markus, *Introduction to the Spectral Theory of Polynomial Operator Pencils*, vol. 71 (Amer. Math. Soci., Providence, RI, 1988)

20. A. S. Markus, V.I. Matsaev, On the convergence of eigenvector expansions for an operator which is close to being self-adjoint, Mat. Issled., (in Russian) **61**, 104–129 (1981)
21. M. Marletta, A. Shkalikov, C. Tretter, Pencils of differential operators containing the eigenvalue parameter in the boundary conditions. Proc. R. Soc. Edinburgh Sect. A **133**(4), 893–917 (2003)
22. B. P. Rao, Optimal energy decay rate in a damped Rayleigh beam, in: S. Cox, I. Lasiecka (eds.), *Optimisation Methods in Partial Differential Equations*, vol. 209 (Contemp. Math., 1997), pp. 221–229
23. M.A. Shubov, Basis property of eigenfunctions of non-self-adjoint operator pencils generated by the equation of nonhomogeneous damped string. Inte. Equa. Oper. Theo. **25**, 289–328 (1996)
24. M.A. Shubov, Asymptotic and spectral analysis of the spatially nonhomogeneous Timoshenko beam model. Math. Nachr. **241**, 125–162 (2002)
25. V. N. Vizitei, A. S. Markus, On convergence of multiple expansions in the eigenvectors and associated vectors of an operator pencil. Mat. Sb. **66**(108), 287–320 (1965); English transl. in Amer. Math. Soc. Trans., **87**(2), 187–227 (1970)
26. C. Wyss, Perturbation theory for Hamiltonian operator matrices and Riccati equations, Ph.D. thesis, University of Bern, (2008)
27. C. Wyss, Riesz bases for p-subordinate perturbations of normal operator. J. Funct. Anal. **258**, 208–240 (2010)
28. C. Wyss, Hamiltonians with Riesz bases of generalised eigenvectors and Riccati Equations, arXiv:1005.5336v1
29. G.Q. Xu, D.X. Feng, Riesz basis property of a Timoshenko beam with boundary feeback and application. IMA J. Appl. Math. **67**(4), 357–370 (2002)
30. G.Q. Xu, S.P. Yung, The expansion of a semigroup and a Riesz basis criterion. J. Diff. Equ. **210**, 1–24 (2005)
31. G.Q. Xu, S.P. Yung, Properties of a class of $\mathscr{C}_0$-semigroups on a Banach spaces and their applications. J. Math. Anal. Appl. **328**, 245–256 (2007)
32. H. Zwart, Riesz basis for strongly continuous groups (2008). arXiv:0808.3447 [math.FA]

Chapter 9
Analytic Operators in Feki-Jeribi-Sfaxi's Sense

The purpose of this chapter is to formulate some new supplements to perturbation theory of linear operators [21] by considering a non-analytic perturbation "analytic operators in Feki-Jeribi-Sfaxi's sense" involving more than one perturbation parameter. On the one hand, under a relative boundedness condition, we show the invariance of the closure and we establish sufficient conditions to give the series expansions of the eigenvalue and the eigenvector of the perturbed operator. The purpose of the present chapter is to formulate some new supplements to perturbation theory of linear operators by considering a non-analytic perturbation involving more than one perturbation parameter.

9.1 Family of Operators Dependent of Several Parameters

Let T_0 be a linear operator on a separable Banach space X such that
$(H1)$ T_0 is closed with domain $\mathscr{D}(T_0)$ dense in X and has isolated discrete eigenvalues.

Let $T_1, T_2, T_3, \ldots$ be some linear operators on X having the same domain $\mathscr{D}$ and satisfying the hypothesis
$(H2)$ $\mathscr{D}(T_0) \subset \mathscr{D}$ and there exist $a, b, q > 0$ and $\beta \in]0, 1]$ such that for all $k \geq 1$

$$\|T_k\varphi\| \leq q^{k-1}(a\|\varphi\| + b\|T_0\varphi\|^{\beta}\|\varphi\|^{1-\beta}), \quad \text{for all } \varphi \in \mathscr{D}(T_0). \tag{9.1.1}$$

In this chapter, we are interested by the perturbed operator

$$T(\xi) := T_0 + \xi_1 T_1 + \xi_2 T_2 + \cdots + \xi_k T_k + \cdots, \tag{9.1.2}$$

where the sequence $(\xi_k)_{k\geq 0}$ satisfy the following condition:
$(H3)$ $\xi := (\xi_k)_{k\geq 0}$ is a sequence of complex numbers verifying $\xi_0 = 1$ and

A. Jeribi, *Perturbation Theory for Linear Operators*,
https://doi.org/10.1007/978-981-16-2528-2_9

$$\tau(\xi) = \sum_{k=1}^{\infty} |\xi_k| < \infty.$$

Let $A(\varepsilon)$ be a perturbed operator on a Banach space X, depending on a complex parameter ε as a convergent power series

$$A(\varepsilon) := A_0 + \varepsilon A_1 + \varepsilon^2 A_2 + \cdots + \varepsilon^k A_k + \cdots , \tag{9.1.3}$$

where $\varepsilon \in \mathbb{C}$ and $A_0, A_1, A_2, A_3, \ldots$ are linear operators on X, having the same domain $\mathscr{D}$ and satisfying the following condition

$$\|A_k\varphi\| \leq q^{k-1}(a\|\varphi\| + b\|A_0\varphi\|) \tag{9.1.4}$$

for all $\varphi \in \mathscr{D}$ and for all $k \geq 1$, with $a, b, q > 0$.

9.2 Invariance of the Closure

The first result of this section is formulated in the following theorem.

Theorem 9.2.1 *Suppose that the assumptions $(H1)$, $(H2)$ with $\beta = q = 1$, and $(H3)$ hold. Then, the series*

$$\sum_{k\geq 0} \xi_k T_k \varphi$$

converges for all $\varphi \in \mathscr{D}(T_0)$. If $T(\xi)\varphi$ denotes its limit, then we define a linear operator $T(\xi)$ with domain $\mathscr{D}(T_0)$. In addition, if $\tau(\xi) < \frac{1}{b}$, then $T(\xi)$, given in (9.1.2), is closed. ◇

Proof Let $\varphi \in \mathscr{D}(T_0)$ and $n \in \mathbb{N}^*$. Using Eq. (9.1.1), we get

$$\begin{aligned}
\left\| \sum_{k=0}^{n} \xi_k T_k \varphi \right\| &\leq \sum_{k=0}^{n} |\xi_k| \|T_k \varphi\| \\
&\leq \|T_0\varphi\| + \sum_{k=1}^{n} |\xi_k| \|T_k \varphi\| \\
&\leq \|T_0\varphi\| + \sum_{k=1}^{n} |\xi_k| \left(a\|\varphi\| + b\|T_0\varphi\|\right) \\
&\leq a\tau(\xi)\|\varphi\| + (1 + b\tau(\xi)) \|T_0\varphi\|.
\end{aligned}$$

Hence, the series

$$\sum_{k\geq 0} \xi_k T_k \varphi$$

is convergent. Setting $T(\xi)\varphi$ its limit, we define a linear operator $T(\xi)$ with domain $\mathscr{D}(T_0)$. Similarly, by Eq. (9.1.1), we obtain

$$\|(T(\xi) - T_0)\varphi\| \leq a\tau(\xi)\|\varphi\| + b\tau(\xi)\|T_0\varphi\|,$$

for all $\varphi \in \mathscr{D}(T_0)$. Since T_0 is a closed operator and $\tau(\xi) < \frac{1}{b}$, we deduce, in view of Lemma 1.4.1, that the operator $T(\xi)$ is also closed. Q.E.D.

Corollary 9.2.1 *Let $A_0, A_1, A_2, A_3, \ldots$ be linear operators on X such that A_0 is closed with domain $\mathscr{D}(A_0)$ dense in X and $A_1, A_2, A_3, \ldots$ are with the same domain $\mathscr{D} \supset \mathscr{D}(A_0)$ and verifying*

$$\|A_k\varphi\| \leq q^{k-1}\left(a\|\varphi\| + b\|A_0\varphi\|^{\beta}\|\varphi\|^{1-\beta}\right), \quad \text{for all } \varphi \in \mathscr{D}(A_0), \tag{9.2.1}$$

where a, b, and q are positive numbers and $\beta \in]0, 1]$. Then, the series

$$\sum_{k\geq 0} \varepsilon^k A_k\varphi$$

converges for all $\varphi \in \mathscr{D}(A_0)$. If $A(\varepsilon)\varphi$ denotes its limit, then $T(\varepsilon)$ is a linear operator with domain $\mathscr{D}(A_0)$. In addition, if $|\varepsilon| < (q + \beta b)^{-1}$, then the operator $A(\varepsilon)$ is closed. ◇

Proof Let $|\varepsilon| < \frac{1}{q}$ and $\varphi \in \mathscr{D}$. It follows from Eq. (9.2.1) that

$$\left\|\sum_{k=0}^{\infty} \varepsilon^k A_k\varphi\right\| \leq \|A_0\varphi\| + \frac{|\varepsilon|}{1 - |\varepsilon|q}(a\|\varphi\| + b\|A_0\varphi\|^{\beta}\|\varphi\|^{1-\beta}).$$

So, the series

$$\sum_{k\geq 0} \varepsilon^k A_k\varphi$$

is convergent. Similarly, using Eq. (9.2.1), we have

$$\|(A(\varepsilon) - A_0)\varphi\| \leq \theta_1(\varepsilon)\|\varphi\| + \theta_2(\varepsilon)\|A_0\varphi\|^{\beta}\|\varphi\|^{1-\beta},$$

for all $\varphi \in \mathscr{D}(A_0)$, where

$$\theta_1(\varepsilon) = \frac{a|\varepsilon|}{1 - |\varepsilon|q}$$

and

$$\theta_2(\varepsilon) = \frac{b|\varepsilon|}{1 - |\varepsilon|q}.$$

If $|\varepsilon| < \frac{1}{q+\beta b}$, we have

$$\theta_2(\varepsilon) < \frac{1}{\beta}.$$

Since A_0 is a closed operator, we deduce, in view of Lemma 1.4.1, that the operator $A(\varepsilon)$ is closed for $|\varepsilon| < \frac{1}{q+\beta b}$. Q.E.D.

Corollary 9.2.2 *Let T_0 be a closed linear operator with domain $\mathscr{D}(T_0)$ dense in a separable Banach space X. Let $T_1, T_2, T_3, \ldots$ be some linear operators on X having the same domain $\mathscr{D}$ and satisfying the hypotheses: $\mathscr{D}(T_0) \subset \mathscr{D}$ and, there exist $a, b > 0$ such that for all $k \geq 1$*

$$\|T_k\varphi\| \leq \xi_k \left(a\|\varphi\| + b\|T_0\varphi\|\right), \quad \textit{for all} \;\; \varphi \in \mathscr{D}(T_0), \tag{9.2.2}$$

where $(\xi_k)_{k\geq 1}$ is a sequence of positive real numbers verifying

$$\xi = \sum_{k=1}^{\infty} \xi_k < \infty.$$

Then, the series

$$\sum_{k\geq 0} T_k\varphi$$

converges for all $\varphi \in \mathscr{D}(T_0)$. If $T(\xi)\varphi$ denotes its limit, then we define a linear operator $T(\xi)$ with domain $\mathscr{D}(T_0)$ and, for $\xi < \frac{1}{b}$, the operator $T(\xi)$ is closed. ◇

Proof Let $\varphi \in \mathscr{D}(T_0)$ and $n \in \mathbb{N}^*$. Using Eq. (9.2.2), we get

$$\begin{aligned}
\left\|\sum_{k=0}^{n} T_k\varphi\right\| &\leq \sum_{k=0}^{n} \|T_k\varphi\| \\
&\leq \|T_0\varphi\| + \sum_{k=1}^{n} \|T_k\varphi\| \\
&\leq \|T_0\varphi\| + \sum_{k=1}^{n} \xi_k(a\|\varphi\| + b\|T_0\varphi\|) \\
&\leq \|T_0\varphi\| + \xi(a\|\varphi\| + b\|T_0\varphi\|).
\end{aligned}$$

Hence, the series

$$\sum_{k\geq 0} T_k\varphi$$

is convergent. Setting $T(\xi)\varphi$ its limit, we define a linear operator $T(\xi)$ with domain $\mathscr{D}(T_0)$. Similarly, using Eq. (9.2.2), we obtain

$$\|(T(\xi) - T_0)\varphi\| \leq a\xi\,\|\varphi\| + b\xi\,\|T_0\varphi\|, \quad \text{for all} \;\; \varphi \in \mathscr{D}(T_0).$$

Since T_0 is a closed operator and $\xi < \frac{1}{b}$, we deduce, in view of Lemma 1.4.1, that the operator $T(\xi)$ is also closed. Q.E.D.

In particular, if we take $\xi_k = \varepsilon^k q^{k-1}$, $T_0 = A_0$ and $T_k = \frac{1}{q^{k-1}} A_k$ $\forall k \geq 1$, where $q > 0$, ε is a complex number and $(A_k)_{k\geq 1}$ are linear operators on X verifying Eq. (9.1.4), we regain the analytic perturbation $A(\varepsilon)$ (see Eq. (9.1.3)) considered by B. Sz. Nagy in [21]. More precisely, we have the following result.

Corollary 9.2.3 *Let $A_0, A_1, A_2, A_3, \ldots$ be linear operators on X such that A_0 is closed with domain $\mathcal{D}(A_0)$ and $A_1, A_2, A_3, \ldots$ are with the same domain $\mathcal{D} \supset \mathcal{D}(A_0)$ and verifying*

$$\|A_k\varphi\| \leq q^{k-1}\left(a\|\varphi\| + b\|A_0\varphi\|\right), \quad \text{for all } \varphi \in \mathcal{D}(A_0), \tag{9.2.3}$$

where a, b, and q are positive numbers. Then, the series

$$\sum_{k\geq 0} \varepsilon^k A_k\varphi$$

converges for all $\varphi \in \mathcal{D}(A_0)$ and for $|\varepsilon| < \frac{1}{q}$. If $A(\varepsilon)\varphi$ denotes its limit, then we define a linear operator $A(\varepsilon)$ with domain $\mathcal{D}(A_0)$. For $|\varepsilon| < \frac{1}{q+b}$, the operator $A(\varepsilon)$ is closed. ◇

Proof We have

$$\begin{aligned}\tau(\xi) &= \sum_{k=1}^{\infty} |\xi_k| \\ &= |\varepsilon| \sum_{k=0}^{\infty} (|\varepsilon| q)^k.\end{aligned}$$

If $|\varepsilon| < \frac{1}{q}$, then

$$\tau(\xi) = \frac{|\varepsilon|}{1 - |\varepsilon| q} < +\infty.$$

Hence, the series

$$\sum_{k\geq 0} \varepsilon^k A_k\varphi$$

converges for all $|\varepsilon| < \frac{1}{q}$. Moreover, $\tau(\xi) < \frac{1}{b}$ if, and only if, $|\varepsilon| < \frac{1}{q+b}$. So, $A(\varepsilon)$ is closed for $|\varepsilon| < \frac{1}{q+b}$. Q.E.D.

9.3 Eigenvalues

Let $n \in \mathbb{N}^*$ and λ_n be the nth isolated eigenvalue of the operator T_0 with multiplicity one. Consider $d_n = d(\lambda_n, \sigma(T_0)\backslash\{\lambda_n\})$ the distance between λ_n and $\sigma(T_0)\backslash\{\lambda_n\}$, and $\mathscr{C}_n = \mathscr{C}(\lambda_n, r_n)$ the closed circle with center λ_n and with radii $r_n = \frac{d_n}{2}$. By using both $(T_0 - z)^{-1}$ is a regular analytic function for $z \in \rho(T_0)$ and $\|(T_0 - z)^{-1}\|$ is a continuous function of z, then we will denote the following by

$$M_n := \max_{z_n \in \mathscr{C}_n} \|(T_0 - z_n)^{-1}\|,$$

$$\begin{aligned}\|T_k(T_0 - z_n)^{-1}g\| N_n &:= \max_{z_n \in \mathscr{C}_n} \|T_0(T_0 - z_n)^{-1}\| \\ &= \max\nolimits_{z_n \in \mathscr{C}_n} \|I + z_n(T_0 - z_n)^{-1}\|\end{aligned}$$

and

$$\alpha_n := aM_n + bN_n,$$

where a and b are the constants given in (9.1.1).

Proposition 9.3.1 *Assume that the assumptions $(H1)$, $(H2)$ with $\beta = q = 1$, and $(H3)$ hold. Let $R_{z_n}(\xi) := (T(\xi) - z_n)^{-1}$. If $\tau(\xi) < \frac{1}{\alpha_n}$, then the resolvent of $T(\xi)$ at $z_n \in \mathscr{C}_n$ is well defined and the following assertions hold:*

(i) the operator $R_{z_n}(\xi)$ can be developed into a series

$$R_{z_n}(\xi) := R_{z_n,0}(\xi) + R_{z_n,1}(\xi) + R_{z_n,2}(\xi) + \cdots + R_{z_n,l}(\xi) + \cdots,$$

where

$$R_{z_n,0}(\xi) := R_{z_n} := (T_0 - z_n)^{-1}, \text{ and for all } l \geq 1$$

$$R_{z_n,l}(\xi) := (-1)^l \sum_{k=l}^{+\infty} \sum_{i_1+i_2+\cdots+i_l=k} \xi_{i_1}\xi_{i_2}\ldots\xi_{i_l} R_{z_n} T_{i_1} R_{z_n} T_{i_2} R_{z_n} \ldots T_{i_l} R_{z_n}, \quad (9.3.1)$$

(ii) we have

$$\|R_{z_n,l}(\xi)\| \leq M_n(\alpha_n \tau(\xi))^l, \text{ for every } l \geq 0. \quad \diamond$$

Proof (i) Let $z_n \in \mathscr{C}_n$. Using Eq. (9.1.1), we infer that for all $g \in X\backslash\{0\}$ and for all $k \geq 1$,

$$\begin{aligned}\|T_k(T_0 - z_n)^{-1}g\| &\leq a\|(T_0 - z_n)^{-1}g\| + b\|T_0(T_0 - z_n)^{-1}g\| \\ &\leq \left(a\|(T_0 - z_n)^{-1}\| + b\|T_0(T_0 - z_n)^{-1}\|\right) \|g\|.\end{aligned}$$

So, for all $k \geq 1$,

$$\|T_k(T_0 - z_n)^{-1}\| \le \alpha_n. \tag{9.3.2}$$

We claim that $\mathscr{C}_n \subset \rho(T(\xi))$. Indeed, for $z_n \in \mathscr{C}_n$, we get

$$\begin{aligned} T(\xi) - z_n &= T_0 + \xi_1 T_1 + \xi_2 T_2 + \cdots - z_n \\ &= \left(I + \xi_1 T_1 (T_0 - z_n)^{-1} + \xi_2 T_2 (T_0 - z_n)^{-1} + \cdots\right)(T_0 - z_n). \end{aligned}$$

Thus,

$$\begin{aligned} T(\xi) - z_n &= \left(I + \sum_{k=1}^{\infty} \xi_k T_k R_{z_n}\right)(T_0 - z_n) \\ &= (I + S)(T_0 - z_n), \end{aligned} \tag{9.3.3}$$

where

$$S = \sum_{k=1}^{\infty} \xi_k T_k R_{z_n}.$$

In view of Eq. (9.3.2), we have if $\tau(\xi) < \frac{1}{\alpha_n}$, then

$$\begin{aligned} \|S\| &\le \sum_{k=1}^{\infty} \|\xi_k T_k R_{z_n}\| \\ &\le \alpha_n \tau(\xi) \\ &< 1. \end{aligned}$$

Hence, $I + S$ is invertible with bounded inverse. Since $z_n \in \mathscr{C}_n \subset \rho(T_0)$, so Eq. (9.3.3) implies that $T(\xi) - z_n$ is also invertible with bounded inverse. Hence, $z_n \in \rho(T(\xi))$, which ends the proof of the claim. Now, Eq. (9.3.3) yields

$$\begin{aligned} R_{z_n}(\xi) &= R_{z_n}(I + S)^{-1} \\ &= R_{z_n} \sum_{\nu=0}^{\infty} (-S)^{\nu} \\ &= R_{z_n} \sum_{\nu=0}^{\infty} \left(\sum_{k=1}^{\infty} - \xi_k T_k R_{z_n}\right)^{\nu} \\ &= R_{z_n} + R_{z_n} \sum_{\nu=1}^{\infty} \left(\sum_{k=1}^{\infty} - \xi_k T_k R_{z_n}\right)^{\nu}. \end{aligned}$$

Since

$$\sum_{k=1}^{\infty} \| - \xi_k T_k R_{z_n}\| < 1,$$

then, in view of Proposition 3.1.2, we get

$$\begin{aligned}R_{z_n}(\xi) &= R_{z_n} + R_{z_n}\sum_{l=1}^{\infty}\sum_{k=l}^{+\infty}\sum_{i_1+i_2+\cdots+i_l=k}\left(-\xi_{i_1}T_{i_1}R_{z_n}\right)\left(-\xi_{i_2}T_{i_2}R_{z_n}\right)\ldots\left(-\xi_{i_l}T_{i_l}R_{z_n}\right)\\ &= R_{z_n} + R_{z_n}\sum_{l=1}^{\infty}\sum_{k=l}^{+\infty}\sum_{i_1+i_2+\cdots+i_l=k}(-1)^l\,\xi_{i_1}\xi_{i_2}\ldots\xi_{i_l}\,T_{i_1}R_{z_n}\,T_{i_2}R_{z_n}\ldots T_{i_l}R_{z_n}\\ &= \sum_{l=0}^{\infty}R_{z_n,l}(\xi),\end{aligned}$$

where $R_{z_n,0}(\xi) := R_{z_n} := (T_0 - z_n)^{-1}$ and for each $l \geq 1$,

$$R_{z_n,l}(\xi) := (-1)^l\sum_{k=l}^{+\infty}\sum_{i_1+i_2+\cdots+i_l=k}\xi_{i_1}\xi_{i_2}\ldots\xi_{i_l}\;R_{z_n}\;T_{i_1}R_{z_n}\;T_{i_2}R_{z_n}\ldots T_{i_l}R_{z_n}.$$

(ii) In view of both the estimates

$$\|R_{z_n}\| \leq M_n$$

and

$$\|T_k R_{z_n}\| \leq \alpha_n$$

for every $k \geq 1$ (see Eq. (9.3.2)) and by using Proposition 3.1.3, we infer that the following estimations hold

$$\begin{aligned}\|R_{z_n,l}(\xi)\| &\leq M_n\sum_{k=l}^{+\infty}\sum_{i_1+i_2+\cdots+i_l=k}\alpha_n^l|\xi_{i_1}|\,|\xi_{i_2}|\ldots|\xi_{i_l}|\\ &\leq M_n\alpha_n^l\sum_{k=l}^{+\infty}\sum_{i_1+i_2+\cdots+i_l=k}|\xi_{i_1}|\,|\xi_{i_2}|\cdots|\xi_{i_l}|\\ &\leq M_n\alpha_n^l(\tau(\xi))^l,\end{aligned}$$

for every $l \geq 1$. This implies that

$$\|R_{z_n,l}(\xi)\| \leq M_n(\alpha_n\tau(\xi))^l,$$

for every $l \geq 0$. The proof is then complete. Q.E.D.

Consider the operator

$$A(\varepsilon) := A_0 + \varepsilon A_1 + \varepsilon^2 A_2 + \varepsilon^3 A_3 + \cdots + \varepsilon^k A_k + \cdots$$

defined in Corollary 9.2.3. Analogously, let $\widetilde{\lambda_n}$ be the nth isolated eigenvalue of the operator A_0 with multiplicity one,

$$\widetilde{d_n} = d(\widetilde{\lambda_n}, \sigma(A_0)\backslash\{\widetilde{\lambda_n}\})$$

be the distance between $\widetilde{\lambda_n}$ and $\sigma(A_0)\backslash\{\widetilde{\lambda_n}\}$, and $\widetilde{\mathscr{C}_n} = \mathscr{C}(\widetilde{\lambda_n}, \widetilde{r_n})$ be the closed circle with center $\widetilde{\lambda_n}$ and radii $\widetilde{r_n} = \frac{\widetilde{d_n}}{2}$. Denoting by $\widetilde{M_n}$, $\widetilde{N_n}$ and $\widetilde{\alpha_n}$ the numbers given by

$$\begin{cases} \widetilde{M_n} := \max\limits_{z_n \in \widetilde{\mathscr{C}}_n} \|(A_0 - z_n)^{-1}\|, \\ \widetilde{N_n} := \max\limits_{z_n \in \widetilde{\mathscr{C}}_n} \|A_0(A_0 - z_n)^{-1}\| = \max\limits_{z_n \in \widetilde{\mathscr{C}}_n} \|I + z_n(A_0 - z_n)^{-1}\|, \text{ and} \\ \widetilde{\alpha_n} := a\widetilde{M_n} + b\widetilde{N_n}, \end{cases}$$

where a and b are the constants given in (9.1.1). We can see the following result:

Corollary 9.3.1 *Suppose that the assumptions of Corollary 9.2.3 hold. Let* $\widetilde{R}_{z_n}(\varepsilon) := (A(\varepsilon) - z_n)^{-1}$. *If* $|\varepsilon| < \frac{1}{q+\widetilde{\alpha_n}}$, *then*

(i) the operator $\widetilde{R}_{z_n}(\varepsilon)$ *can be developed into an entire series*

$$\widetilde{R}_{z_n}(\varepsilon) := \widetilde{R}_{z_n,0} + \varepsilon\widetilde{R}_{z_n,1} + \varepsilon^2\widetilde{R}_{z_n,2} + \cdots + \varepsilon^l\widetilde{R}_{z_n,l} + \cdots,$$

where

$$\widetilde{R}_{z_n,0} := \widetilde{R}_{z_n} := (A_0 - z_n I)^{-1}$$

$$\widetilde{R}_{z_n,l} := \sum_{l=l}^{k}(-1)^l \sum_{i_1+i_2+\cdots+i_l=k} \widetilde{R}_{z_n} A_{i_1}\widetilde{R}_{z_n} A_{i_2}\widetilde{R}_{z_n} \ldots A_{i_l}\widetilde{R}_{z_n}, \quad \forall\, l \geq 1, \quad (9.3.4)$$

(ii) for every $l \geq 1$,

$$\|\widetilde{R}_{z_n,l}\| \leq \widetilde{M}_n\widetilde{\alpha_n}(q + \widetilde{\alpha_n})^{l-1}.$$

◇

Proof (i) Let $\xi_k = \varepsilon^k q^{k-1}$, $T_0 = A_0$ and $T_k = \frac{1}{q^{k-1}} A_k$, $\forall k \geq 1$. In view of Proposition 9.3.1, we have

$$
\begin{aligned}
\widetilde{R}_{z_n}(\xi) &= \widetilde{R}_{z_n} + \sum_{l=1}^{+\infty}(-1)^l \sum_{k=l}^{+\infty} \sum_{i_1+i_2+\cdots+i_l=k} \xi_{i_1}\xi_{i_2}\ldots\xi_{i_l}\, \widetilde{R}_{z_n}\, T_{i_1}\widetilde{R}_{z_n}\, T_{i_2}\widetilde{R}_{z_n} \ldots T_{i_l}\widetilde{R}_{z_n} \\
&= \widetilde{R}_{z_n} + \sum_{l=1}^{+\infty}(-1)^l \sum_{k=l}^{+\infty} \sum_{i_1+i_2+\cdots+i_l=k} \varepsilon^{i_1} q^{i_1-1}\, \varepsilon^{i_2} q^{i_2-1} \ldots \varepsilon^{i_l} q^{i_l-1}\, \widetilde{R}_{z_n} \\
&\qquad \times \frac{A_{i_1}\widetilde{R}_{z_n}}{q^{i_1-1}} \frac{A_{i_2}\widetilde{R}_{z_n}}{q^{i_2-1}} \cdots \frac{A_{i_l}\widetilde{R}_{z_n}}{q^{i_l-1}} \\
&= \widetilde{R}_{z_n} + \sum_{l=1}^{+\infty}(-1)^l \sum_{k=l}^{+\infty} \varepsilon^k \sum_{i_1+i_2+\cdots+i_l=k} \widetilde{R}_{z_n}\, A_{i_1}\widetilde{R}_{z_n}\, A_{i_2}\widetilde{R}_{z_n} \ldots A_{i_l}\widetilde{R}_{z_n} \\
&= \widetilde{R}_{z_n} + \sum_{k=1}^{+\infty} \varepsilon^k \sum_{l=1}^{k}(-1)^l \sum_{i_1+i_2+\cdots+i_l=k} \widetilde{R}_{z_n}\, A_{i_1}\widetilde{R}_{z_n}\, A_{i_2}\widetilde{R}_{z_n} \ldots A_{i_l}\widetilde{R}_{z_n} \\
&= \sum_{k=0}^{+\infty} \varepsilon^k \widetilde{R}_{z_n,k}\,,
\end{aligned}
$$

where

$$
\begin{cases}
\widetilde{R}_{z_n,0} := \widetilde{R}_{z_n} := (A_0 - z_n)^{-1} \\
\widetilde{R}_{z_n,k} := \displaystyle\sum_{l=1}^{k}(-1)^l \sum_{i_1+i_2+\cdots+i_l=k} \widetilde{R}_{z_n}\, A_{i_1}\widetilde{R}_{z_n}\, A_{i_2}\widetilde{R}_{z_n} \ldots A_{i_l}\widetilde{R}_{z_n}, \quad \forall\, k \geq 1.
\end{cases}
$$

(ii) In view of Eq. (9.2.3), we get

$$
\|A_k R_{z_n}\| \leq q^{k-1}\widetilde{\alpha_n}.
$$

So, in view of Lemma 3.1.3, we have for every $k \geq 1$

$$
\begin{aligned}
\|\widetilde{R}_{z_n,k}\| &\leq \sum_{l=1}^{k} \sum_{i_1+i_2+\cdots+i_l=k} \widetilde{M_n}\left(q^{i_1-1}\widetilde{\alpha_n}\right)\left(q^{i_2-1}\widetilde{\alpha_n}\right) \ldots \left(q^{i_1-1}\widetilde{\alpha_n}\right) \\
&\leq \widetilde{M_n}\, q^k \sum_{l=1}^{k} \sum_{i_1+i_2+\cdots+i_l=k} \left(\frac{\widetilde{\alpha_n}}{q}\right)^l \\
&\leq \widetilde{M_n}\, q^k \sum_{l=1}^{k} \left(\frac{\widetilde{\alpha_n}}{q}\right)^l \sum_{i_1+i_2+\cdots+i_l=k} 1 \\
&\leq \widetilde{M_n}\, q^k\, \frac{\widetilde{\alpha_n}}{q}\left(1+\frac{\widetilde{\alpha_n}}{q}\right)^{k-1} \\
&\leq \widetilde{M_n}\widetilde{\alpha_n}\,(q+\widetilde{\alpha_n})^{k-1}.
\end{aligned}
$$

The proof is then complete. Q.E.D.

Having obtained these results, we are now ready to investigate under the spectral properties of the perturbed operator $T(\xi)$. In that line our first result asserts:

Theorem 9.3.1 *Suppose that the hypotheses* $(H1)$, $(H2)$ *with* $\beta = q = 1$, *and* $(H3)$ *hold. Let* φ_n *(respectively,* φ_n^**) be an eigenvector of* T_0 *(respectively,* T_0^**) associated to the eigenvalue* λ_n *(respectively,* $\overline{\lambda_n}$*) such that* $\|\varphi_n\| = \|\varphi_n^*\| = \omega_n$ *and* $\varphi_n^*(\varphi_n) = 1$. *Then,*

(i) if $\tau(\xi) < \frac{1}{\alpha_n(1+r_n M_n)}$, *then* $T(\xi)$ *has a unique point* $\lambda_n(\xi)$ *of its spectrum in the neighborhood of* λ_n *and this point is also with multiplicity one,*

(ii) if $\tau(\xi) < \frac{1}{\alpha_n(1+\omega_n^2 r_n M_n)}$, *then the eigenvalue* $\lambda_n(\xi)$ *can be developed into a series*

$$\lambda_n(\xi) := \lambda_n + \lambda_{n,1}(\xi) + \lambda_{n,2}(\xi) + \cdots + \lambda_{n,i}(\xi) + \cdots, \tag{9.3.5}$$

where

$$\lambda_{n,i}(\xi) = \sum_{l=0}^{i-1} b_{n,l}(\xi)\, c_{n,i-l}(\xi) \text{ for all } i \geq 1,$$

with

$$\begin{aligned}
b_{n,0}(\xi) &= 1, \\
b_{n,l}(\xi) &= \sum_{k=l}^{\infty} \sum_{i_1+i_2+\cdots+i_l=k} \beta_{i_1}\, \beta_{i_2} \ldots \beta_{i_l}, \quad \forall l \geq 1 \\
\beta_k &= -\varphi_n^*\left(P_{n,k}(\xi)\varphi_n\right), \quad \forall\, k \geq 1, \\
P_{n,k}(\xi) &= \frac{-1}{2i\pi} \int_{\mathscr{C}_n} R_{z_n,k}(\xi)\, \mathrm{d}z_n, \quad \forall\ k \geq 1, \\
c_{n,k}(\xi) &= \varphi_n^*\left(B_{n,k}(\xi)\varphi_n\right), \quad \forall\, k \geq 1, \\
B_{n,k}(\xi) &= \frac{-1}{2i\pi} \int_{\mathscr{C}_n} (z_n - \lambda_n) R_{z_n,k}(\xi)\, \mathrm{d}z_n, \quad \forall\ k \geq 1,
\end{aligned} \tag{9.3.6}$$

and $\left(R_{z_n,k}(\xi)\right)_{k\geq 1}$ *is defined in Eq. (9.3.1), and*

(iii) for all $i \geq 1$,

$$|\lambda_{n,i}(\xi)| \leq \frac{\omega_n^2 r_n^2 M_n\,(1-\alpha_n\tau(\xi))}{\alpha_n\tau(\xi) - 1 + \omega_n^2 r_n M_n} \left(\left(\frac{\omega_n^2 r_n M_n \alpha_n \tau(\xi)}{1-\alpha_n\tau(\xi)} \right)^i - (\alpha_n\tau(\xi))^i \right). \ \diamond$$

Proof (i) Let $P_n(\xi)$ (respectively, P_n) be the spectral projection of $T(\xi)$ (respectively, T_0) corresponding to the eigenvalue $\lambda_n(\xi)$ (respectively, λ_n) and $\mathscr{R}_n(\xi)$ (respectively, $\mathscr{R}_n$) be the eigenspace for $\lambda_n(\xi)$ (respectively, λ_n). In view of Proposition 3.2.1, we get

$$\begin{aligned} P_n(\xi) &= \frac{-1}{2\pi i} \int\limits_{\mathscr{C}_n} R_{z_n}(\xi)\, \mathrm{d}z_n \\ &= \frac{-1}{2\pi i} \int\limits_{\mathscr{C}_n} \left(R_{z_n} + \sum_{k=1}^{\infty} R_{z_n,k}(\xi) \right) \mathrm{d}z_n. \end{aligned}$$

Thus,

$$P_n(\xi) = P_n + \sum_{k=1}^{\infty} P_{n,k}(\xi),$$

where

$$P_{n,k}(\xi) = \frac{-1}{2\pi i} \int\limits_{\mathscr{C}_n} R_{z_n,k}(\xi)\, \mathrm{d}z_n, \quad \forall\, k \geq 1.$$

Since

$$\begin{aligned} \|P_{n,k}(\xi)\| &\leq \frac{1}{2\pi} \int\limits_{\mathscr{C}_n} \|R_{z_n,k}(\xi)\|\, \mathrm{d}z_n \\ &\leq r_n M_n (\alpha_n \tau(\xi))^k, \end{aligned}$$

then for $\tau(\xi) < \frac{1}{\alpha_n(1+r_n M_n)}$, we have

$$\begin{aligned} \|P_n(\xi) - P_n\| &= \left\| \sum_{k=1}^{\infty} P_{n,k}(\xi) \right\| \\ &\leq \sum_{k=1}^{\infty} \|P_{n,k}(\xi)\| \\ &\leq \sum_{k=1}^{\infty} r_n M_n (\alpha_n \tau(\xi))^k \\ &\leq \frac{r_n M_n \alpha_n \tau(\xi)}{1 - \alpha_n \tau(\xi)} \\ &< 1. \end{aligned}$$

Consequently,

$$\dim \mathscr{R}_n(\xi) = \dim \mathscr{R}_n = 1.$$

So, $T(\xi)$ has a unique eigenvalue $\lambda_n(\xi)$ in $\mathscr{C}_n$, for $\tau(\xi) < \frac{1}{\alpha_n(1+r_n M_n)}$. Moreover, $\lambda_n(\xi)$ is with multiplicity one.

(ii) Let

$$\varphi_n(\xi) = \frac{P_n(\xi)\varphi_n}{[\varphi_n^*(P_n(\xi)\varphi_n)]^{\frac{1}{2}}}$$

be an eigenvector of $T(\xi)$ associated to the eigenvalue $\lambda_n(\xi)$. We have

$$\begin{aligned}\lambda_n(\xi) - \lambda_n &= (\lambda_n(\xi) - \lambda_n)\,\frac{\varphi_n^*\,(\varphi_n(\xi))}{\varphi_n^*\,(\varphi_n(\xi))}\\ &= \frac{\varphi_n^*\,((\lambda_n(\xi) - \lambda_n)\,\varphi_n(\xi))}{\varphi_n^*\,(\varphi_n(\xi))}.\end{aligned}$$

Thus,

$$\begin{aligned}\lambda_n(\xi) - \lambda_n &= \frac{\varphi_n^*\,(T(\xi)\varphi_n(\xi) - \lambda_n\varphi_n(\xi))}{\varphi_n^*\,(\varphi_n(\xi))}\\ &= \frac{\varphi_n^*\,((T(\xi) - \lambda_n I)P_n(\xi)\varphi_n)}{\varphi_n^*(P_n(\xi)\varphi_n)}.\end{aligned} \tag{9.3.7}$$

In view of $\varphi_n^*(P_n\varphi_n) = \varphi_n^*(\varphi_n) = 1$, then Eq. (9.3.7) implies that

$$\lambda_n(\xi) - \lambda_n = \frac{\varphi_n^*\,((T(\xi) - \lambda_n I)P_n(\xi)\varphi_n)}{1 + \varphi_n^*((P_n(\xi) - P_n)\varphi_n)}. \tag{9.3.8}$$

Furthermore, for $\tau(\xi) < \frac{1}{\alpha_n(1+r_n M_n\omega_n^2)}$, we get

$$\begin{aligned}\|\varphi_n^*((P_n(\xi) - P_n)\varphi_n)\| &\leq \|\varphi_n^*\|\;\|(P_n(\xi) - P_n)\varphi_n\|\\ &\leq \|P_n(\xi) - P_n\|\;\|\varphi_n\|\;\|\varphi_n^*\|\\ &\leq \frac{r_n M_n \alpha_n \tau(\xi)}{1 - \alpha_n\tau(\xi)}\,\omega_n^2\\ &< 1.\end{aligned}$$

Hence,

$$\frac{1}{1 + \varphi_n^*((P_n(\xi) - P_n)\varphi_n)} = \sum_{\nu=0}^{\infty}(-1)^\nu\left[\varphi_n^*\,((P_n(\xi) - P_n)\varphi_n)\right]^\nu. \tag{9.3.9}$$

On the other hand, we obtain

$$\begin{aligned}(T(\xi)-\lambda_n I)\,P_n(\xi) &= T(\xi)P_n(\xi)-\lambda_n P_n(\xi)\\ &= \frac{-1}{2\pi i}\int_{\mathscr{C}_n} z_n R_{z_n}(\xi)\,\mathrm{d}z_n-\lambda_n\left(\frac{-1}{2\pi i}\right)\int_{\mathscr{C}_n} R_{z_n}(\xi)\,\mathrm{d}z_n\\ &= \frac{-1}{2\pi i}\int_{\mathscr{C}_n}(z_n-\lambda_n)R_{z_n}(\xi)\,\mathrm{d}z_n\\ &= \frac{-1}{2\pi i}\int_{\mathscr{C}_n}(z_n-\lambda_n)\left(R_{z_n}+\sum_{k=1}^{\infty}R_{z_n,k}(\xi)\right)\mathrm{d}z_n\\ &= \frac{-1}{2\pi i}\int_{\mathscr{C}_n}(z_n-\lambda_n)R_{z_n}\,\mathrm{d}z_n+\sum_{k=1}^{\infty}\left(\frac{-1}{2\pi i}\int_{\mathscr{C}_n}(z_n-\lambda_n)R_{z_n,k}(\xi)\,\mathrm{d}z_n\right)\\ &= \sum_{k=1}^{\infty}B_{n,k}(\xi),\end{aligned} \tag{9.3.10}$$

where

$$B_{n,k}(\xi)=\frac{-1}{2\pi i}\int_{\mathscr{C}_n}(z_n-\lambda_n)R_{z_n,k}(\xi)\,\mathrm{d}z_n.$$

Consequently, Eqs. (9.3.8), (9.3.9) and (9.3.10) imply that

$$\begin{aligned}\lambda_n(\xi)-\lambda_n &= \sum_{\nu=0}^{\infty}(-1)^{\nu}\left[\varphi_n^*\left((P_n(\xi)-P_n)\varphi_n\right)\right]^{\nu}\ \varphi_n^*\left(\sum_{l=1}^{\infty}B_{n,l}(\xi)\varphi_n\right)\\ &= \sum_{\nu=0}^{\infty}(-1)^{\nu}\left(\sum_{k=1}^{\infty}\varphi_n^*\left(P_{n,k}(\xi)\varphi_n\right)\right)^{\nu}\ \sum_{l=1}^{\infty}\varphi_n^*\left(B_{n,l}(\xi)\varphi_n\right).\end{aligned}$$

Thus,

$$\lambda_n(\xi)-\lambda_n=\sum_{\nu=0}^{\infty}\left(\sum_{k=1}^{\infty}-\varphi_n^*\left(P_{n,k}(\xi)\varphi_n\right)\right)^{\nu}\ \sum_{l=1}^{\infty}\varphi_n^*\left(B_{n,l}(\xi)\varphi_n\right). \tag{9.3.11}$$

Denoting by $\beta_k=-\varphi_n^*\left(P_{n,k}(\xi)\varphi_n\right)$, $\forall\, k\geq 1$, we can easily check that if $\tau(\xi)<\frac{1}{\alpha_n(1+r_nM_n\omega_n^2)}$, then

$$\sum_{k=1}^{\infty}|\beta_k|<1$$

and

$$\sum_{k=0}^{\infty}\widetilde{\beta}_k^*<+\infty,$$

where

$$\widetilde{\beta}_k^* = \sum_{l=1}^{k} \sum_{i_1+i_2+\ldots+i_l=k} |\beta_{i_1}|\,|\beta_{i_2}|\,\ldots|\beta_{i_l}|.$$

Hence, in view of Proposition 3.1.3, we get

$$\sum_{\nu=0}^{\infty}\left(\sum_{k=1}^{\infty} -\varphi_n^*\left(P_{n,k}(\xi)\varphi_n\right)\right)^{\nu} = \sum_{l=0}^{\infty} b_{n,l}(\xi),$$

where

$$b_{n,0}(\xi) = 1,$$

and

$$b_{n,l}(\xi) = \sum_{k=l}^{\infty} \sum_{i_1+i_2+\cdots+i_l=k} \beta_{i_1}\,\beta_{i_2}\ldots\beta_{i_l}, \quad \forall l \geq 1.$$

So, using Eq. (9.3.11), we get

$$\lambda_n(\xi) - \lambda_n = \sum_{l=0}^{\infty} b_{n,l}(\xi) \sum_{k=1}^{\infty} c_{n,k}(\xi),$$

where $c_{n,k}(\xi) = \varphi_n^*\left(B_{n,k}(\xi)\varphi_n\right)$, for all $k \geq 1$. By the Cauchy product of two series, we get

$$\lambda_n(\xi) - \lambda_n = \sum_{i=1}^{\infty} \lambda_{n,i}(\xi),$$

where

$$\lambda_{n,i}(\xi) = \sum_{l=0}^{i-1} b_{n,l}(\xi)\, c_{n,i-l}(\xi).$$

(iii) In order to estimate the coefficients $(|\lambda_{n,i}(\xi)|)_{i=1}^{\infty}$, we shall first estimate $(|b_{n,l}(\xi)|)_{l\geq 0}$ and $(|c_{n,k}(\xi)|)_{k\geq 1}$. Indeed, in view of Proposition 3.1.3, we have for $l \geq 1$

$$\begin{aligned}|b_{n,l}(\xi)| &\leq \sum_{k=l}^{\infty} \sum_{i_1+i_2+\cdots+i_l=k} |\beta_{i_1}|\,|\beta_{i_2}|\,\ldots|\beta_{i_l}| \\ &\leq \left(\sum_{k=1}^{\infty}|\beta_k|\right)^{l}.\end{aligned}$$

Since

$$\begin{aligned}|\beta_k| &\leq \|\varphi_n^*\|\|P_{n,k}(\xi)\|\,\|\varphi_n\| \\ &\leq \omega_n^2 M_n r_n (\alpha_n \tau(\xi))^k, \quad \forall k \geq 1,\end{aligned}$$

we obtain

$$|b_{n,l}(\xi)| \leq \left(\frac{\omega_n^2 M_n r_n \alpha_n \tau(\xi)}{1-\alpha_n \tau(\xi)} \right)^l, \quad \forall l \geq 1.$$

In view of $b_{n,0}(\xi) = 1$, we obtain

$$|b_{n,l}(\xi)| \leq \left(\frac{\omega_n^2 M_n r_n \alpha_n \tau(\xi)}{1-\alpha_n \tau(\xi)} \right)^l, \quad \forall l \geq 0. \tag{9.3.12}$$

On the other hand, we have for $l \geq 1$

$$\begin{aligned} |c_{n,l}(\xi)| &\leq \|\varphi_n^*\| \|B_{n,l}(\xi)\| \ \|\varphi_n\| \\ &\leq \omega_n^2 \frac{1}{2\pi} \left\| \int_{\mathscr{C}_n} (z_n - \lambda_n) R_{z_n,k}(\xi) \mathrm{d}z_n \right\| \\ &\leq \omega_n^2 r_n^2 M_n (\alpha_n \tau(\xi))^l. \end{aligned} \tag{9.3.13}$$

Consequently, combining both Eqs. (9.3.12) and (9.3.13), we get

$$\begin{aligned} |\lambda_{n,i}(\xi)| &\leq \sum_{l=0}^{i-1} |b_{n,l}(\xi)| \ |c_{n,i-l}(\xi)| \\ &\leq \omega_n^2 r_n^2 M_n \sum_{l=0}^{i-1} \left(\frac{\omega_n^2 M_n r_n \alpha_n \tau(\xi)}{1-\alpha_n \tau(\xi)} \right)^l (\alpha_n \tau(\xi))^{i-l}. \end{aligned}$$

Since $M_n \geq \frac{1}{r_n}$ and $\omega_n^2 \geq 1$, we get

$$1 - \omega_n^2 M_n r_n \leq 0.$$

Hence,

$$\tau(\xi) \neq \frac{1}{\alpha_n}(1 - \omega_n^2 M_n r_n).$$

So,

$$\frac{\omega_n^2 M_n r_n \alpha_n \tau(\xi)}{1-\alpha_n \tau(\xi)} \neq \alpha_n \tau(\xi).$$

Thus,

$$
\begin{aligned}
|\lambda_{n,i}(\xi)| &\le \omega_n^2 r_n^2 M_n \alpha_n \tau(\xi) \frac{\left(\frac{\omega_n^2 r_n M_n \alpha_n \tau(\xi)}{1-\alpha_n \tau(\xi)}\right)^i - (\alpha_n \tau(\xi))^i}{\frac{\omega_n^2 r_n M_n \alpha_n \tau(\xi)}{1-\alpha_n \tau(\xi)} - \alpha_n \tau(\xi)} \\
&\le \frac{\omega_n^2 r_n^2 M_n (1-\alpha_n \tau(\xi))}{\alpha_n \tau(\xi) - 1 + \omega_n^2 r_n M_n} \left(\left(\frac{\omega_n^2 r_n M_n \alpha_n \tau(\xi)}{1-\alpha_n \tau(\xi)}\right)^i - (\alpha_n \tau(\xi))^i \right),
\end{aligned}
$$

which completes the proof. Q.E.D.

We have the analogue of Theorem 9.3.1 if we consider the analytic operator

$$
A(\varepsilon) := A_0 + \varepsilon A_1 + \varepsilon^2 A_2 + \cdots + \varepsilon^k A_k + \cdots .
$$

Corollary 9.3.2 *Suppose that the assumptions of Corollary 9.2.3 are satisfied. Let $\widetilde{\varphi}_n$ (respectively, $\widetilde{\varphi}_n^*$) be an eigenvector of A_0 (respectively, A_0^*) associated to the eigenvalue $\widetilde{\lambda}_n$ (respectively, $\overline{\widetilde{\lambda}_n}$) such that $\|\widetilde{\varphi}_n\| = \|\widetilde{\varphi}_n^*\| = \widetilde{\omega}_n$ and $\widetilde{\varphi}_n^*(\widetilde{\varphi}_n) = 1$. Then,*

(i) for $|\varepsilon| < \frac{1}{q+\widetilde{\alpha}_n+\widetilde{r}_n\widetilde{M}_n\widetilde{\alpha}_n}$, the operator $A(\varepsilon)$ has a unique point $\widetilde{\lambda}_n(\varepsilon)$ of its spectrum in the neighborhood of $\widetilde{\lambda}_n$ and this point is also with multiplicity one,
(ii) if $|\varepsilon| < \frac{1}{q+\widetilde{\alpha}_n+\widetilde{\omega}_n^2\widetilde{r}_n\widetilde{M}_n\widetilde{\alpha}_n}$, then the value $\widetilde{\lambda}_n(\varepsilon)$ can be developed into an entire series

$$
\widetilde{\lambda}_n(\varepsilon) := \widetilde{\lambda}_n + \varepsilon \widetilde{\lambda}_{n,1} + \varepsilon^2 \widetilde{\lambda}_{n,2} + \cdots + \varepsilon^k \widetilde{\lambda}_{n,k} + \cdots ,
$$

where

$$
\widetilde{\lambda}_{n,k} = \sum_{i=0}^{k-1} \widetilde{b}_{n,i}\, \widetilde{c}_{n,k-i}, \quad \textit{for all } k \ge 1,
$$

with

$$
\begin{aligned}
\widetilde{b}_{n,k} &= \sum_{l=0}^{k} (-1)^l \sum_{i_1+i_2+\cdots+i_l=k} \widetilde{\beta}_{i_1}\, \widetilde{\beta}_{i_2} \ldots \widetilde{\beta}_{i_l}, \quad \forall\ k \ge 1, \\
\widetilde{\beta}_k &= \widetilde{\varphi}_n^* \left(\widetilde{P}_{n,k} \widetilde{\varphi}_n\right), \quad \forall\ k \ge 1, \\
\widetilde{P}_{n,k} &= \frac{-1}{2i\pi} \int_{\widetilde{\mathscr{C}}_n} \widetilde{R}_{z_n,k}\, dz_n, \quad \forall\ k \ge 1, \\
\widetilde{c}_{n,k} &= \widetilde{\varphi}_n^* \left(\widetilde{B}_{n,k} \widetilde{\varphi}_n\right), \quad \forall\ k \ge 1, \\
\widetilde{B}_{n,k} &= \frac{-1}{2i\pi} \int_{\widetilde{\mathscr{C}}_n} (z_n - \widetilde{\lambda}_n) \widetilde{R}_{z_n,k}\, dz_n, \quad \forall\ k \ge 1,
\end{aligned}
\tag{9.3.14}
$$

and $\left(\widetilde{R}_{z_n,k}\right)_{k \ge 1}$ is defined in Eq. (9.3.4), and

(iii) for all $k \geq 1$*, we have*

$$|\widetilde{\lambda}_{n,k}| \leq \widetilde{\omega}_n^2 \widetilde{r}_n^2 \widetilde{M}_n \widetilde{\alpha}_n \left((q + \widetilde{\alpha}_n + \widetilde{\omega}_n^2 \widetilde{r}_n \widetilde{M}_n \widetilde{\alpha}_n)^{k-1} - (q + \widetilde{\alpha}_n)^{k-1} \right). \qquad \diamond$$

Proof (i) The result is an immediate consequence of Theorem 9.3.1, it suffices to take $T_0 = A_0$, $T_k = \frac{1}{q^{k-1}} A_k$ and $\xi_k = \varepsilon^k q^{k-1}$, $\forall k \geq 1$. In this case, we have

$$\tau(\xi) = \frac{|\varepsilon|}{1 - |\varepsilon| q} < \frac{1}{\widetilde{\alpha}_n (1 + \widetilde{r}_n \widetilde{M}_n)} \text{ if, and only if, } |\varepsilon| < \frac{1}{q + \widetilde{\alpha}_n + \widetilde{r}_n \widetilde{M}_n \widetilde{\alpha}_n}.$$

(ii) Let $\widetilde{P}_n(\varepsilon)$ (respectively, $\widetilde{P}_n$) be the spectral projection of $A(\varepsilon)$ (respectively, A_0) corresponding to the eigenvalue $\widetilde{\lambda}_n(\varepsilon)$. Consider

$$\widetilde{\varphi}_n(\varepsilon) = \frac{\widetilde{P}_n(\varepsilon) \widetilde{\varphi}_n}{[\widetilde{\varphi}_n^*(\widetilde{P}_n(\varepsilon) \widetilde{\varphi}_n)]^{\frac{1}{2}}}$$

an eigenvector of $A(\varepsilon)$ associated to the eigenvalue $\widetilde{\lambda}_n(\varepsilon)$. Making the same reasoning as the one in the proof of Theorem 9.3.1(ii), we obtain

$$\widetilde{\lambda}_n(\varepsilon) - \widetilde{\lambda}_n = \frac{\widetilde{\varphi}_n^* \left((A(\varepsilon) - \widetilde{\lambda}_n I) \, \widetilde{P}_n(\varepsilon) \widetilde{\varphi}_n \right)}{1 + \widetilde{\varphi}_n^* ((\widetilde{P}_n(\varepsilon) - \widetilde{P}_n) \widetilde{\varphi}_n)}. \tag{9.3.15}$$

If $|\varepsilon| < \frac{1}{q + \widetilde{\alpha}_n + \widetilde{\omega}_n^2 \widetilde{r}_n \widetilde{M}_n \widetilde{\alpha}_n}$, we have

$$\|\widetilde{\varphi}_n^* ((\widetilde{P}_n(\varepsilon) - \widetilde{P}_n) \widetilde{\varphi}_n)\| < 1.$$

So,

$$\begin{aligned} \frac{1}{1 + \widetilde{\varphi}_n^* ((\widetilde{P}_n(\varepsilon) - \widetilde{P}_n) \widetilde{\varphi}_n)} &= \sum_{\nu=0}^{\infty} (-1)^{\nu} \left[\widetilde{\varphi}_n^* \left((\widetilde{P}_n(\varepsilon) - \widetilde{P}_n) \widetilde{\varphi}_n \right) \right]^{\nu} \\ &= \sum_{\nu=0}^{\infty} \left(\sum_{k=1}^{\infty} -\varepsilon^k \widetilde{\varphi}_n^* (\widetilde{P}_{n,k} \varphi_n) \right)^{\nu}. \end{aligned}$$

Since

$$\sum_{k=1}^{\infty} \left| \varepsilon^k \widetilde{\varphi}_n^* \left(\widetilde{P}_{n,k} \widetilde{\varphi}_n \right) \right| < 1,$$

for $|\varepsilon| < \frac{1}{q + \widetilde{\alpha}_n + \widetilde{\omega}_n^2 \widetilde{r}_n \widetilde{M}_n \widetilde{\alpha}_n}$, we obtain by using Proposition 3.1.1,

$$\sum_{\nu=0}^{\infty}\left(\sum_{k=1}^{\infty}-\varepsilon^k\widetilde{\varphi}_n^*\left(\widetilde{P}_{n,k}\widetilde{\varphi}_n\right)\right)^{\nu}=\sum_{k=0}^{\infty}\varepsilon^k\widetilde{b}_{n,k}, \tag{9.3.16}$$

where

$$\widetilde{b}_{n,0}=1,$$

$$\widetilde{b}_{n,k}=\sum_{l=1}^{k}(-1)^l\sum_{i_1+i_2+\cdots+i_l=k}\beta_{i_1}\,\beta_{i_2}\,\ldots\beta_{i_l}$$

and, for all $k\geq 1$,

$$\beta_k=\widetilde{\varphi}_n^*\left(\widetilde{P}_{n,k}\widetilde{\varphi}_n\right).$$

On the other hand, in view of Corollary 9.3.1 and proceeding as in Theorem 9.3.1, we get

$$\left(A(\varepsilon)-\widetilde{\lambda}_n I\right)\widetilde{P}_n(\varepsilon)=\sum_{k=1}^{\infty}\varepsilon^k\widetilde{B}_{n,k},$$

where

$$\widetilde{B}_{n,k}=\frac{-1}{2\pi i}\int_{\widetilde{\mathscr{C}}_n}(z_n-\widetilde{\lambda}_n)\widetilde{R}_{z_n,k}\,\mathrm{d}z_n,\quad\forall k\geq 1.$$

So, we deduce that

$$\widetilde{\varphi}_n^*\left(\left(A(\varepsilon)-\widetilde{\lambda}_n I\right)\widetilde{P}_n(\varepsilon)\widetilde{\varphi}_n\right)=\sum_{k=1}^{\infty}\varepsilon^k\widetilde{c}_{n,k}, \tag{9.3.17}$$

where $\widetilde{c}_{n,k}=\widetilde{\varphi}_n^*\left(\widetilde{B}_{n,k}\widetilde{\varphi}_n\right)$, for all $k\geq 1$. Consequently, Eqs. (9.3.15), (9.3.16) and (9.3.17) yield

$$\begin{aligned}\widetilde{\lambda}_n(\varepsilon)-\widetilde{\lambda}_n&=\sum_{k=0}^{\infty}\varepsilon^k\widetilde{b}_{n,k}\sum_{k=1}^{\infty}\varepsilon^k\widetilde{c}_{n,k}\\&=\sum_{k=1}^{\infty}\varepsilon^k\widetilde{\lambda}_{n,k},\end{aligned}$$

where

$$\widetilde{\lambda}_{n,k}=\sum_{i=0}^{k-1}\widetilde{b}_{n,i}\,\widetilde{c}_{n,k-i}.$$

(iii) We have

$$|\widetilde{b}_{n,k}| \leq \sum_{l=1}^{k} \sum_{i_1+i_2+\cdots+i_l=k} \left|\beta_{i_1}\right| \, \left|\beta_{i_2}\right| \, \ldots \left|\beta_{i_l}\right|.$$

Or,

$$\begin{aligned} |\beta_k| &= \left|\widetilde{\varphi}_n^* \left(\widetilde{P}_{n,k}\widetilde{\varphi}_n\right)\right| \\ &\leq \widetilde{\omega}_n^2 \widetilde{r}_n \widetilde{M}_n \widetilde{\alpha}_n (q+\widetilde{\alpha}_n)^{k-1}. \end{aligned}$$

Hence,

$$\begin{aligned} |\widetilde{b}_{n,k}| &\leq \widetilde{\omega}_n^2 \widetilde{r}_n \widetilde{M}_n \widetilde{\alpha}_n \sum_{l=1}^{k} \sum_{i_1+i_2+\cdots+i_l=k} (q+\widetilde{\alpha}_n)^{i_1-1}(q+\widetilde{\alpha}_n)^{i_2-1} \ldots (q+\widetilde{\alpha}_n)^{i_l-1} \\ &\leq \widetilde{\omega}_n^2 \widetilde{r}_n \widetilde{M}_n \widetilde{\alpha}_n \sum_{l=1}^{k} \sum_{i_1+i_2+\cdots+i_l=k} (q+\widetilde{\alpha}_n)^{k-l} \\ &\leq \widetilde{\omega}_n^2 \widetilde{r}_n \widetilde{M}_n \widetilde{\alpha}_n (q+\widetilde{\alpha}_n)^k \sum_{l=1}^{k} \frac{1}{(q+\widetilde{\alpha}_n)^l} \sum_{i_1+i_2+\cdots+i_l=k} 1. \end{aligned}$$

Using Lemma 3.1.3, we deduce that

$$\left|\widetilde{b}_{n,k}\right| \leq \widetilde{\omega}_n^2 \widetilde{r}_n \widetilde{M}_n \widetilde{\alpha}_n (1+q+\widetilde{\alpha}_n)^{k-1}. \tag{9.3.18}$$

Furthermore, we obtain

$$\left|\widetilde{c}_{n,k}\right| \leq \left|\widetilde{\varphi}_n^* \left(\widetilde{B}_{n,k}\widetilde{\varphi}_n\right)\right| \leq \widetilde{\omega}_n^2 \widetilde{r}_n^2 \widetilde{M}_n \widetilde{\alpha}_n (q+\widetilde{\alpha}_n)^{k-1}. \tag{9.3.19}$$

Consequently, Eqs. (9.3.18) and (9.3.19) imply

$$\begin{aligned} |\widetilde{\lambda}_{n,k}| &\leq \sum_{i=1}^{k-1} |\widetilde{b}_{n,i}| \, |\widetilde{c}_{n,k-i}| \\ &\leq \widetilde{\omega}_n^4 \widetilde{r}_n^3 \widetilde{M}_n^2 \widetilde{\alpha}_n^2 \sum_{i=1}^{k-1} \left(q+\widetilde{\alpha}_n+\widetilde{\omega}_n^2 \widetilde{r}_n \widetilde{M}_n \widetilde{\alpha}_n\right)^{i-1} (q+\widetilde{\alpha}_n)^{k-1-i} \\ &\leq \widetilde{\omega}_n^4 \widetilde{r}_n^3 \widetilde{M}_n^2 \widetilde{\alpha}_n^2 \, \frac{\left(q+\widetilde{\alpha}_n+\widetilde{\omega}_n^2 \widetilde{r}_n \widetilde{M}_n \widetilde{\alpha}_n\right)^{k-1} - (q+\widetilde{\alpha}_n)^{k-1}}{\widetilde{\omega}_n^2 \widetilde{r}_n \widetilde{M}_n \widetilde{\alpha}_n} \\ &\leq \widetilde{\omega}_n^2 \widetilde{r}_n^2 \widetilde{M}_n \widetilde{\alpha}_n \left(\left(q+\widetilde{\alpha}_n+\widetilde{\omega}_n^2 \widetilde{r}_n \widetilde{M}_n \widetilde{\alpha}_n\right)^{k-1} - (q+\widetilde{\alpha}_n)^{k-1}\right). \end{aligned}$$

The proof is then complete. Q.E.D.

9.4 Eigenvectors

Now, setting $\varphi_n(\xi)$ an eigenvector of $T(\xi)$ associated to the eigenvalue $\lambda_n(\xi)$, we prove the analogue of Theorem 9.4.1 for $\varphi_n(\xi)$.

Theorem 9.4.1 *Assume that the hypotheses* $(H1)$, $(H2)$ *with* $\beta = q = 1$, *and* $(H3)$ *hold. Let* φ_n *(respectively,* φ_n^**) be an eigenvector of* T_0 *(respectively,* T_0^**) associated to the eigenvalue* λ_n *(respectively,* $\overline{\lambda_n}$*) such that* $\|\varphi_n\| = \|\varphi_n^*\|$ *and* $\varphi_n^*(\varphi_n) = 1$. *Then, for* $\tau(\xi) < \frac{1}{\alpha_n(1+r_n M_n \omega_n^2)}$, *we have*

(i) the eigenvector $\varphi_n(\xi)$ *can be developed into a series*

$$\varphi_n(\xi) := \varphi_n + \varphi_{n,1}(\xi) + \varphi_{n,2}(\xi) + \cdots + \varphi_{n,i}(\xi) + \cdots, \tag{9.4.1}$$

where

$$\begin{cases} \varphi_{n,i}(\xi) := \displaystyle\sum_{k=1}^{i} d_{n,k}(\xi) P_{n,i-k}(\xi)\varphi_n, \quad \forall\, i \geq 1 \\ d_{n,0}(\xi) = 1, \\ d_{n,l}(\xi) = \displaystyle\sum_{k=l}^{\infty} C^l_{-\frac{1}{2}} \sum_{i_1+i_2+\cdots+i_l=k} \gamma_{i_1}\gamma_{i_2}\ldots\gamma_{i_l}, \quad \forall\ l \geq 1, \\ \gamma_l = \varphi_n^*(P_{n,l}(\xi)\,\varphi_n), \quad \forall\ l \geq 1 \end{cases}$$

and $\left(P_{n,k}(\xi)\right)_{k\geq 1}$ *is defined in Eq. (9.3.6),*

(ii) the coefficients $\left(\varphi_{n,i}(\xi)\right)_{i=1}^{\infty}$ *satisfy*

$$\left\|\varphi_{n,i}(\xi)\right\| \leq \frac{\omega_n r_n M_n\ (1-\alpha_n\tau(\xi))}{\alpha_n\tau(\xi)(\omega_n^2 r_n M_n - 1)}\left[\left(\frac{\omega_n^2 r_n M_n \alpha_n \tau(\xi)}{1-\alpha_n\tau(\xi)}\right)^{i+1} - (\alpha_n\tau(\xi))^{i+1}\right],$$

(iii) the coefficients $\left(\varphi_{n,i}(\xi)\right)_{i=1}^{\infty}$ *satisfy*

$$\left\|\varphi_{n,i}(\xi)\right\| \leq \frac{\omega_n^3 r_n^2 M_n^2}{\omega_n^2 r_n M_n - 1 + \alpha_n\tau(\xi)}\left[\left(\frac{\omega_n^2 r_n M_n \alpha_n \tau(\xi)}{1-\alpha_n\tau(\xi)}\right)^{i} - (\alpha_n\tau(\xi))^{i}\right]. \ \diamond$$

Proof (i) We have

$$\begin{aligned}\varphi_n(\xi) &= \frac{P_n(\xi)\varphi_n}{[\varphi_n^*(P_n(\xi)\varphi_n)]^{\frac{1}{2}}} \\ &= [\varphi_n^*(P_n(\xi)\varphi_n)]^{-\frac{1}{2}} P_n(\xi)\varphi_n.\end{aligned}$$

Thus,

$$\varphi_n(\xi) = \left(1 + \varphi_n^*((P_n(\xi) - P_n)\varphi_n))\right)^{-\frac{1}{2}} \sum_{l=0}^{\infty} P_{n,l}(\xi)\varphi_n,$$

where $(P_{n,k}(\xi))_{k\geq 1}$ is defined in Eq. (9.3.6). In view of $\tau(\xi) < \frac{1}{\alpha_n(1+r_n M_n \omega_n^2)}$, we get

$$\|\varphi_n^*((P_n(\xi) - P_n)\varphi_n)\| < 1.$$

So,

$$\left(1 + \varphi_n^*((P_n(\xi) - P_n)\varphi_n)\right)^{-\frac{1}{2}} = \sum_{\nu=0}^{\infty} C_{-\frac{1}{2}}^{\nu} \left[\varphi_n^* \left((P_n(\xi) - P_n)\varphi_n\right)\right]^{\nu}.$$

Hence,

$$\begin{aligned}\varphi_n(\xi) &= \sum_{\nu=0}^{\infty} C_{-\frac{1}{2}}^{\nu} \left(\varphi_n^* \left((P_n(\xi) - P_n)\varphi_n\right)\right)^{\nu} \sum_{l=0}^{\infty} P_{n,l}(\xi)\varphi_n \\ &= \sum_{\nu=0}^{\infty} C_{-\frac{1}{2}}^{\nu} \left(\varphi_n^* \left(\sum_{k=1}^{\infty} P_{n,k}(\xi)\varphi_n\right)\right)^{\nu} \sum_{l=0}^{\infty} P_{n,l}(\xi)\varphi_n.\end{aligned}$$

Setting $\gamma_k = \varphi_n^* \left(P_{n,k}(\xi)\varphi_n\right), \forall k \geq 1$, we can easily check that if $\tau(\xi) < \frac{1}{\alpha_n(1+r_n M_n \omega_n^2)}$, then

$$\sum_{k=1}^{\infty} |\gamma_k| < 1$$

and

$$\sum_{k=0}^{\infty} \widetilde{\gamma}_k^* < +\infty,$$

where

$$\widetilde{\gamma}_k^* = \sum_{l=1}^{k} \sum_{i_1+i_2+\cdots+i_l=k} |\gamma_{i_1}|\,|\gamma_{i_2}| \ldots |\gamma_{i_l}|.$$

Consequently, in view of Theorem 3.1.1, we get

$$\sum_{\nu=0}^{\infty} C_{-\frac{1}{2}}^{\nu} \left(\sum_{k=1}^{\infty} \varphi_n^*(P_{n,k}(\xi)\varphi_n)\right)^{\nu} = \sum_{l=0}^{\infty} d_{n,l}(\xi),$$

where

$$\begin{cases} d_{n,0}(\xi) = 1, \\ d_{n,l}(\xi) = C^{l}_{-\frac{1}{2}} \sum_{k=l}^{\infty} \sum_{i_1+i_2+\cdots+i_l=k} \gamma_{i_1}\gamma_{i_2}\ldots\gamma_{i_l}, \quad \forall\ l \geq 1. \end{cases}$$

By the Cauchy product of the two series, we get

$$\begin{aligned} \varphi_n(\xi) &= \sum_{l=0}^{\infty} d_{n,l}(\xi) \sum_{l=0}^{\infty} P_{n,l}(\xi)\varphi_n \\ &= \sum_{i=0}^{\infty} \varphi_{n,i}(\xi), \end{aligned}$$

where

$$\begin{cases} \varphi_{n,0}(\xi) := \varphi_n, \\ \varphi_{n,i}(\xi) := \sum_{k=1}^{i} d_{n,k}(\xi) P_{n,i-k}(\xi)\varphi_n, \quad \forall\, i \geq 1. \end{cases}$$

(ii) Since $\left|C^{\nu}_{-\frac{1}{2}}\right| < \frac{1}{2}, \ \forall\ \nu \geq 1$, then for every $l \geq 1$, we have

$$\left|d_{n,l}(\xi)\right| \leq \frac{1}{2}\sum_{k=l}^{\infty} \sum_{i_1+i_2+\cdots+i_l=k} |\gamma_{i_1}|\,|\gamma_{i_2}|\ldots|\gamma_{i_l}|.$$

Denoting that for all $l \geq 1$

$$\begin{aligned} |\gamma_l| &= \left|\varphi_n^*(P_{n,l}(\xi)\ \varphi_n)\right| \\ &\leq \omega_n^2 r_n M_n\, (\alpha_n \tau(\xi))^l\,, \end{aligned}$$

we get, in view of Proposition 3.1.3,

$$\begin{aligned} \left|d_{n,l}(\xi)\right| &\leq \frac{1}{2}\sum_{k=l}^{\infty} \sum_{i_1+i_2+\cdots+i_l=k} (\omega_n^2 r_n M_n)^l\, (\alpha_n\tau(\xi))^k \\ &\leq \frac{1}{2}(\omega_n^2 r_n M_n)^l \sum_{k=l}^{\infty} \sum_{i_1+i_2+\cdots+i_l=k} (\alpha_n\tau(\xi))^{i_1+i_2+\cdots+i_l} \\ &\leq \frac{1}{2}(\omega_n^2 r_n M_n)^l \left(\sum_{k=l}^{\infty} (\alpha_n\tau(\xi))^k\right)^l \\ &\leq \frac{1}{2}(\omega_n^2 r_n M_n)^l \left(\frac{\alpha_n\tau(\xi)}{1-\alpha_n\tau(\xi)}\right)^l. \end{aligned}$$

Hence,

$$\left|d_{n,l}(\xi)\right| \le \frac{1}{2}\left(\frac{\omega_n^2 r_n M_n \alpha_n \tau(\xi)}{1-\alpha_n \tau(\xi)}\right)^l, \quad \text{for all } l \ge 0. \tag{9.4.2}$$

On the other hand, we have

$$\left\| P_{n,i-k}(\xi)\varphi_n \right\| \le \omega_n r_n M_n \left(\alpha_n \tau(\xi)\right)^{i-k}, \quad \forall i \ge 1 \text{ and } \forall k \ge 0. \tag{9.4.3}$$

Consequently, Eqs. (9.4.2) and (9.4.3) imply that for all $i \ge 1$

$$\begin{aligned}
\|\varphi_{n,i}(\xi)\| &\le \sum_{k=0}^{i} |d_{n,k}(\xi)| \| P_{n,i-k}(\xi)\varphi_n \| \\
&\le \sum_{k=0}^{i} \left(\frac{\omega_n^2 r_n M_n \alpha_n \tau(\xi)}{1-\alpha_n \tau(\xi)}\right)^k \omega_n r_n M_n \left(\alpha_n \tau(\xi)\right)^{i-k} \\
&\le \omega_n r_n M_n \sum_{k=0}^{i} \left(\frac{\omega_n^2 r_n M_n \alpha_n \tau(\xi)}{1-\alpha_n \tau(\xi)}\right)^k \left(\alpha_n \tau(\xi)\right)^{i-k}.
\end{aligned} \tag{9.4.4}$$

Now, making the same reasoning as the one in the proof of Theorem 9.3.1, we have $\frac{\omega_n^2 r_n M_n \alpha_n \tau(\xi)}{1-\alpha_n \tau(\xi)} \neq \alpha_n \tau(\xi)$. So, Eq. (9.4.4) yields for every $i \ge 1$,

$$\begin{aligned}
\|\varphi_{n,i}(\xi)\| &\le \omega_n r_n M_n \frac{\left(\frac{\omega_n^2 r_n M_n \alpha_n \tau(\xi)}{1-\alpha_n \tau(\xi)}\right)^{i+1} - \left(\alpha_n \tau(\xi)\right)^{i+1}}{\frac{\omega_n^2 r_n M_n \alpha_n \tau(\xi)}{1-\alpha_n \tau(\xi)} - \alpha_n \tau(\xi)} \\
&\le \frac{\left(1-\alpha_n \tau(\xi)\right) \omega_n r_n M_n}{\alpha_n \tau(\xi)(\omega_n^2 r_n M_n - 1)} \left(\left(\frac{\omega_n^2 r_n M_n \alpha_n \tau(\xi)}{1-\alpha_n \tau(\xi)}\right)^{i+1} - \left(\alpha_n \tau(\xi)\right)^{i+1}\right).
\end{aligned}$$

(iii) Using Eqs. (9.4.2) and (9.4.3) imply that for all $i \ge 1$,

$$\begin{aligned}
\|\varphi_{n,i}(\xi)\| &\le \sum_{k=1}^{i} |d_{n,k}(\xi)| \| P_{n,i-k}(\xi)\varphi_n \| \\
&\le \sum_{k=1}^{i} \left(\frac{\omega_n^2 r_n M_n \alpha_n \tau(\xi)}{1-\alpha_n \tau(\xi)}\right)^k \omega_n r_n M_n \left(\alpha_n \tau(\xi)\right)^{i-k} \\
&\le \omega_n r_n M_n \sum_{k=1}^{i} \left(\frac{\omega_n^2 r_n M_n \alpha_n \tau(\xi)}{1-\alpha_n \tau(\xi)}\right)^k \left(\alpha_n \tau(\xi)\right)^{i-k} \\
&\le \frac{\omega_n^3 r_n^2 M_n^2 \alpha_n \tau(\xi)}{1-\alpha_n \tau(\xi)} \sum_{k=1}^{i} \left(\frac{\omega_n^2 r_n M_n \alpha_n \tau(\xi)}{1-\alpha_n \tau(\xi)}\right)^{k-1} \left(\alpha_n \tau(\xi)\right)^{i-k}.
\end{aligned} \tag{9.4.5}$$

Now, making the same reasoning as the one in the proof of Theorem 9.3.1, we have

$$\frac{\omega_n^2 r_n M_n \alpha_n \tau(\xi)}{1-\alpha_n \tau(\xi)} \neq \alpha_n \tau(\xi).$$

So, Eq. (9.4.5) yields for every $i \geq 1$

$$\begin{aligned}\|\varphi_{n,i}(\xi)\| &\leq \frac{\omega_n^3 r_n^2 M_n^2 \alpha_n \tau(\xi)}{1-\alpha_n \tau(\xi)} \frac{\left(\frac{\omega_n^2 r_n M_n \alpha_n \tau(\xi)}{1-\alpha_n \tau(\xi)}\right)^i - (\alpha_n \tau(\xi))^i}{\frac{\omega_n^2 r_n M_n \alpha_n \tau(\xi)}{1-\alpha_n \tau(\xi)} - \alpha_n \tau(\xi)} \\ &\leq \frac{\omega_n^3 r_n^2 M_n^2}{\alpha_n \tau(\xi) - 1 + \omega_n^2 r_n M_n} \left(\left(\frac{\omega_n^2 r_n M_n \alpha_n \tau(\xi)}{1-\alpha_n \tau(\xi)} \right)^i - (\alpha_n \tau(\xi))^i \right).\end{aligned}$$

The proof is then complete. Q.E.D.

Let us consider the analytic perturbation $A(\varepsilon)$ (see Eq. (9.1.3)) and let $\widetilde{\varphi}_n(\varepsilon)$ be an eigenvector of $A(\varepsilon)$ associated to the eigenvalue $\widetilde{\lambda}_n(\varepsilon)$. Then, we can see the following result:

Corollary 9.4.1 *Assume that the hypotheses $(H1)$, $(H2)$ with $\beta = q = 1$, and $(H3)$ hold. Let $\widetilde{\varphi}_n$ (respectively, $\widetilde{\varphi}_n^*$) be an eigenvector of A_0 (respectively, A_0^*) associated to the eigenvalue $\widetilde{\lambda}_n$ (respectively, $\overline{\widetilde{\lambda}_n}$) such that $\|\widetilde{\varphi}_n\| = \|\widetilde{\varphi}_n^*\|$ and $\widetilde{\varphi}_n^*(\widetilde{\varphi}_n) = 1$. Then, for $|\varepsilon| < \frac{1}{q+\widetilde{\alpha}_n+\widetilde{\omega}_n^2\widetilde{r}_n\widetilde{M}_n\widetilde{\alpha}_n}$, we have*

(i) the eigenvector $\widetilde{\varphi}_n(\varepsilon)$ can be developed into an entire series

$$\widetilde{\varphi}_n(\varepsilon) := \widetilde{\varphi}_n + \varepsilon\widetilde{\varphi}_{n,1} + \varepsilon^2\widetilde{\varphi}_{n,2} + \cdots + \varepsilon^k\widetilde{\varphi}_{n,k} + \cdots,$$

where

$$\begin{cases} \widetilde{\varphi}_{n,i} := \displaystyle\sum_{k=1}^{i} \widetilde{d}_{n,k}\, \widetilde{P}_{n,i-k}\widetilde{\varphi}_n, & \forall\, i \geq 1, \\ \widetilde{d}_{n,k} = C^k_{-\frac{1}{2}} \displaystyle\sum_{l=1}^{k} \sum_{i_1+i_2+\cdots+i_l=k} \widetilde{\gamma}_{i_1}\widetilde{\gamma}_{i_2}\ldots\widetilde{\gamma}_{i_l}, & \forall\ k \geq 1, \\ \widetilde{\gamma}_k = \widetilde{\varphi}_n^*(\widetilde{P}_{n,k}\ \widetilde{\varphi}_n), \ \ \forall\ k \geq 1 \end{cases}$$

and $\left(\widetilde{P}_{n,k}\right)_{k\geq 1}$ is defined in Eq. (9.3.14), and

(ii) the coefficients $\left(\widetilde{\varphi}_{n,i}\right)_{i=1}^{\infty}$ satisfy

$$\|\varphi_{n,i}\| \leq \widetilde{r}_n\widetilde{M}_n\widetilde{\alpha}_n \left[\frac{\left(q+\widetilde{\alpha}_n+\widetilde{\omega}_n^2\widetilde{r}_n\widetilde{M}_n\widetilde{\alpha}_n\right)^i - (q+\widetilde{\alpha}_n)^i}{q+\widetilde{\alpha}_n} \right]. \tag{9.4.6}$$

◇

Proof (i) We can write $\widetilde{\varphi}_n(\varepsilon)$ in this form

$$\widetilde{\varphi}_n(\varepsilon) = \frac{\widetilde{P}_n(\varepsilon)\widetilde{\varphi}_n}{[\widetilde{\varphi}_n^*(\widetilde{P}_n(\varepsilon)\widetilde{\varphi}_n)]^{\frac{1}{2}}}.$$

Making the same reasoning as the one in the proof of Theorem 9.4.1 and by using Theorem P2ii, we get

$$\left(1 + \widetilde{\varphi}_n^*((\widetilde{P}_n(\varepsilon) - \widetilde{P}_n)\widetilde{\varphi}_n))\right)^{-\frac{1}{2}} = \sum_{k=0}^{\infty} \varepsilon^k \widetilde{d}_{n,k},$$

where

$$\begin{cases} \widetilde{d}_{n,0} = 1, \\ \widetilde{d}_{n,k} = \displaystyle\sum_{l=1}^{k} C^l_{-\frac{1}{2}} \sum_{i_1+i_2+\cdots+i_l=k} \widetilde{\gamma}_{i_1}\widetilde{\gamma}_{i_2}\ldots\widetilde{\gamma}_{i_l}, \quad \forall\ k \geq 1, \\ \widetilde{\gamma}_k = \widetilde{\varphi}_n^*(\widetilde{P}_{n,k}\ \widetilde{\varphi}_n), \quad \forall\ k \geq 1, \\ \widetilde{P}_{n,k} = \dfrac{-1}{2i\pi} \displaystyle\int_{\widetilde{\mathscr{C}}_n} \widetilde{R}_{z_n,k}\, \mathrm{d}z_n, \quad \forall\ k \geq 1. \end{cases}$$

So, we have

$$\begin{aligned} \widetilde{\varphi}_n(\varepsilon) &= \sum_{k=0}^{\infty} \varepsilon^k \widetilde{d}_{n,k} \sum_{k=0}^{\infty} \varepsilon^k \widetilde{P}_{n,k}\widetilde{\varphi}_n \\ &= \sum_{i=0}^{\infty} \varepsilon^i \widetilde{\varphi}_{n,i}, \end{aligned}$$

where

$$\begin{cases} \widetilde{\varphi}_{n,0} := \widetilde{\varphi}_n, \\ \widetilde{\varphi}_{n,i} := \displaystyle\sum_{k=1}^{i} \widetilde{d}_{n,k}\ \widetilde{P}_{n,i-k}\widetilde{\varphi}_n, \quad \forall\, i \geq 1. \end{cases}$$

(ii) We have, for all $k \geq 1$,

$$\left|\widetilde{d}_{n,k}\right| = \sum_{l=1}^{k} \sum_{i_1+i_2+\cdots+i_l=k} |\widetilde{\gamma}_{i_1}|\,|\widetilde{\gamma}_{i_2}|\ldots|\widetilde{\gamma}_{i_l}|.$$

Or, for all $k \geq 1$,

$$|\widetilde{\gamma}_k| \leq \widetilde{\omega}_n^2 \widetilde{r}_n \widetilde{M}_n \widetilde{\alpha}_n (q + \widetilde{\alpha}_n)^{k-1}.$$

So, using Proposition 3.1.3, we deduce for all $k \geq 1$

$$\begin{aligned}
\left|d_{n,k}\right| &\leq \sum_{l=1}^{k} (\widetilde{\omega}_n^2 \widetilde{r}_n \widetilde{M}_n \widetilde{\alpha}_n)^l \sum_{i_1+i_2+\cdots+i_l=k} (q+\widetilde{\alpha}_n)^{k-l} \\
&\leq (q+\widetilde{\alpha}_n)^k \sum_{l=1}^{k} \left(\frac{\widetilde{\omega}_n^2 \widetilde{r}_n \widetilde{M}_n \widetilde{\alpha}_n}{q+\widetilde{\alpha}_n}\right)^l \sum_{i_1+i_2+\cdots+i_l=k} 1 \\
&\leq (q+\widetilde{\alpha}_n)^k \frac{\widetilde{\omega}_n^2 \widetilde{r}_n \widetilde{M}_n \widetilde{\alpha}_n}{q+\widetilde{\alpha}_n} \left(1+\frac{\widetilde{\omega}_n^2 \widetilde{r}_n \widetilde{M}_n \widetilde{\alpha}_n}{q+\widetilde{\alpha}_n}\right)^{k-1} \\
&\leq \widetilde{\omega}_n^2 \widetilde{r}_n \widetilde{M}_n \widetilde{\alpha}_n \left(q+\widetilde{\alpha}_n+\widetilde{\omega}_n^2 \widetilde{r}_n \widetilde{M}_n \widetilde{\alpha}_n\right)^{k-1}.
\end{aligned} \tag{9.4.7}$$

Noting that

$$\left\| \widetilde{P}_{n,k} \right\| \leq \widetilde{r}_n \widetilde{M}_n \widetilde{\alpha}_n (q+\widetilde{\alpha}_n)^{k-1}, \quad \forall k \geq 1,$$

we get, in view of Eq. (9.4.7), that for all $i \geq 1$,

$$\begin{aligned}
\|\widetilde{\varphi}_{n,i}\| &\leq \omega_n^2 r_n^2 M_n^2 \alpha_n^2 \sum_{k=1}^{i} \left(q+\widetilde{\alpha}_n+\widetilde{\omega}_n^2 \widetilde{r}_n \widetilde{M}_n \widetilde{\alpha}_n\right)^{k-1} (q+\widetilde{\alpha}_n)^{i-k-1} \\
&\leq \frac{\omega_n^2 r_n^2 M_n^2 \alpha_n^2}{q+\widetilde{\alpha}_n} \sum_{k=1}^{i} \left(q+\widetilde{\alpha}_n+\widetilde{\omega}_n^2 \widetilde{r}_n \widetilde{M}_n \widetilde{\alpha}_n\right)^{k-1} (q+\widetilde{\alpha}_n)^{i-k} \\
&\leq \frac{\omega_n^2 r_n^2 M_n^2 \alpha_n^2}{q+\widetilde{\alpha}_n} \frac{\left(q+\widetilde{\alpha}_n+\widetilde{\omega}_n^2 \widetilde{r}_n \widetilde{M}_n \widetilde{\alpha}_n\right)^i - (q+\widetilde{\alpha}_n)^i}{\widetilde{\omega}_n^2 \widetilde{r}_n \widetilde{M}_n \widetilde{\alpha}_n} \\
&\leq \widetilde{r}_n \widetilde{M}_n \widetilde{\alpha}_n \left[\frac{\left(q+\widetilde{\alpha}_n+\widetilde{\omega}_n^2 \widetilde{r}_n \widetilde{M}_n \widetilde{\alpha}_n\right)^i - (q+\widetilde{\alpha}_n)^i}{q+\widetilde{\alpha}_n}\right].
\end{aligned}$$

The proof is then complete. Q.E.D.

9.5 Notes and Remarks

- Perturbation theory of linear operators has been pioneered by Rayleigh [22] and Schrödinger ([24, 25]) and has been the object of much research until now [16]. In [22], L. Rayleigh has given a formula for computing the natural frequencies and modes of a vibrating system deviating slightly from a simpler system which admits a complete determination of the frequencies and modes. Mathematically speaking, the method was equivalent to an approximation solution of the eigenvalue problem for a linear operator slightly different from a simpler operator for which the problem is completely solved. Schrödinger [24, 25] has developed a similar method, with more generality and systematization, for the eigenvalue problems that appear in quantum mechanics. Later, Jeribi [18], Kato [19], and Rellich [23] have been mainly concerned with the regular perturbation of self-adjoint operators of a Hilbert space, while some attempts have also been made toward the treatment of non-regular cases which are not less important in physics and applications.

- Notice here that if in particular we take $\xi_k = \varepsilon^k q^{k-1}$ and $T_k = \varepsilon^k A_k$ in Eqs (9.1.2) and (9.1.1) ($\beta = k = 1$), we recognize the B. Sz. Nagy's perturbation model of linear operators (see [20]–[21]). Moreover, we regain the spectral study of the operator $A(\varepsilon)$ (see Eqs (9.1.3) and (9.1.4)) investigated in [21] see also [18]. The main novelty in this chapter is that we give the exact expressions of the coefficients $(\lambda_{n,i})_{i\geq 1}$ and $(\varphi_{n,i})_{i\geq 1}$ in Eqs (9.3.5) and (9.4.1) (see Theorems 9.3.1 and 9.4.1). The key tool here was the following identity,

$$\sum_{n=1}^{+\infty} a_n \left(\sum_{k=1}^{+\infty} s_k\right)^n = \sum_{n=1}^{+\infty} a_n \sum_{l=n}^{+\infty} \sum_{i_1+i_2+\ldots+i_n=l} s_{i_1} s_{i_2} \ldots s_{i_n},$$

where $(s_k)_{k\geq 0}$ and $(a_k)_{k\geq 0}$ are given complex sequences satisfying some convergence conditions.
- Most of the research on the perturbations of linear operators concerns the autoadjoint operators of Hilbert space, which is quite natural given the importance of their applications to physics. It is perhaps not without interest to examine to some extent the reasoning used in this research is adapted to the case of linear operators of the more general types, of Hilbert space, or even of a Banach space.
- Among basic results already obtained by B. Sz. Nagy [21], we focus our attention on the following:
(α) The series

$$A_0\varphi + \varepsilon A_1\varphi + \varepsilon^2 A_2\varphi + \cdots + \varepsilon^k A_k\varphi + \cdots$$

converges for all $\varphi \in \mathscr{D}$ and for all $|\varepsilon| < q^{-1}$. Setting $A(\varepsilon)\varphi$ its limit, we have $A(\varepsilon)$ is a linear operator with domain $\mathscr{D}$.
(β) If A_0 is closed, then $A(\varepsilon)$ is also closed, for $|\varepsilon| < (q+b)^{-1}$.
(γ) Suppose that the unperturbed operator A_0 has an isolated eigenvalue λ_n with multiplicity one. Then, $A(\varepsilon)$ has a unique eigenvalue $\lambda_n(\varepsilon)$ in the neighborhood of λ_n for sufficiently small $|\varepsilon|$ and, this eigenvalue can be expanded into a convergent series

$$\lambda_n(\varepsilon) = \lambda_n + \varepsilon\lambda_{n,1} + \varepsilon^2\lambda_{n,2} + \cdots.$$

The associated eigenvector $\varphi_n(\varepsilon)$ of $A(\varepsilon)$ can also be chosen as a convergent series

$$\varphi_n(\varepsilon) = \varphi_n + \varepsilon\varphi_{n,1} + \varepsilon^2\varphi_{n,2} + \cdots,$$

where φ_n is an eigenvector of A_0 associated to the eigenvalue λ_n.
- Recently, in [15] I. Feki, A. Jeribi and R. Sfaxi ask what conclusions can be drawn if they consider a more general type of non-analytic perturbation involving more than one perturbation parameter? Based on rather long and explicit calculations, they investigate under this question and they give new variants on B. Sz. Nagy's results [21]. To this end, they consider the operator (9.1.2), where $\xi = (\xi_k)_{k\geq 1}$ is

a sequence of complex numbers such that

$$\tau(\xi) := \sum_{k=1}^{\infty} |\xi_k| < \infty,$$

T_0 is a closed linear operator on a Banach space X with domain $\mathscr{D}(T_0)$ and $(T_k)_{k\geq 1}$ are linear operators on X, having the same domain $\mathscr{D}$ such that $\mathscr{D}(T_0) \subset \mathscr{D}$ and (9.1.1), where a, $b > 0$ and $k = \beta = 1$. They emphasize the fact that under those considerations, they give an essential improvement to the results developed by B. Sz. Nagy [21] since they deal with a non-analytic perturbation involving more than one perturbation parameter and covering cases where the results developed in [21] can not be applied.

- Based on the analysis started in see also [1–14, 17, 21], they study the behavior of spectral properties of $T(\xi)$. More precisely, they prove that (θ) The series

$$T_0\varphi + \xi_1 T_1\varphi + \xi_2 T_2\varphi + \cdots + \xi_k T_k\varphi + \cdots$$

converges for all $\varphi \in \mathscr{D}(T_0)$. If the quantities $T(\xi)\varphi$ denotes its limit, then $T(\xi)$ is a linear operator with domain $\mathscr{D}(T_0)$. Moreover, $T(\xi)$ is closed if $\tau(\xi) < \frac{1}{b}$. (η) Let λ_n be the nth eigenvalue of T_0. If λ_n is isolated with multiplicity one, then for ξ small enough, there exists a neighborhood of λ_n in which there is a unique eigenvalue $\lambda_n(\xi)$ of the operator $T(\xi)$ with multiplicity one. Setting $\varphi_n(\xi)$ an eigenvector of $T(\xi)$ associated to $\lambda_n(\xi)$, then one can develop $\lambda_n(\xi)$ and $\varphi_n(\xi)$ into series

$$\lambda_n(\xi) := \lambda_n + \lambda_{n,1}(\xi) + \lambda_{n,2}(\xi) + \cdots + \lambda_{n,i}(\xi) + \cdots \tag{9.5.1}$$

and

$$\varphi_n(\xi) := \varphi_n + \varphi_{n,1}(\xi) + \varphi_{n,2}(\xi) + \cdots + \varphi_{n,i}(\xi) + \cdots . \tag{9.5.2}$$

If, in particular, they take $\xi_k = \varepsilon^k q^{k-1}$ and $T_k = \varepsilon^k A_k$ in Eqs. (9.1.2) and (9.1.1), they recognize the B. Sz. Nagy's perturbation model of linear operators (see [20, 21]). Moreover, they regain the spectral study of the operator $A(\varepsilon)$ (see Eqs. (9.1.3) and (9.1.4)) investigated in [21]. The main novelty in this part is that the authors in [15] give the exact expressions of the coefficients $(\lambda_{n,i})_{i=1}^{\infty}$ and $(\varphi_{n,i})_{i=1}^{\infty}$ in Eqs. (9.5.1) and (9.5.2).

(i) In Corollary 9.4.1, we give some supplements to [21, p. 136]. Indeed, in [21] (see also [18]) has proved that the eigenvector $\tilde{\lambda}_n(\varepsilon)$ of $A(\varepsilon)$ can be developed into an entire series of ε and he has given the estimation of the coefficients $(\tilde{\varphi}_{n,i})_{i\geq 1}$, by comparing them to those of an other series. In our considerations, we estimate the coefficients $(\tilde{\varphi}_{n,i})_{i=1}^{\infty}$ by means of their expression. The obtained estimations are more precise than the one given by B. Sz. Nagy since Eq. (9.4.6) implies that

$$\begin{aligned}\|\varphi_{n,i}\| &\leq \tilde{r}_n \tilde{M}_n \tilde{\alpha}_n \frac{\left(q + \tilde{\alpha}_n + \tilde{\omega}_n^2 \tilde{r}_n \tilde{M}_n \tilde{\alpha}_n\right)^i}{q + \tilde{\alpha}_n} \\ &\leq \tilde{r}_n \tilde{M}_n \left(q + \tilde{\alpha}_n + \tilde{\omega}_n^2 \tilde{r}_n \tilde{M}_n \tilde{\alpha}_n\right)^i \\ &\leq \tilde{\omega}_n \tilde{r}_n \tilde{M}_n \left(q + \tilde{\alpha}_n + \tilde{\omega}_n^2 \tilde{r}_n \tilde{M}_n \tilde{\alpha}_n\right)^i.\end{aligned}$$

(ii) The results of this chapter is no longer valid in general if the nth isolated eigenvalue of the operator T_0 has a multiplicity greater than 1. Here are three simple examples where these are linear operators of the complex two-dimensional Euclidean space, operators which we present by their matrices:

$$(a)\ \begin{pmatrix} 0 & 1 \\ \varepsilon & 0 \end{pmatrix}, \qquad (b)\ \begin{pmatrix} 0 & \varepsilon^2 \\ \varepsilon & 0 \end{pmatrix}, \qquad (c)\ \begin{pmatrix} 0 & \varepsilon \\ 0 & 0 \end{pmatrix}$$

In all these examples, the spectrum of the unperturbed operator ($\varepsilon = 0$) is made up of the only point $\lambda = 0$ and this is of principal multiplicity 2. Its proper multiplicity is equal to 1 in the example (a), and 2 in the examples (b) and (c). The spectrum of the perturbed operator ($\varepsilon \neq 0$) consists, in the example (a), of the points $\lambda = \pm\varepsilon^{\frac{1}{2}}$, and in the example (b), of the points $\lambda = \pm\varepsilon^{\frac{3}{2}}$. None of them is an analytical function of ε at the point $\varepsilon = 0$. In the example (c), the spectrum of the perturbed operator consists of the only point $\lambda = 0$, but this will be of proper multiplicity 1.

References

1. N. Ben Ali, A. Jeribi, On the Riesz basis of a family of analytic operators in the sens of Kato and application to the problem of radiation of a vibriting structure in a light fluid, J. Math. Anal. Appl. **320**(1), 78–94 (2006)
2. S. Charfi, A. Damergi, A. Jeribi, On a Riesz basis of finite-dimensional invariant subspaces and application to Gribov operator in Bargmann space. Line. Mult. Alge. **61**(11), 1577–1591 (2013)
3. S. Charfi, A. Intissar, A. Jeribi, Expansion of solution in terms of generalized eigenvectors for a rectilinear transport equation. Transport Theory Statist. Phys. **38**, 1–16 (2009)
4. S. Charfi, A. Jeribi, I. Walha, Riesz basis property of families of nonharmonic exponential and application to a problem of a radiation of a vibrating structure in a light fluid. Numerical Functional Analysis and Optimization, **32**(4), 370–382, (2011)
5. H. Ellouz, I. Feki, A. Jeribi, On a Riesz basis of exponentials related to the eigenvalues of an analytic operator and application to a non-self-adjoint problem deduced from a perturbation method, J. Math. Phys. **54**, 112101 (2013)
6. H. Ellouz, I. Feki, A. Jeribi, On the asymptotic behavior of the eigenvalues of an analytic operator in the sense of Kato and applications, Serdica J. Math. **45**(1), 55–88 (2019)
7. H. Ellouz, I. Feki, A. Jeribi, On a Riesz basis of exponentials related to a family of analytic operators and application. J. Pseudo-Differ. Oper. Appl. **10**(4), 999–1014 (2019)

8. H. Ellouz, I. Feki, A. Jeribi, Non-orthogonal Fusion Frames of an Analytic Operator and Application to a One-Dimensional Wave Control System. Mediterr. J. Math. **16**(2), 16:52 (2019)
9. H. Ellouz, I. Feki, A. Jeribi, On the basis property of root vectors related to a non-self-adjoint analytic operator and applications, (2021)
10. I. Feki, A. Jeribi, R. Sfaxi, On an unconditionl basis of generalized eigenvectors of an analytic operator and application to a problem of radiation of a vibrating structure in a light fluid, J. Math. Anal. Appl. **375** 261–269 (2011)
11. I. Feki, A. Jeribi, R. Sfaxi, On a Schauder basis related to the eigenvectors of a family of non-self-adjoint analytic operators and applications, Anal. Math. Phys. **3**(4), 311–331 (2013)
12. I. Feki, A. Jeribi, R. Sfaxi, On a Riesz basis of eigenvectors of a non-self-adjoint analytic operator and applications. Line. Mult. Alge. **62**(8), 1049–1068 (2014)
13. I. Feki, A. Jeribi, R. Sfaxi, On a Riesz basis of eigenvectors of a non-self-adjoint perturbation and application to a Gribov operator in Bargmann space, preprint (2021)
14. A. Intissar, A. Jeribi, I. Walha, Riesz basis of exponential family for a hyperbolic system. J. Appl. Anal. **25**(1), 13–23 (2019)
15. I. Feki, A. Jeribi, R. Sfaxi, On a Riesz basis of eigenvectors of a non-self-adjoint perturbation and application to a Gribov operator in Bargmann space, preprint (2021)
16. A. Jeribi, *Spectral Theory and Applications of Linear Operators and block operator Matrices* (Springer, New-York, 2015)
17. A. Jeribi, A. Intissar, On an Riesz basis of generalized eigenvectors of the nonself-adjoint problem deduced from a perturbation method for sound radiation by a vibrating plate in a light fluid. J. Math. Anal. Appl. **292**, 1–16 (2004)
18. A. Jeribi, *Denseness, Bases and Frames in Banach Spaces and Applications* (de Gruyter, Berlin, 2018)
19. T. Kato, *Perturbation Theory for Linear Operators* (Springer, Berlin, 1980)
20. B. Sz, Nagy, Perturbations des transformations autoadjointes dans l'espace de Hilbert. Comment. Math. Helv. **19**, 347–366 (1947)
21. B. Sz, Nagy, Perturbations des transformations linéaires fermées. Acta Sci. Math. Szeged. **14**, 125–137 (1951)
22. L. Rayleigh, The theory of sound **1** (1927)
23. F. Rellich, Storungstheorie der Spektral zerlegung. I. Mitt. Math. Ann. **113**, 600–619 (1937)
24. E. Schrödinger, Quantisierung als eigenwertproblem. Ann. Physik **80**, 437–490 (1926)
25. E. Schrödinger, *Collected Papers on Wave Mechanics* (London and Glasgow, 1928)

Chapter 10
On a Schauder and Riesz Bases of Eigenvectors of an Analytic Operator

This chapter deals with Schauder and Riesz bases of eigenvectors of a family of non-normal operators. Indeed, we generalize some results due to Nagy in [29] and we extend the main result in [7] to Schauder basis in a separable Banach space. Second, we investigate under sufficient conditions assuring the existence of a Riesz basis in a Hilbert space X, formed by eigenvectors of the perturbed operator $T(\varepsilon)$. We also study the existence of a finitely spectral Riesz basis of a family of non-normal operators including the spectral Riesz basis of subspaces and the Riesz basis of subspaces. The aim of this chapter is to prove that the system of eigenvectors of the perturbed operator $T(\xi)$, given in (9.1.2), forms a Schauder and Riesz bases in a separable Banach space X.

10.1 Completeness of the System of Root Vectors of $T(\varepsilon)$

Consider, in this section, the following operator:

$$T(\varepsilon) := T_0 + \varepsilon T_1 + \varepsilon^2 T_2 + \cdots + \varepsilon^k T_k + \cdots , \qquad (10.1.1)$$

where $\varepsilon \in \mathbb{C}$ and $T_0, T_1, T_2, \ldots$ are linear transformations acting from a Banach space X into a Banach space Y.

10.1.1 In Banach Space

In this section, we prove that the system of root vectors of the operator $T(\varepsilon)$, given in (10.1.1), is complete in a Banach space X. Now, we are ready to state the first result of this section:

A. Jeribi, *Perturbation Theory for Linear Operators*,
https://doi.org/10.1007/978-981-16-2528-2_10

Theorem 10.1.1 *Assume that the hypotheses* $(H1)$ *and* $(H2)$ *hold. In addition, suppose that the resolvent of* T_0 *is quasiconservative and belongs to the Carleman-class* $\mathscr{C}_p(X)$, $0 < p < \infty$. *If* T_i *is* T_0*-compact for all* $i \in \mathbb{N}^*$, *then for* $|\varepsilon|$ *small enough, the system of root vectors of the operator* $T(\varepsilon)$, *given in (10.1.1), is complete in* X. ◇

Proof Since the resolvent of T_0 is quasiconservative, then there exists $\lambda_0 \in \rho(T_0)$ such that $(T_0 - \lambda_0)^{-1}$ is quasiconservative. We have

$$\begin{aligned} T(\varepsilon) - \lambda_0 &= T_0 - \lambda_0 + \varepsilon T_1 + \varepsilon^2 T_2 + \cdots \\ &= \left(I + \varepsilon T_1 (T_0 - \lambda_0)^{-1} + \varepsilon^2 T_2 (T_0 - \lambda_0)^{-1} + \cdots\right)(T_0 - \lambda_0) \\ &= (I + S)(T_0 - \lambda_0), \end{aligned} \tag{10.1.2}$$

where

$$S := \sum_{k=1}^{\infty} \varepsilon^k T_k (T_0 - \lambda_0)^{-1}.$$

Using Eq. (9.1.1), we get

$$\begin{aligned} \|S\| &\leq \sum_{k=1}^{\infty} |\varepsilon|^k \|T_k (T_0 - \lambda_0)^{-1}\| \\ &\leq M(\lambda_0, \beta, a, b) \sum_{k=1}^{\infty} |\varepsilon|^k q^{k-1}, \end{aligned}$$

where

$$M(\lambda_0, \beta, a, b) = a\|(T_0 - \lambda_0)^{-1}\| + b\|I + \lambda_0 (T_0 - \lambda_0)^{-1}\|^{\beta} \|(T_0 - \lambda_0)^{-1}\|^{1-\beta}.$$

Since $|\varepsilon| < q^{-1}$, then

$$\|S\| \leq \frac{M(\lambda_0, \beta, a, b)|\varepsilon|}{1 - |\varepsilon| q}$$

and therefore, for $|\varepsilon| < \frac{1}{q + M(\lambda_0, \beta, a, b)}$, we have

$$\|S\| < 1$$

and $I + S$ is invertible. So, Eq. (10.1.2) implies that $T(\varepsilon) - \lambda_0$ is invertible and

$$\begin{aligned} (T(\varepsilon) - \lambda_0)^{-1} &= (T_0 - \lambda_0)^{-1}(I + S)^{-1} \\ &= (T_0 - \lambda_0)^{-1}(I + K), \end{aligned}$$

where

$$K = -S + S^2 - S^3 + \cdots + (-1)^n S^n + \cdots.$$

Since the resolvent of T_0 is compact and T_k is T_0-compact for all $k \geq 1$, then due to (1.7.3), $T_k(T_0 - \lambda_0)^{-1}$ is compact for all $k \geq 1$. So, the operator K is compact on X. Consequently, the result follows from Theorem 6.1.5. Q.E.D.

10.1.2 In Hilbert Space

Similarly, if X is a separable Hilbert space, then making the same reasoning as the one in the proof of Theorem 10.1.1 and using Keldysh theorem (see Theorem 6.1.3), we have the following result:

Corollary 10.1.1 *Let X be a separable Hilbert space. Assume that the hypotheses $(H1)$ and $(H2)$ hold. In addition, suppose that T_0 is self-adjoint and its resolvent belongs to the Carleman-class $\mathscr{C}_p(X)$, $0 < p < \infty$. If T_i is T_0-compact for all $i \in \mathbb{N}^*$, then for $|\varepsilon|$ small enough, the system of root vectors of the operator $T(\varepsilon)$, given in (10.1.1), is complete in X.* ◇

10.2 On Riesz Bases in a Separable Hilbert Space

Theorem 10.2.1 *Let $T(\varepsilon)$ be the perturbed operator given in (10.1.1). Assume that the resolvent of T_0 belongs to Carleman-class $\mathscr{C}_p$, $p > 0$ and, the hypotheses $(H1)$ and $(H2)$ with $\beta = 1$ are satisfied. Then, for some $\lambda \in \rho(T_0)$ and for $|\varepsilon| < \frac{1}{q+M(\lambda,a,b)}$, the operator $(\lambda - T(\varepsilon))^{-1}$ belongs to Carleman-class $\mathscr{C}_p$, $p > 0$, where*

$$M(\lambda, a, b) = a\|(\lambda - T_0)^{-1}\| + b[1 + |\lambda|\|(\lambda - T_0)^{-1}\|].$$ ◇

Proof Let $\lambda \in \rho(T_0)$ and $\psi \in \mathscr{D}(T_0)$, we have

$$\begin{aligned}(\lambda - T(\varepsilon))\psi &= \lambda\psi - T_0\psi - \varepsilon T_1\psi - \varepsilon^2 T_2\psi - \cdots \\ &= [I - \varepsilon T_1(\lambda - T_0)^{-1} - \varepsilon^2 T_2(\lambda - T_0)^{-1} - \cdots](\lambda - T_0)\psi.\end{aligned}$$

For $\varphi \in R(T_0)$, consider

$$S\varphi = T_1(\lambda - T_0)^{-1}\varphi + \varepsilon T_2(\lambda - T_0)^{-1}\varphi + \cdots.$$

So,

$$(\lambda - T(\varepsilon))\psi = (I - \varepsilon S)(\lambda - T_0)\psi. \tag{10.2.1}$$

For every $\varphi \in R(T_0)$ such that $\varphi \neq 0$, we have the following estimate:

$$\|S\varphi\| \leq \sum_{k=1}^{+\infty} |\varepsilon|^{k-1} \|T_k(\lambda - T_0)^{-1}\varphi\|. \tag{10.2.2}$$

Using the hypothesis $(H2)$, we have

$$\begin{aligned}\|T_k(\lambda - T_0)^{-1}\varphi\| &\le q^{k-1}[a\|(\lambda - T_0)^{-1}\varphi\| + b\|T_0(\lambda - T_0)^{-1}\varphi\|] \\ &\le q^{k-1}[a\|(\lambda - T_0)^{-1}\varphi\| + b\| - \varphi + \lambda(\lambda - T_0)^{-1}\varphi\|] \\ &\le q^{k-1}[a\|(\lambda - T_0)^{-1}\varphi\| + b(\|\varphi\| + |\lambda|\|(\lambda - T_0)^{-1}\varphi\|)].\end{aligned}$$

So,

$$\|T_k(\lambda - T_0)^{-1}\varphi\| \le q^{k-1}[a\|(\lambda - T_0)^{-1}\| + b(1 + |\lambda|\|(\lambda - T_0)^{-1}\|)]\|\varphi\|. \tag{10.2.3}$$

Substituting the inequality (10.2.3) in (10.2.2), we get

$$\|S\varphi\| \le \sum_{k=1}^{+\infty}(|\varepsilon|q)^{k-1}[a\|(\lambda - T_0)^{-1}\| + b(1 + |\lambda|\|(\lambda - T_0)^{-1}\|)]\|\varphi\|.$$

Therefore,

$$\frac{\|S\varphi\|}{\|\varphi\|} \le \sum_{k=1}^{+\infty}(|\varepsilon|q)^{k-1}[a\|(\lambda - T_0)^{-1}\| + b(1 + |\lambda|\|(\lambda - T_0)^{-1}\|)].$$

So,

$$\|S\| \le M(\lambda, a, b)\sum_{k=1}^{+\infty}(|\varepsilon|q)^{k-1},$$

where

$$M(\lambda, a, b) = a\|(\lambda - T_0)^{-1}\| + b(1 + |\lambda|\|(\lambda - T_0)^{-1}\|).$$

Since $|\varepsilon|q < 1$, then

$$\|S\| \le \frac{M(\lambda, a, b)}{1 - |\varepsilon|q},$$

and therefore, for $|\varepsilon| < \frac{1}{q + M(\lambda, a, b)}$, we have $|\varepsilon|\|S\| < 1$ and $I - \varepsilon S$ is invertible. Using Eq. (10.2.1), we infer that $\lambda - T(\varepsilon)$ is invertible,

$$\rho(T_0) \subset \rho(T(\varepsilon))$$

and

$$(\lambda - T(\varepsilon))^{-1} = (\lambda - T_0)^{-1}(I - \varepsilon S)^{-1}. \tag{10.2.4}$$

Since the resolvent of T_0 belongs to Carleman-class $\mathscr{C}_p$, $p > 0$, and the linear operator $(I - \varepsilon S)^{-1}$ is bounded, Eq. (10.2.4) and Proposition 4.4.2 leads to the fact that the resolvent of $T(\varepsilon)$ belongs to Carleman-class $\mathscr{C}_p$, $p > 0$. Q.E.D.

Proposition 10.2.1 *Let X be a separable Hilbert space and let T_0 be a self-adjoint operator whose resolvent belongs to Carleman-class $\mathscr{C}_p$, $p > 0$. Suppose that T_0 satisfies the hypotheses $(H1)$ and $(H2)$ with $\beta = 1$ and that, for all $\delta > 0$, there exists a constant $c_\delta > 0$ such that*

$$\|T_i\varphi\| \leq \delta\|T_0\varphi\| + c_\delta\|\varphi\|$$

for all $\varphi \in \mathscr{D}(T_0)$ and for all $i \in \mathbb{N}^*$. Then,

(i) T_i is T_0-compact for all $i \in \mathbb{N}^*$, and
(ii) for a sufficiently small $|\varepsilon|$, the system of generalized eigenvectors of the operator $T(\varepsilon)$ is complete in X. ◇

Proof (i) This assertion is an immediate consequence of the definition.
(ii) Let $\lambda \in \rho(T_0)\bigcap\mathbb{R}$ and

$$M(\lambda, a, b) = a\|(\lambda - T_0)^{-1}\| + b[1 + |\lambda|\|(\lambda - T_0)^{-1}\|].$$

If $|\varepsilon| < \frac{1}{q+M(\lambda,a,b)}$, then $\lambda \in \rho(T(\varepsilon))$ and, using Eq. (10.2.1), we have

$$\begin{aligned}(\lambda - T(\varepsilon))^{-1} &= (\lambda - T_0)^{-1}(I - \varepsilon S)^{-1} \\ &= (\lambda - T_0)^{-1}(I + \varepsilon S + \varepsilon^2 S^2 + \cdots).\end{aligned} \tag{10.2.5}$$

Next, using the assertion (i), we infer that the operator

$$\varepsilon S + \varepsilon^2 S^2 + \cdots + \varepsilon^n S^n + \cdots$$

is a compact operator on X. The result of (ii) follows from both Eq. (10.2.5) and the Keldysh theorem (see Theorem 6.1.3). Q.E.D.

We suppose that
$(H4)$ the eigenvectors $\mathscr{R} := (\varphi_n)_n$ of the operator T_0 form a Riesz basis of X.
Let $\varepsilon \in \mathbb{C}^*$ and consider the eigenvector problem

$$\begin{cases} T_0\varphi + \varepsilon T_1\varphi + \varepsilon^2 T_2\varphi + \cdots + \varepsilon^k T_k\varphi + \cdots = \lambda\varphi \\ \varphi \in \mathscr{D}(T_0). \end{cases}$$

Theorem 10.2.2 *Let X be a separable Hilbert space. Assume that $(H1)$, $(H2)$ with $\beta = 1$ and $(H4)$ hold. If the eigenvalues $(\lambda_n)_{n=1}^{\infty}$ of the operator T_0 are with simple multiplicity and, there exists a positive constant r, independent of n, such that*

$$\{z \in \mathbb{C} \text{ such that } |z - \lambda_n| < r\}\bigcap\sigma(T_0) = \{\lambda_n\} \text{ for all } n \in \mathbb{N}^*, \tag{10.2.6}$$

then there exists a sequence of complex $(\varepsilon_n)_{n=1}^{\infty}$ and a sequence of eigenvectors of the operator $T(\varepsilon)$, having the following form

$$\varphi_n(\varepsilon) = \varphi_n + \varepsilon\varphi_n^1 + \varepsilon^2\varphi_n^2 + \cdots,$$

such that the system $\{\varphi_n(\varepsilon_n)\}_{n=1}^{\infty}$ *forms a Riesz basis of* X. ◇

Proof Let $n \in \mathbb{N}^*$, we note $\sigma_0 = \{\lambda_n\}$ and $\sigma_1 = \sigma(T_0)\backslash\sigma_0$, where λ_n is the nth eigenvalue, with principal multiplicity one, of the operator T_0. Consider $\mathscr{C}_n = \mathscr{C}(\lambda_n, r_n)$ the closed circle with center λ_n and with radii r_n, lying in the resolvent set of T_0, and having σ_0 inside and σ_1 outside. The fact that $(\lambda - T_0)^{-1}$ is a regular analytic function for $\lambda \in \rho(T_0)$, we infer that $\|(\lambda - T_0)^{-1}\|$ is a continuous function. Thus, we denote M_n, N_n, α_n, and ω_n, respectively, by

$$M_n = \sup_{\lambda\in\mathscr{C}_n} \|(\lambda - T_0)^{-1}\|,$$

$$\begin{aligned} N_n &= \sup_{\lambda\in\mathscr{C}_n} \|T_0(\lambda - T_0)^{-1}\| \\ &= \sup_{\lambda\in\mathscr{C}_n} \| - I + \lambda(\lambda - T_0)^{-1}\|, \end{aligned}$$

$$\alpha_n = aM_n + bN_n,$$

and

$$\omega_n = \|\varphi_n\|,$$

where φ_n is an eigenvector of T_0 associated to the eigenvalue λ_n. Due to Corollary 9.3.2, for $|\varepsilon| < \frac{1}{q+\alpha_n+\omega_n^2 r_n M_n\alpha_n}$, the perturbed transformation $T(\varepsilon)$ will have a unique point of the spectrum inside the circle $\mathscr{C}_n = \mathscr{C}(\lambda_n, r_n)$ and, this point $\lambda_n(\varepsilon)$ will also be with principal multiplicity 1 (see Corollary 9.3.2(i)). The eigenvalue $\lambda_n(\varepsilon)$ and the corresponding eigenvector $\varphi_n(\varepsilon)$ of the perturbed operator $T(\varepsilon)$ can be developed as entire series of ε and, we have

$$\lambda_n(\varepsilon) = \lambda_n + \varepsilon\lambda_n^1 + \varepsilon^2\lambda_n^2 + \cdots \tag{10.2.7}$$

$$\varphi_n(\varepsilon) = \varphi_n + \varepsilon\varphi_n^1 + \varepsilon^2\varphi_n^2 + \cdots, \tag{10.2.8}$$

where λ_n^i and φ_n^i satisfy the following estimations:

$$\begin{aligned} |\lambda_n^i| &\leq \omega_n^2 r_n^2 M_n\alpha_n(q + \alpha_n + \omega_n^2 r M_n\alpha_n)^{i-1} \\ \|\varphi_n^i\| &\leq \omega_n r_n M_n(q + \alpha_n + \omega_n^2 r_n M_n\alpha_n)^i. \end{aligned} \tag{10.2.9}$$

On the other hand, by hypothesis $(H4)$, the eigenvectors $(\varphi_n)_{n=1}^{\infty}$ of the operator T_0 form a Riesz basis of the space X. Hence, in view of Proposition 7.6.2, there are $A > 0$ and $B > 0$ such that

$$A\|u\|^2 \leq \sum_{n=1}^{+\infty} |\langle\varphi_n, u\rangle|^2 \leq B\|u\|^2$$

for all $u \in X$. Let $n \in \mathbb{N}^*$. For each eigenvalue λ_n of T_0, we fix an $\varepsilon_n \in \mathbb{C}$ such that

$$|\varepsilon_n| \in \left]0, \frac{A}{(\beta\omega_n r_n M_n n\sqrt{B} + A)(q + \alpha_n + \omega_n^2 r_n M_n \alpha_n)}\right[,$$

where

$$\beta^2 = \sum_{k=1}^{\infty} \frac{1}{k^2}.$$

Since

$$\left]0, \frac{A}{(\beta\omega_n r_n M_n n\sqrt{B} + A)(q + \alpha_n + \omega_n^2 r_n M_n \alpha_n)}\right[\subset \left]0, \frac{1}{q + \alpha_n + \omega_n^2 r_n M_n \alpha_n}\right[,$$

then by using both Eqs. (10.2.7) and (10.2.8), the eigenvalue $\lambda_n(\varepsilon_n)$ and the eigenvector $\varphi_n(\varepsilon_n)$ can be developed as entire series of ε_n and, we have

$$\lambda_n(\varepsilon_n) = \lambda_n + \varepsilon_n \lambda_n^1 + \varepsilon_n^2 \lambda_n^2 + \cdots$$

$$\varphi_n(\varepsilon_n) = \varphi_n + \varepsilon_n \varphi_n^1 + \varepsilon_n^2 \varphi_n^2 + \cdots .$$

Using the last equation and the estimate (10.2.9), we obtain the following estimation

$$\begin{aligned}
\|\varphi_n(\varepsilon_n) - \varphi_n\| &\le \sum_{k=1}^{+\infty} |\varepsilon_n|^k \omega_n r_n M_n (q + \alpha_n + \omega_n^2 r_n M_n \alpha_n)^k \\
&\le \omega_n r_n M_n \sum_{k=1}^{+\infty} [|\varepsilon_n|(q + \alpha_n + \omega_n^2 r_n M_n \alpha_n)]^k \\
&\le \frac{\omega_n r_n M_n |\varepsilon_n| \, (q + \alpha_n + \omega_n^2 r_n M_n \alpha_n)}{1 - |\varepsilon_n|(q + \alpha + \omega_n^2 r_n M_n \alpha_n)}.
\end{aligned}$$

Thus,

$$\|\varphi_n(\varepsilon_n) - \varphi_n\| < \frac{A}{\beta n\sqrt{B}}.$$

We deduce that

$$\begin{aligned}
\sum_{n=1}^{+\infty} \|\varphi_n(\varepsilon_n) - \varphi_n\|^2 &< \sum_{n=1}^{+\infty} \frac{A^2}{\beta^2 n^2 B} \\
&= \frac{A^2}{B}.
\end{aligned}$$

In view of Theorem 7.6.4, we infer that the system $(\varphi_n(\varepsilon_n))_{n=1}^{\infty}$ forms a Riesz basis of the space X. This achieves the proof. Q.E.D.

Arguing as the proof of the last theorem and by using Theorem 1.15.1, we have a more general result than Theorem 10.2.2.

Theorem 10.2.3 *Let X be a separable Hilbert space. Assume that $(H1)$, $(H2)$ with $\beta = 1$ and $(H4)$ hold. If the operator $P_j T_{1_{|E_j}}$ has at least one eigenvalue σ_j^1 with simple multiplicity, where E_j is the eigenspace with m_j dimension associated to the eigenvalue λ_j of T_0 and P_j is the spectral projector on E_j, then*

(i) for $|\varepsilon|$ smooth, the eigenvalues $\{\lambda_n\}_{n=1}^{\infty}$ and the eigenvectors $\{\varphi_n\}_{n=1}^{\infty}$ of $T(\varepsilon)$ can be developed as entire series of ε

$$\begin{aligned}\lambda_n(\varepsilon) &= \lambda_n + \varepsilon\sigma_n^1 + \varepsilon^2\lambda_n^2 + \cdots \\ \varphi_n(\varepsilon) &= \varphi_n + \varepsilon\varphi_n^1 + \varepsilon^2\varphi_n^2 + \cdots,\end{aligned}$$

(ii) there exists a sequence of complex $(\varepsilon_n)_{n=1}^{\infty}$ and a sequence of eigenvectors of the operator $T(\varepsilon)$ having the following form:

$$\varphi_n(\varepsilon_n) = \varphi_n + \varepsilon_n\varphi_n^1 + \varepsilon_n^2\varphi_n^2 + \cdots$$

such that the system $\{\varphi_n(\varepsilon_n)\}_{n=1}^{\infty}$ forms a Riesz Basis of X. ◇

We have the following result:

Theorem 10.2.4 *Let X be a separable Hilbert space and assume that the eigenvalues $(\lambda_n)_{n=1}^{\infty}$ of T_0 are isolated with multiplicity one. Assume further that the hypotheses $(H1)$, $(H2)$ and $(H4)$ hold. Then, there are a sequence of complex $(\varepsilon_n)_{n=1}^{\infty}$ and a sequence $\{\varphi_n(\varepsilon_n)\}_{n=1}^{\infty}$ having the form*

$$\varphi_n(\varepsilon_n) = \varphi_n + \varepsilon_n\varphi_n^1 + \varepsilon_n^2\varphi_n^2 + \cdots$$

such that the system $\{\varphi_n\}_1^N \bigcup \{\varphi_n(\varepsilon_n)\}_{N+1}^{\infty}$ $(N > 1)$ forms a Riesz basis of X. ◇

Proof Let $n \in \mathbb{N}^*$ and let λ_n be the nth eigenvalue of T_0. For each eigenvalue λ_n of T_0, we fix an $\varepsilon_n \in \mathbb{C}$ such that

$$|\varepsilon_n| \in \left]0, \frac{r(\mathscr{R})}{(\eta\omega_n r_n M_n n + r(\mathscr{R}))(q + \alpha_n + \omega_n^2 r_n M_n \alpha_n)}\right[,$$

where $r(\mathscr{R})$ is given in (7.6.1) and

$$\eta^2 = \sum_{n=N+1}^{\infty} \frac{1}{n^2}.$$

Since

$$\left]0, \frac{r(\mathscr{R})}{(\eta\omega_n r_n M_n n + r(\mathscr{R}))(q + \alpha_n + \omega_n^2 r_n M_n \alpha_n)}\right[\subset \left]0, \frac{1}{q + \alpha_n + \omega_n^2 r_n M_n \alpha_n}\right[,$$

then using Corollary 9.4.1, we can easily see that

$$\|\varphi_n(\varepsilon_n) - \varphi_n\| < \frac{r(\mathscr{R})}{\eta n}.$$

So,

$$\sum_{n=N+1}^{\infty} \|\varphi_n(\varepsilon_n) - \varphi_n\|^2 < r(\mathscr{R})^2.$$

Consequently, in view of Theorem 7.6.3 allows us to conclude that the system $\{\varphi_n\}_1^N \bigcup \{\varphi_n(\varepsilon_n)\}_{N+1}^{\infty}$ $(N > 1)$ forms a Riesz basis of the space X. Q.E.D.

Now, using A. Schueller's theorem (see Theorem 7.6.4), we get the following results:

Theorem 10.2.5 *Let X be a separable Hilbert space and assume that the eigenvalues $(\lambda_n)_{n=1}^{\infty}$ of T_0 are isolated with multiplicity one. Assume further that the hypotheses $(H1)$, $(H2)$, and $(H4)$ hold, then there are a sequence of complex $(\varepsilon_n)_{n=1}^{\infty}$ and a sequence $\{\varphi_n(\varepsilon_n)\}_{n=1}^{\infty}$ having the form*

$$\varphi_n(\varepsilon_n) = \varphi_n + \varepsilon_n \varphi_n^1 + \varepsilon_n^2 \varphi_n^2 + \cdots$$

such that the system $\{\varphi_n(\varepsilon_n)\}_{n=1}^{\infty}$ forms a Riesz basis of X. ◇

Proof Let $n \in \mathbb{N}^*$ and let λ_n be the nth eigenvalue of T_0. Since the eigenvectors $(\varphi_n)_{n=1}^{\infty}$ of the operator T_0 form a Riesz basis of X, then in view of Proposition 7.6.2, there are numbers A, $B > 0$ such that

$$A\|u\|^2 \leq \sum_{n=1}^{\infty} |\langle \varphi_n, u\rangle|^2 \leq B\|u\|^2$$

for all $u \in X$. For all eigenvalue λ_n of the operator T_0, we fix an $\varepsilon_n \in \mathbb{C}$ such that

$$|\varepsilon_n| \in \left]0, \frac{A}{(\beta\omega_n r_n M_n n\sqrt{B} + A)(q + \alpha_n + \omega_n^2 r_n M_n \alpha_n)}\right[,$$

where r_n, α_n, and ω_n are defined by the same way as in the proof of Theorem 10.2.2 and

$$\beta^2 = \sum_{n=1}^{\infty} \frac{1}{n^2}.$$

Making the same reasoning as the one of Theorem 10.2.4, we get

$$\sum_{j=1}^{\infty} \|\varphi_n(\varepsilon_n) - \varphi_n\|^2 < \frac{A^2}{B}.$$

Consequently, in view of Theorem 7.6.4 allows us to conclude that the system $\{\varphi_n(\varepsilon_n)\}_{n=1}^{\infty}$ forms a Riesz basis of X. Q.E.D.

Corollary 10.2.1 *Let X be a separable Hilbert space and assume that the eigenvalues $(\lambda_n)_{n=1}^{\infty}$ of T_0 are isolated with multiplicity one. Assume, further, that the hypotheses $(H1)$, $(H2)$ and $(H4)$ hold, then there are a constant $C_N > 0$, $(N > 1)$, a sequence of complex $(\varepsilon_n)_{n=1}^{\infty}$, and a sequence of eigenvectors of the operator $T(\varepsilon)$ having the following form:*

$$\varphi_n(\varepsilon) = \varphi_n + \varepsilon\varphi_n^1 + \varepsilon^2\varphi_n^2 + \cdots$$

such that for all $|\varepsilon| \in]0, C_N]$, the system $\{\varphi_n(\varepsilon)\}_1^N \bigcup \{\varphi_n(\varepsilon_n)\}_{N+1}^{\infty}$ $(N > 1)$ forms a Riesz basis of X. ◇

Proof Consider

$$C_N = \min_{n \in [1,N]} \frac{A}{(\beta\omega_n r_n M_n n\sqrt{B} + A)(q + \alpha_n + \omega_n^2 r_n M_n \alpha_n)},$$

where A, B, β, r_n, α_n, ω_n are defined by the same way as in the proof of Theorem 10.2.5. Now, the proof follows from Theorem 7.6.4. Q.E.D.

Corollary 10.2.2 *Let X be a separable Hilbert space and assume that the eigenvalues $(\lambda_n)_{n=1}^{\infty}$ of T_0 are isolated with multiplicity one. Assume, further, that the hypotheses $(H1)$, $(H2)$ and $(H4)$ hold, then there exist a constant $C_N > 0$ $(N > 1)$ and a sequence of eigenvectors of $T(\varepsilon)$ having the form*

$$\varphi_n(\varepsilon) = \varphi_n + \varepsilon\varphi_n^1 + \varepsilon^2\varphi_n^2 + \cdots$$

such that for all $|\varepsilon| \in]0, C_N]$, the system $\{\varphi_n(\varepsilon)\}_1^N \bigcup \{\varphi_n\}_{N+1}^{\infty}$ forms a Riesz Basis of X. ◇

Proof It suffices to choose C_N the same constant as in the proof of Corollary 10.2.1 and to apply Theorem 7.6.4. Q.E.D.

The bases defined in this section are Riesz bases of the Hilbert space X. To answer the question raised by Filippi et al. in [19] and to give a Riesz basis of X formed by eigenvectors of the operator $T(\varepsilon)$, we shall add the following hypothesis:

$(H5)$ For all $n \in \mathbb{N}^*$, there is a sequence $(r_n)_{n\geq 1}$ in $\mathbb{R}^*_{\geq 0}$ satisfying

$$\begin{cases} \{z \in \mathbb{C} \text{ such that } |z - \lambda_n| \leq r_n\} \bigcap \sigma(T_0) = \{\lambda_n\}, \quad \text{for all } n \geq 1 \\ \sup\limits_{n\geq 1} \ (q + \alpha_n + w_n^2 r_n M_n \alpha_n) < \infty, \ \text{ and} \\ \sum\limits_{n=1}^{\infty} (\omega_n r_n M_n)^2 < \infty. \end{cases}$$

Theorem 10.2.6 *Let X be a separable Hilbert space and assume that the eigenvalues $(\lambda_n)_{n=1}^{\infty}$ of T_0 are isolated with multiplicity one. Assume, further, that the hypotheses $(H1)$, $(H2)$, $(H4)$, and $(H5)$ hold, then there exist a constant $D > 0$ and a sequence of eigenvectors of the operator $T(\varepsilon)$ having the following form*

$$\varphi_n(\varepsilon) = \varphi_n + \varepsilon\varphi_n^1 + \varepsilon^2\varphi_n^2 + \varepsilon^3\varphi_n^3 + \cdots$$

such that for all $|\varepsilon| \in]0, D[$, the system $\{\varphi_n(\varepsilon)\}_{n=1}^{\infty}$ forms a Riesz basis in X. ◇

Proof Let $n \in \mathbb{N}^*$ and consider λ_n the nth eigenvalue of T_0. Let's pose

$$\theta_0 = \frac{A}{A + \left(B \sum\limits_{n=1}^{+\infty} (w_n r_n M_n)^2\right)^{1/2}}$$

and

$$\tau = \frac{1}{\sup\limits_{n\geq 1}(q + \alpha_n + w_n^2 r_n M_n \alpha_n)},$$

where $A, B, \beta^2, r_n, \alpha_n, \omega_n$ are defined by the same way as in the proof of Theorem 10.2.5. For $\theta \in]0, \theta_0[$, $D = \theta\tau$ and $|\varepsilon| \in]0, D[$, we have

$$\begin{aligned} |\varepsilon| &< \theta\tau \\ &< \theta_0\tau \\ &< \tau \\ &< \frac{1}{q + \alpha_n + w_n^2 r_n M_n \alpha_n}. \end{aligned}$$

In view of Corollary 9.4.1, we have

$$\|\varphi_n(\varepsilon) - \varphi_n\| \leq w_n r_n M_n \sum_{k=1}^{+\infty} \left(|\varepsilon|(q + \alpha_n + w_n^2 r M_n \alpha_n)\right)^k.$$

Moreover,

$$|\varepsilon|(q+\alpha_n+w_n^2r_nM_n\alpha_n) < \theta\frac{(q+\alpha_n+w_n^2r_nM_n\alpha_n)}{\sup\limits_{n\geq 1}(q+\alpha_n+w_n^2r_nM_n\alpha_n)} \leq \theta < 1.$$

Hence,

$$\|\varphi_n(\varepsilon)-\varphi_n\| \leq w_nr_nM_n\frac{|\varepsilon|(q+\alpha_n+w_n^2r_nM_n\alpha_n)}{1-|\varepsilon|(q+\alpha_n+w_n^2r_nM_n\alpha_n)}.$$

Since the function

$$x \longrightarrow \frac{x}{1-x}$$

is strictly increasing in $]0, 1[$, then

$$\|\varphi_n(\varepsilon)-\varphi_n\| < w_nr_nM_n\frac{\theta}{1-\theta}.$$

On the other hand, taking into account the hypothesis $(H5)$, we have

$$\begin{aligned}\sum_{n=1}^{+\infty}\|\varphi_n(\varepsilon)-\varphi_n\|^2 &< \left(\frac{\theta}{1-\theta}\right)^2\sum_{n=1}^{+\infty}(w_nr_nM_n)^2\\ &< \left(\frac{\theta_0}{1-\theta_0}\right)^2\sum_{n=1}^{+\infty}(w_nr_nM_n)^2\\ &< \frac{A^2}{B}.\end{aligned}$$

Thus, Theorem 7.6.4 allows us to conclude that for all $|\varepsilon| \in]0, D[$, the system $\{\varphi_n(\varepsilon)\}_{n=1}^{\infty}$ forms a Riesz basis of the space X. Q.E.D.

10.3 On a Finitely Spectral Riesz Basis of a Family of Non-normal Operators

We investigate the existence of finitely spectral Riesz basis of the perturbed operator $T(\varepsilon)$, given in (10.1.1), where the operators $T_0, T_1, T_2, T_3, \ldots$ satisfy the following
$(H6)$ T_0 is a normal operator on a separable Hilbert space X,
$(H7)$ $T_1, T_2, T_3, \ldots$ are linear operators on X with the same domain $\mathscr{D} \subseteq \mathscr{D}(T_0)$ and, for every sufficiently small η, the following inequalities are satisfied:

$$\|T_k\varphi\| \leq q^{k-1}(\eta\|T_0\varphi\|^{\beta}\|\varphi\|^{1-\beta}+\Phi_\eta(\varphi)), \quad \text{for all } \varphi \in \mathscr{D}(T_0) \text{ and } k \geq 1, \tag{10.3.1}$$

where $\beta \in]0, 1[$ and Φ_η is a continuous and sublinear functional.

10.3.1 Spectrum of $T(\varepsilon)$

In this section, we suppose that the hypotheses $(H1)$, $(H7)$, $(H6)$, and $(H7)$ are fulfilled. Consider the eigenvector problem

$$\begin{cases} T_0\varphi + \varepsilon T_1\varphi + \varepsilon^2 T_2\varphi + \cdots + \varepsilon^k T_k\varphi + \cdots = \lambda\varphi \\ \varphi \in \mathscr{D}(T_0). \end{cases}$$

Theorem 10.3.1 *If $|\varepsilon| < \frac{1}{q}$, then the series*

$$\sum_{k\geq 0} \varepsilon^k T_k\varphi$$

converges for all $\varphi \in \mathscr{D}(T_0)$. ◇

Proof Let $\varphi \in \mathscr{D}(T_0)$. If $|\varepsilon| < \frac{1}{q}$, then it follows from Eq. (10.3.1) that

$$\begin{aligned} \left\| \sum_{k\geq 0} \varepsilon^k T_k\varphi \right\| &\leq \sum_{k\geq 0} \|\varepsilon^k T_k\varphi\| \\ &\leq \sum_{k\geq 0} |\varepsilon|^k \|T_k\varphi\| \\ &\leq \|T_0\varphi\| + \sum_{k\geq 1} |\varepsilon|^k \|T_k\varphi\|. \end{aligned}$$

For every sufficiently small η, we have

$$\begin{aligned} \left\| \sum_{k\geq 0} \varepsilon^k T_k\varphi \right\| &\leq \|T_0\varphi\| + \sum_{k\geq 1} |\varepsilon|^k q^{k-1} (\|\Phi_\eta\| \|\varphi\| + \eta \|T_0\varphi\|^\beta \|\varphi\|^{1-\beta}) \\ &\leq \|T_0\varphi\| + \frac{|\varepsilon|}{1-|\varepsilon|q} (\|\Phi_\eta\| \|\varphi\| + \eta \|T_0\varphi\|^\beta \|\varphi\|^{1-\beta}). \end{aligned}$$

So, the series

$$\sum_{k=0}^{n} \varepsilon^k T_k\varphi$$

is convergent. Q.E.D.

Next, $T(\varepsilon)$ denotes the limit, as $n \to \infty$, of the series

$$\sum_{k=0}^{n} \varepsilon^k T_k\varphi.$$

In order to prove that the operator $T(\varepsilon)$ is closed, we need the following lemma:

Theorem 10.3.2 *Let $\eta > 0$. If $|\varepsilon| < (q+\beta\eta)^{-1}$, then $T(\varepsilon)$ is closed.* ◇

Proof Let $\varphi \in \mathcal{D}(T_0)$ and we set $S := T(\varepsilon) - T_0$. By using condition $(H7)$, we obtain for $\eta > 0$,

$$\begin{aligned}\|S\varphi\| &= \left\|\sum_{k\geq 1}\varepsilon^k T_k\varphi\right\| \\ &\leq \sum_{k\geq 1}|\varepsilon|^k\|T_k\varphi\| \\ &\leq \frac{\|\Phi_\eta\||\varepsilon|}{1-|\varepsilon|q}\|\varphi\| + \frac{\eta|\varepsilon|}{1-|\varepsilon|q}\|T_0\varphi\|^\beta\|\varphi\|^{1-\beta}.\end{aligned}$$

If $|\varepsilon| < \frac{1}{q+\beta\eta}$, then

$$\frac{\eta|\varepsilon|}{1-|\varepsilon|q} < \frac{1}{\beta}.$$

Now, in view of Lemma 2.7.2 and by using the fact that T_0 is closed, implies that $T(\varepsilon)$ is closed. This achieves the proof. Q.E.D.

Now, the following lemma is needed:

Lemma 10.3.1 *Assume that $\sigma(T_0)\bigcap\Omega(2\varphi_-, 2\varphi_+) \subset \mathbb{R}_{\geq 0}$ with $-\pi \leq \varphi_- < 0 < \varphi_+ \leq \pi$ and $|\varepsilon|$ small enough. Let $\eta > 0$, then for $\alpha > \frac{\eta|\varepsilon|}{1-|\varepsilon|q} > 0$, $\frac{\eta|\varepsilon|}{(1-|\varepsilon|q)\alpha} < \theta < 1$ and $0 < \psi < \min\{-\varphi_-, \varphi_+, \pi/2\}$, there exists $r_0 > 0$ such that the sets*

$$\begin{aligned}\rho_1 &= \{z \in \overline{\Omega(\varphi_-, \varphi_+)} \text{ such that } |z| \geq r_0, z \notin \Omega(\psi)\},\\ \rho_2 &= \{z = x+iy \in \overline{\Omega(\psi)} \text{ such that } |z| \geq r_0, |y| \geq \alpha x^\beta\},\end{aligned}$$

and

$$\rho_3 = \{z = x+iy \in \overline{\Omega(\psi)} \text{ such that } |z| \geq r_0, |y| \leq \alpha x^\beta \leq dist(z, \sigma(T_0))\}$$

satisfy $\rho_1\bigcup\rho_2\bigcup\rho_3 \subset \rho(T(\varepsilon))$ and for $z \in \rho_1\bigcup\rho_2\bigcup\rho_3$, we have

$$||S(T_0-z)^{-1}|| \leq \theta,$$

$$||(T(\varepsilon)-z)^{-1}|| \leq \frac{(1-\theta)^{-1}}{dist(z, \sigma(T_0))}$$

and

$$||S(T(\varepsilon)-z)^{-1}|| \leq \frac{\theta}{1-\theta},$$

where $S := T(\varepsilon) - T_0$. Furthermore, there is a constant $M > 0$ such that

$$||(T(\varepsilon) - z)^{-1}|| \leq M,$$

for all $z \in \rho_1 \bigcup \rho_2 \bigcup \rho_3$. ◇

Proof Let $z \in \rho_1 \bigcup \rho_2 \bigcup \rho_3$. By using the spectral theorem for normal operators (see [27]), and writing $d = \text{dist}(z, \sigma(T_0))$, we obtain

$$\begin{aligned} ||(T_0 - z)^{-1}|| &= \sup_{\lambda \in \sigma(T_0)} \frac{1}{|\lambda - z|} \\ &= \frac{1}{d} \end{aligned}$$

and

$$\begin{aligned} ||T_0(T_0 - z)^{-1}|| &= ||I + z(T_0 - z)^{-1}|| \\ &\leq 1 + \frac{|z|}{d}. \end{aligned}$$

For $|\varepsilon| < \frac{1}{q+\eta\beta}$ and by virtue of the proof of Theorem 10.3.2, we deduce that the operator $S := T(\varepsilon) - T_0$ satisfies

$$\|S\varphi\| \leq \frac{\eta|\varepsilon|}{1 - |\varepsilon|q}\|T_0\varphi\|^{\beta}\|\varphi\|^{1-\beta} + \frac{\|\Phi_\eta\||\varepsilon|}{1 - |\varepsilon|q}\|\varphi\|, \quad \forall \varphi \in \mathscr{D}(T_0). \tag{10.3.2}$$

So, the following inequalities hold

$$\begin{aligned} ||S(T_0 - z)^{-1}u|| &\leq \frac{\eta|\varepsilon|}{1 - |\varepsilon|q}||T_0(T_0 - z)^{-1}u||^{\beta}||(T_0 - z)^{-1}u||^{1-\beta} + \frac{\|\Phi_\eta\||\varepsilon|}{1 - |\varepsilon|q}||(T_0 - z)^{-1}u|| \\ &\leq \left[\frac{\eta|\varepsilon|}{1 - |\varepsilon|q}\left(1 + \frac{|z|}{d}\right)^{\beta}\frac{1}{d^{1-\beta}} + \frac{\|\Phi_\eta\||\varepsilon|}{d(1 - |\varepsilon|q)}\right]||u|| \end{aligned}$$

for every $u \in X$. To obtain the desired estimates, it suffices to show that

$$C = \frac{\eta|\varepsilon|}{1 - |\varepsilon|q}\left(1 + \frac{|z|}{d}\right)^{\beta}\frac{1}{d^{1-\beta}} + \frac{\|\Phi_\eta\||\varepsilon|}{d(1 - |\varepsilon|q)} \leq \theta.$$

Note that, in the case $\beta = 0$, we can suppose without loss of generality that $\|\Phi_\eta\| = 0$. Consider $z = x + iy$, the analysis of the geometry of the situation leads to the following:

$$\varphi_- \leq arg(z) \leq -\pi/2 \text{ or } \pi/2 \leq arg(z) \leq \varphi_+ \Rightarrow d \geq |z|, \tag{10.3.3}$$

$$\max\{\varphi_-, -\pi/2\} \leq arg(z) \leq \min\{\varphi_+, \pi/2\} \Rightarrow d \geq |y|. \tag{10.3.4}$$

Moreover, we obtain

$$\psi \leq |arg(z)| \leq \pi/2 \Rightarrow |y| \geq |z| \sin\psi, \tag{10.3.5}$$

$$|arg(z)| \leq \psi \Rightarrow x \geq |z| \cos\psi. \tag{10.3.6}$$

In what follows, three cases are discussed:

First case $z \in \rho_1$: If (10.3.3) is satisfied, together with the fact that $d \geq |z|$, for r_0 large enough, we obtain

$$C \leq 2^{\beta} \frac{\eta|\varepsilon|}{1 - |\varepsilon|q} |z|^{\beta - 1} + \frac{\|\Phi_\eta\||\varepsilon|}{(1 - |\varepsilon|q)|z|} \leq \theta.$$

If $\psi \leq |arg(z)| \leq \pi/2$, then by using both (10.3.4) and (10.3.5), we have $d \geq |z| \sin \psi$. So, we conclude, for r_0 large enough, that

$$C \leq \frac{\eta|\varepsilon|}{1 - |\varepsilon|q} \left(1 + \frac{1}{\sin \psi}\right)^{\beta} \frac{1}{(|z| \sin \psi)^{1-\beta}} + \frac{\|\Phi_\eta\||\varepsilon|}{(1 - |\varepsilon|q)|z| \sin \psi} \leq \theta.$$

Second case $z \in \rho_2$: From both (10.3.4), (10.3.6) and the inequality $|y| \geq \alpha x^{\beta}$, it is easy to see that $d \geq \alpha x^{\beta}$. On the one hand, in the case $\beta = 0$, then $d \geq \alpha$ and

$$C = \frac{\eta|\varepsilon|}{(1 - |\varepsilon|q)d} \leq \frac{\eta|\varepsilon|}{(1 - |\varepsilon|q)\alpha} < \theta.$$

Moreover, in the case $\beta > 0$, the Minkowski inequality yields

$$\begin{aligned} \left(1 + \frac{|z|}{d}\right)^{\beta} &\leq \left(1 + \frac{x + |y|}{d}\right)^{\beta} \\ &\leq 1 + \frac{x^{\beta} + |y|^{\beta}}{d^{\beta}}. \end{aligned}$$

Thus,

$$\begin{aligned} \left(1 + \frac{|z|}{d}\right)^{\beta} &\leq 1 + \frac{\alpha^{-1}d + d^{\beta}}{d^{\beta}} \\ &= 2 + \frac{1}{\alpha} d^{1-\beta}. \end{aligned}$$

i.e.,

$$C \leq 2\frac{\eta|\varepsilon|}{1 - |\varepsilon|q} d^{\beta - 1} + \frac{\|\Phi_\eta\||\varepsilon|}{1 - |\varepsilon|q} d^{-1} + \frac{\eta|\varepsilon|}{(1 - |\varepsilon|q)\alpha}.$$

Since $\frac{\eta|\varepsilon|}{(1-|\varepsilon|q)\alpha} < \theta$ and taking account of the fact that

$$d \geq \alpha(|z| \cos \psi)^{\beta},$$

we obtain $C \leq \theta$ for r_0 sufficiently large.

Third case $z \in \rho_3$: In this case, (10.3.4), (10.3.6) apply and, we obtain

$$d \geq \alpha x^{\beta}.$$

So, the argument used in the second case adapts easily.
To complete the proof of the lemma, we mention that if $z \in \rho_1$, then

$$d \geq |z| \geq r_0 > 0$$

or

$$d \geq |z| \sin \psi \geq r_0 \sin \psi > 0.$$

Furthermore, if $z \in \rho_2 \bigcup \rho_3$, then

$$\begin{aligned} d &\geq \alpha(|z| \cos \psi)^\beta \\ &\geq \alpha(r_0 \cos \psi)^\beta \\ &> 0. \end{aligned}$$

Hence, d^{-1} is bounded independently of z. So, we infer that $||(T(\varepsilon) - z)^{-1}||$ is uniformly bounded. Q.E.D.

Using the results of the previous lemma (see Lemma 10.3.1), the description of the spectrum of $T(\varepsilon)$ is given by the following.

Theorem 10.3.3 *Assume that the spectrum of T_0 lies on finitely many rays $e^{i\theta_j}\mathbb{R}_{\geq 0}$ with $0 \leq \theta_j < 2\pi$, $j = 1, \ldots, n$ and let $\eta > 0$, then for every $\alpha > \frac{\eta|\varepsilon|}{1-|\varepsilon|q}$, there exists $r_0 > 0$ such that the spectrum of $T(\varepsilon)$, $\sigma(T(\varepsilon))$, satisfies*

$$\sigma(T(\varepsilon)) \subset B_{r_0}(0) \bigcup \bigcup_{j=1}^{n} \{e^{i\theta_j}(x + iy);\ x \geq 0 \ \text{and} \ |y| \leq \alpha x^\beta\}$$

for $|\varepsilon|$ small enough. Moreover, $T(\varepsilon)$ has a compact resolvent. ◇

The proof of this theorem can be checked in the same ways as Theorem 8.1.1.

10.3.2 Riesz Basis of Subspaces

Theorem 10.3.4 *If the spectrum of T_0 lies on finitely many rays $e^{i\theta_j}\mathbb{R}_{\geq 0}$ with $0 \leq \theta_j < 2\pi$, $j = 1, \ldots, n$ and*

$$\liminf_{r \longrightarrow \infty} \frac{n(\mathbb{D}(0, r), T_0)}{r^{1-\beta}} < \infty,$$

then for $|\varepsilon|$ small enough, $T(\varepsilon)$ has a finitely spectral Riesz basis of subspaces $(V_k)_{k \in \mathbb{N}}$. ◇

Proof To prove this theorem, we shall distinguish the following two cases:

First case: $N(T_0) = \{0\}$: According to the first item of Proposition 2.7.2 and Eq. (10.3.2), we infer that S is β-subordinate to T_0. Hence, by using Sect. 8.2.5, we conclude that $T(\varepsilon)$ has a finitely spectral Riesz basis of subspaces $\{V_k\}_{k\in\mathbb{N}}$.
Second case: $N(T_0) \neq \{0\}$: Let P be the orthogonal projection onto $N(T_0)$ and, consider the operator

$$G_0 := T_0 + P.$$

This operator G_0 is normal with compact resolvent and satisfies also the growth condition. Consider now the operator

$$T(\varepsilon) = T_0 + S = G_0 + S - P.$$

The second item of Proposition 2.7.2 allows us to infer that $S - P$ is β-subordinate to G_0. Thus, Sect. 8.2.5 guarantees that $T(\varepsilon)$ has a finitely spectral Riesz basis of subspaces $\{V_k\}_{k\in\mathbb{N}}$. This achieves the proof. Q.E.D.

10.4 Riesz Basis in $L_2(0, T)$

In the next, consider the eigenvector problem of Eq. (9.1.2)

$$\begin{cases} T_0\varphi + \xi_1 T_1\varphi + \xi_2 T_2\varphi + \cdots + \xi_k T_k\varphi + \cdots = \lambda\varphi \\ \varphi \in \mathscr{D}(T_0). \end{cases}$$

We have the following.

Theorem 10.4.1 *Suppose that the assumptions* $(H1)$, $(H3)$ *and* $(H7)$ *with* $q = \beta = 1$ *hold. Then, the series*

$$\sum_{k\geq 0} \xi_k T_k\varphi$$

converges for every $\varphi \in \mathscr{D}(T_0)$. *If* $T(\xi)\varphi$ *denotes its limit, then we define a linear operator* $T(\xi)$ *with domain* $\mathscr{D}(T_0)$. *In addition, if* $\tau(\xi) < \frac{1}{\eta}$, *then* $T(\xi)$ *is closed.* ◇

Proof Let $\varphi \in \mathscr{D}(T_0)$ and $n \in \mathbb{N}^*$. By using Eq. (10.3.1), we obtain

$$\begin{aligned}\left\|\sum_{k=0}^{n}\xi_k T_k\varphi\right\| &\le \sum_{k=0}^{n}|\xi_k|\|T_k\varphi\| \\ &\le \|T_0\varphi\| + \sum_{k=1}^{n}|\xi_k|\|T_k\varphi\| \\ &\le \|T_0\varphi\| + \sum_{k=1}^{n}|\xi_k|\left(\eta\|T_0\varphi\| + \Phi_\eta(\varphi)\right) \\ &\le \tau(\xi)\Phi_\eta(\varphi) + (1+\eta\tau(\xi))\,\|T_0\varphi\|.\end{aligned}$$

So, the series

$$\sum_{k\ge 0}\xi_k T_k\varphi$$

is convergent. Setting $T(\xi)\varphi$ its limit, we define a linear operator $T(\xi)$ with domain $\mathscr{D}(T_0)$. Similarly, by using Eq. (10.3.1), we get

$$\|(T(\xi)-T_0)\varphi\| \le \tau(\xi)\Phi_\eta(\varphi) + \eta\tau(\xi)\|T_0\varphi\|, \quad \text{for all } \varphi \in \mathscr{D}(T_0).$$

Since T_0 is closed and $\tau(\xi) < \frac{1}{\eta}$, we infer, in view of Lemma 2.7.2, that $T(\xi)$ is also closed. Q.E.D.

We suppose the following:

$(H8)$ The family $\{e^{i\lambda_n t}\}_1^\infty$ associated to the operator T_0 forms a Riesz basis in $L_2(0, T)$, for some $T > 0$.

The main result in this section is formulated as follows.

Theorem 10.4.2 *Suppose that the hypotheses $(H1)$, $(H3)$, $(H7)$ with $q = \beta = 1$, and $(H8)$ hold. If $\{\lambda_n\}_1^\infty$ are isolated eigenvalues of T_0 with multiplicity one, then the exponential family*

$$\{e^{i\lambda_n(\xi)t}\}_1^\infty$$

forms a Riesz basis in the space $L_2(0, T)$, for some $T > 0$, where $(\lambda_n(\xi))_1^\infty$ is a sequence of eigenvalues of the perturbed transformation $T(\xi)$. ◇

Proof Let $n \in \mathbb{N}^*$, we denote by $\sigma_0 = \{\lambda_n\}$ and by $\sigma_1 = \sigma(T_0)\backslash\{\sigma_0\}$, where λ_n is the nth isolated eigenvalue, which multiplicity one, of the operator T_0. Let $\mathscr{C}_n = \mathscr{C}(\lambda_n, r_n)$ be the closed circle with center λ_n and radii r_n, lying in the resolvent set of T_0, and having σ_0 inside and σ_1 outside. Since $(\lambda_n - T_0)^{-1}$ is a regular analytic function of $\lambda_n \in \rho(T_0)$, then $\|(\lambda_n - T_0)^{-1}\|$ is continuous. Let's pose

$$M_n := \max_{z_n\in\mathscr{C}_n}\|(T_0 - z_n I)^{-1}\|,$$

$$\begin{aligned}N_n &:= \max_{z_n\in\mathscr{C}_n}\|T_0(T_0 - z_n I)^{-1}\| \\ &= \max_{z_n\in\mathscr{C}_n}\|I + z_n(T_0 - z_n I)^{-1}\|\end{aligned}$$

$$\alpha_n := \|\Phi_\eta\| M_n + \eta N_n,$$

and

$$w := \|\varphi_n\|,$$

where φ_n is an eigenvector of T_0 associated to the eigenvalue λ_n. If

$$\tau(\xi) < \frac{1}{\alpha_n(1 + r_n M_n \omega_n^2)},$$

then in view of Theorem 9.3.1, the perturbed transformation $T(\xi)$ will have a unique point of the spectrum inside the circle $\mathscr{C}_n = \mathscr{C}(\lambda_n, r_n)$ and the corresponding $\lambda_n(\xi)$ have the same multiplicity as λ_n. The eigenvalue $\lambda_n(\xi)$ can be developed into a series

$$\lambda_n(\xi) := \lambda_n + \lambda_{n,1}(\xi) + \lambda_{n,2}(\xi) + \cdots + \lambda_{n,i}(\xi) + \cdots, \tag{10.4.1}$$

where $\lambda_{n,i}$ satisfy the following estimation:

$$|\lambda_{n,i}(\xi)| \leq \frac{\omega_n^2 r_n^2 M_n (1 - \alpha_n \tau(\xi))}{1 - \omega_n^2 r_n M_n - \alpha_n \tau(\xi)} \left((\alpha_n \tau(\xi))^i - \left(\frac{\omega_n^2 r_n M_n \alpha_n \tau(\xi)}{1 - \alpha_n \tau(\xi)} \right)^i \right). \tag{10.4.2}$$

Since the family $\{e^{i\lambda_n t}\}_1^\infty$ forms a Riesz basis in $L_2(0, T)$, then in view of Proposition 7.6.1, we deduce that there exist two constants $A > 0$ and $B > 0$ such that

$$A\|u\|^2 \leq \sum_{n=1}^{+\infty} |\langle e^{i\lambda_n t}, u\rangle|^2 \leq B\|u\|^2, \quad \forall u \in L_2(0, T).$$

Furthermore, it follows from Theorem 7.6.1 that the sequence $\{\lambda_n\}_1^\infty$ lies in the strip parallel to the real axis. Hence, there exists a positive constant h such that for all $n \geq 1$, we have

$$|Im\lambda_n| \leq h,$$

where $h := \sup |Im\lambda_n|$. Now, let $n \in \mathbb{N}^*$. For all eigenvalue λ_n of T_0, we fix an element $\tau(\xi) \in \mathbb{C}$ such that

$$\tau(\xi) \in \left]0, \frac{(1 - \omega_n^2 r_n M_n)A}{\beta w_n^2 r_n^2 M_n \alpha_n n \sqrt{BT} t e^{t r_1} + \alpha_n A}\right[,$$

where $\beta^2 := \sum_{k=1}^{\infty} \frac{1}{k^2}$ and $r_n' := r_n + h$. It is easy to see that

$$\left]0, \frac{(1 - \omega_n^2 r_n M_n)A}{\beta w_n^2 r_n^2 M \alpha_n n \sqrt{BT} t e^{t r_n'} + \alpha_n A}\right[\subset \left]0, \frac{1}{\alpha_n(1 + w_n^2 r_n M_n)}\right[.$$

In view of Eq. (10.4.1), we get

$$\begin{aligned}
|e^{i\lambda_n(\xi)t} - e^{i\lambda_n t}| &= \left|e^{i\lambda_n t}(e^{i(\lambda_n(\xi)-\lambda_n)t} - 1)\right| \\
&\leq \left|e^{i\lambda_n t}\right| \left|e^{i(\lambda_n(\xi)-\lambda_n)t} - 1\right| \\
&\leq e^{|Im\lambda_n|t} \left|e^{i(\lambda_n(\xi)-\lambda_n)\frac{t}{2}}\right| \left|e^{i(\lambda_n(\xi)-\lambda_n)\frac{t}{2}} - e^{-i(\lambda_n(\xi)-\lambda_n)\frac{t}{2}}\right| \\
&\leq 2e^{|Im\lambda_n|t} e^{-Im((\lambda_n(\xi)-\lambda_n)\frac{t}{2})} \left|\sin\left((\lambda_n(\xi) - \lambda_n)\frac{t}{2}\right)\right| \\
&\leq 2e^{ht} e^{|Im((\lambda_n(\xi)-\lambda_n)t)|} \left|(\lambda_n(\xi) - \lambda_n)\frac{t}{2}\right| \\
&\leq e^{ht} e^{|Im((\lambda_n(\xi)-\lambda_n)t)|} |\lambda_n(\xi) - \lambda_n| t.
\end{aligned}$$

Since the eigenvalue $\lambda_n(\xi)$ of $T(\xi)$ is inside the circle C_n, then

$$e^{|Im((\lambda_n(\xi)-\lambda_n)t)|} \leq e^{tr_n}.$$

Now, the estimate (10.4.2) allows us to conclude the following

$$\begin{aligned}
|e^{i\lambda_n(\xi)t} - e^{i\lambda_n t}| &\leq te^{tr'_n} \sum_{i=1}^{\infty} |\lambda_{n,i}(\xi)| \\
&\leq te^{tr'_n} \frac{\omega_n^2 r_n^2 M_n (1 - \alpha_n \tau(\xi))}{1 - \omega_n^2 r_n M_n - \alpha_n \tau(\xi)} \sum_{i=1}^{\infty} (\alpha_n \tau(\xi))^i - \left(\frac{\omega_n^2 r_n M_n \alpha_n \tau(\xi)}{1 - \alpha_n \tau(\xi)}\right)^i \\
&\leq te^{tr'_n} \frac{\omega_n^2 r_n^2 M_n (1 - \alpha_n \tau(\xi))}{1 - \omega_n^2 r_n M_n - \alpha_n \tau(\xi)} \left(\sum_{i=1}^{\infty} (\alpha_n \tau(\xi))^i - \sum_{i=1}^{\infty} \left(\frac{\omega_n^2 r_n M_n \alpha_n \tau(\xi)}{1 - \alpha_n \tau(\xi)}\right)^i\right) \\
&\leq te^{tr'_n} \frac{\omega_n^2 r_n^2 M_n (1 - \alpha_n \tau(\xi))}{1 - \omega_n^2 r_n M_n - \alpha_n \tau(\xi)} \left(\frac{\alpha_n \tau(\xi)}{1 - \alpha_n \tau(\xi)} - \frac{\omega_n^2 r_n M_n \alpha_n \tau(\xi)}{1 - \alpha_n \tau(\xi) - \omega_n^2 r_n M_n \alpha_n \tau(\xi)}\right) \\
&\leq \frac{te^{tr'_n} \omega_n^2 r_n^2 M_n \alpha_n \tau(\xi)}{1 - \omega_n^2 r_n M_n - \alpha_n \tau(\xi)} \\
&< \frac{A}{\beta n \sqrt{TB}}.
\end{aligned}$$

Thus,

$$\begin{aligned}
\|e^{i\lambda_n(\xi)t} - e^{i\lambda_n t}\|^2 &= \int_0^T |e^{i\lambda_n(\xi)t} - e^{i\lambda_n t}|^2 \, dt \\
&< \int_0^T \frac{A^2}{\beta^2 n^2 TB} \, dt \\
&= \frac{A^2}{\beta^2 n^2 B}.
\end{aligned}$$

Then, we infer that

$$\sum_{n=1}^{\infty} \|e^{i\lambda_n(\xi)t} - e^{i\lambda_n t}\|^2 < \sum_{n=1}^{\infty} \frac{A^2}{\beta^2 n^2 B} = \frac{A^2}{B}.$$

Now, by using Theorem 7.6.4, we can deduce that the family $\{e^{i\lambda_n(\xi)t}\}_1^\infty$ forms a Riesz basis in the space $L_2(0, T)$ for some $T > 0$. Q.E.D.

Taking account of the Notes and Remarks 10.5 (iv), given at the end of this chapter, we are going to check the Riesz basis exponential family of the perturbed transformation $T(\xi)$ by applying the general results stated in Theorem 10.4.2.

Theorem 10.4.3 *Suppose that hypotheses $(H1)$, $(H3)$, $(H7)$ with $q = \beta = 1$ hold and $\{\lambda_n\}_1^\infty$ are isolated eigenvalues of T_0 with multiplicity one. If the family $\{e^{if(\lambda_n)t}\}_1^\infty$ forms a Riesz basis in the space $L_2(0, T)$, $T > 0$, then the exponential family*

$$\{e^{if(\lambda_n(\xi))t}\}_1^\infty$$

forms a Riesz basis in the space $L_2(0, T)$, for some $T > 0$, where $(\lambda_n(\xi))_1^\infty$ is a sequence of eigenvalues of the perturbed transformation $T(\xi)$. ◇

Proof First, note that

$$\begin{aligned} |e^{if(\lambda_n(\xi))t} - e^{if(\lambda_n)t}| &= |e^{if(\lambda_n)t}(e^{i(f(\lambda_n(\xi))-f(\lambda_n))t} - 1)| \\ &\le t|f(\lambda_n(\xi)) - f(\lambda_n)| \\ &\le tH|\lambda_n(\xi) - \lambda_n|. \end{aligned}$$

Choosing

$$\left]0, \frac{(1-\omega_n^2 r_n M_n)A}{\beta w_n^2 r_n^2 M_n H\alpha_n n\sqrt{BT} t e^{tr_1} + \alpha_n A}\right[\subset \left]0, \frac{1}{\alpha_n(1 + w_n^2 r_n M_n)}\right[,$$

and by using the same reasoning as the proof of Theorem 10.4.2, we get the desired result. Q.E.D.

10.5 Notes and Remarks

- During the second half of the twentieth and early twenty-first centuries, bases in Banach spaces, called Schauder basis, drew a lot of attention. One of the reasons is its connection with solutions of the problems of heat conductivity and the theory of oscillations. Bari [5, 6] placed for consideration a class of bases in the Hilbert space and called them the Riesz bases. Grothendieck [21] investigated Banach spaces that have approximation property. Enflo [15] constructed reflexive separable Banach space, where basis does not exist. Kostyuchenko (1953) demonstrated Riesz's basisness in $L_2[0, \pi]$ of the system of eigenvectors of Sturm-Liouville operator. Ilyin (1986) started to study necessary and sufficient conditions for Riesz's basisness of the system of eigen and adjoint functions for ordinary differential operators of any order.

- Later, the basis of eigenvectors property was confirmed for the operator $A(\varepsilon)$ in order to describe the radiation of a vibrating structure in a light fluid, and to study the strong interactions in the context of the Reggeon theory (see [1–4, 7–10, 12–14, 16–18, 22–25]). After that [11], it has been proved that the family of exponentials associated with the eigenvalues of the operator $A(\varepsilon)$ forms a Riesz basis. This result was of importance for application to a context of non-self-adjoint problem deduced from perturbation method for sound radiation.
- In this context, Jeribi and Intissar in [23] have exploited the results obtained by Nagy [29] to give a positive answer to some positive open questions posed by Filippi et al. in [19] by studying the existence of a Riesz basis in the Hilbert space $L_2\big(]-L, L[\big)$, formed by the eigenvectors of the following integrodifferential operator which will be more detailed in the last part of this chapter:

$$\big(I + \varepsilon K\big)^{-1} \frac{d^4}{dx^4},$$

where $L > 0$ and K is an integral operator with kernel the Hankel function of the first kind and order 0.

- Later, Ben Ali and Jeribi [7] pursued the analysis started in [26] aiming at extending it to the general case by proving the Riesz basis property of the eigenvectors of the system operator

$$T(\varepsilon) := T_0 + \varepsilon T_1 + \varepsilon^2 T_2 + \cdots + \varepsilon^k T_k + \cdots,$$

where $\varepsilon \in \mathbb{C}$ and $T_0, T_1, T_2, \ldots$ are linear operators on X, having the same domain $\mathscr{D}$ and satisfying the following condition:

$$\|T_k\varphi\| \leq q^{k-1}(a\|\varphi\| + b\|T_0\varphi\|)$$

for all $\varphi \in \mathscr{D}$ and for all $k \geq 1$, where $a, b, q > 0$. More precisely, they proved that the family $(\varphi_n(\varepsilon_n))_{n=1}^{\infty}$, corresponding as a sequence of eigenvectors of $T(\varepsilon_n)$ and can be developed as an entire series of ε, forms a Riesz basis in X. As application, they have studied the problem of radiation produced by a vibrating structure in a light fluid, given in [28].

- Recently, Feki et al. in [17, 18] have tried to formulate some supplements to the results developed in [7, 29] which occur in application to a problem of radiation of a vibrating structure in a light fluid and in Reggeon field theory due to Gribov [20]. More specially, they have studied the operator $T(\varepsilon)$, where $\varepsilon \in \mathbb{C}$, T_0 is a closed linear operator on a separable Banach space X with dense domain $\mathscr{D}(T_0)$, while T_1, T_2, $T_3 \dots$ are linear operators on X with the same domain $\mathscr{D} \supset \mathscr{D}(T_0)$ and satisfying the following inequality

$$\|T_k\varphi\| \le q^{k-1}(a\|\varphi\| + b\|T_0\varphi\|^\beta \|\varphi\|^{1-\beta})$$

for all $\varphi \in \mathscr{D}(T_0)$ and for all $k \ge 1$, with $\beta \in]0, 1]$ and a, b, $q > 0$. This new condition enables the authors in [17] to give a considerable improvement to some estimations related to the eigenvalues and the eigenvectors of the perturbed operator.

- Based on the obtained estimations, they have investigated under sufficient conditions assuring the completeness of the system of eigenvectors of the operator $T(\varepsilon)$. Furthermore, they have extended the main result in [7] to the existence of Schauder basis in any Banach space.

(i) If the eigenvalues $(\lambda_n)_{n=1}^{\infty}$ of the operator T_0 are with simple multiplicity ordered as an ascending sequence in modulus and, satisfying the condition

$$\lim_{n\to+\infty} |\lambda_{n+1}| - |\lambda_n| = +\infty,$$

then the condition (10.2.6) is verified.

(ii) The Riesz basis $\{\varphi_n(\varepsilon)\}_{n=1}^{\infty}$ obtained in Theorem 10.2.6 is formed by eigenvectors of the operator $T(\varepsilon)$, for all $|\varepsilon| \in]0, D[$.

(iii) Theorem 10.2.6 provides a positive answer to the open problem posed in [19] by Filippi et al.

(iv) In many applications, one expects that the hypothesis $(H8)$ is not generally true and we don't order to check the Riesz basis property of the exponential family $\{e^{i\lambda_n t}\}_1^\infty$ in $L_2(0, T)$, for some $T > 0$. More precisely, in some applications, it may happen that the family $\{f(\lambda_n)\}_1^\infty$ is the set of zeros of a function of sine type, where f is a Lipschitz function with a positive constant H, i.e., for each x and y elements of $[c, +\infty[$, $c > 0$, there is a positive constant H such that

$$|f(x) - f(y)| \le H|x - y|.$$

References

1. M.T. Aimar, A. Intissar, J.M. Paoli, Densité des vecteurs propres généralisés d'une classe d'opérateurs compacts non auto-adjoints et applications. Commun. Math. Phys. **156**, 169–177 (1993)
2. M.T. Aimar, A. Intissar, J.M. Paoli, Quelques nouvelles propriétés de régularité de l'opérateur de Gribov. Commun. Math. Phys. **172**(3), 461–466 (1995)
3. M.T. Aimar, A. Intissar, J.M. Paoli, Critères de complétude des vecteurs propres généralisés d'une classe d'opérateurs non auto-adjoints compacts ou à résolvante compacte et applications. Publ. Res. Inst. Math. Sci. **32**(2), 191–205 (1996)
4. M.T. Aimar, A. Intissar, A. Jeribi, On an unconditional basis of generalized eigenvectors of the non-self-adjoint Gribov operator in Bargmann space. J. Math. Anal. Appl. **231**, 588–602 (1999)

5. N.K. Bari, Sur les bases dans l'espace de Hilbert. Dokl. Akad. Nauk. SSSR **54**, 379–382 (1946)
6. N.K. Bari, Biorthogonal systems and bases in Hilbert space. Uchen. Zap. Moskov. Gos. Univ. **148**(4), 69–107 (1951) (in Russian); Math. Rev. **14**, 289 (1953)
7. N. Ben Ali and A. Jeribi, On the Riesz basis of a family of analytic operators in the sens of Kato and application to the problem of radiation of a vibriting structure in a light fluid. J. Math. Anal. Appl. **320**(1), 78-94 (2006)
8. S. Charfi, A. Damergi, A. Jeribi, On a Riesz basis of finite-dimensional invariant subspaces and application to Gribov operator in Bargmann space. Line. Mult. Alge. **61**(11), 1577–1591 (2013)
9. S. Charfi, A. Intissar, A. Jeribi, Expansion of solution in terms of generalized eigenvectors for a rectilinear transport equation. Transport Theory Statist. Phys. **38**, 1–16 (2009)
10. S. Charfi, A. Jeribi, I. Walha, Riesz basis property of families of nonharmonic exponential and application to a problem of a radiation of a vibrating structure in a light fluid. Numerical Functional Analysis and Optimization, **32**(4), 370–382 (2011)
11. H. Ellouz, I. Feki, A. Jeribi, On a Riesz basis of exponentials related to the eigenvalues of an analytic operator and application to a non-self-adjoint problem deduced from a perturbation method. J. Math. Phys. **54**, 112101 (2013)
12. H. Ellouz, I. Feki, A. Jeribi, On the asymptotic behavior of the eigenvalues of an analytic operator in the sense of Kato and applications, Serdica J. Math. **45**(1), 55–88 (2019)
13. H. Ellouz, I. Feki, A. Jeribi, On a Riesz basis of exponentials related to a family of analytic operators and application. J. Pseudo-Differ. Oper. Appl. **10**(4), 999–1014 (2019)
14. H. Ellouz, I. Feki, A. Jeribi, Non-orthogonal Fusion Frames of an Analytic Operator and Application to a One-Dimensional Wave Control System. Mediterr. J. Math. **16**(2), 16:52 (2019)
15. P. Enflo, A counterexample to the approximation problem in Banach spaces. Acta Math. **130**(3–4), 309–317 (1973)
16. I. Feki, A. Jeribi, R. Sfaxi, On an unconditionl basis of generalized eigenvectors of an analytic operator and application to a problem of radiation of a vibrating structure in a light fluid. J. Math. Anal. Appl. **375**, 261–269 (2011)
17. I. Feki, A. Jeribi, R. Sfaxi, On a Schauder basis related to the eigenvectors of a family of non-self-adjoint analytic operators and applications. Anal. Math. Phys. **3**(4), 311–331 (2013)
18. I. Feki, A. Jeribi, R. Sfaxi, On a Riesz basis of eigenvectors of a non-self-adjoint analytic operator and applications. Line. Mult. Alge. **62**(8), 1049–1068 (2014)
19. P.J.T. Filippi, O. Lagarrigue, P.O. Mattei, Perturbation method for sound radiation by a vibrating plate in a light fluid: comparison with the exact solution. J. Sound Vib. **177**, 259–275 (1994)
20. V.N. Gribov, A Reggon diagram technique. J. E. T. P. Sov. Phys. **26**, 414–423 (1968)
21. A. Grothendieck, Produits tensoriels topologiques et espaces nucleaires. Mem. Am. Math. Soc. **16**, 13–30 (1955)
22. A. Intissar, Sur une méthode de perturbation, C. R. Acad. Sci. Paris, Série 1, Mathématique **304**(4), 95–98 (1987)
23. A. Intissar, Analyse fonctionnelle et théorie spectrale pour les opérateurs compacts non-autoadjoints, Editions CEPADUES (1997)
24. A. Intissar, A. Jeribi, I. Walha, Riesz basis of exponential family for a hyperbolic system. J. Appl. Anal. **25**(1), 13–23 (2019)
25. A. Jeribi, *Denseness, Bases and Frames in Banach Spaces and Applications* (de Gruyter, Berlin, 2018)
26. A. Jeribi, A. Intissar, On an Riesz basis of generalized eigenvectors of the non-self-adjoint problem deduced from a perturbation method for sound radiation by a vibrating plate in a light fluid. J. Math. Anal. Appl. **292**, 1–16 (2004)
27. T. Kato, *Perturbation Theory for Linear Operators* (Springer, Berlin, 1980)
28. P.M. Morse, H. Feshbach, *Methods of Theorical Physics* (Book co. Inc, New-York-Toronto-London, 1953)
29. B. Sz, Nagy, Perturbations des transformations linéaires fermées. Acta Sci. Math. Szeged. **14**, 125–137 (1951)

Chapter 11
On the Asymptotic Behavior of the Eigenvalues of an Analytic Operator in the Sense of Kato

The basic idea of this chapter is to derive a precise description, on a separable Hilbert space X, to the behavior of the spectrum of a self-adjoint operator T_0 after a perturbation by an infinite sum of operators

$$B(\varepsilon) := \varepsilon T_1 + \varepsilon^2 T_2 + \cdots + \varepsilon^k T_k + \cdots ,$$

where $\varepsilon \in \mathbb{C}$ and $T_1, T_2, T_3 \ldots$ are linear operators on the space X having the same domain $\mathscr{D} \supset \mathscr{D}(T_0)$ and satisfying a specific growing inequality, while the spectrum of T_0 is discrete and its eigenvalues are not condensed.

11.1 Perturbation of T_0

Consider the self-adjoint operator with compact resolvent T_0 and

$$T_0 f = \sum_{n=1}^{\infty} \lambda_n \langle f, \varphi_n \rangle \varphi_n$$

its spectral decomposition, where λ_n is the nth eigenvalue of T_0 associated to the eigenvector φ_n. Denoting by

$$K_r := 2l \sum_{r-2l<\lambda_n<r} \langle \cdot, \varphi_n \rangle \varphi_n - 2l \sum_{r\leq\lambda_n<r+2l} \langle \cdot, \varphi_n \rangle \varphi_n,$$

with $r > 4l > 0$, and $l \in \mathbb{N}^*$. Then, K_r is a finite rank operator with

A. Jeribi, *Perturbation Theory for Linear Operators*,
https://doi.org/10.1007/978-981-16-2528-2_11

$$rk(K_r) \leq 4lp.$$

Let us consider $T_r := T_0 - K_r$.

Lemma 11.1.1 *The operator T_r is positive and self-adjoint.* ◇

Proof First, we have $\mathscr{D}(T_r) = \mathscr{D}(T_0)$, and let $f \in \mathscr{D}(T_0)$. Then,

$$\begin{aligned}\langle T_r f, f\rangle &= \langle (T_0 - K_r)f, f\rangle \\ &= \sum_{n=1}^{\infty} \lambda_n \langle f, \varphi_n\rangle\langle \varphi_n, f\rangle - 2l \sum_{r-2l<\lambda_n<r} \langle f, \varphi_n\rangle\langle \varphi_n, f\rangle + 2l \sum_{r\leq\lambda_n<r+2l} \langle f, \varphi_n\rangle\langle \varphi_n, f\rangle \\ &= \sum_{n=1}^{\infty} \lambda_n |\langle f, \varphi_n\rangle|^2 - 2l \sum_{r-2l<\lambda_n<r} |\langle f, \varphi_n\rangle|^2 + 2l \sum_{r\leq\lambda_n<r+2l} |\langle f, \varphi_n\rangle|^2.\end{aligned}$$

Hence,

$$\begin{aligned}\langle T_r f, f\rangle = &\sum_{\lambda_n\leq r-2l} \lambda_n |\langle f, \varphi_n\rangle|^2 + \sum_{r-2l<\lambda_n<r} (\lambda_n - 2l)|\langle f, \varphi_n\rangle|^2 + \\ &\sum_{r\leq\lambda_n<r+2l} (\lambda_n + 2l)|\langle f, \varphi_n\rangle|^2 + \sum_{\lambda_n\geq r+2l} \lambda_n |\langle f, \varphi_n\rangle|^2 \\ &> 0 \ \text{(since } r > 4l).\end{aligned}$$

So, T_r is a positive operator. Moreover, let $f, g \in \mathscr{D}(T_0)$. Since T_0 is self-adjoint, we obtain

$$\begin{aligned}\langle T_r f, g\rangle &= \langle (T_0 - K_r)f, g\rangle \\ &= \sum_{n=1}^{\infty} \lambda_n \langle f, \varphi_n\rangle\langle \varphi_n, g\rangle - 2l \sum_{r-2l<\lambda_n<r} \langle f, \varphi_n\rangle\langle \varphi_n, g\rangle + 2l \sum_{r\leq\lambda_n<r+2l} \langle f, \varphi_n\rangle\langle \varphi_n, g\rangle \\ &= \sum_{n=1}^{\infty} \lambda_n \langle f, \varphi_n\rangle\overline{\langle g, \varphi_n\rangle} - 2l \sum_{r-2l<\lambda_n<r} \langle f, \varphi_n\rangle\overline{\langle g, \varphi_n\rangle} + 2l \sum_{r\leq\lambda_n<r+2l} \langle f, \varphi_n\rangle\overline{\langle g, \varphi_n\rangle} \\ &= \langle f, T_r g\rangle.\end{aligned}$$

Consequently, T_r is a self-adjoint operator. Q.E.D.

On the other hand, T_r preserves the system of eigenvectors $(\varphi_n)_n$ and the eigenvalues $(\lambda_n)_n$ lying outside the interval $]r - 2l, r + 2l[$ and changes only the $(\lambda_n)_n$ lying inside the interval (see Fig. 11.1).

More precisely, if we denote by $(\nu_n)_n$ the eigenvalues of T_r, numbered in the increasing order, we have

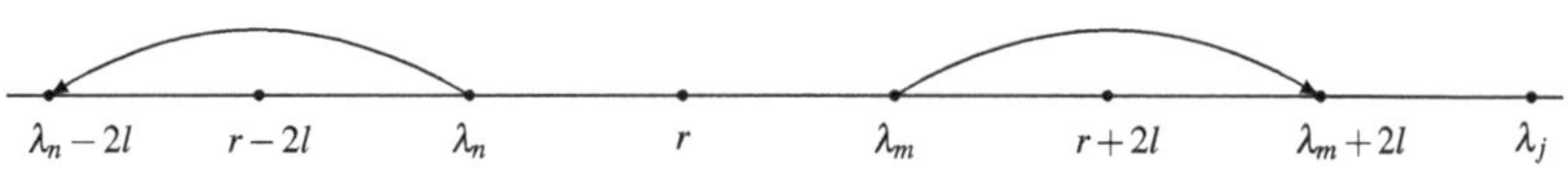

Fig. 11.1 Intervals of λ_n

$$\nu_n = \begin{cases} \lambda_n - 2l & \text{if} \quad \lambda_n \in]r-2l, r[\\ \lambda_n + 2l & \text{if} \quad \lambda_n \in [r, r+2l[\\ \lambda_n & \text{if} \quad \lambda_n \notin]r-2l, r+2l[. \end{cases}$$

So, $]r-2l, r+2l[$ doesn't contain any eigenvalues ν_n of the operator T_r.

Theorem 11.1.1 *The eigenvalues* $(\nu_n)_n$ *of the operator* T_r *verify the following* $\nu_{n+2p} - \nu_n \geq 1$, $n = 1, 2, \ldots$. ◇

Proof Let $k \in \mathbb{N}$. We will discuss three cases.

(i) First case: $[k, k+1[\bigcap]r-2l, r+2l[= \emptyset$.
(ii) Second case: $[k, k+1[\bigcap]r-2l, r+2l[\neq \emptyset$.
(iii) Third case: $[k, k+1[\bigcap]r-2l, r+2l[= [k, k+1[$.

(i) First case $[k, k+1[\bigcap]r-2l, r+2l[= \emptyset$: The fact that $[k, k+1[\bigcap]r-2l, r+2l[= \emptyset$, we have $[k, k+1[$ contains at most p eigenvalues of the operator T_r having the form $\nu_n = \lambda_n$.
(ii) Second case $[k, k+1[\bigcap]r-2l, r+2l[\neq \emptyset$: First, case where $[k, k+1[\bigcap]r-2l, r+2l[= [k, r+2l[$. The interval $[k, k+1[$ contains at most p eigenvalues of T_0, then we will denote by p_1, (respectively, p_2) the number of eigenvalues of T_0 contained in the interval $[k, r+2l[$ (respectively, $[r+2l, k+1[$) such that $p_1 + p_2 \leq p$. The fact that $[k, r+2l[$ contains at most p_1 eigenvalues of T_0, then

$$[k+2l, r+4l[\subset [k+2l, k+2l+1[$$

contains also at most p_1 eigenvalues of T_r having the following form $\nu_n = \lambda_n + 2l$. Moreover, as $[k+2l, k+2l+1[$ contains at most p eigenvalues of T_0 and $[k+2l, k+2l+1[\bigcap]r-2l, r+2l[= \emptyset$, then $[k+2l, k+2l+1[$ contains at most p eigenvalues of T_r having the following form $\nu_n = \lambda_n$. Hence, the interval $[k+2l, k+2l+1[$ contains at most $p + p_1 \leq 2p$ eigenvalues of T_r. Moreover, it suffices to notice that $[r+2l, k+1[\bigcap]r-2l, r+2l[= \emptyset$ and $[r+2l, k+1[\subset [r+2l, r+2l+1[$ to conclude that $[r+2l, r+2l+1[$ contains at most $p \geq p_2$ eigenvalues of T_r having the following form $\nu_n = \lambda_n$. Second, case where $[k, k+1[\bigcap]r-2l, r+2l[=]r-2l, k+1[$. We make the same reasoning as above allows us to conclude.
(iii) Third case $[k, k+1[\bigcap]r-2l, r+2l[= [k, k+1[$: Case where $[k, k+1[\subset] r-2l, r]$. The fact that $[k, k+1[$ contains at most p eigenvalues of T_0, then the interval $[k-2l, k-2l+1[$ contains also at most p eigenvalues of the operator T_r having the form $\nu_n = \lambda_n - 2l$. Furthermore, as $[k-2l, k-2l+1[$ contains at most p eigenvalues of T_0 and $[k-2l, k-2l+1[\bigcap]r-2l, r+2l[= \emptyset$, we have $[k-2l, k-2l+1[$ contains at most p eigenvalues of T_r having the following form $\nu_n = \lambda_n$. Hence, the interval $[k-2l, k-2l+1[$ contains at most $2p$ eigenvalues of T_r. For the case of $[k, k+1[\subset [r, r+2l[$, we use the same reasoning as above with $[k+2l, k+2l+1[$ instead of $[k-2l, k-2l+1[$ and $\nu_n = \lambda_n + 2l$ instead of $\nu_n = \lambda_n - 2l$. If $r \in [k, k+1[\subset]r-2l, r+2l[$, then

we designate by p_1, (respectively, p_2) the number of eigenvalues of T_0 contained in the interval $[k, r[$, (respectively, $[r, k+1[$) such that $p_1 + p_2 \leq p$ since $[k, k+1[$ contains at most p eigenvalues of T_0. As $[r, k+1[$ contains at most p_2 eigenvalues of T_0, we have

$$[r + 2l, k + 1 + 2l[\subset [r + 2l, r + 2l + 1[$$

contains also at most p_2 eigenvalues of T_r having the following form $\nu_n = \lambda_n + 2l$. Moreover, the fact that $[r + 2l, r + 2l + 1[$ contains at most p eigenvalues of T_r having the form $\nu_n = \lambda_n$, then the interval $[r + 2l, r + 2l + 1[$ contains at most $p + p_2 \leq 2p$ eigenvalues of the operator T_r. For the other case, we use the same reasoning as above with p_1 instead of p_2, the interval $[k - 2l, r - 2l[$ instead of the interval $[r + 2l, k + 1 + 2l[$, the interval $[r - 2l - 1, r - 2l[$ instead of the interval $[r + 2l, r + 2l + 1[$ and the element $\nu_n = \lambda_n - 2l$ instead of the element $\nu_n = \lambda_n + 2l$. Q.E.D.

11.2 Behavior of the Spectrum of Perturbed Operator $T(\varepsilon)$ Under a Finite Rank Perturbation

Let X be a separable Hilbert space and let T_0 be an operator satisfying the following hypotheses:
$(H9)$ T_0 is self-adjoint, positive and with domain $\mathscr{D}(T_0)$ dense on X.
$(H10)$ The eigenvalues $(\lambda_n)_{n=1}^{\infty}$ of T_0 do not condense, i.e., $\lambda_{n+p} - \lambda_n \geq 1$, for some $p \in \mathbb{N}^*$.

11.2.1 Discrete Spectrum

In the sequel, we will denote by

$$B(\varepsilon) := \sum_{i \geq 1} \varepsilon^i T_i. \tag{11.2.1}$$

The first result of this section is formulated in the following proposition.

Proposition 11.2.1 *Assume that hypotheses $(H9)$, $(H10)$ and $(H2)$ with $\beta \in]0, \frac{1}{2}[$ hold. Then, for $|\varepsilon| < \frac{1}{q+\beta b}$, we have*

(i) the operator $B(\varepsilon)$ is T_0-compact. Further, the operator $T(\varepsilon)$ is with compact resolvent, where $T(\varepsilon)$ is defined in Eq. (10.1.1), and
(ii) the operator $B(\varepsilon)$ is T_r-compact. Moreover, the operator $T(\varepsilon) - K_r$ is with compact resolvent. ◇

Proof (i) Let $n \in \mathbb{N}^*$ and let λ_n be the nth eigenvalue of T_0. By using the hypothesis $(H2)$, we obtain

$$\begin{aligned}\|B(\varepsilon)\varphi_n\| &= \|(\varepsilon T_1 + \varepsilon^2 T_2 + \cdots)\varphi_n\| \\ &\le \sum_{i=1}^{\infty} \|\varepsilon^i T_i \varphi_n\| \\ &\le \sum_{i=1}^{\infty} |\varepsilon|^i q^{i-1}(a\|\varphi_n\| + b\|T_0\varphi_n\|^{\beta}\|\varphi_n\|^{1-\beta}) \\ &\le \sum_{i=1}^{\infty} |\varepsilon|^i q^{i-1}(a + b\lambda_n^{\beta}). \end{aligned} \tag{11.2.2}$$

Hence, for $|\varepsilon| < \frac{1}{q}$, Eq. (11.2.2) implies that

$$\|B(\varepsilon)\varphi_n\| \le \frac{|\varepsilon|}{1 - |\varepsilon|q}\,(a + b\lambda_n^{\beta}). \tag{11.2.3}$$

So, for $|\varepsilon| < \frac{1}{q}$, we get

$$\frac{\|B(\varepsilon)\varphi_n\|^2}{\lambda_n^2} \le \frac{|\varepsilon|^2}{(1 - |\varepsilon|q)^2}\frac{(a + b\lambda_n^{\beta})^2}{\lambda_n^2}.$$

Taking into account hypothesis $(H10)$, we infer that

$$\begin{aligned}\lambda_{n+1} - \lambda_1 &= \underbrace{\lambda_{n+1} - \lambda_{(n+1)-p}}_{\ge 1} + \underbrace{\lambda_{(n+1)-p} - \lambda_{(n+1)-2p}}_{\ge 1} + \cdots + \underbrace{\lambda_{1+p} - \lambda_{(n+1)-\frac{n}{p}p(=1)}}_{\ge 1} \\ &\ge \frac{n}{p}. \end{aligned} \tag{11.2.4}$$

Consequently, Eq. (11.2.4) yields

$$\lambda_n \ge \frac{n-1}{p} + \lambda_1.$$

On the other hand, writing $\frac{(a+b\lambda_n^{\beta})^2}{\lambda_n^2}$ as

$$\frac{(a + b\lambda_n^{\beta})^2}{\lambda_n^2} = \frac{a^2}{\lambda_n^2} + \frac{2ab\lambda_n^{\beta}}{\lambda_n^2} + \frac{b^2\lambda_n^{2\beta}}{\lambda_n^2},$$

we conclude that the series

$$\sum_n \frac{2ab\lambda_n^{\beta}}{\lambda_n^2}$$

and

$$\sum_n \frac{b^2\lambda_n^{2\beta}}{\lambda_n^2}$$

are convergent since $\beta \in]0, \frac{1}{2}[$. Hence,

$$\sum_n \frac{\| B(\varepsilon)\varphi_n \|^2}{\lambda_n^2}$$

is also convergent. So,

$$\sum_{n=1}^{\infty} \frac{\| B(\varepsilon)\varphi_n \|^2}{\lambda_n^2} < \infty. \tag{11.2.5}$$

In view of both Eqs. (11.2.5) and (1.15.3), the operator $B(\varepsilon)$ is T_0-compact and, from Eq. (1.15.4), the spectrum of the operator $T(\varepsilon)$ is discrete for $|\varepsilon| < \frac{1}{q+\beta b}$. Furthermore, since T_0 is self-adjoint with compact resolvent and its spectrum is positive, then by using Lemma 1.13.3, the operator $T(\varepsilon)$ has compact resolvent.

(ii) Making the same reasoning as the proof in (i), we get $B(\varepsilon)$ is T_r-compact and $T(\varepsilon) - K_r$ has compact resolvent. Q.E.D.

Corollary 11.2.1 *Assume that hypotheses* $(H2)$ *with* $\beta \in]0, \frac{1}{2}[$ *and* $(H9)$ *are verified. If*

$$\lambda_{n+p}^{1-\alpha} - \lambda_n^{1-\alpha} \geq 1, \tag{11.2.6}$$

where $0 \leq \alpha < 1$. *Then, for* $\beta \in]0, 1 + \frac{\alpha-1}{2}[$ *and* $|\varepsilon| < \frac{1}{q+\beta b}$, *the spectrum of the operator* $T(\varepsilon)$ *is discrete.* ◇

Proof By using Eq. (11.2.6), we get

$$\lambda_n^{1-\alpha} \geq \frac{n-1}{p} + \lambda_1^{1-\alpha}.$$

Then, the series

$$\sum_n \frac{2ab\lambda_n^{\beta}}{\lambda_n^2}$$

and

$$\sum_n \frac{b^2\lambda_n^{2\beta}}{\lambda_n^2}$$

are convergent since $\beta \in]0, 1 + \frac{\alpha-1}{2}[$. Moreover, making the same reasoning as the proof of Proposition 11.2.1, we infer that the spectrum of the operator $T(\varepsilon)$ is discrete for $|\varepsilon| < \frac{1}{q+\beta b}$. Q.E.D.

11.2.2 Estimate Norm

Lemma 11.2.1 *For $|\varepsilon| < \frac{1}{q}$ and $r - l \leq Re\lambda \leq r + l$, there exists a positive constant $M(\varepsilon, a, 2p, q, l)$ such that*

$$\sum_{n=1}^{\infty} \frac{\|B(\varepsilon)\varphi_n\|^2}{|\lambda - \nu_n|^2} < M(\varepsilon, a, 2p, q, l),$$

where ν_n is the nth eigenvalue of T_r. ◇

Proof Let $n \in \mathbb{N}^*$ and $\sigma = Re\lambda$. Then, there exists $k \in \mathbb{N}^*$ such that $\nu_{k-1} \leq \sigma$ and $\nu_k > \sigma$. Hence, we get

$$\nu_k - \sigma > \nu_k^{\beta}\left(\nu_k^{1-\beta} - \sigma^{1-\beta}\right) \geq C_1 \nu_k^{\beta} \;\; (C_1 > 0) \tag{11.2.7}$$

and

$$|\lambda - \nu_{k-1}| \geq ||\lambda| - \nu_{k-1}| > |\lambda|^{\beta}\left(|\lambda|^{1-\beta} - \nu_{k-1}^{1-\beta}\right) \geq C_2 \nu_{k-1}^{\beta} \;\; (C_2 > 0). \tag{11.2.8}$$

Moreover, we obtain

$$\sigma - \nu_n > \nu_{k-1} - \nu_n \;\; (n < k-1), \quad \nu_n - \sigma > \nu_n - \nu_k \;\; (n > k).$$

So, we have

$$\nu_{k-1} - \nu_n \geq \nu_n^{\beta}(\nu_{k-1}^{1-\beta} - \nu_n^{1-\beta}) \;\; (n < k-1), \quad \nu_n - \nu_k \geq \nu_n^{\beta}(\nu_n^{1-\beta} - \nu_k^{1-\beta}) \;\; (n > k). \tag{11.2.9}$$

Consequently, by using Eqs. (11.2.7), (11.2.8) and (11.2.9), we get

$$\begin{aligned} \sum_{n=1}^{\infty} \frac{\lambda_n^{2\beta}}{|\lambda - \nu_n|^2} &\leq \sum_{n=1}^{\infty} \frac{\lambda_n^{2\beta}}{|\sigma - \nu_n|^2} \\ &< \frac{c}{C_1^2} + \frac{c}{C_2^2} + \sum_{n<k-1} \frac{\lambda_n^{2\beta}}{\nu_n^{2\beta}\left(\nu_{k-1}^{1-\beta} - \nu_n^{1-\beta}\right)^2} + \sum_{n>k} \frac{\lambda_n^{2\beta}}{\nu_n^{2\beta}\left(\nu_n^{1-\beta} - \nu_k^{1-\beta}\right)^2} \\ &\leq \frac{2c}{C_3^2} + \sum_{n<k-1} \frac{c}{\left(\nu_{k-1}^{1-\beta} - \nu_n^{1-\beta}\right)^2} + \sum_{n>k} \frac{c}{\left(\nu_n^{1-\beta} - \nu_k^{1-\beta}\right)^2}, \end{aligned} \tag{11.2.10}$$

where $0 < c \leq 1 + \frac{2l}{\nu_1}$ and $C_3 := \min\{C_1, C_2\}$. As $\beta \in]0, \frac{1}{2}[$, we have

$$\nu_{k-1}^{1-\beta} - \nu_n^{1-\beta} \geq (1-\beta)(\nu_{k-1} - \nu_n)\nu_{k-1}^{-\beta} \geq c_1(1-\beta)(\nu_{k-1} - \nu_n)^{1-\beta}, \;\; 0 < c_1 < 1 \tag{11.2.11}$$

and

$$\nu_n^{1-\beta} - \nu_k^{1-\beta} \geq (1-\beta)(\nu_n - \nu_k)\nu_n^{-\beta} \geq c_2(1-\beta)(\nu_n - \nu_k)^{1-\beta}, \quad 0 < c_2 < 1. \tag{11.2.12}$$

Thus, in view of Eqs. (11.2.10), (11.2.11) and (11.2.12) that

$$\begin{aligned}\sum_{n=1}^{\infty} \frac{\lambda_n^{2\beta}}{|\lambda - \nu_n|^2} &< \frac{2c}{C_3^2} + \sum_{n<k-1} \frac{c}{c_1^2(1-\beta)^2\,(\nu_{k-1} - \nu_n)^{2(1-\beta)}} + \\ &\quad \sum_{n>k} \frac{c}{c_2^2(1-\beta)^2\,(\nu_n - \nu_k)^{2(1-\beta)}} \\ &\leq \frac{2c}{C_3^2} + \sum_{n<k-1} \frac{c}{c'^2(1-\beta)^2\,(\nu_{k-1} - \nu_n)^{2(1-\beta)}} + \\ &\quad \sum_{n>k} \frac{c}{c'^2(1-\beta)^2\,(\nu_n - \nu_k)^{2(1-\beta)}},\end{aligned} \tag{11.2.13}$$

where $c' := \min\{c_1, c_2\}$. Consequently, taking into account that $\nu_n \geq \frac{n-1}{2p} + \nu_1$, Eq. (11.2.13) yields

$$\begin{aligned}\sum_{n=1}^{\infty} \frac{\lambda_n^{2\beta}}{|\lambda - \nu_n|^2} &< \frac{2c}{C_3^2} + \sum_{n<k-1} \frac{c(2p)^{2(1-\beta)}}{c'^2(1-\beta)^2\,(k-1-n)^{2(1-\beta)}} + \sum_{n>k} \frac{c(2p)^{2(1-\beta)}}{c'^2(1-\beta)^2\,(n-k)^{2(1-\beta)}} \\ &\leq \frac{2c}{C_3^2} + \frac{c(2p)^{2(1-\beta)}}{c'^2(1-\beta)^2} \left(\sum_{n<k-1} \frac{1}{(k-1-n)^{2(1-\beta)}} + \sum_{m=1}^{\infty} \frac{1}{m^{2(1-\beta)}} \right) \\ &< \frac{2c}{C_3^2} + \frac{2c(2p)^{2(1-\beta)}}{c'^2(1-\beta)^2} \sum_{m=1}^{\infty} \frac{1}{m^{2(1-\beta)}} =: \xi_1 < \infty.\end{aligned}$$

Furthermore, we get

$$\sum_{n=1}^{\infty} \frac{\lambda_n^{\beta}}{|\lambda - \nu_n|^2} < \frac{2c}{C_3^2} + \frac{8c(2p)^{2-\beta}}{c'^2(2-\beta)^2} \sum_{m=1}^{\infty} \frac{1}{m^{2-\beta}} =: \xi_2 < \infty.$$

Indeed, it suffices to make the same reasoning as above with the element $\frac{\beta}{2}$ instead of β. So, the series

$$\sum_n \frac{2ab\lambda_n^{\beta}}{|\lambda - \nu_n|^2}$$

and

$$\sum_n \frac{b^2\lambda_n^{2\beta}}{|\lambda - \nu_n|^2}$$

are convergent. Hence, let ξ denotes a positive constant such that

$$b^2\xi_1 + 2ab\xi_2 = \xi.$$

Then, we obtain

$$\sum_{n=1}^{\infty} \frac{2ab\lambda_n^{\beta}}{|\lambda - \nu_n|^2} + \sum_{n=1}^{\infty} \frac{b^2\lambda_n^{2\beta}}{|\lambda - \nu_n|^2} < \xi. \tag{11.2.14}$$

On the other hand, since the interval $]r - 2l, r + 2l[$ does not contain any eigenvalue ν_n of T_r, then there exists $k \in \mathbb{N}^*$ such that $\nu_{k-1} \leq r - 2l$ and $\nu_k \geq r + 2l$. Moreover, by using Theorem 11.1.1 and for any $\sigma \in [r - l, r + l]$, we obtain

$$\nu_{k+j+2sp} - \sigma \geq s + \nu_{k+j} - \sigma \geq s + r + 2l - r - l = l + s$$

and

$$\sigma - \nu_{k-j-2sp-1} \geq \sigma + s - \nu_{k-j-1} \geq r - l + s - r + 2l = l + s.$$

Consequently,

$$\nu_{k+j+2sp} - \sigma \geq l + s \text{ and } \sigma - \nu_{k-j-2sp-1} \geq l + s, \tag{11.2.15}$$

where $j = 0, 1, \ldots, 2p - 1$ and $s = 0, 1, \ldots$. In view of hypothesis $(H2)$ and Eq. (11.2.3), we get for $|\varepsilon| < \frac{1}{q}$

$$\begin{aligned} \frac{\|B(\varepsilon)\varphi_n\|^2}{|\lambda - \nu_n|^2} &\leq \frac{|\varepsilon|^2}{(1 - |\varepsilon|q)^2} \frac{(a + b\lambda_n^{\beta})^2}{|\lambda - \nu_n|^2} \\ &\leq \frac{|\varepsilon|^2}{(1 - |\varepsilon|q)^2} \left(\frac{a^2}{|\lambda - \nu_n|^2} + \frac{2ab\lambda_n^{\beta}}{|\lambda - \nu_n|^2} + \frac{b^2\lambda_n^{2\beta}}{|\lambda - \nu_n|^2} \right). \end{aligned}$$

Hence, for $|\varepsilon| < \frac{1}{q}$, Eq. (11.2.14) yields

$$\sum_{n=1}^{\infty} \frac{\|B(\varepsilon)\varphi_n\|^2}{|\lambda - \nu_n|^2} < \frac{|\varepsilon|^2}{(1 - |\varepsilon|q)^2} \left(\xi + \sum_{n=1}^{\infty} \frac{a^2}{|\sigma - \nu_n|^2} \right). \tag{11.2.16}$$

Now, let $\sigma \in [r - l, r + l]$. Using Eqs. (11.2.15) and (11.2.16), we obtain

$$\sum_{n=1}^{\infty}\frac{\|B(\varepsilon)\varphi_n\|^2}{|\lambda-\nu_n|^2} < \frac{|\varepsilon|^2}{(1-|\varepsilon|q)^2}\left(\xi + a^2\left(\sum_{n=1}^{k-1}\frac{1}{|\sigma-\nu_n|^2}+\sum_{n=k}^{\infty}\frac{1}{|\nu_n-\sigma|^2}\right)\right)$$
$$\leq \frac{|\varepsilon|^2}{(1-|\varepsilon|q)^2}\left(\xi + a^2\left(\sum_{s=0}^{\frac{k}{2p}-1}\frac{2p}{(l+s)^2}+\sum_{s=0}^{\infty}\frac{2p}{(l+s)^2}\right)\right)$$
$$\leq \frac{|\varepsilon|^2}{(1-|\varepsilon|q)^2}\left(\xi + 2a^2\left(\sum_{s=1}^{\infty}\frac{2p}{(l+s)^2}+\frac{2p}{l^2}\right)\right).$$

As a consequence, it is easy to see that for $|\varepsilon| < \frac{1}{q}$, we have

$$\sum_{n=1}^{\infty}\frac{\|B(\varepsilon)\varphi_n\|^2}{|\lambda-\nu_n|^2} < \frac{|\varepsilon|^2}{(1-|\varepsilon|q)^2}\left(\xi + 4pa^2\left(\int_0^{\infty}\frac{dx}{(l+x)^2}+\frac{1}{l^2}\right)\right)$$
$$\leq \frac{|\varepsilon|^2}{(1-|\varepsilon|q)^2}\left(\xi + 4pa^2\frac{l+1}{l^2}\right) =: M(\varepsilon, a, 2p, q, l).$$

The proof is then complete. Q.E.D.

Lemma 11.2.2 *Let τ be an arbitrary positive number. If $|Im\lambda| \geq \tau$, then for $|\varepsilon| < \frac{1}{q}$, there exists a positive constant $N(\varepsilon, a, 2p, q, \tau)$ such that*

$$\sum_{n=1}^{\infty}\frac{\|B(\varepsilon)\varphi_n\|^2}{|\lambda-\nu_n|^2} < N(\varepsilon, a, 2p, q, \tau). \tag{11.2.17}$$

If $Re\lambda \leq -\tau$, then for $|\varepsilon| < \frac{1}{q}$, there exists also a positive constant $N_1(\varepsilon, a, 2p, q, \tau)$ such that

$$\sum_{n=1}^{\infty}\frac{\|B(\varepsilon)\varphi_n\|^2}{|\lambda-\nu_n|^2} < N_1(\varepsilon, a, 2p, q, \tau). \tag{11.2.18}$$

◇

Proof Let $n \in \mathbb{N}^*$, $\sigma = Re\lambda$ and $\lambda = \sigma \pm i\tau$, where $\tau > 0$. Then, there exists $k \in \mathbb{N}^*$ such that $\nu_{k-1} \leq \sigma$ and $\nu_k > \sigma$. In view of both hypothesis $(H2)$ and Eq. (11.2.3), we get for $|\varepsilon| < \frac{1}{q}$

$$\frac{\|B(\varepsilon)\varphi_n\|^2}{|\lambda-\nu_n|^2} \leq \frac{|\varepsilon|^2}{(1-|\varepsilon|q)^2}\left(\frac{a^2}{|\lambda-\nu_n|^2}+\frac{2ab\lambda_n^{\beta}}{|\lambda-\nu_n|^2}+\frac{b^2\lambda_n^{2\beta}}{|\lambda-\nu_n|^2}\right). \tag{11.2.19}$$

- Let us consider $|Im\lambda| \geq \tau$. In view of Theorem 11.1.1, we obtain

$$\nu_{k+j+2sp} - \sigma \geq s \text{ and } \sigma - \nu_{k-j-2sp-1} \geq s, \tag{11.2.20}$$

where $j = 0, 1, \ldots, 2p-1$ and $s = 0, 1, \ldots$. Hence, for $|\varepsilon| < \frac{1}{q}$, Eqs. (11.2.14), (11.2.19) and (11.2.20) yield

$$\begin{aligned}
\sum_{n=1}^{\infty} \frac{\|B(\varepsilon)\varphi_n\|^2}{|\lambda-\nu_n|^2} &< \frac{|\varepsilon|^2}{(1-|\varepsilon|q)^2}\left(\xi + a^2\sum_{n=1}^{\infty}\frac{1}{(\nu_n-\sigma)^2+\tau^2}\right)\\
&\leq \frac{|\varepsilon|^2}{(1-|\varepsilon|q)^2}\left(\xi + a^2\left(\sum_{n=1}^{k-1}\frac{1}{(\sigma-\nu_n)^2+\tau^2} + \sum_{n=k}^{\infty}\frac{1}{(\nu_n-\sigma)^2+\tau^2}\right)\right)\\
&\leq \frac{|\varepsilon|^2}{(1-|\varepsilon|q)^2}\left(\xi + 2a^2 p\left(\sum_{s=0}^{\frac{k}{2p}-1}\frac{1}{s^2+\tau^2} + \sum_{s=0}^{\infty}\frac{1}{s^2+\tau^2}\right)\right)\\
&\leq \frac{|\varepsilon|^2}{(1-|\varepsilon|q)^2}\left(\xi + 4pa^2\left(\sum_{s=1}^{\infty}\frac{1}{s^2+\tau^2} + \frac{1}{\tau^2}\right)\right)\\
&< \frac{|\varepsilon|^2}{(1-|\varepsilon|q)^2}\left(\xi + 4pa^2\left(\frac{1}{\tau^2} + \int_0^{\infty}\frac{dx}{x^2+\tau^2}\right)\right)\\
&\leq \frac{|\varepsilon|^2}{(1-|\varepsilon|q)^2}\left(\xi + 4pa^2\left(\frac{1}{\tau^2} + \frac{1}{\tau}\left[\arctan\frac{x}{\tau}\right]_0^{\infty}\right)\right)\\
&\leq \frac{|\varepsilon|^2}{(1-|\varepsilon|q)^2}\left(\xi + 2a^2\frac{p}{\tau}\left(\pi + \frac{2}{\tau}\right)\right)\\
&=: N(\varepsilon, a, 2p, q, \tau).
\end{aligned}$$

- Now, if $Re\lambda \leq -\tau$. By using Theorem 11.1.1, we obtain

$$\nu_{1+j+2sp} - \sigma \geq s + \nu_{1+j} - \sigma,$$

where $j = 0, \ldots, 2p-1$ and $s = 0, 1, \ldots$. Since $\nu_{1+j} > 0$, we have

$$\nu_{1+j+2sp} - \sigma \geq s - \sigma.$$

On the other hand, since $Re\lambda \leq -\tau$, we get

$$\nu_{1+j+2sp} - \sigma \geq s - \sigma \geq s + \tau, \tag{11.2.21}$$

where $j = 0, \ldots, 2p-1$ and $s = 0, 1, \ldots$.
Consequently, Eqs. (11.2.14), (11.2.19) and (11.2.21) imply that for $|\varepsilon| < \frac{1}{q}$

$$\begin{aligned}
\sum_{n=1}^{\infty} \frac{\|B(\varepsilon)\varphi_n\|^2}{|\lambda-\nu_n|^2} &< \frac{|\varepsilon|^2}{(1-|\varepsilon|q)^2}\left(\xi + a^2\sum_{n=1}^{\infty}\frac{1}{|\nu_n-\sigma|^2}\right)\\
&\leq \frac{|\varepsilon|^2}{(1-|\varepsilon|q)^2}\left(\xi + 2pa^2\left(\sum_{s=0}^{\infty}\frac{1}{(\tau+s)^2}\right)\right)\\
&< \frac{|\varepsilon|^2}{(1-|\varepsilon|q)^2}\left(\xi + 2pa^2\left(\int_0^{\infty}\frac{dx}{(\tau+x)^2} + \frac{1}{\tau^2}\right)\right)\\
&\leq \frac{|\varepsilon|^2}{(1-|\varepsilon|q)^2}\left(\xi + 2pa^2\left(\frac{1}{\tau^2} + \left[-\frac{1}{\tau+x}\right]_0^{\infty}\right)\right)\\
&\leq \frac{|\varepsilon|^2}{(1-|\varepsilon|q)^2}\left(\xi + 2a^2\frac{p}{\tau}\left(1 + \frac{1}{\tau}\right)\right)\\
&=: N_1(\varepsilon, a, 2p, q, \tau).
\end{aligned}$$

The proof is then complete. Q.E.D.

Now, let us estimate the relation between $\|B(\varepsilon)(\lambda - T_r)^{-1}\|$ and

$$\sum_{n=1}^{\infty} \frac{\|B(\varepsilon)\varphi_n\|^2}{|\lambda - \nu_n|^2}.$$

Theorem 11.2.1 *We have*

$$\|B(\varepsilon)(\lambda - T_r)^{-1}\|^2 \le \sum_{n=1}^{\infty} \frac{\|B(\varepsilon)\varphi_n\|^2}{|\lambda - \nu_n|^2}.$$ ◇

Proof Let $f \in X$ such that $\|f\| = 1$. Then, we have

$$\begin{aligned}
\|B(\varepsilon)(\lambda - T_r)^{-1} f\|^2 &= \left\| \sum_{n=1}^{\infty} \frac{\langle f, \varphi_n\rangle B(\varepsilon)\varphi_n}{\lambda - \nu_n} \right\|^2 \\
&\le \left(\sum_{n=1}^{\infty} |\langle f, \varphi_n\rangle| \frac{\|B(\varepsilon)\varphi_n\|}{|\lambda - \nu_n|} \right)^2 \\
&\le \left(\sum_{n=1}^{\infty} |\langle f, \varphi_n\rangle|^2 \right) \left(\sum_{n=1}^{\infty} \frac{\|B(\varepsilon)\varphi_n\|^2}{|\lambda - \nu_n|^2} \right) \\
&\le \|f\|^2 \left(\sum_{n=1}^{\infty} \frac{\|B(\varepsilon)\varphi_n\|^2}{|\lambda - \nu_n|^2} \right).
\end{aligned}$$

Consequently, we obtain

$$\|B(\varepsilon)(\lambda - T_r)^{-1}\|^2 \le \sum_{n=1}^{\infty} \frac{\|B(\varepsilon)\varphi_n\|^2}{|\lambda - \nu_n|^2}.$$ Q.E.D.

11.2.3 Sum of Multiplicities of All Eigenvalues of $T(\varepsilon) - K_r$

We denote by $\Omega_{r,h}$ the intersection of S_h with the half-plane $Re\lambda < r$, where

$$S_h := \{\lambda \in \mathbb{C} \text{ such that } |Im\lambda| < h \text{ and } Re\lambda > -h\}$$

and $h > 3l$ (see Fig. 11.2).

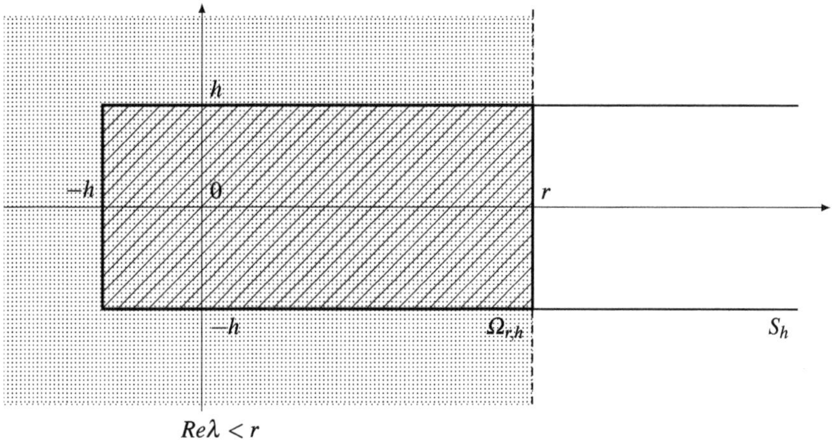

Fig. 11.2 The domain $\Omega_{r,h}$

Now, we are ready to prove the following result.

Proposition 11.2.2 *Assume that hypotheses* $(H2)$ *with* $\beta \in]0, \frac{1}{2}[$, $(H9)$ *and* $(H10)$ *are verified. Then, for* $|\varepsilon|$ *small enough, we have*

$$n(\overline{\mathbb{D}}(0,r), T(\varepsilon) - K_r) = n(\overline{\mathbb{D}}(0,r), T_0). \qquad \diamond$$

Proof Since the interval $]r - 2l, r + 2l[$ does not contain any eigenvalue ν_n of T_r and T_r is positive, we have

$$\sigma(T_r) \bigcap \partial\Omega_{r,h} = \emptyset.$$

Now, let $\lambda \in \partial\Omega_{r,h}$ and $t \in [0, 1]$, we have

$$\lambda - T_r - tB(\varepsilon) = [I - tB(\varepsilon)(\lambda - T_r)^{-1}](\lambda - T_r). \tag{11.2.22}$$

Due to both Lemma 11.2.1 and Theorem 11.2.1, we have for $Re\lambda \in [r - l, r + l]$ and $|\varepsilon| < \frac{1}{q}$,

$$\|B(\varepsilon)(\lambda - T_r)^{-1}\|^2 < M(\varepsilon, a, 2p, q, l).$$

Then, for $Re\lambda \in [r - l, r + l]$ and $|\varepsilon| < \frac{1}{q+\sqrt{\xi+4pa^2\frac{l+1}{l^2}}}$, we have

$$\|B(\varepsilon)(\lambda - T_r)^{-1}\| < 1. \tag{11.2.23}$$

Similarly, by Eq. (11.2.17) and Theorem 11.2.1, we get for $|Im\lambda| \geq h$ and $|\varepsilon| < \frac{1}{q}$,

$$\|B(\varepsilon)(\lambda - T_r)^{-1}\|^2 < N(\varepsilon, a, 2p, q, h).$$

So, for $|Im\lambda| \geq h$ and $|\varepsilon| < \frac{1}{q+\sqrt{\xi+2a^2\frac{p}{h}(\pi+\frac{2}{h})}}$, we have

$$\|B(\varepsilon)(\lambda - T_r)^{-1}\| < 1. \tag{11.2.24}$$

Furthermore, for $Re\lambda \leq -h$, $|\varepsilon| < \frac{1}{q}$, Eq. (11.2.18) and Theorem 11.2.1 imply that

$$\|B(\varepsilon)(\lambda - T_r)^{-1}\|^2 < N_1(\varepsilon, a, 2p, q, h).$$

Hence, for $Re\lambda \leq -h$ and $|\varepsilon| < \frac{1}{q+\sqrt{\xi+2a^2\frac{p}{h}(1+\frac{1}{h})}}$, we have

$$\|B(\varepsilon)(\lambda - T_r)^{-1}\| < 1. \tag{11.2.25}$$

Consequently, since $h > 3l$, Eqs. (11.2.23), (11.2.24) and (11.2.25) yield, for $|\varepsilon| < \frac{1}{q+\sqrt{\xi+4pa^2\frac{l+1}{l^2}}}$,

$$\|tB(\varepsilon)(\lambda - T_r)^{-1}\|_{\partial\Omega_{r,h}} \leq \|B(\varepsilon)(\lambda - T_r)^{-1}\|_{\partial\Omega_{r,h}} < 1.$$

Hence, $I - tB(\varepsilon)(\lambda - T_r)^{-1}$ is invertible with bounded inverse. So, Eq. (11.2.22) implies that $\lambda - T_r - tB(\varepsilon)$ is invertible with bounded inverse. Then,

$$\sigma(T_r + tB(\varepsilon)) \bigcap \partial\Omega_{r,h} = \emptyset.$$

On the other hand, T_r is self-adjoint with discrete spectrum, then due to [3, p. 146] T_r is with compact resolvent. Moreover, by using Proposition 11.2.1(ii), the operator $B(\varepsilon)$ is T_r-compact. Hence, Proposition 1.15.1 allows us to conclude that

$$n(\overline{\mathbb{D}}(0, r), T(\varepsilon) - K_r) = n(\overline{\mathbb{D}}(0, r), T_r).$$

In view of Notes and Remarks 11.4(i), given at the end of this chapter, we obtain

$$n(\overline{\mathbb{D}}(0, r), T(\varepsilon) - K_r) = n(\overline{\mathbb{D}}(0, r), T_0).$$

Q.E.D.

11.3 Behavior of the Spectrum of Perturbed Operator $T(\varepsilon)$

In this section, we will study the comportment of the eigenvalues of the operator $T(\varepsilon)$ in terms of the one of T_0. To this end, we shall emphasize both Lemma 2.5.4 and Proposition 11.2.2 imply the following

$$n(\overline{\mathbb{D}}(0, r), T(\varepsilon)) = n(\overline{\mathbb{D}}(0, r), T_0) + \frac{1}{2\pi}[\arg D_\varepsilon(\lambda)]_{\partial\Omega_{r,h}},$$

where

$$D_\varepsilon(\lambda) := \det(I - K_\varepsilon(\lambda))$$

and

$$K_\varepsilon(\lambda) := K_r(\lambda - T(\varepsilon) + K_r)^{-1}.$$

11.3.1 Argument of the Function $D_\varepsilon(\lambda)$

In this subsection, we give an estimate of the increment of the argument of $D_\varepsilon(\lambda)$ along the boundary of $\Omega_{r,h}$. To this interest, we need the following results enabling us to obtain such estimates.

Proposition 11.3.1 *Assume that hypotheses $(H2)$ with $\beta \in]0, \frac{1}{2}[$, $(H9)$ and $(H10)$ hold. Then,*

(i) in the strip $Re\lambda \in [r-l, r+l]$ and for $|\varepsilon|$ small enough, the function $D_\varepsilon(\lambda)$ is holomorphic and is estimated as the following

$$|D_\varepsilon(\lambda)| \leq (1 + 2(1 - \sqrt{M(\varepsilon, a, 2p, q, l)})^{-1})^{4lp},$$

(ii) in the same strip, for $Im\lambda = \pm h$ and for $|\varepsilon|$ small enough, the following lower estimate holds

$$|D_\varepsilon(\lambda)| \geq \left(1 - \frac{2l}{h}(1 - \sqrt{N(\varepsilon, a, 2p, q, h)})^{-1}\right)^{4lp}. \qquad \diamond$$

Proof (i) Let $\lambda \in \mathbb{C}$ such that $Re\lambda \in [r-l, r+l]$. The fact that the interval $[r-l, r+l]$ does not contain any eigenvalue of T_r, then

$$\lambda - T_r - B(\varepsilon) = [I - B(\varepsilon)(\lambda - T_r)^{-1}](\lambda - T_r).$$

Moreover, combining Lemma 11.2.1 together with Theorem 11.2.1, we get for $|\varepsilon| < \frac{1}{q+\sqrt{\xi+4pa^2\frac{l+1}{l^2}}}$,

$$\|B(\varepsilon)(\lambda - T_r)^{-1}\| < \sqrt{M(\varepsilon, a, 2p, q, l)} < 1.$$

So, $I - B(\varepsilon)(\lambda - T_r)^{-1}$ is invertible with bounded inverse. Thus, for $|\varepsilon| < \frac{1}{q+\sqrt{\xi+4pa^2\frac{l+1}{l^2}}}$, the operator $K_\varepsilon(\lambda)$ can be written as the following

$$K_\varepsilon(\lambda) := K_r(\lambda - T_r)^{-1}[I - B(\varepsilon)(\lambda - T_r)^{-1}]^{-1}. \tag{11.3.1}$$

Taking into account the fact that $D_\varepsilon(\lambda)$ is meromorphic and the strip $Re\lambda \in [r-l, r+l]$ does not contain any eigenvalue of T_r, we get $D_\varepsilon(\lambda)$ is holomorphic in the strip $Re\lambda \in [r-l, r+l]$. Now, we are going to estimate $|D_\varepsilon(\lambda)|$ for $Re\lambda \in [r-l, r+l]$ and $|\varepsilon| < \frac{1}{q+\sqrt{\xi+4pa^2\frac{l+1}{l^2}}}$. Due to Proposition 2.5.1(i), we have

$$|D_\varepsilon(\lambda)| \le (1 + \|K_\varepsilon(\lambda)\|)^{rk(K_\varepsilon)}. \tag{11.3.2}$$

On the other hand, Eq. (11.3.1) implies that

$$\|K_\varepsilon(\lambda)\| \le \|K_r\| \, \|(\lambda - T_r)^{-1}\| \, \|[I - B(\varepsilon)(\lambda - T_r)^{-1}]^{-1}\|. \tag{11.3.3}$$

As

$$\|K_r\| \le 2l,$$

$$\|(\lambda - T_r)^{-1}\| \le \frac{1}{\text{dist}(\lambda, \sigma(T_r))} \le \frac{1}{l}$$

and

$$\left\|\left[I - B(\varepsilon)(\lambda - T_r)^{-1}\right]^{-1}\right\| \le \left(1 - \sqrt{M(\varepsilon, a, 2p, q, l)}\right)^{-1},$$

we obtain, in view of Eqs. (11.3.2) and (11.3.3), that

$$\begin{aligned} |D_\varepsilon(\lambda)| &\le \left(1 + 2\left(1 - \sqrt{M(\varepsilon, a, 2p, q, l)}\right)^{-1}\right)^{rk(K_\varepsilon(\lambda))} \\ &\le \left(1 + 2\left(1 - \sqrt{M(\varepsilon, a, 2p, q, l)}\right)^{-1}\right)^{4lp}. \end{aligned}$$

(ii) Let $\lambda \in \mathbb{C}$ such that $Re\lambda \in [r-l, r+l]$ and $Im\lambda = \pm h$. In view of Eq. (11.2.17) and Theorem 11.2.1, we have for $|Im\lambda| \ge h$ and $|\varepsilon| < \frac{1}{q+\sqrt{\xi+2a^2\frac{p}{h}(\pi+\frac{2}{h})}}$,

$$\|B(\varepsilon)(\lambda - T_r)^{-1}\| < \sqrt{N(\varepsilon, a, 2p, q, h)} < 1.$$

So, for $|Im\lambda| \ge h$ and $|\varepsilon| < \frac{1}{q+\sqrt{\xi+4pa^2\frac{l+1}{l^2}}}$, we get

$$\left\|\left[I - B(\varepsilon)(\lambda - T_r)^{-1}\right]^{-1}\right\| \le \left(1 - \sqrt{N(\varepsilon, a, 2p, q, h)}\right)^{-1}.$$

Moreover, we have

$$\|(\lambda - T_r)^{-1}\| \le \frac{1}{|Im\lambda|} = \frac{1}{h}$$

and

$$\|K_r\| \leq 2l.$$

Hence, for $Im\lambda = \pm h$ and $|\varepsilon| < \frac{1}{q+\sqrt{\xi+4pa^2\frac{l+1}{l^2}}}$, Eq. (11.3.1) implies that

$$\|K_\varepsilon(\lambda)\| \leq \frac{2l}{h}\left(1-\sqrt{N(\varepsilon,a,2p,q,h)}\right)^{-1}. \tag{11.3.4}$$

Consequently, by using Notes and Remark 2.8(iv), we obtain for $Im\lambda = \pm h$ and $|\varepsilon| < \frac{1}{q+\sqrt{\xi+4pa^2\frac{l+1}{l^2}}}$, that

$$\left|D_\varepsilon(\lambda)\right| \geq \left|1-\frac{2l}{h}\left(1-\sqrt{N(\varepsilon,a,2p,q,h)}\right)^{-1}\right|^{rk(K_\varepsilon(\lambda))}.$$

Since $h > 3l$, we have

$$\frac{2l}{h}\left(1-\sqrt{N(\varepsilon,a,2p,q,h)}\right)^{-1} < 1, \quad \text{for } |\varepsilon| < W, \tag{11.3.5}$$

where

$$W = \min\left\{\frac{1}{q+\sqrt{\xi+4pa^2\frac{l+1}{l^2}}}, \frac{1}{q+\frac{h}{h-2l}\sqrt{\xi+2a^2\frac{p}{h}(\pi+\frac{2}{h})}}\right\}.$$

So, for $|\varepsilon| < W$,

$$\left|D_\varepsilon(\lambda)\right| \geq \left(1-\frac{2l}{h}\left(1-\sqrt{N(\varepsilon,a,2p,q,h)}\right)^{-1}\right)^{4lp}. \quad \text{Q.E.D.}$$

A key tool in this section is the following proposition.

Proposition 11.3.2 *Assume that hypotheses $(H2)$ with $\beta \in]0, \frac{1}{2}[$, $(H9)$ and $(H10)$ are verified. Suppose that the line $Re\lambda = r$ does not contain any eigenvalue of the operator $T(\varepsilon)$. Then, for $|\varepsilon|$ small enough the variation of the argument of the function $D_\varepsilon(\lambda)$ along the boundary of $\Omega_{r,h}$ satisfies the estimate*

$$\left|[\arg D_\varepsilon(\lambda)]_{\partial\Omega_{r,h}}\right| \leq C_{1,\varepsilon},$$

where $C_{1,\varepsilon}$ is a constant. ◇

Proof • For $Re\lambda \in [r-l, r+l]$, $Im\lambda = \pm h$ and $|\varepsilon| < W$, Eqs. (11.3.4) and (11.3.5) imply that

$$\|K_\varepsilon(\lambda)\| \leq \frac{2l}{h}\left(1-\sqrt{N(\varepsilon,a,2p,q,h)}\right)^{-1} < 1. \tag{11.3.6}$$

- On the other hand, for $Re\lambda = -h$ and $|\varepsilon| < \frac{1}{q+\sqrt{\xi+4pa^2\frac{l+1}{l^2}}}$, it follows from Eq. (11.2.18) and Theorem 11.2.1 that

$$\|B(\varepsilon)(\lambda - T_r)^{-1}\| < \sqrt{N_1(\varepsilon, a, 2p, q, h)} < 1.$$

Moreover, we have

$$\begin{aligned} \|(\lambda - T_r)^{-1}\| &\leq \frac{1}{\operatorname{dist}(\lambda, \sigma(T_r))} \\ &\leq \frac{1}{|Re\lambda - \lambda_n'|}, \qquad \lambda_n' \in \sigma(T_r) \\ &\leq \frac{1}{|-h - \lambda_n'|}, \qquad Re\lambda = -h \\ &< \frac{1}{h}. \end{aligned}$$

So, for $Re\lambda = -h$ and $|\varepsilon| < \frac{1}{q+\sqrt{\xi+4pa^2\frac{l+1}{l^2}}}$, Eq. (11.3.1) implies that

$$\|K_\varepsilon(\lambda)\| \leq \frac{2l}{h}\left(1 - \sqrt{N_1(\varepsilon, a, 2p, q, h)}\right)^{-1}.$$

As $h > 3l$, we have

$$\frac{2l}{h}\left(1 - \sqrt{N_1(\varepsilon, a, 2p, q, h)}\right)^{-1} < 1, \quad \text{for } |\varepsilon| < W_1, \tag{11.3.7}$$

where

$$W_1 = \min\left\{\frac{1}{q + \sqrt{\xi + 4pa^2\frac{l+1}{l^2}}}, \frac{1}{q + \frac{h}{h-2l}\sqrt{\xi + 2a^2\frac{p}{h}(1 + \frac{1}{h})}}\right\}.$$

Hence, Eqs. (11.3.6) and (11.3.7) yield

$$\|K_\varepsilon(\lambda)\|_{\partial S_h} < 1, \text{ for } |\varepsilon| < W.$$

Since $rk(K_\varepsilon(\lambda)) \leq 4lp$, then in virtue of Lemma 2.5.3, we obtain

$$|[\arg\det(I - K_\varepsilon(\lambda))]_{\partial S_h}| \leq 4lp\pi, \quad \text{for } |\varepsilon| < W. \tag{11.3.8}$$

To complete the proof, we apply Lemma 2.5.2. We consider the rectangle $\Gamma_{l,h}$ bounded by the straight lines $Re\lambda = r \pm l$ and $Im\lambda = \pm 2h$ and the second contracted rectangle $\Gamma'_{l,h}$ with the same center r and bounded by the lines $Re\lambda = r \pm \frac{l}{2}$ and $Im\lambda = \pm h$ (see Fig. 11.3). In view of Proposition 11.3.1, we have $D_\varepsilon(\lambda)$ is a bounded and holomorphic function in the strip $Re\lambda \in [r - l, r + l]$. We denote by

$$m_1 = \sup_{\lambda \in \Gamma_{l,h}} D_\varepsilon(\lambda)$$

and

$$m_2 = \sup_{\lambda \in \Gamma'_{l,h}} D_\varepsilon(\lambda).$$

Taking into account that the line $Re\lambda = r$ does not contain any eigenvalue of $T(\varepsilon)$ i.e., does not contain zeros of

$$D_\varepsilon(\lambda) := \det\big((\lambda - T(\varepsilon))(\lambda - T(\varepsilon) + K_r)^{-1}\big).$$

We get, in view of Lemma 2.5.2(iii), that

$$|[\arg D_\varepsilon(\lambda)]_{Re\lambda=r}| \leq C'(\log m_1 - \log m_2), \tag{11.3.9}$$

where $C' > 0$. Hence, using Proposition 11.3.1, we obtain for $|\varepsilon| < W$

$$\log m_1 - \log m_2 \leq \left[\log\left(\left(1 + 2\left(1 - \sqrt{M(\varepsilon, a, 2p, q, l)}\right)^{-1}\right)^{4lp}\right) - \log\left(\left(1 - \frac{2l}{h}\left(1 - \sqrt{N(\varepsilon, a, 2p, q, h)}\right)^{-1}\right)^{4lp}\right)\right]. \tag{11.3.10}$$

So, both Eqs. (11.3.9) and (11.3.10) imply that, for $|\varepsilon| < W$

$$|[\arg D_\varepsilon(\lambda)]_{Re\lambda=r}| \leq C'\left[\log\left(\left(1 + 2\left(1 - \sqrt{M(\varepsilon, a, 2p, q, l)}\right)^{-1}\right)^{4lp}\right) - \log\left(\left(1 - \frac{2l}{h}\left(1 - \sqrt{N(\varepsilon, a, 2p, q, h)}\right)^{-1}\right)^{4lp}\right)\right].$$

Hence,

$$|[\arg D_\varepsilon(\lambda)]_{Re\lambda=r}| \leq C'_\varepsilon, \tag{11.3.11}$$

where

$$C'_\varepsilon = 4lp\, C' \log\left(\frac{1 + 2\left(1 - \sqrt{M(\varepsilon, a, 2p, q, l)}\right)^{-1}}{1 - \frac{2l}{h}\left(1 - \sqrt{N(\varepsilon, a, 2p, q, h)}\right)^{-1}}\right) \text{ and } |\varepsilon| < W.$$

Consequently, for $|\varepsilon| < W$, Eqs. (11.3.8) and (11.3.11) yield

$$|[\arg D_\varepsilon(\lambda)]_{\partial\Omega_{r,h}}| \leq C_{1,\varepsilon},$$

where $C_{1,\varepsilon} := \min\{4lp\pi, C'_\varepsilon\}$. Q.E.D.

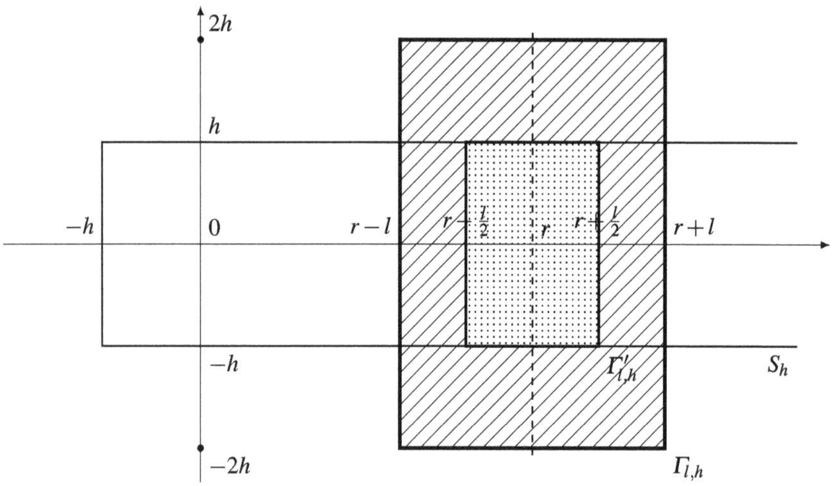

Fig. 11.3 The rectangles $\Gamma_{l,h}$ and $\Gamma'_{l,h}$

11.3.2 Sum of Multiplicities of All Eigenvalues of $T(\varepsilon)$

Using the results described above, we can now prove the following:

Theorem 11.3.1 *Assume that hypotheses $(H9)$, $(H10)$ and $(H2)$ with $\beta \in]0, \frac{1}{2}[$ are satisfied. Then, for $|\varepsilon|$ small enough, the spectrum of the operator $T(\varepsilon)$ is constituted by isolated eigenvalues satisfying*

$$n(\overline{\mathbb{D}}(0, r), T(\varepsilon)) = n(\overline{\mathbb{D}}(0, r), T_0) + O(1)$$

i.e.,

$$|n(\overline{\mathbb{D}}(0, r), T(\varepsilon)) - n(\overline{\mathbb{D}}(0, r), T_0)| < C \text{ as } r \to \infty,$$

where C is a constant. ◇

Proof Due to Proposition 11.2.1, the operators $T(\varepsilon)$ and $T(\varepsilon) - K_r$ are with compact resolvent, for $|\varepsilon| < \frac{1}{q+\beta b}$. Further, making the same reasoning as the one in the proof of Proposition 11.2.2, we have for $|\varepsilon| < \frac{1}{q+\sqrt{\xi+4pa^2\frac{l+1}{l^2}}}$,

$$\sigma(T(\varepsilon) - K_r) \bigcap \partial\Omega_{r,h} = \emptyset.$$

Consequently, Lemma 2.5.4 implies that, for $|\varepsilon| < \frac{1}{q+\sqrt{\xi+4pa^2\frac{l+1}{l^2}}}$,

$$n(\overline{\mathbb{D}}(0, r), T(\varepsilon)) - n(\overline{\mathbb{D}}(0, r), T(\varepsilon) - K_r) = \frac{1}{2\pi}[\arg D_\varepsilon(\lambda)]_{\partial\Omega_{r,h}}.$$

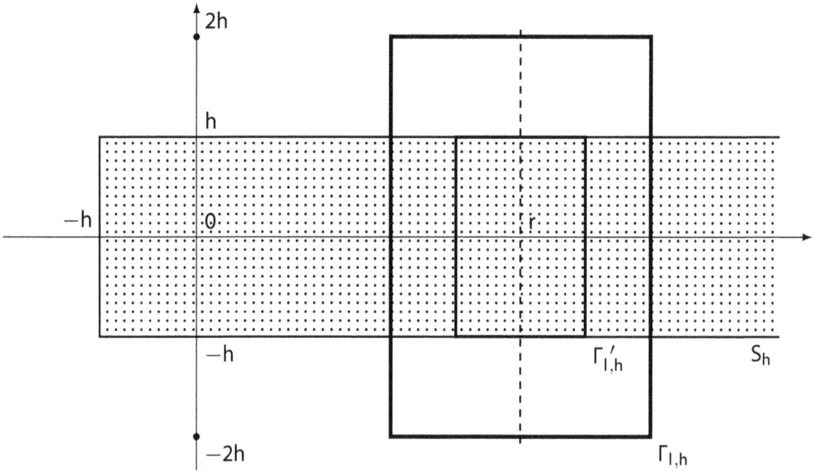

Fig. 11.4 The domains S_h, $\Gamma_{l,h}$ and $\Gamma'_{l,h}$

Taking into account Proposition 11.2.2, we obtain, for $|\varepsilon| < \frac{1}{q+\sqrt{\xi+4pa^2\frac{l+1}{l^2}}}$,

$$n(\overline{\mathbb{D}}(0,r), T(\varepsilon)) - n(\overline{\mathbb{D}}(0,r), T_0) = \frac{1}{2\pi}[\arg D_\varepsilon(\lambda)]_{\partial\Omega_{r,h}}.$$

On the other hand, due to Proposition 11.3.2, we have $[\arg D_\varepsilon(\lambda)]$ is bounded for any r by $C_{1,\varepsilon}$ when λ varies along the boundary of $\Omega_{r,h}$, for $|\varepsilon| < W$. As a consequence, for $|\varepsilon| < W$, we have

$$|n(\overline{\mathbb{D}}(0,r), T(\varepsilon)) - n(\overline{\mathbb{D}}(0,r), T_0)| \leq \frac{1}{2\pi}C_{1,\varepsilon} = O(1) \text{ as } r \to \infty.$$

But, we obtain this result when we exclude the values of r for which the straight line $Re\lambda = r$ passes through the zeros of $D_\varepsilon(\lambda)$. However, all such zeros lie in the rectangle $\Gamma'_{l,h}$ (see Fig. 11.4).
Therefore, in view of Lemma 2.5.2(i), we have

$$n_{D_\varepsilon(\lambda)}(\Gamma'_{l,h}) \leq C''\left[\log\left(\left(1+2\left(1-\sqrt{M(\varepsilon,a,2p,q,l)}\right)^{-1}\right)^{4lp}\right) - \log\left(\left(1-\frac{2l}{h}\left(1-\sqrt{N(\varepsilon,a,2p,q,h)}\right)^{-1}\right)^{4lp}\right)\right],$$

for $|\varepsilon| < W$ and $C'' > 0$. So, we get

$$n_{D_\varepsilon(\lambda)}(\Gamma'_{l,h}) \leq C''_\varepsilon,$$

where

$$C''_\varepsilon = 4lp\, C'' \log\left(\frac{1+2\left(1-\sqrt{M(\varepsilon,a,2p,q,l)}\right)^{-1}}{1-\frac{2l}{h}\left(1-\sqrt{N(\varepsilon,a,2p,q,h)}\right)^{-1}}\right) \quad \text{and } |\varepsilon| < W.$$

Hence, the number of zeros of $D_\varepsilon(\lambda)$ inside $\Gamma'_{l,h}$ does not exceed C''_ε i.e., the eigenvalues of $T(\varepsilon)$ does not exceed C''_ε. Consequently, the relation remains valid for all $r \to \infty$, i.e.,

$$n(\overline{\mathbb{D}}(0,r), T(\varepsilon)) - n(\overline{\mathbb{D}}(0,r), T_0) = O(1). \qquad \text{Q.E.D.}$$

Corollary 11.3.1 *Assume that hypotheses $(H2)$ with $\beta \in]0, \frac{1}{2}[$, and $(H9)$ and Eq. (11.2.6) hold. Then, for $\beta \in]0, 1+\frac{\alpha-1}{2}[$ and $|\varepsilon|$ small enough, the spectrum of the operator $T(\varepsilon)$ is constituted by isolated eigenvalues satisfying*

$$n(\overline{\mathbb{D}}(0,r), T(\varepsilon)) = n(\overline{\mathbb{D}}(0,r), T_0) + O(1)$$

i.e.,

$$|n(\overline{\mathbb{D}}(0,r), T(\varepsilon)) - n(\overline{\mathbb{D}}(0,r), T_0)| < C \text{ as } r \to \infty,$$

where C is a constant. ◇

Proof By using $(H2)$ and making the same reasoning as the proof developed in Proposition 11.2.1, we have for $|\varepsilon| < \frac{1}{q}$

$$\frac{\|B(\varepsilon)\varphi_n\|^2}{|\lambda-\nu_n|^2} \leq \frac{|\varepsilon|^2}{(1-|\varepsilon|q)^2}\left(\frac{a^2}{|\lambda-\nu_n|^2} + \frac{2ab\lambda_n^\beta}{|\lambda-\nu_n|^2} + \frac{b^2\lambda_n^{2\beta}}{|\lambda-\nu_n|^2}\right).$$

Let $\sigma = Re\lambda$. Then, there is $k \in \mathbb{N}^*$ such that $\nu_{k-1} \leq \sigma$ and $\nu_k > \sigma$. Furthermore, the fact that $\nu_n^{1-\alpha} \geq \frac{n-1}{2p} + \nu_1^{1-\alpha}$, we have

$$\nu_n \geq \left(\frac{n-1}{2p} + \nu_1^{1-\alpha}\right)^{\frac{1}{1-\alpha}}.$$

So, Eq. (11.2.10) yields

$$\sum_{n=1}^{\infty} \frac{\lambda_n^{2\beta}}{|\lambda-\nu_n|^2} < \frac{2c}{C_3^2} + \sum_{n<k-1} \frac{c}{\left(\left(\frac{k-2}{2p}+\nu_1^{1-\alpha}\right)^{\frac{1-\beta}{1-\alpha}} - \left(\frac{n-1}{2p}+\nu_1^{1-\alpha}\right)^{\frac{1-\beta}{1-\alpha}}\right)^2} + \sum_{n>k} \frac{c}{\left(\left(\frac{n-1}{2p}+\nu_1^{1-\alpha}\right)^{\frac{1-\beta}{1-\alpha}} - \left(\frac{k-1}{2p}+\nu_1^{1-\alpha}\right)^{\frac{1-\beta}{1-\alpha}}\right)^2}. \qquad (11.3.12)$$

If $\beta \in]0, \alpha]$, then

$$\frac{1-\beta}{1-\alpha} \geq 1.$$

Hence, it view from Eq. (11.3.12) that

$$\begin{aligned}\sum_{n=1}^{\infty} \frac{\lambda_n^{2\beta}}{|\lambda - \nu_n|^2} &< \frac{2c}{C_3^2} + \sum_{n<k-1} \frac{c(2p)^{\frac{2(1-\beta)}{1-\alpha}}}{(k-1-n)^{\frac{2(1-\beta)}{1-\alpha}}} + \sum_{n>k} \frac{c(2p)^{\frac{2(1-\beta)}{1-\alpha}}}{(n-k)^{\frac{2(1-\beta)}{1-\alpha}}} \\ &\leq \frac{2c}{C_3^2} + c(2p)^{\frac{2(1-\beta)}{1-\alpha}} \left(\sum_{n<k-1} \frac{1}{(k-1-n)^{\frac{2(1-\beta)}{1-\alpha}}} + \sum_{m=1}^{\infty} \frac{1}{m^{\frac{2(1-\beta)}{1-\alpha}}} \right) \\ &< \frac{2c}{C_3^2} + 2c(2p)^{\frac{2(1-\beta)}{1-\alpha}} \sum_{m=1}^{\infty} \frac{1}{m^{\frac{2(1-\beta)}{1-\alpha}}} =: \xi_1' < \infty.\end{aligned}$$

If $\beta \in]\alpha, 1 + \frac{\alpha-1}{2}[$, then we get

$$0 < \frac{-\alpha+\beta}{1-\alpha} < \frac{1}{2}$$

and

$$\frac{1}{2} < 1 - \frac{-\alpha+\beta}{1-\alpha} = \frac{1-\beta}{1-\alpha} < 1.$$

Thus, Eqs. (11.2.11), (11.2.12) and (11.3.12) imply that

$$\begin{aligned}\sum_{n=1}^{\infty} \frac{\lambda_n^{2\beta}}{|\lambda - \nu_n|^2} &< \frac{2c}{C_3^2} + \sum_{n<k-1} \frac{c(2p)^{\frac{2(1-\beta)}{1-\alpha}}}{c_1^2\left(\frac{1-\beta}{1-\alpha}\right)^2 (k-1-n)^{\frac{2(1-\beta)}{1-\alpha}}} + \sum_{n>k} \frac{c(2p)^{\frac{2(1-\beta)}{1-\alpha}}}{c_2^2\left(\frac{1-\beta}{1-\alpha}\right)^2 (n-k)^{\frac{2(1-\beta)}{1-\alpha}}} \\ &\leq \frac{2c}{C_3^2} + \frac{c(2p)^{\frac{2(1-\beta)}{1-\alpha}}}{c'^2\left(\frac{1-\beta}{1-\alpha}\right)^2} \left(\sum_{n<k-1} \frac{1}{(k-1-n)^{\frac{2(1-\beta)}{1-\alpha}}} + \sum_{m=1}^{\infty} \frac{1}{m^{\frac{2(1-\beta)}{1-\alpha}}} \right) \\ &< \frac{2c}{C_3^2} + \frac{2c(2p)^{\frac{2(1-\beta)}{1-\alpha}}}{c'^2\left(\frac{1-\beta}{1-\alpha}\right)^2} \sum_{m=1}^{\infty} \frac{1}{m^{\frac{2(1-\beta)}{1-\alpha}}} =: \xi_2' < \infty,\end{aligned}$$

where $c' := \min\{c_1, c_2\}$. Consequently, for $\beta \in]0, 1 + \frac{\alpha-1}{2}[$, we have

$$\sum_{n=1}^{\infty} \frac{\lambda_n^{2\beta}}{|\lambda - \nu_n|^2} < \max\{\xi_1', \xi_2'\} =: \xi_2'.$$

Moreover, following some ideas of above with $\frac{\beta}{2}$, we get

$$\sum_{n=1}^{\infty} \frac{\lambda_n^{\beta}}{|\lambda - \nu_n|^2} < \frac{2c}{C_3^2} + \frac{2c(2p)^{\frac{2-\beta}{1-\alpha}}}{c'^2\left(\frac{1-\frac{\beta}{2}}{1-\alpha}\right)^2} \sum_{m=1}^{\infty} \frac{1}{m^{\frac{2-\beta}{1-\alpha}}} =: \xi_3' < \infty.$$

So, the series

$$\sum_n \frac{2ab\lambda_n^{\beta}}{|\lambda - \nu_n|^2}$$

and

$$\sum_n \frac{b^2\lambda_n^{2\beta}}{|\lambda - \nu_n|^2}$$

are convergent. Hence, let ξ' be a positve constant such that

$$b^2\xi_2' + 2ab\xi_3' = \xi'.$$

Then,

$$\sum_{n=1}^{\infty} \frac{2ab\lambda_n^{\beta}}{|\lambda - \nu_n|^2} + \sum_{n=1}^{\infty} \frac{b^2\lambda_n^{2\beta}}{|\lambda - \nu_n|^2} < \xi'.$$

Consequently, by a similar reasoning as the proof of Theorem 11.3.1, we have for $|\varepsilon| < W_2$

$$n(\overline{\mathbb{D}}(0,r), T(\varepsilon)) = n(\overline{\mathbb{D}}(0,r), T_0) + O(1),$$

where

$$W_2 := \min\left\{ \frac{1}{q + \sqrt{\xi' + 4pa^2\frac{l+1}{l^2}}}, \frac{1}{q + \frac{h}{h-2l}\sqrt{\xi' + 2a^2\frac{p}{h}(\pi + \frac{2}{h})}} \right\}.$$

Q.E.D.

11.4 Notes and Remarks

- For non-self-adjoint perturbations, in a Banach space, of a self-adjoint operator, the crucial problem is the study of some spectral properties. For instance, the existence of a basis (possibly with parentheses) of root vectors is, in this case, an important property. In order to show the existence of such a basis, several authors studied the comportment of the eigenvalues of this problem and established different conditions in terms of the spectrum (see [4–7]). In fact, many non-self-adjoint ordinary differential operators can be considered as a perturbation $T + B$ of a leading self-adjoint component T by its subordinate B. In [5], A. S. Markus claimed that

$$G = T + B$$

admits an unconditional basis with parentheses of root vectors if B is p-subordinate to T and the eigenvalue-counting function of T satisfy a certain asymptotic growth condition. One might ask whether we can construct a basis if the p-subordinate condition is relaxed. A positive answer is given by Shkalikov [6]. He assumed that T is positive, self-adjoint with discrete spectrum and its eigenvalues $(\mu_n)_{n=1}^{\infty}$ are not condense, i.e.,

$$\mu_{n+q} - \mu_n \geq 1, \quad \text{for some } q \in \mathbb{N}^*. \tag{11.4.1}$$

Further, he required that B verify

$$\|B\psi_n\| \leq b, \tag{11.4.2}$$

where $(\psi_n)_{n=1}^{\infty}$ is an orthonormal system of eigenvectors associated to the eigenvalues $(\mu_n)_{n=1}^{\infty}$ of T. Under these assumptions, he established an asymptotic relation between the eigenvalue-counting functions of G and T and he claimed that the system of root vectors of G forms a basis with parentheses in $\mathscr{H}$. More precisely, he proved the existence of a spectral condition

$$n(\overline{\mathbb{D}}(0,r), G) - n(\overline{\mathbb{D}}(0,r), T) = O(1),$$

under which he guarantees the existence of a basis with parentheses of root vectors (see [6, Theorem 2]).

- The asymptotic behavior of the eigenvalues of self-adjoint differential operators and their perturbations has long been studied by many authors. Apparently, the first important results for ordinary differential operators were obtained by Birkhoff [8], and for partial differential operators, by Weyl [9] (see the monograph by Naimark [10] for more details). Keldysh [11] proved the first results for general operators in a Hilbert space using the Tauberian technique. The most complete and the sharpest results for compact perturbations and for the so-called p-subordinate perturbations of self-adjoint operators are due to Markus and Matsaev [3]. Mainly in [3], the authors often considered an unbounded operator G on a separable Hilbert space written as a perturbation of a leading normal component T by its p-subordinate B, i.e.,

$$\|B\varphi\| \leq b\|T\varphi\|^p\|\varphi\|^{1-p} \quad \text{for every } \varphi \in \mathscr{D}(T), \tag{11.4.3}$$

where $p \in [0, 1]$ and b is a strictly positive constant.
- It should be noted here that the sharpest results in the comparison of the spectrum which are due to Markus and Matsaev [3] dealt with the subordinate condition (11.4.3). In fact, under Eq. (11.4.3), the authors in [3, 5] developed several approaches on the distribution of the eigenvalues of the perturbed operator G. More precisely, they established asymptotic relations between the eigenvalue-counting functions of G and T and provided sufficient conditions to prove the existence of bases.
- Later, making Eq. (11.4.3) much weaker, Shkalikov [6] obtained an estimate about the distribution of the eigenvalues of G in terms of the one of T. Indeed, he

replaced Eq. (11.4.3) by (11.4.2), where $(\psi_n)_{n=1}^{\infty}$ is an orthonormal system of eigenvectors of T associated to the eigenvalues $(\mu_n)_{n=1}^{\infty}$ and assumed that T is positive and self-adjoint and the spectrum of T is discrete and its eigenvalues $(\mu_n)_{n=1}^{\infty}$ are not condense. Using Eqs. (11.4.2) and (11.4.1), he gave a precise description of the localization of the spectrum of the perturbed operator G. More precisely, he constructed a half-strip containing all the eigenvalues of T and G. As a consequence, he claimed that the difference between the eigenvalue-counting functions of G and T is bounded by a constant, i.e.,

$$n(\overline{\mathbb{D}}(0, r), G) - n(\overline{\mathbb{D}}(0, r), T) = O(1), \tag{11.4.4}$$

where $n(\overline{\mathbb{D}}(0, r), T)$ (respectively, $n(\overline{\mathbb{D}}(0, r), G)$) denotes the sum of multiplicities of each eigenvalues of the operator T (respectively, G) contained in the disk $\overline{\mathbb{D}}(0, r) := \{\lambda \in \mathbb{C} \text{ such that } |\lambda| \leq r\}$ (see [6, Theorem 1]).

- It is interesting to note that this outcome is of importance on itself. In fact, it is quite hard to control the jump of the eigenvalue-counting function of G. For instance, the main results stated in [3] guarantee the equivalence between the eigenvalue-counting functions of G and T if the operator B is T-compact and the growth of $n(\overline{\mathbb{D}}(0, r), T)$ is regular. Further, [3, Theorem 9.2] characterize the behavior of $n(\overline{\mathbb{D}}(0, r), G)$ in terms of $n(\overline{\mathbb{D}}(0, r), T)$ if Eq. (11.4.3) holds. However, [6, Theorem 1] improves these theorems in the sense that Eqs. (11.4.2) and (11.4.1), under which Eq. (11.4.4) holds, are much weaker.
- Although, in certain applications, H. Ellouz, I. Feki, and A. Jeribi show in [12] that it is quite hard to verify Eq. (11.4.2) and all the obtained results in literature are restrictive. Among these applications, the authors cite the Gribov operator and the integro-differential operator suggested by Filippi et al. [14] to describe the radiation of a vibrating structure in a light fluid.
- As a tentative approach to grapple with such difficulty, the authors in [12] have the idea to extend the outcomes obtained by Shkalikov [6] to an abstract analytic operator introduced by Nagy in [15] and considered later in [1, 2, 4, 13, 16–20]. Their results improve the ones on [4, Theorem 3.1] since the spectral analysis method developed in [4] is based on the fact that the eigenvalues $(\lambda_n)_{n=1}^{\infty}$ of T_0 are with multiplicity one, whereas in [4], the authors control the jump of $n(\overline{\mathbb{D}}(0, r), T(\varepsilon))$ in terms of $n(\overline{\mathbb{D}}(0, r), T_0)$ by assuming that the eigenvalues $(\lambda_n)_{n=1}^{\infty}$ of T_0 are with finite multiplicity and do not condense.
- Moreover, the results of [12] extend the work of [6, Theorem 1] to an analytic operator instead of the sum of two operators; and this generalization leads to the study the asymptotic behavior of the eigenvalues of some analytic operators where the criteria of Shkalikov [6] can not be applied.

(i) As the spectrum of T_0 is discrete so is the spectrum of T_r. Furthermore, we have

$$n(\overline{\mathbb{D}}(0, r), T_r) = n(\overline{\mathbb{D}}(0, r), T_0),$$

where $n(\overline{\mathbb{D}}(0,r), T_r)$, (respectively, $n(\overline{\mathbb{D}}(0,r), T_0)$) denotes the sum of multiplicities of each eigenvalues of the operator T_r (respectively, T_0) contained in the closed disk $\overline{\mathbb{D}}(0,r)$.

(ii) $D_\varepsilon(\lambda)$ is well defined for $\lambda \notin \sigma(T(\varepsilon) - K_r)$. In fact, due to [21, p. 160], $K_\varepsilon(\lambda)$ is with finite rank for $\lambda \notin \sigma(T(\varepsilon) - K_r)$. Furthermore, $D_\varepsilon(\lambda)$ is a meromorphic function. Indeed, let $\lambda, \lambda_1 \in \mathbb{C}\backslash\sigma(T(\varepsilon) - K_r)$.

$$\lim_{\lambda \to \lambda_1} \frac{R(\lambda, T(\varepsilon) - K_r) - R(\lambda_1, T(\varepsilon) - K_r)}{\lambda - \lambda_1}$$

$$= \lim_{\lambda \to \lambda_1} \frac{(\lambda - \lambda_1) R(\lambda, T(\varepsilon) - K_r) R(\lambda_1, T(\varepsilon) - K_r)}{\lambda - \lambda_1}$$
$$= R(\lambda_1, T(\varepsilon) - K_r)^2.$$

Then, $R(\lambda, T(\varepsilon) - K_r)$ is holomorphic in $\mathbb{C}\backslash\sigma(T(\varepsilon) - K_r)$. Thus, if $\lambda \in \sigma(T(\varepsilon) - K_r)$, then λ is a pole for $R(\lambda, T(\varepsilon) - K_r)$. So, $R(\lambda, T(\varepsilon) - K_r)$ is meromorphic. Consequently, by Proposition 2.5.1(ii), $D_\varepsilon(\lambda)$ is a meromorphic function.

References

1. S. Charfi, A. Damergi, A. Jeribi, On a Riesz basis of finite-dimensional invariant subspaces and application to Gribov operator in Bargmann space. Line. Mult. Alge. **61**(11), 1577–1591 (2013)
2. S. Charfi, A. Jeribi, I. Walha, Riesz basis property of families of nonharmonic exponential and application to a problem of a radiation of a vibrating structure in a light fluid. Numer. Funct. Anal. Optim. **32**(4), 370–382 (2011)
3. A.S. Markus, V.I. Matsaev, Comparison theorems for spectra of linear operators, and spectral asymptotics. Tr. Mosk. Mat. Obshch. **45**, 133–181 (1982) [Trans. Moscow Math. Soc. (1), 139–187 (1984)]
4. I. Feki, A. Jeribi, R. Sfaxi, On a Schauder basis related to the eigenvectors of a family of non-self-adjoint analytic operators and applications. Anal. Math. Phys. **3**(4), 311–331 (2013)
5. A.S. Markus, *Introduction to the Spectral Theory of Polynomial Operator Pencils*, vol. 71 (Amer. Math. Soci., Providence, RI, 1988)
6. A.A. Shkalikov, On the basis property of root vectors of a perturbed self-adjoint operator. Proc. Steklov Inst. Math. **269**, 284–298 (2010)
7. C. Wyss, Riesz bases for p-subordinate perturbations of normal operator. J. Funct. Anal. **258**, 208–240 (2010)
8. G.D. Birkhoff, Boundary value and expansion problems of ordinary linear differential equations. Trans. Am. Math. Soc. **9**, 373–395 (1908)
9. H. Weyl, Das asymptotische verteilungsgesetz der eigenwerte linearer partieller differentialgleichungen. Math. Ann. **71**, 441–479 (1912)
10. M.A. Naimark, *Linear Differential Operators* (Nauka, Russian, Moscow, 1969)
11. M.V. Keldysh, On eigenvalues and eigenfunctions of some classes of non-self-adjoint equations. Dokl. Akad. Nauk SSSR **77**, 11–14 (In Russian) (1951)
12. H. Ellouz, I. Feki, A. Jeribi, On the asymptotic behavior of the eigenvalues of an analytic operator in the sense of Kato and applications. Serdica Math. J. **45**(1), 55–88 (2019)

13. H. Ellouz, I. Feki, A. Jeribi, Non-orthogonal fusion frames of an analytic operator and application to a one-dimensional wave control system. Mediterr. J. Math. **16**(2), 16:52 (2019)
14. P.J.T. Filippi, O. Lagarrigue, P.O. Mattei, Perturbation method for sound radiation by a vibrating plate in a light fluid: comparison with the exact solution. J. Sound Vib. **177**, 259–275 (1994)
15. B. Sz, Nagy, Perturbations des transformations linéaires fermées. Acta Sci. Math. Szeged. **14**, 125–137 (1951)
16. N. Ben Ali, A. Jeribi, On the Riesz basis of a family of analytic operators in the sens of Kato and application to the problem of radiation of a vibriting structure in a light fluid. J. Math. Anal. Appl. **320**(1), 78–94 (2006)
17. I. Feki, A. Jeribi, R. Sfaxi, On an unconditionl basis of generalized eigenvectors of an analytic operator and application to a problem of radiation of a vibrating structure in a light fluid. J. Math. Anal. Appl. **375**, 261–269 (2011)
18. I. Feki, A. Jeribi, R. Sfaxi, On a Riesz basis of eigenvectors of a non-self-adjoint analytic operator and applications. Line. Mult. Alge. **62**(8), 1049–1068 (2014)
19. H. Ellouz, I. Feki, A. Jeribi, On a Riesz basis of exponentials related to the eigenvalues of an analytic operator and application to a non-self-adjoint problem deduced from a perturbation method. J. Math. Phys. **54**, 112101 (2013)
20. A. Jeribi, A. Intissar, On an Riesz basis of generalized eigenvectors of the nonself-adjoint problem deduced from a perturbation method for sound radiation by a vibrating plate in a light fluid. J. Math. Anal. Appl. **292**, 1–16 (2004)
21. T. Kato, *Perturbation Theory for Linear Operators* (Springer, Berlin, 1980)

Chapter 12
On the Basis Property of Root Vectors Related to a Non-self-adjoint Analytic Operator

In this chapter, we focus on the study of the asymptotic behavior of the eigenvalues of an analytic operator in the sense of Kato. More precisely, we investigate the behavior of the spectrum of the perturbed operator $T(\varepsilon)$ under a finite rank perturbation and we develop perturbation theory for these new type conditions. We concentrate exclusively on the study of the basis property of root vectors related to a non-self-adjoint analytic operator. We replace the p-subordination condition used in [8] by a weaker one which is more suitable for applications in the theory of differential operators and we develop perturbation theory for these new type conditions (see [10–21]). The novelty allows us to prove some results about the completeness of the system of root vectors of the perturbed operator, the existence of a basis with parentheses of root vectors and to give an idea about the location of the spectrum of the considered operators.

12.1 Completeness of the System of Root Vectors of $T(\varepsilon)$

The aim of this part is to establish the completeness of the system of root vectors of the perturbed operator $T(\varepsilon)$ in X.

Theorem 12.1.1 *Suppose that hypotheses $(H2)$ with $\beta \in]0, \frac{1}{2}[$, $(H9)$ and $(H10)$ are verified. If the resolvent of the operator T_0 is compact, then for $|\varepsilon| < \frac{1}{q+\beta b}$, the system of root vectors of $T(\varepsilon)$, given in (10.1.1), is complete in X.* ◇

Proof In view of both hypotheses, $(H9)$ and $(H10)$, we get that the operator T_0 is self-adjoint with compact resolvent. Furthermore, by using hypothesis $(H10)$, we have

A. Jeribi, *Perturbation Theory for Linear Operators*,
https://doi.org/10.1007/978-981-16-2528-2_12

$$\lambda_{n+1} - \lambda_1 = \underbrace{\lambda_{n+1} - \lambda_{(n+1)-p}}_{\geq 1} + \underbrace{\lambda_{(n+1)-p} - \lambda_{(n+1)-2p}}_{\geq 1} + \cdots + \underbrace{\lambda_{1+p} - \lambda_{(n+1)-\frac{n}{p}p(=1)}}_{\geq 1}$$
$$\geq \frac{n}{p}. \tag{12.1.1}$$

Hence, Eq. (12.1.1) yields

$$\lambda_n \geq \frac{n-1}{p} + \lambda_1.$$

Thus, there is $P > 1$ such that the series

$$\sum_{n=1}^{\infty} \left(\frac{1}{\lambda_n} \right)^P$$

is convergent. As a consequence, the resolvent of the operator T_0 belongs to the Carleman-class $\mathscr{C}_P$. Furthermore, in virtue of Proposition 11.2.1, the operator $B(\varepsilon)$, given in (11.2.1), is T_0-compact for $|\varepsilon| < \frac{1}{q+\beta b}$. Hence, Theorem 6.1.4 implies that for $|\varepsilon| < \frac{1}{q+\beta b}$, the system of root vectors of the operator $T(\varepsilon)$ is complete in X. This achieves the proof.

Corollary 12.1.1 *Assume that hypotheses* $(H2)$ *with* $\beta \in]0, \frac{1}{2}[$ *and* $(H9)$ *are verified. Forthermore, suppose that*

$$\lambda_{n+p}^{1-\alpha} - \lambda_n^{1-\alpha} \geq 1, \tag{12.1.2}$$

where $0 \leq \alpha < 1$. *Then, for* $\beta \in]0, 1 + \frac{\alpha-1}{2}[$ *and* $|\varepsilon| < \frac{1}{q+\beta b}$, *the system of root vectors of the operator* $T(\varepsilon)$ *is complete in* X. ◇

Proof By using Eq. (11.2.6), we have

$$\lambda_n^{1-\alpha} \geq \frac{n-1}{p} + \lambda_1^{1-\alpha}.$$

Thus, there is $P > 1 - \alpha$ such that the series

$$\sum_{n=1}^{\infty} \frac{1}{\lambda_n^P}$$

is convergent. Since T_0 is self-adjoint with compact resolvent, then the resolvent of the operator T_0 belongs to the Carleman-class $\mathscr{C}_P$. Furthermore, due to Proposition 11.2.1, the operator $B(\varepsilon)$, given in (11.2.1), is T_0-compact for $|\varepsilon| < \frac{1}{q+\beta b}$. So, according to Theorem 6.1.4, we infer that the system of root vectors of the operator $T(\varepsilon)$ is complete in X for $|\varepsilon| < \frac{1}{q+\beta b}$. Q.E.D.

12.2 Basis with Parentheses of Root Vectors of $T(\varepsilon)$

In Theorem 12.1.1, we have showed that the system of root vectors of $T(\varepsilon)$ is complete. The question that occurs is whether this system forms a basis in X. In other words, if

$$P_{n,\varepsilon} = \frac{1}{2\pi i}\int_{\partial\Delta_n}(\lambda - T(\varepsilon))^{-1}\mathrm{d}\lambda$$

designates the spectral projection corresponding to the spectrum of $T(\varepsilon)$ inside Δ_n, where Δ_n is a bounded closed isolated part of the spectrum of the operator $T(\varepsilon)$, then is the series

$$\sum_n P_{n,\varepsilon} f$$

convergent and its sum is it f. To give a positive answer to this question, we shall show some preliminary results.

12.2.1 *Localization of the Spectrum of $T(\varepsilon)$*

Lemma 12.2.1 *Let τ be an arbitrary positive number. If $|Im\lambda| \geq \tau$, then for $|\varepsilon| < \frac{1}{q}$, there is a positive constant $N(\varepsilon, a, p, q, \tau)$ such that*

$$\sum_{n=1}^{\infty}\frac{\|B(\varepsilon)\varphi_n\|^2}{|\lambda - \lambda_n|^2} < N(\varepsilon, a, p, q, \tau), \tag{12.2.1}$$

where $B(\varepsilon)$ is given in (11.2.1). If $Re\lambda \leq -\tau$, then for $|\varepsilon| < \frac{1}{q}$, there is also a positive constant $N_1(\varepsilon, a, p, q, \tau)$ such that

$$\sum_{n=1}^{\infty}\frac{\|B(\varepsilon)\varphi_n\|^2}{|\lambda - \lambda_n|^2} < N_1(\varepsilon, a, p, q, \tau). \tag{12.2.2}$$

◇

Proof Let $n \in \mathbb{N}^*$ and let λ_n be the nth eigenvalue of T_0. Then,

$$\begin{aligned}\|B(\varepsilon)\varphi_n\| &= \|(\varepsilon T_1 + \varepsilon^2 T_2 + \cdots)\varphi_n\| \\ &\leq \sum_{i=1}^{\infty}\|\varepsilon^i T_i\varphi_n\|.\end{aligned}$$

By using the hypothesis $(H2)$, we get

$$\|B(\varepsilon)\varphi_n\| \leq \sum_{i=1}^{\infty} |\varepsilon|^i q^{i-1}(a\|\varphi_n\| + b\|T_0\varphi_n\|^{\beta}\|\varphi_n\|^{1-\beta})$$
$$\leq \sum_{i=1}^{\infty} |\varepsilon|^i q^{i-1}(a + b\lambda_n^{\beta}). \tag{12.2.3}$$

Thus, for $|\varepsilon| < \frac{1}{q}$, and in view from Eq. (12.2.3), we have

$$\frac{\|B(\varepsilon)\varphi_n\|^2}{|\lambda - \lambda_n|^2} \leq \frac{|\varepsilon|^2}{(1-|\varepsilon|q)^2}\left(\frac{a^2}{|\lambda-\lambda_n|^2} + \frac{2ab\lambda_n^{\beta}}{|\lambda-\lambda_n|^2} + \frac{b^2\lambda_n^{2\beta}}{|\lambda-\lambda_n|^2}\right). \tag{12.2.4}$$

Now, let $\sigma = Re\lambda$ and $\lambda = \sigma \pm i\tau$, where $\tau > 0$. So, there is $k \in \mathbb{N}^*$ such that $\lambda_{k-1} \leq \sigma$ and $\lambda_k > \sigma$. Hence,

$$\lambda_k - \sigma > \lambda_k^{\beta}\left(\lambda_k^{1-\beta} - \sigma^{1-\beta}\right) \geq C_1\lambda_k^{\beta} \ \ (C_1 > 0) \tag{12.2.5}$$

and

$$|\lambda - \lambda_{k-1}| \geq ||\lambda| - \lambda_{k-1}| > |\lambda|^{\beta}\left(|\lambda|^{1-\beta} - \lambda_{k-1}^{1-\beta}\right) \geq C_2\lambda_{k-1}^{\beta} \ \ (C_2 > 0). \tag{12.2.6}$$

So, both Eqs. (12.2.5) and (12.2.6) imply that

$$\sum_{n=1}^{\infty}\frac{\lambda_n^{2\beta}}{|\lambda_n-\sigma|^2} < \frac{1}{C_1^2} + \frac{1}{C_2^2} + \sum_{n<k-1}\frac{\lambda_n^{2\beta}}{|\lambda_n-\sigma|^2} + \sum_{n>k}\frac{\lambda_n^{2\beta}}{|\lambda_n-\sigma|^2}$$
$$\leq \frac{2}{C^2} + \sum_{n<k-1}\frac{\lambda_n^{2\beta}}{|\lambda_n-\sigma|^2} + \sum_{n>k}\frac{\lambda_n^{2\beta}}{|\lambda_n-\sigma|^2}, \tag{12.2.7}$$

where $C := \min\{C_1, C_2\}$. Furthermore, since $\beta \in]0, \frac{1}{2}[$, and for $n < k-1$, we get

$$\begin{aligned}\sigma - \lambda_n &> \lambda_{k-1} - \lambda_n \\ &> (1-\beta)\lambda_n^{\beta}\left(\lambda_{k-1} - \lambda_n\right)^{1-\beta}\left(\lambda_{k-1} - \lambda_n\right)^{\beta}\lambda_n^{-\beta} \\ &\geq \gamma_1(1-\beta)\lambda_n^{\beta}\left(\lambda_{k-1} - \lambda_n\right)^{1-\beta}, \ \ 0 < \gamma_1 < 1.\end{aligned} \tag{12.2.8}$$

So, for $n > k$, we get

$$\begin{aligned}\lambda_n - \sigma &> \lambda_n - \lambda_k \\ &> (1-\beta)\lambda_n^{\beta}\left(\lambda_n - \lambda_k\right)^{1-\beta}\left(\lambda_n - \lambda_k\right)^{\beta}\lambda_n^{-\beta} \\ &\geq \gamma_2(1-\beta)\lambda_n^{\beta}\left(\lambda_n - \lambda_k\right)^{1-\beta}, \ \ 0 < \gamma_2 < 1.\end{aligned} \tag{12.2.9}$$

Hence, Eqs. (12.2.7), (12.2.8) and (12.2.9) yield

$$\sum_{n=1}^{\infty} \frac{\lambda_n^{2\beta}}{|\lambda_n - \sigma|^2} < \frac{2}{C^2} + \sum_{n<k-1} \frac{1}{\gamma_1^2 (1-\beta)^2 (\lambda_{k-1} - \lambda_n)^{2(1-\beta)}} + \sum_{n>k} \frac{1}{\gamma_2^2 (1-\beta)^2 (\lambda_n - \lambda_k)^{2(1-\beta)}}.$$

As a consequence, if we designate by $\gamma := \min\{\gamma_1, \gamma_2\}$, we get

$$\sum_{n=1}^{\infty} \frac{\lambda_n^{2\beta}}{|\lambda_n - \sigma|^2} < \frac{2}{C^2} + \sum_{n<k-1} \frac{1}{\gamma^2 (1-\beta)^2 (\lambda_{k-1} - \lambda_n)^{2(1-\beta)}} + \sum_{n>k} \frac{1}{\gamma^2 (1-\beta)^2 (\lambda_n - \lambda_k)^{2(1-\beta)}}. \tag{12.2.10}$$

As $\lambda_n \geq \frac{n-1}{p} + \lambda_1$, and Eq. (12.2.10) yields

$$\begin{aligned}
\sum_{n=1}^{\infty} \frac{\lambda_n^{2\beta}}{|\lambda - \lambda_n|^2} &< \sum_{n=1}^{\infty} \frac{\lambda_n^{2\beta}}{|\lambda_n - \sigma|^2} \\
&< \frac{2}{C^2} + \sum_{n<k-1} \frac{p^{2(1-\beta)}}{\gamma^2 (1-\beta)^2 (k-1-n)^{2(1-\beta)}} + \sum_{n>k} \frac{p^{2(1-\beta)}}{\gamma^2 (1-\beta)^2 (n-k)^{2(1-\beta)}} \\
&\leq \frac{2}{C^2} + \sum_{n<k-1} \frac{p^{2(1-\beta)}}{\gamma^2 (1-\beta)^2 (k-1-n)^{2(1-\beta)}} + \sum_{m=1}^{\infty} \frac{p^{2(1-\beta)}}{\gamma^2 (1-\beta)^2 m^{2(1-\beta)}} \\
&< \frac{2}{C^2} + \frac{2p^{2(1-\beta)}}{\gamma^2 (1-\beta)^2} \sum_{m=1}^{\infty} \frac{1}{m^{2(1-\beta)}} =: \xi_1 < \infty.
\end{aligned}$$

Furthermore, if we make the same reasoning as above with $\frac{\beta}{2}$, we have

$$\sum_{n=1}^{\infty} \frac{\lambda_n^{\beta}}{|\lambda - \lambda_n|^2} < \frac{2}{C^2} + \frac{8p^{2-\beta}}{\gamma^2 (2-\beta)^2} \sum_{m=1}^{\infty} \frac{1}{m^{2-\beta}} =: \xi_2 < \infty.$$

Hence, the series

$$\sum_n \frac{2ab\lambda_n^{\beta}}{|\lambda - \lambda_n|^2}$$

and

$$\sum_n \frac{b^2\lambda_n^{2\beta}}{|\lambda-\lambda_n|^2}$$

are convergent. Thus, there is $\xi > 0$ verifying

$$\sum_{n=1}^{\infty} \frac{2ab\lambda_n^{\beta}}{|\lambda-\lambda_n|^2} + \sum_{n=1}^{\infty} \frac{b^2\lambda_n^{2\beta}}{|\lambda-\lambda_n|^2} < b^2\xi_1 + 2ab\xi_2 =: \xi. \tag{12.2.11}$$

- Consider $|Im\lambda| \geq \tau$. For $|\varepsilon| < \frac{1}{q}$, it follows from Eqs. (12.2.4) and (12.2.11) that

$$\sum_{n=1}^{\infty} \frac{\|B(\varepsilon)\varphi_n\|^2}{|\lambda-\lambda_n|^2} < \frac{|\varepsilon|^2}{(1-|\varepsilon|q)^2}\left(\xi + a^2\sum_{n=1}^{\infty}\frac{1}{(\lambda_n-\sigma)^2+\tau^2}\right).$$

  Moreover, hypothesis $(H10)$ implies that

$$\lambda_{k+j+sp} - \sigma \geq s$$

  and

$$\sigma - \lambda_{k-j-sp-1} \geq s,$$

  where $j = 0, \ldots, p-1$ and $s = 0, 1, \ldots$. So, for $|\varepsilon| < \frac{1}{q}$, we get

$$\sum_{n=1}^{\infty} \frac{\|B(\varepsilon)\varphi_n\|^2}{|\lambda-\lambda_n|^2} < \frac{|\varepsilon|^2}{(1-|\varepsilon|q)^2}\left(\xi + a^2 p\left(\sum_{s=0}^{\frac{k}{p}-1}\frac{1}{s^2+\tau^2} + \sum_{s=0}^{\infty}\frac{1}{s^2+\tau^2}\right)\right).$$

  Hence, for $|\varepsilon| < \frac{1}{q}$, we get

$$\begin{aligned}\sum_{n=1}^{\infty} \frac{\|B(\varepsilon)\varphi_n\|^2}{|\lambda-\lambda_n|^2} &< \frac{|\varepsilon|^2}{(1-|\varepsilon|q)^2}\left(\xi + 2pa^2\left(\sum_{s=1}^{\infty}\frac{1}{s^2+\tau^2} + \frac{1}{\tau^2}\right)\right)\\ &< \frac{|\varepsilon|^2}{(1-|\varepsilon|q)^2}\left(\xi + 2pa^2\left(\frac{1}{\tau^2} + \int_0^{\infty}\frac{dx}{x^2+\tau^2}\right)\right)\\ &\leq N(\varepsilon, a, p, q, \tau),\end{aligned}$$

  where

$$N(\varepsilon, a, p, q, \tau) := \frac{|\varepsilon|^2}{(1-|\varepsilon|q)^2}\left(\xi + a^2\frac{p}{\tau}\left(\pi + \frac{2}{\tau}\right)\right).$$

- Now, if $Re\lambda \leq -\tau$, then using hypothesis $(H10)$, we have

$$\lambda_{1+j+sp} - \sigma \geq s - \sigma \geq s + \tau, \tag{12.2.12}$$

where $j = 0, \dots, p-1$ and $s = 0, 1, \dots$, since $\lambda_{1+j} > 0$ and $\lambda_{1+j+sp} - \sigma \geq s + \lambda_{1+j} - \sigma$. Thus, Eqs. (12.2.4), (12.2.11) and (12.2.12) imply that for $|\varepsilon| < \frac{1}{q}$, we get

$$\begin{aligned}\sum_{n=1}^{\infty} \frac{\|B(\varepsilon)\varphi_n\|^2}{|\lambda - \lambda_n|^2} &< \frac{|\varepsilon|^2}{(1-|\varepsilon|q)^2}\left(\xi + a^2 \sum_{n=1}^{\infty} \frac{1}{|\sigma - \lambda_n|^2}\right) \\ &\leq \frac{|\varepsilon|^2}{(1-|\varepsilon|q)^2}\left(\xi + pa^2\left(\sum_{s=0}^{\infty} \frac{1}{(\tau+s)^2}\right)\right) \\ &< \frac{|\varepsilon|^2}{(1-|\varepsilon|q)^2}\left(\xi + pa^2\left(\int_0^{\infty} \frac{\mathrm{d}x}{(\tau+x)^2} + \frac{1}{\tau^2}\right)\right) \\ &\leq N_1(\varepsilon, a, p, q, \tau),\end{aligned}$$

where

$$N_1(\varepsilon, a, p, q, \tau) := \frac{|\varepsilon|^2}{(1-|\varepsilon|q)^2}\left(\xi + a^2\frac{p}{\tau}\left(1 + \frac{1}{\tau}\right)\right).$$

Now, let us give the relation between $\|B(\varepsilon)(\lambda - T_0)^{-1}\|$ and

$$\sum_{n=1}^{\infty} \frac{\|B(\varepsilon)\varphi_n\|^2}{|\lambda - \lambda_n|^2}.$$

Proposition 12.2.1 *We have the following estimation*

$$\|B(\varepsilon)(\lambda - T_0)^{-1}\|^2 \leq \sum_{n=1}^{\infty} \frac{\|B(\varepsilon)\varphi_n\|^2}{|\lambda - \lambda_n|^2}.$$ ◇

Proof Let $f \in X$ such that $\|f\| = 1$. Since T_0 is self-adjoint with compact resolvent, then

$$T_0 f = \sum_{n=1}^{\infty} \lambda_n \langle f, \varphi_n \rangle \varphi_n$$

be its spectral decomposition, where λ_n is the nth eigenvalue of the operator T_0 associated to the eigenvector φ_n. Hence,

$$\begin{aligned}\|B(\varepsilon)(\lambda - T_0)^{-1} f\|^2 &= \left\| \sum_{n=1}^{\infty} \frac{\langle f, \varphi_n \rangle B(\varepsilon)\varphi_n}{\lambda - \lambda_n} \right\|^2 \\ &\leq \left(\sum_{n=1}^{\infty} |\langle f, \varphi_n \rangle| \frac{\|B(\varepsilon)\varphi_n\|}{|\lambda - \lambda_n|} \right)^2 \\ &\leq \left(\sum_{n=1}^{\infty} |\langle f, \varphi_n \rangle|^2 \right) \left(\sum_{n=1}^{\infty} \frac{\|B(\varepsilon)\varphi_n\|^2}{|\lambda - \lambda_n|^2} \right) \\ &\leq \sum_{n=1}^{\infty} \frac{\|B(\varepsilon)\varphi_n\|^2}{|\lambda - \lambda_n|^2}.\end{aligned}$$

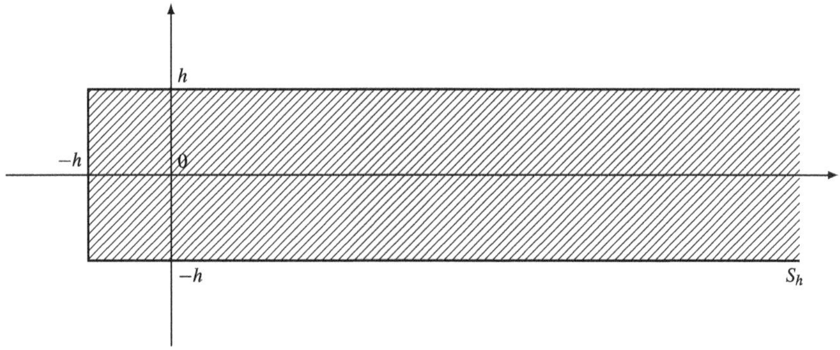

Fig. 12.1 The domain S_h

As a consequence, we get

$$\|B(\varepsilon)(\lambda - T_0)^{-1}\|^2 \leq \sum_{n=1}^{\infty} \frac{\|B(\varepsilon)\varphi_n\|^2}{|\lambda - \lambda_n|^2}.$$

This completes the proof. Q.E.D.

We denote by

$$S_h := \{\lambda \in \mathbb{C} \text{ such that } |Im\lambda| < h \text{ and } Re\lambda > -h\},$$

with $h > 0$ (see Fig. 12.1).

Proposition 12.2.2 *For $|\varepsilon|$ small enough, the spectrum, $\sigma(T(\varepsilon))$, of the operator $T(\varepsilon)$ lies in the half-strip S_h.* ◇

Proof Let $\lambda \in \mathbb{C}$ such that $|Im\lambda| \geq h$ or $Re\lambda \leq -h$. The fact that T_0 is self-adjoint and positive, then

$$\lambda - T(\varepsilon) = [I - B(\varepsilon)(\lambda - T_0)^{-1}](\lambda - T_0). \tag{12.2.13}$$

Further, combining Eq. (12.2.1) together with Proposition 12.2.1, we obtain for $|Im\lambda| \geq h$ and $|\varepsilon| < \frac{1}{q}$,

$$\|B(\varepsilon)(\lambda - T_0)^{-1}\|^2 < N(\varepsilon, a, p, q, h). \tag{12.2.14}$$

So, for $|Im\lambda| \geq h$ and $|\varepsilon| < \frac{1}{q+\sqrt{\xi + a^2\frac{p}{h}(\pi + \frac{2}{h})}}$,

$$\|B(\varepsilon)(\lambda - T_0)^{-1}\| < 1.$$

On the other hand, for $Re\lambda \leq -h$ and $|\varepsilon| < \frac{1}{q}$, Eq. (12.2.2) and Proposition 12.2.1 yield

$$\|B(\varepsilon)(\lambda - T_0)^{-1}\|^2 < N_1(\varepsilon, a, p, q, h). \tag{12.2.15}$$

Hence, for $Re\lambda \leq -h$ and $|\varepsilon| < \frac{1}{q+\sqrt{\xi+a^2\frac{p}{h}(1+\frac{1}{h})}}$,

$$\|B(\varepsilon)(\lambda - T_0)^{-1}\| < 1.$$

Consequently, for

$$\begin{aligned} |\varepsilon| &< \min\left\{\frac{1}{q+\sqrt{\xi+a^2\frac{p}{h}(\pi+\frac{2}{h})}}, \frac{1}{q+\sqrt{\xi+a^2\frac{p}{h}(1+\frac{1}{h})}}\right\} \\ &= \frac{1}{q+\sqrt{\xi+a^2\frac{p}{h}(\pi+\frac{2}{h})}}, \end{aligned}$$

we have

$$\|B(\varepsilon)(\lambda - T_0)^{-1}\| < 1$$

for $|Im\lambda| \geq h$ or $Re\lambda \leq -h$. Hence, $I - B(\varepsilon)(\lambda - T_0)^{-1}$ is invertible with bounded inverse outside S_h. So, Eq. (12.2.13) implies that $\lambda - T(\varepsilon)$ is invertible with bounded inverse and, we obtain

$$(\lambda - T(\varepsilon))^{-1} = (\lambda - T_0)^{-1}[I - B(\varepsilon)(\lambda - T_0)^{-1}]^{-1}. \tag{12.2.16}$$

Consequently, $\lambda \in \rho(T(\varepsilon))$. Thus, the spectrum of the operator $T(\varepsilon)$ lies in the half-strip S_h. Q.E.D.

12.2.2 Basis with Parentheses

These results are of importance to prove the aim of this subsection.

Theorem 12.2.1 *Suppose that hypotheses $(H2)$ with $\beta \in]0, \frac{1}{2}[$, $(H9)$ and $(H10)$ hold. If the resolvent of the operator T_0 is compact, then for $|\varepsilon|$ small enough, the system of root vectors of the operator $T(\varepsilon)$ forms a basis with parentheses in X.* ◇

Proof Let $\lambda \in \mathbb{C}$. By using Proposition 12.2.2, the spectrum, $\sigma(T(\varepsilon))$, of $T(\varepsilon)$ lies in the half-strip

$$S_h := \{\lambda \in \mathbb{C} \text{ such that } |Im\lambda| < h \text{ and } Re\lambda > -h\},$$

for

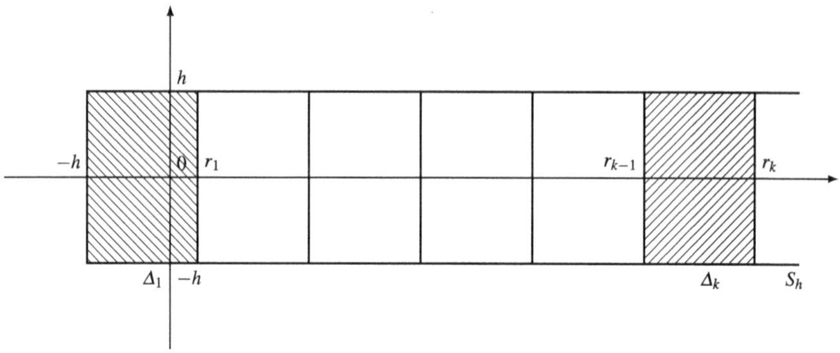

Fig. 12.2 The rectangles $(\Delta_k)_k$

$$|\varepsilon| < \frac{1}{q + \sqrt{\xi + a^2 \frac{p}{h}(\pi + \frac{2}{h})}}.$$

Hence, let $(\Delta_k)_{k\geq 1}$ be the rectangles bounded by the straight lines $Im\lambda = \pm h$, $Re\lambda = r_k$ and $Re\lambda = r_{k-1}$, where $r_0 = -h$ and $r_k \to \infty$ (see Fig. 12.2). We note here that the numbers r_k are chosen in such a way that the boundary $\partial \Delta_k$ of any rectangle Δ_k does not pass through the eigenvalues of the operator $T(\varepsilon)$. Then, for

$$|\varepsilon| < \frac{1}{q + \sqrt{\xi + a^2 \frac{p}{h}(\pi + \frac{2}{h})}},$$

we get

$$\sum_{k=1}^{n} \frac{1}{2\pi i} \int_{\partial \Delta_k} (\lambda - T(\varepsilon))^{-1} \mathrm{d}\lambda = \sum_{k=1}^{n} P_k(\varepsilon),$$

where $P_k(\varepsilon)$ designates the spectral projection corresponding to the spectrum of the operator $T(\varepsilon)$ inside Δ_k. To show our result, it suffices to prove that

$$\sup_n \left| \sum_{k=1}^{n} \frac{1}{2\pi i} \int_{\partial \Delta_k} (\lambda - T(\varepsilon))^{-1} \mathrm{d}\lambda \right| < \infty. \tag{12.2.17}$$

To this interest, we are going first to estimate of the norm $\|(\lambda - T(\varepsilon))^{-1}\|$ for
(i) $|Im\lambda| = \tau \geq h$ and $|\varepsilon| < \dfrac{1}{q + \sqrt{\xi + a^2 \frac{p}{h}(\pi + \frac{2}{h})}}$.

(ii) $Re\lambda = -\tau$ and $|\varepsilon| < \dfrac{1}{q+\sqrt{\xi + a^2\frac{p}{h}(1+\frac{1}{h})}}$.

For this purpose, let us consider $\lambda = \sigma + i\tau$.
(i) From Eq. (12.2.14), we get

$$\|B(\varepsilon)(\lambda - T_0)^{-1}\| < \sqrt{N(\varepsilon, a, p, q, \tau)} < 1, \tag{12.2.18}$$

where

$$N(\varepsilon, a, p, q, \tau) = \frac{|\varepsilon|^2}{(1-|\varepsilon|q)^2}\left(\xi + a^2\frac{p}{\tau}\left(\pi + \frac{2}{\tau}\right)\right).$$

Furthermore,

$$\|(\lambda - T_0)\|^{-1} \le \frac{1}{|Im\lambda|} \le \frac{1}{\tau}. \tag{12.2.19}$$

Thus, both Eqs. (12.2.18) and (12.2.19) yield

$$\begin{aligned}
\left\|(\lambda - T(\varepsilon))^{-1}\right\| &= \left\|(\lambda - T_0)^{-1}\left[I - B(\varepsilon)(\lambda - T_0)^{-1}\right]^{-1}\right\| \\
&\le \left\|(\lambda - T_0)^{-1}\right\| \left\|\left[I - B(\varepsilon)(\lambda - T_0)^{-1}\right]^{-1}\right\| \\
&\le \frac{1}{\tau}\left(1 - \sqrt{N(\varepsilon, a, p, q, \tau)}\right)^{-1}.
\end{aligned}$$

(ii) Equation (12.2.15) implies that

$$\|B(\varepsilon)(\lambda - T_0)^{-1}\| < \sqrt{N_1(\varepsilon, a, p, q, \tau)} < 1, \tag{12.2.20}$$

where

$$N_1(\varepsilon, a, p, q, \tau) = \frac{|\varepsilon|^2}{(1-|\varepsilon|q)^2}\left(\xi + a^2\frac{p}{\tau}\left(1 + \frac{1}{\tau}\right)\right).$$

Further, we get

$$\begin{aligned}
\|(\lambda - T_0)^{-1}\| &\le \frac{1}{\text{dist}(\lambda, \sigma(T_0))} \\
&\le \frac{1}{|Re\lambda - \lambda_n|}, \qquad \lambda_n \in \sigma(T_0) \\
&\le \frac{1}{|-\tau - \lambda_n|}, \qquad Re\lambda = -\tau \\
&< \frac{1}{\tau}.
\end{aligned} \tag{12.2.21}$$

Consequently, in view of both Eqs. (12.2.20) and (12.2.21), we get

$$\begin{aligned}\|(\lambda - T(\varepsilon))^{-1}\| &= \left\|(\lambda - T_0)^{-1}\left[I - B(\varepsilon)(\lambda - T_0)^{-1}\right]^{-1}\right\| \\ &\leq \frac{1}{\tau}\left(1 - \sqrt{N_1(\varepsilon, a, p, q, \tau)}\right)^{-1}.\end{aligned}$$

Now, to show Eq. (12.2.17), it remains to prove the existence of vertical segments in the half-strip S_h that tend to infinity and on which $(\lambda - T(\varepsilon))^{-1}$ is uniformly bounded (see [8, p. 292]). Let us begin with the boundedness of $\|(\lambda - T(\varepsilon))^{-1}\|$. Let $f, g \in X$ and, consider the scalar function $F_\varepsilon(\lambda)$ defined as

$$F_\varepsilon(\lambda) = \langle(\lambda - T(\varepsilon))^{-1} f, g\rangle.$$

It is easy to see that

* $F_\varepsilon(\lambda)$ is meromorphic and belongs to the Carleman-class $\mathscr{C}_P$, $P > 1$. In fact, the set of the Carleman-class $\mathscr{C}_P$ is a two-sided ideal of the algebra of bounded operators $\mathscr{L}(X)$. Further, the resolvent of T_0 belongs to the Carleman-class $\mathscr{C}_P$ (see the proof of Theorem 12.1.1). Hence, in view of Eq. (12.2.16), the resolvent of $T(\varepsilon)$ belongs to the Carleman-class $\mathscr{C}_P$.

* The poles of $F_\varepsilon(\lambda)$ lie in the strip $|Im\lambda| < h$. Indeed, in view of [6, p. 38], the poles of $(\lambda - T(\varepsilon))^{-1}$ are exactly the eigenvalues of $T(\varepsilon)$ which lies in the half-strip S_h.

* $|F_\varepsilon(\lambda)| \leq \frac{1}{\tau}\left(1 - \sqrt{N(\varepsilon, a, p, q, \tau)}\right)^{-1}$, for $|Im\lambda| = \tau = h + \delta$, $\delta > 0$ and $|\varepsilon| < \frac{1}{q + \sqrt{\xi + a^2 \frac{p}{h}(\pi + \frac{2}{h})}}$. Hence, in view of Lemma 4.4.1, we have

$$\log|F_\varepsilon(\lambda)| \leq C'\left(\frac{1}{\tau}\left(1 - \sqrt{N(\varepsilon, a, p, q, \tau)}\right)^{-1} + \sup_{|t-r|\leq r^\eta} (n(t+1, F_\varepsilon) - n(t, F_\varepsilon))\right),$$

for $|Im\lambda| \leq \tau$ and $Re\lambda = r_n \to \infty$ outside an exceptional set of disks $\mathscr{D}$, with $r = |\lambda|$. On the other hand, in virtue of Theorem 11.3.1, there exists a positive constant W such that for $|\varepsilon| < W$, we have

$$n(\overline{\mathbb{D}}(0, r), T(\varepsilon)) = n(\overline{\mathbb{D}}(0, r), T_0) + O(1). \tag{12.2.22}$$

Hence, hypothesis $(H10)$ and Eq. (12.2.22) imply that, for $|\varepsilon| < W$

$$\begin{aligned}n(\overline{\mathbb{D}}(0, t+1), F_\varepsilon) - n(\overline{\mathbb{D}}(0, t), F_\varepsilon) &= n(\overline{\mathbb{D}}(0, t+1), T(\varepsilon)) - n(\overline{\mathbb{D}}(0, t), T(\varepsilon)) \\ &= [n(\overline{\mathbb{D}}(0, t+1), T_0) + O(1)] - [n(\overline{\mathbb{D}}(0, t), T_0) + O(1)] \\ &= O(1) + [n(\overline{\mathbb{D}}(0, t+1), T_0) - n(\overline{\mathbb{D}}(0, t), T_0)] \\ &\leq O(1) + p = p'.\end{aligned}$$

Consequently, for $|\varepsilon| < V$,

$$|F_\varepsilon(\lambda)| \leq C_\varepsilon,$$

where C_ε is a constant independent of f, g and

$$V = \min\left\{W, \frac{1}{q+\sqrt{\xi + a^2\frac{p}{h}(\pi + \frac{2}{h})}}\right\}.$$

Therefore, for $|\varepsilon| < V$, we obtain

$$\|(\lambda - T(\varepsilon))^{-1}\| \leq C'_{\varepsilon},$$

where $|Im\lambda| \leq \tau$ and $Re\lambda = r_n \to \infty$ outside an exceptional set of disks $\mathcal{D}$. So, for $|\varepsilon| < V$, we have

$$\left|\sum_{k=1}^{n} \frac{1}{2\pi i} \int_{\partial\Delta_k} (\lambda - T(\varepsilon))^{-1}\, \mathrm{d}\lambda\right| < \infty.$$

Hence,

$$\sup_n \left|\sum_{k=1}^{n} \frac{1}{2\pi i} \int_{\partial\Delta_k} (\lambda - T(\varepsilon))^{-1}\, \mathrm{d}\lambda\right| < \infty.$$

Thus,

$$\sup_n \left\|\sum_{k=1}^{n} P_k(\varepsilon)\right\| < \infty.$$

As a consequence, due to Theorem 7.6.6, we claim that the family $(R(P_k(\varepsilon)))_{k\geq 1}$ forms a basis in X which means that the family of root vectors of $T(\varepsilon)$ forms a basis with parentheses in X. To complete the proof of our result, it suffices to show the existence of the vertical segments that do not pass through the eigenvalues of the operator $T(\varepsilon)$. Indeed, in each rectangle bounded by the straight lines $Re\lambda = n$, $Re\lambda = n+1$ and $Im\lambda = \pm h$, there are at most p' points of the eigenvalues $\lambda_k(\varepsilon)$. Then, for $|\varepsilon| < W$ and $d < \frac{1}{2p'}$ (where d is the total radii of the disks from $\mathcal{D}$ inside each rectangle), the projection of the disks from $\mathcal{D}$ onto the real axis do not fill the interval $[n, n+1]$. Therefore, there exists a vertical segment in this rectangle that does not intersect $\mathcal{D}$ (see Fig. 12.3).

Moreover, the vertical segments can be chosen in such a way that only points $\lambda_k(\varepsilon)$ with

$$|Re(\lambda_k(\varepsilon)) - Re(\lambda_j(\varepsilon))| < d$$

fall between the neighboring segments. Q.E.D.

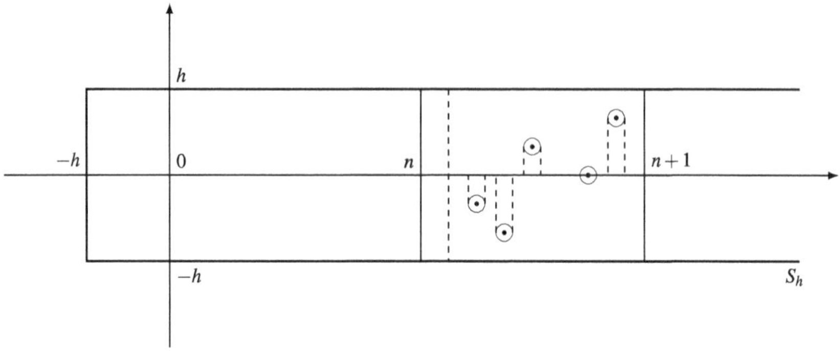

Fig. 12.3 Location of the eigenvalues of $T(\varepsilon)$

Corollary 12.2.1 *Suppose that hypotheses* $(H2)$ *with* $\beta \in]0, \frac{1}{2}[$ *and* $(H9)$ *and Eq. (12.1.2) hold. Then, for* $\beta \in]0, 1 + \frac{\alpha-1}{2}[$ *and* $|\varepsilon|$ *small enough, the system of root vectors of the operator* $T(\varepsilon)$ *forms a basis with parentheses in* X. ◇

Proof By using the hypothesis $(H2)$ and making the same reasoning as the one developed in the proof of Lemma 12.2.1, we obtain for $|\varepsilon| < \frac{1}{q}$

$$\frac{\|B(\varepsilon)\varphi_n\|^2}{|\lambda - \lambda_n|^2} \leq \frac{|\varepsilon|^2}{(1-|\varepsilon|q)^2}\left(\frac{a^2}{|\lambda-\lambda_n|^2} + \frac{2ab\lambda_n^{\beta}}{|\lambda-\lambda_n|^2} + \frac{b^2\lambda_n^{2\beta}}{|\lambda-\lambda_n|^2}\right).$$

Now, let $\sigma = Re\lambda$. Then, there is $k \in \mathbb{N}^*$ such that $\lambda_{k-1} \leq \sigma$ and $\lambda_k > \sigma$. Since $\lambda_n^{1-\alpha} \geq \frac{n-1}{p} + \lambda_1^{1-\alpha}$, then

$$\lambda_n \geq \left(\frac{n-1}{p} + \lambda_1^{1-\alpha}\right)^{\frac{1}{1-\alpha}}.$$

Hence, for $n < k-1$, we get

$$\begin{aligned}\sigma - \lambda_n &> \lambda_{k-1} - \lambda_n \\ &\geq \lambda_n^{\beta}(\lambda_{k-1}^{1-\beta} - \lambda_n^{1-\beta}) \\ &> \lambda_n^{\beta}\left(\left(\frac{k-2}{p} + \lambda_1^{1-\alpha}\right)^{\frac{1-\beta}{1-\alpha}} - \left(\frac{n-1}{p} + \lambda_1^{1-\alpha}\right)^{\frac{1-\beta}{1-\alpha}}\right). \qquad (12.2.23)\end{aligned}$$

Furthermore, for $n > k$, we obtain

$$\begin{aligned}\lambda_n - \sigma &> \lambda_n - \lambda_k \\ &\geq \lambda_n^{\beta}(\lambda_n^{1-\beta} - \lambda_k^{1-\beta}) \\ &> \lambda_n^{\beta}\left(\left(\frac{n-1}{p} + \lambda_1^{1-\alpha}\right)^{\frac{1-\beta}{1-\alpha}} - \left(\frac{k-1}{p} + \lambda_1^{1-\alpha}\right)^{\frac{1-\beta}{1-\alpha}}\right). \quad (12.2.24)\end{aligned}$$

Two cases are presented: If $\beta \in]0, \alpha]$, then

$$\frac{1-\beta}{1-\alpha} \geq 1.$$

So, Eq. (12.2.23) implies that

$$\sigma - \lambda_n > \lambda_n^{\beta} \frac{(k-1-n)^{\frac{1-\beta}{1-\alpha}}}{p^{\frac{1-\beta}{1-\alpha}}} \quad (12.2.25)$$

and, Eq. (12.2.24) yields

$$\lambda_n - \sigma > \lambda_n^{\beta} \frac{(n-k)^{\frac{1-\beta}{1-\alpha}}}{p^{\frac{1-\beta}{1-\alpha}}}. \quad (12.2.26)$$

Consequently, it follows from Eqs. (12.2.7), (12.2.25) and (12.2.26) that

$$\begin{aligned}&\sum_{n=1}^{\infty} \frac{\lambda_n^{2\beta}}{|\lambda_n - \sigma|^2} \\ &< \frac{2}{C^2} + p^{\frac{2(1-\beta)}{1-\alpha}} \left(\sum_{n<k-1} \frac{1}{(k-1-n)^{\frac{2(1-\beta)}{1-\alpha}}} + \sum_{n>k}^{\infty} \frac{1}{(n-k)^{\frac{2(1-\beta)}{1-\alpha}}} \right) \quad (12.2.27) \\ &\leq \frac{2}{C^2} + p^{\frac{2(1-\beta)}{1-\alpha}} \left(\sum_{n<k-1} \frac{1}{(k-1-n)^{\frac{2(1-\beta)}{1-\alpha}}} + \sum_{m=1}^{\infty} \frac{1}{m^{\frac{2(1-\beta)}{1-\alpha}}} \right) \\ &< \frac{2}{C^2} + 2p^{\frac{2(1-\beta)}{1-\alpha}} \sum_{m=1}^{\infty} \frac{1}{m^{\frac{2(1-\beta)}{1-\alpha}}} =: \xi_1' < \infty.\end{aligned}$$

Now, if $\beta \in]\alpha, 1 + \frac{\alpha-1}{2}[$. Then,

$$0 < \frac{-\alpha+\beta}{1-\alpha} < \frac{1}{2}$$

and

$$\frac{1}{2} < 1 - \frac{-\alpha+\beta}{1-\alpha} = \frac{1-\beta}{1-\alpha} < 1.$$

Hence, by using Eq. (12.2.23), we obtain

$$
\begin{aligned}
\sigma-\lambda_n &> \lambda_n^{\beta}\frac{1-\beta}{1-\alpha}\left(\frac{k-1-n}{p}\right)\left(\frac{k-2}{p}+\lambda_1^{1-\alpha}\right)^{\frac{\alpha-\beta}{1-\alpha}}\\
&= \lambda_n^{\beta}\frac{\left(\frac{1-\beta}{1-\alpha}\right)(k-1-n)^{\frac{1-\beta}{1-\alpha}}}{p^{\frac{1-\beta}{1-\alpha}}}\frac{(k-1-n)^{\frac{-\alpha+\beta}{1-\alpha}}}{p^{\frac{-\alpha+\beta}{1-\alpha}}}\frac{\left(k-2+p\lambda_1^{1-\alpha}\right)^{\frac{\alpha-\beta}{1-\alpha}}}{p^{\frac{\alpha-\beta}{1-\alpha}}}\\
&= \lambda_n^{\beta}\frac{\left(\frac{1-\beta}{1-\alpha}\right)(k-1-n)^{\frac{1-\beta}{1-\alpha}}}{p^{\frac{1-\beta}{1-\alpha}}}\frac{(k-1-n)^{\frac{-\alpha+\beta}{1-\alpha}}}{\left(k-2+p\lambda_1^{1-\alpha}\right)^{\frac{-\alpha+\beta}{1-\alpha}}}\\
&\geq \gamma'\lambda_n^{\beta}\frac{\left(\frac{1-\beta}{1-\alpha}\right)(k-1-n)^{\frac{1-\beta}{1-\alpha}}}{p^{\frac{1-\beta}{1-\alpha}}},\quad 0<\gamma'<1.
\end{aligned}
\tag{12.2.28}
$$

Similarly, based on Eq. (12.2.24), we get

$$
\lambda_n-\sigma > \gamma''\lambda_n^{\beta}\frac{\left(\frac{1-\beta}{1-\alpha}\right)(n-k)^{\frac{1-\beta}{1-\alpha}}}{p^{\frac{1-\beta}{1-\alpha}}},\quad 0<\gamma''<1.
\tag{12.2.29}
$$

So, if we denote by $\gamma_1' := \min\{\gamma',\gamma''\}$, Eqs. (12.2.7), (12.2.28) and (12.2.29) imply that

$$
\begin{aligned}
\sum_{n=1}^{\infty}\frac{\lambda_n^{2\beta}}{|\lambda_n-\sigma|^2} &< \frac{2}{C^2}+\frac{p^{\frac{2(1-\beta)}{1-\alpha}}}{\gamma_1'^2\left(\frac{1-\beta}{1-\alpha}\right)^2}\left(\sum_{n<k-1}\frac{1}{(k-1-n)^{\frac{2(1-\beta)}{1-\alpha}}}+\right.\\
&\qquad\left.\sum_{n>k}^{\infty}\frac{1}{(n-k)^{\frac{2(1-\beta)}{1-\alpha}}}\right)
\end{aligned}
\tag{12.2.30}
$$

$$
\begin{aligned}
&\leq \frac{2}{C^2}+\frac{p^{\frac{2(1-\beta)}{1-\alpha}}}{\gamma_1'^2\left(\frac{1-\beta}{1-\alpha}\right)^2}\left(\sum_{n<k-1}\frac{1}{(k-1-n)^{\frac{2(1-\beta)}{1-\alpha}}}+\sum_{m=1}^{\infty}\frac{1}{m^{\frac{2(1-\beta)}{1-\alpha}}}\right)\\
&< \frac{2}{C^2}+\frac{2p^{\frac{2(1-\beta)}{1-\alpha}}}{\gamma_1'^2\left(\frac{1-\beta}{1-\alpha}\right)^2}\sum_{m=1}^{\infty}\frac{1}{m^{\frac{2(1-\beta)}{1-\alpha}}} =: \xi_2' < \infty.
\end{aligned}
$$

Consequently, for $\beta\in]0,1+\frac{\alpha-1}{2}[$, we obtain

$$
\sum_{n=1}^{\infty}\frac{\lambda_n^{2\beta}}{|\lambda-\lambda_n|^2} < \sum_{n=1}^{\infty}\frac{\lambda_n^{2\beta}}{|\sigma-\lambda_n|^2} < \max\{\xi_1',\xi_2'\} =: \xi_2'.
$$

On the other hand, we get

$$\sum_{n=1}^{\infty} \frac{\lambda_n^{\beta}}{|\lambda - \lambda_n|^2} < \frac{2}{C^2} + \frac{2p^{\frac{2-\beta}{1-\alpha}}}{\gamma_1'^2 \left(\frac{1-\frac{\beta}{2}}{1-\alpha}\right)^2} \sum_{m=1}^{\infty} \frac{1}{m^{\frac{2-\beta}{1-\alpha}}} =: \xi_3' < \infty.$$

In fact, it suffices to replace β by $\frac{\beta}{2}$. Thus, the series

$$\sum_n \frac{2ab\lambda_n^{\beta}}{|\sigma - \lambda_n|^2}$$

and

$$\sum_n \frac{b^2\lambda_n^{2\beta}}{|\lambda - \lambda_n|^2}$$

are convergent. Hence, let ξ' be a positive constant satisfying

$$\sum_{n=1}^{\infty} \frac{2ab\lambda_n^{\beta}}{|\lambda - \lambda_n|^2} + \sum_{n=1}^{\infty} \frac{b^2\lambda_n^{2\beta}}{|\lambda - \lambda_n|^2} < b^2\xi_2' + 2ab\xi_3' =: \xi'.$$

Further, in view of Corollary 12.2.1, we have for $|\varepsilon|$ small enough and $\beta \in]0, 1 + \frac{\alpha-1}{2}[$, we have

$$n(\overline{\mathbb{D}}(0,r), T(\varepsilon)) = n(\overline{\mathbb{D}}(0,r), T_0) + O(1).$$

To get the desired result, we advise that the rest of the proof is similar to the proof of Theorem 10.4.2. Q.E.D.

12.3 Notes and Remarks

- Notice that in classical perturbation theorems, in a Banach space, for bases or Riesz bases, the authors always required that the eigenvalues of the operator T are with multiplicity one (see, for instance, [1], [2, Theorem XIX.2.7], [5] and [6, Theorem V.4.15a]). Although, by assuming that the eigenvalues are with finite multiplicity, several authors such as Markus [7] and Wyss [9] proved the existence of bases with parentheses or unconditional bases with parentheses.
- At the same time, Shkalikov [8, Theorem 2] ameliorates the result stated in [7]. Indeed, he obtained a basis with parentheses under Eqs. (11.4.1) and (11.4.2) which are much weaker.
- Among this direction, Ellouz et al. [3] have the idea to extend [8, Theorem 2] to an abstract setting. More precisely, they continue the analysis started in [4] and they

focus on the property of bases with parentheses of the analytic operator (10.1.1) acting on a separable Hilbert space.

- Furthermore, it is interesting to note here that in [3], the authors derive a precise description to the localization of the spectrum of the perturbed operator (10.1.1) and they prove an asymptotic relation between the eigenvalue-counting functions of T_0 and $T(\varepsilon)$. In other words, they claim that the difference between the eigenvalue-counting functions of T_0 and $T(\varepsilon)$ is bounded by a constant. This generalization is of much importance. In fact, it allows them to control the jump of the eigenvalue-counting function of some analytic operators where the criteria of Shkalikov [8] can not be applied.

References

1. C. Clark, On relatively bounded perturbations of ordinary differential operators. Pac. J. Math. **25**, 59–70 (1968)
2. N. Dunford, J.T. Schwartz, *Linear Operators* (Part III, Wiley-Interscience, New-York, 1971)
3. H. Ellouz, I. Feki, A. Jeribi, On the basis property of root vectors related to a non-self-adjoint analytic operator and applications (2020)
4. I. Feki, A. Jeribi, R. Sfaxi, On a Schauder basis related to the eigenvectors of a family of non-self-adjoint analytic operators and applications. Anal. Math. Phys. **3**(4), 311–331 (2013)
5. A. Jeribi, *Denseness, Bases and Frames in Banach Spaces and Applications* (de Gruyter, Berlin, 2018)
6. T. Kato, *Perturbation Theory for Linear Operators* (Springer, Berlin, 1980)
7. A.S. Markus, V.I. Matsaev, On the convergence of eigenvector expansions for an operator which is close to being self-adjoint. Mat. Issled. (in Russian) **61**, 104–129 (1981)
8. A.A. Shkalikov, On the basis property of root vectors of a perturbed self-adjoint operator. Proc. Steklov Inst. Math. **269**, 284–298 (2010)
9. C. Wyss, Riesz bases for p-subordinate perturbations of normal operator. J. Funct. Anal. **258**, 208–240 (2010)
10. N. Ben Ali, A. Jeribi, On the Riesz basis of a family of analytic operators in the sens of Kato and application to the problem of radiation of a vibriting structure in a light fluid. J. Math. Anal. Appl. **320**(1), 78–94 (2006)
11. S. Charfi, A. Damergi, A. Jeribi, On a Riesz basis of finite-dimensional invariant subspaces and application to Gribov operator in Bargmann space. Line. Mult. Alge. **61**(11), 1577–1591 (2013)
12. S. Charfi, A. Intissar, A. Jeribi, Expansion of solution in terms of generalized eigenvectors for a rectilinear transport equation. Trans. Theo. Stati. Phys. **38**, 1–16 (2009)
13. S. Charfi, A. Jeribi, I. Walha, Riesz basis property of families of nonharmonic exponential and application to a problem of a radiation of a vibrating structure in a light fluid. Numer. Funct. Anal. Optim. **32**(4), 370–382 (2011)
14. H. Ellouz, I. Feki, A. Jeribi, On a Riesz basis of exponentials related to the eigenvalues of an analytic operator and application to a non-self-adjoint problem deduced from a perturbation method. J. Math. Phys. **54**, 112101 (2013)
15. H. Ellouz, I. Feki, A. Jeribi, On the asymptotic behavior of the eigenvalues of an analytic operator in the sense of Kato and applications. Serdica Math. J. **45**(1), 55–88 (2019)
16. H. Ellouz, I. Feki, A. Jeribi, On a Riesz basis of exponentials related to a family of analytic operators and application. J. Pseudo-Differ. Oper. Appl. **10**(4), 999–1014 (2019)
17. H. Ellouz, I. Feki, A. Jeribi, Non-orthogonal fusion frames of an analytic operator and application to a one-dimensional wave control system. Mediterr. J. Math. **16**(2), 16:52 (2019)

18. H. Ellouz, I. Feki, A. Jeribi, On the basis property of root vectors related to a non-self-adjoint analytic operator and applications (2021)
19. I. Feki, A. Jeribi, R. Sfaxi, On an unconditionl basis of generalized eigenvectors of an analytic operator and application to a problem of radiation of a vibrating structure in a light fluid. J. Math. Anal. Appl. **375**, 261–269 (2011)
20. I. Feki, A. Jeribi, R. Sfaxi, On a Schauder basis related to the eigenvectors of a family of non-self-adjoint analytic operators and applications. Anal. Math. Phys. **3**(4), 311–331 (2013)
21. I. Feki, A. Jeribi, R. Sfaxi, On a Riesz basis of eigenvectors of a non-self-adjoint analytic operator and applications. Line. Mult. Alge. **62**(8), 1049–1068 (2014)

Chapter 13
Perturbation Method for Sound Radiation by a Vibrating Plate in a Light Fluid

In this chapter, we apply the results of Chaps. 6–12 to two examples: to the problem of radiation of a vibrating structure in a light fluid and perturbation method for sound radiation by a vibrating plate in a light fluid. This chapter contains two sections. However, it is worth mentioning that each section has its own equations, notations, and symbols. In other words, the reader should remember that the same symbol doesn't have the same meaning or significance from one section to another.

13.1 Perturbation Method for Sound Radiation by a Vibrating Plate in a Light Fluid

Let's consider a one-dimensional clamped thin plate, extended by an infinite perfectly rigid baffle, which is excited by a harmonic force

$$F(x)e^{-iwt}.$$

It occupies the region $-L < x < L$ of the $z = 0$ plane. The two half-spaces $z < 0$ and $z > 0$ are filled with gas. The mechanical parameters of the plate are E the Young modulus, ν the Poisson ratio, m the surface density, h the thickness of the plate, and

$$D = \frac{Eh^3}{12(1-\nu^2)}$$

the rigidity. The fluid is characterized by ρ_0 the density, c the sound speed, and $k = \frac{w}{c}$ the wavenumber.

A. Jeribi, *Perturbation Theory for Linear Operators*,
https://doi.org/10.1007/978-981-16-2528-2_13

13.1.1 Position of the Problem

Let u denote the displacement of the baffled plate ($u > 0$ in the direction $z > 0$) and p the radiated pressure. These functions fulfilled the system

$$\left(\frac{\mathrm{d}^4}{\mathrm{d}x^4} - \frac{m\omega^2}{D}\right) u(x) = \frac{1}{D}(F(x) - P(x)) \;\; \forall x \in (-L, L), \tag{13.1.1}$$

where

$$P(x) = \lim_{\eta \to 0^+} [p(x, \eta) - p(x, -\eta)], \tag{13.1.2}$$

$$(\Delta + k^2) p(x, z) = 0 \;\; \forall x \in (-\infty, +\infty) \text{ and } \forall z < 0 \text{ or } z > 0, \tag{13.1.3}$$

$$\lim_{z \to 0} \frac{\partial p(x, z)}{\partial z} \omega^2 \rho_0 u(x) \;\; \forall x \in]-L, L[, \;\; \lim_{z \to 0} \frac{\partial p(x, z)}{\partial z} = 0 \;\; \forall x \notin]-L, L[, \tag{13.1.4}$$

$$\lim_{r \to +\infty} \left(\frac{\partial p}{\partial r} - ikp\right) = 0, \tag{13.1.5}$$

$$u(x) = \frac{\partial u(x)}{\partial x} = 0 \qquad \text{for } x = -L \text{ and } x = L. \tag{13.1.6}$$

It is easy to see that the system (13.1.3), (13.1.4) does not have a unique solution $p(x, z)$. Hence, the equations given in the paper [5] lack a condition at infinity (13.1.5) which is an appropriate and very well-known radiation condition given by Sommerfeld [14]. See also [1, 2, 7, 11]. Hence, the equations introduced in [5] are insufficient. Now, let us consider the Hankel function, $H_0(\cdot)$, of the first kind and order 0 (see [12]) defined by

$$H_0(z) = -\frac{1}{\pi} e^{i(z - \frac{\pi}{4})} \int_0^{+\infty} \frac{e^{-zt}}{\sqrt{t(1 + i\frac{t}{2})}} \mathrm{d}t. \tag{13.1.7}$$

In the next, we claim that for all $x, \; x' \in]-L, L[$ such that $|x - x'| \geq \varepsilon$ (with $\varepsilon > 0$), there exists a positive constant $c = c(\varepsilon) > 0$ depending on ε such that

$$|H_0(k|x - x'|)| \leq c. \tag{13.1.8}$$

Indeed, let $x, x' \in]-L, L[$ such that $|x - x'| \geq \varepsilon$. Then, in view of Eq. (13.1.7), we get

$$
\begin{aligned}
|H_0(k|x-x'|)| &\leq \frac{1}{\pi}\int_0^{+\infty} \frac{e^{-k|x-x'|t}}{\sqrt{t\sqrt{1+\frac{t^2}{4}}}}\mathrm{d}t \\
&\leq \frac{1}{\pi}\int_0^{1} \frac{e^{-k|x-x'|t}}{\sqrt{t\sqrt{1+\frac{t^2}{4}}}}\mathrm{d}t + \frac{1}{\pi}\int_1^{+\infty} \frac{e^{-k|x-x'|t}}{\sqrt{t\sqrt{1+\frac{t^2}{4}}}}\mathrm{d}t \\
&\leq \frac{1}{\pi}\int_0^{1} \frac{\mathrm{d}t}{\sqrt{t}} + \frac{1}{\pi}\int_1^{+\infty} e^{-k|x-x'|t}\mathrm{d}t \\
&\leq \frac{1}{\pi}\left(2+\frac{1}{k|x-x'|}\right) \\
&\leq \frac{1}{\pi}\left(2+\frac{1}{k\varepsilon}\right).
\end{aligned}
$$

This completes the proof of the claim. The solution of (13.1.3), (13.1.4) and (13.1.5) is given by the following:

$$
p(x,z) = -\mathrm{sgn}(z)\, i\, \frac{\omega^2\rho_0}{2}\int_{-L}^{L} H_0\left[k\sqrt{(x-x')^2+z^2}\right]u(x')\mathrm{d}x' \;\; \forall z<0 \text{ or } z>0. \tag{13.1.9}
$$

The expression of Eqs. (13.1.1), (13.1.2), (13.1.6) and (13.1.9) lead to the following boundary value Problem:

$$
\left(\frac{\mathrm{d}^4}{\mathrm{d}x^4} - \frac{m\omega^2}{D}\right)u(x) - i\frac{\omega^2\rho_0}{D}\int_{-L}^{L} H_0\left[k|x-x'|\right]u(x')\mathrm{d}x' = \frac{F(x)}{D} \;\; \forall x \in]-L,L[, \tag{13.1.10}
$$

$$
u(x) = \frac{\partial u(x)}{\partial x} = 0 \quad \text{for } x=-L \text{ and } x=L. \tag{13.1.11}
$$

We need, in the following, the operators

$$
\begin{cases}
T_0 : \mathscr{D}(T_0) \subset L_2(]-L,L[) \longrightarrow L_2(]-L,L[) \\
\qquad \psi \longrightarrow T_0\psi(x) = \dfrac{\mathrm{d}^4\psi}{\mathrm{d}x^4}(x) \\
\mathscr{D}(T_0) = H_0^2(]-L,L[)\bigcap H^4(]-L,L[)
\end{cases}
$$

and

$$\begin{cases} K : L_2(]-L, L[) \longrightarrow L_2(]-L, L[) \\ \qquad \psi \longrightarrow K\psi(x) = \dfrac{i}{2}\displaystyle\int_{-L}^{L} H_0(k|x-x'|)\psi(x')\mathrm{d}x'. \end{cases} \tag{13.1.12}$$

Based on (13.1.10) and (13.1.11), Filippi et al. [4] have considered the following eigenvalue problem:
Finding the values $\lambda \in \mathbb{C}$ for which there is a solution $u \in H_0^2(]-L, L[) \bigcap H^4(]-L, L[)$, $u \neq 0$ for the equation

$$T_0 u = \lambda(I + \varepsilon K)u. \tag{13.1.13}$$

According to the definition, introduced in reference [10, Chap. 9, Sect. 4], λ is the eigenvalue and u is the eigenmode. Note that λ and u each depends on the value of ε. Hence, we denote in the following $\lambda := \lambda(\varepsilon)$ and $u := u(\varepsilon)$.

13.1.2 Open Questions Introduced in [5]

The aim of this section is to give a positive answer for the following hypotheses (introduced in [5]):

(i) There exists an infinite sequence of eigenvalues with a finite order of multiplicity, this sequence being denoted by $\lambda_n(\varepsilon)$, an eigenvalue of order p appearing p times with different indices.
(ii) To each eigenvalue corresponds to an eigenmode $u_n(\varepsilon)$.
(iii) The set $u_n(\varepsilon)$, $n = 1, 2, \ldots, \infty$ is a basis of the space $H_0^2(]-L, L[)$.

Furthermore, the authors in [5] assume that the eigenvalue $\lambda_n(\varepsilon)$ and the eigenvector $u_n(\varepsilon)$ of the problem (13.1.13) can be developed as an entire series of ε and so, they can be written as

$$\lambda_n(\varepsilon) = \lambda_n + \varepsilon\lambda_n^1 + \varepsilon^2\lambda_n^2 + \cdots$$

$$u_n(\varepsilon) = u_n + \varepsilon u_n^1 + \varepsilon^2 u_n^2 + \cdots,$$

where $\lambda_n(0) := \lambda_n$ (respectively, $u_n(0) := u_n$) is the eigenvalue (respectively, eigenvector) of the operator T_0.

In this case, the classical spectral theory can't be applied since the involved operator is not self-adjoint. Nevertheless, assuming (i), (ii) and (iii) is reasonable.

13.1.3 Spectral Properties of the Operator T_0

We will give in this section a precise description of the spectrum of the operator T_0.

Lemma 13.1.1 *T_0 is a closed operator.* ◇

Proof According to the closed graph theorem, we will prove that, $G(T_0)$, of T_0 is a closed set. Let $\lambda \in \rho(T_0)$, and $(\psi_n, T_0\psi_n) \in G(T_0)$ be sequence which converges to (ψ, φ). Our aim is to prove that $\psi \in H_0^2(]-L, L[) \bigcap H^4(]-L, L[)$ and

$$\varphi = T_0\psi.$$

In fact, the sequence $(\psi_n)_n$ converges, in $H^2(]-L, L[)$, to ψ and $\psi_n \in H_0^2(]-L, L[)$ thus $\psi \in H_0^2(]-L, L[)$. Since $(\psi_n)_n$ converges, in $H^2(]-L, L[)$, to ψ, then $(\psi_n)_n$ converges, in $L_2(]-L, L[)$, to ψ. Hence, we conclude that $(\lambda\psi_n - T_0\psi_n)_n$ converges, in $L_2(]-L, L[)$, to $\lambda\psi - \varphi$. The fact that $\lambda \in \rho(T_0)$, implies that $(\psi_n)_n$ converges, in $L_2(]-L, L[)$, to $(\lambda - T_0)^{-1}(\lambda\psi - \varphi)$. Using the fact that $(\psi_n)_n$ converges, in $L_2(]-L, L[)$, to ψ, we deduce that

$$\psi = (\lambda - T_0)^{-1}(\lambda\psi - \varphi)$$

and

$$(\lambda - T_0)\psi = \lambda\psi - \varphi$$

which leads to

$$T_0\psi = \varphi.$$

Since $\varphi \in L_2(]-L, L[)$, we have $T_0\psi \in L_2(]-L, L[)$ and $\psi \in H^2(]-L, L[)$. Then, $\psi \in H^4(]-L, L[)$. This achieves the proof. Q.E.D.

Obviously, we have

Lemma 13.1.2 *(i) The injection from $\mathscr{D}(T_0)$ into $L_2(]-L, L[)$ is compact.*
(ii) T_0 is a self-adjoint operator with compact resolvent.
(iii) The continuous spectrum and the residual spectrum of T_0 are empty. ◇

Lemma 13.1.3 *The resolvent set of T_0 is not empty. In fact, $0 \in \rho(T_0)$.* ◇

Proof Let $\varphi \in N(T_0)$. Since φ is continuous and T_0 is V-elliptic, then $\varphi = 0$. This leads to

$$N(A) = \{0\}.$$

Let $\psi \in L_2(]-L, L[)$, we will find $\varphi \in H_0^2(]-L, L[) \bigcap H^4(]-L, L[)$ such that the equality

$$-T_0\varphi = \psi$$

hold. In particular, it will be satisfied for all $\varphi \in H_0^2(]-L, L[)$. Based on the fundamental lemma of the calculus of variations, there exist four constants α, β, μ, and δ such that

$$\varphi(x) = \alpha x^3 + \beta x^2 + \mu x + \delta + \int_{-L}^{x}\int_{-L}^{\alpha}\int_{-L}^{\beta}\int_{-L}^{\mu} \psi(s)\mathrm{d}s\mathrm{d}\mu\mathrm{d}\beta\mathrm{d}\alpha$$

$$\varphi = \frac{\partial \varphi}{\partial x} = 0 \text{ for } x = -L \text{ and } x = L.$$

We obtain a linear system of four equations with four unknowns,

$$\begin{pmatrix} -L^3 & L_2 & -L & 1 \\ 3L_2 & -2L & 1 & 0 \\ L^3 & L_2 & L & 1 \\ 3L_2 & 2L & 1 & 0 \end{pmatrix} \begin{pmatrix} \alpha \\ \beta \\ \mu \\ \delta \end{pmatrix} \begin{pmatrix} 0 \\ 0 \\ -\eta \\ -\gamma \end{pmatrix}, \tag{13.1.14}$$

where

$$\eta = \int_{-L}^{L}\int_{-L}^{\alpha}\int_{-L}^{\beta}\int_{-L}^{\mu} \psi(s)\mathrm{d}s\mathrm{d}\mu\mathrm{d}\beta\mathrm{d}\alpha$$

and

$$\gamma = \int_{-L}^{L}\int_{-L}^{\beta}\int_{-L}^{\mu} \psi(s)\mathrm{d}s\mathrm{d}\mu\mathrm{d}\beta.$$

Since the matrix's determinant is nonzero, the system (13.1.14) has a unique solution. A simple calculation shows the following operators are bounded and satisfy the following estimates

$$|\eta| \leq (2L)^4 \|\psi\|$$

$$|\gamma| \leq (2L)^3 \|\psi\|$$

$$\left| \int_{-L}^{x}\int_{-L}^{\alpha}\int_{-L}^{\beta}\int_{-L}^{\mu} \psi(s)\mathrm{d}s\mathrm{d}\mu\mathrm{d}\beta\mathrm{d}\alpha \right| \leq (2L)^4 \|\psi\|.$$

Thus,

$$|\varphi(x)| \leq L^3|\alpha| + L_2|\beta| + L|\mu| + |\delta| + \left| \int_{-L}^{x} \int_{-L}^{\alpha} \int_{-L}^{\beta} \int_{-L}^{\mu} \psi(s) \mathrm{d}s \mathrm{d}\mu \mathrm{d}\beta \mathrm{d}\alpha \right|.$$

So,

$$|\varphi(x)| \leq c\|\psi\|$$

and

$$\|(0 - T_0)^{-1}\| = \sup_{\|\psi\| \leq 1} \|(0 - T_0)^{-1}\psi\| \leq c.$$

This completes the proof of lemma. Q.E.D.

Lemma 13.1.4 *The spectrum of T_0 is constituted only of point spectrums which are positive, denumerable and of which the multiplicity is one and which has no finite limit points and satisfies*

$$0 < \lambda_1 \leq \lambda_2 \leq \cdots \leq \lambda_n \rightarrow +\infty \text{ as } n \rightarrow +\infty.$$

Further,

$$\left(\frac{(2n+1)\pi}{4L}\right)^4 \leq \lambda_n \leq \left(\frac{(2n+3)\pi}{4L}\right)^4$$

i.e.,

$$\lambda_n \underset{+\infty}{\sim} \left(\frac{n\pi}{2L}\right)^4.$$ ◇

Proof Let's consider the eigenvalue problem

$$\begin{cases} T_0\varphi = \lambda\varphi \\ \quad \varphi \in \mathscr{D}(T_0). \end{cases} \tag{13.1.15}$$

The solution of the problem (13.1.15) is, formally, given by

$$\varphi(x) = \mu \exp(\sqrt[4]{\lambda_m}x) + \eta \exp(-\sqrt[4]{\lambda_m}x) + \theta \exp(i\sqrt[4]{\lambda_m}x) + \delta \exp(-i\sqrt[4]{\lambda_m}x) \tag{13.1.16}$$

and satisfies the following boundary conditions

$$\varphi(-L) = \varphi(L) = \varphi'(-L) = \varphi'(L) = 0.$$

Hence, we obtain a linear system of four equations with four unknowns and which equal zero. To avoid a trivial solution, it suffices that the determinant of this system be zero:

$$\det\begin{pmatrix} e^{-\sqrt[4]{\lambda_m}L} & e^{\sqrt[4]{\lambda_m}L} & e^{-i\sqrt[4]{\lambda_m}L} & e^{i\sqrt[4]{\lambda_m}L} \\ e^{\sqrt[4]{\lambda_m}L} & e^{-\sqrt[4]{\lambda_m}L} & e^{i\sqrt[4]{\lambda_m}L} & e^{-i\sqrt[4]{\lambda_m}L} \\ \sqrt[4]{\lambda_m}e^{-\sqrt[4]{\lambda_m}L} & -\sqrt[4]{\lambda_m}e^{\sqrt[4]{\lambda_m}L} & i\sqrt[4]{\lambda_m}e^{-i\sqrt[4]{\lambda_m}L} & -i\sqrt[4]{\lambda_m}e^{i\sqrt[4]{\lambda_m}L} \\ \sqrt[4]{\lambda_m}e^{\sqrt[4]{\lambda_m}L} & -\sqrt[4]{\lambda_m}e^{-\sqrt[4]{\lambda_m}L} & i\sqrt[4]{\lambda_m}e^{i\sqrt[4]{\lambda_m}L} & -i\sqrt[4]{\lambda_m}e^{-i\sqrt[4]{\lambda_m}L} \end{pmatrix} = 0. \tag{13.1.17}$$

So, the equality (13.1.17) is equivalent to

$$\cos(2L\sqrt[4]{\lambda_m})\cosh(2L\sqrt[4]{\lambda_m}) = 1.$$

We must discuss two cases:
If n is even, then

$$\left](2n+1)\frac{\pi}{2},(2n+3)\frac{\pi}{2}\right[\bigcap\left\{x \text{ such that } \cosh(x) = \frac{1}{\cos(x)}\right\} = \emptyset.$$

If n is odd, then

$$\left](2n+1)\frac{\pi}{2},(2n+3)\frac{\pi}{2}\right[\bigcap\left\{x \text{ such that } \cosh(x) = \frac{1}{\cos(x)}\right\} = \{x_n^1, x_n^2\}.$$

Thus, we obtain

$$(2n+1)\frac{\pi}{2} \leq 2L\sqrt[4]{\lambda_m} \leq (2n+3)\frac{\pi}{2}.$$

This achieves the proof. Q.E.D.

Using both Lemma 13.1.4 and Notes and Remark 1.16 (iv), we have the following:

Lemma 13.1.5 $\sigma_j(T_0) = \sigma_c(T_0) = \sigma_r(T_0) = \emptyset$ *and* $\sigma(T_0) = \sigma_p(T_0)$. ◇

Due to Lemma 13.1.2, T_0 is a self-adjoint operator and has compact resolvent. So, let

$$T_0\varphi = \sum_{n=1}^{\infty} \lambda_n \langle\varphi, \varphi_n\rangle\varphi_n \tag{13.1.18}$$

its spectral decomposition, where

$$\lambda_n = \kappa n^4$$

is the nth eigenvalue of T_0 associated to the following eigenvector

$$\varphi_n(x) = \mu e^{\sqrt[4]{\lambda_n}x} + \eta e^{-\sqrt[4]{\lambda_n}x} + \theta e^{i\sqrt[4]{\lambda_n}x} + \delta e^{\sqrt[4]{\lambda_n}x}$$

(see Eq. (13.1.16)). So, for $\gamma > 0$, we consider the operator T_0^γ by

$$\begin{cases} T_0^\gamma : \mathscr{D}(T_0^\gamma) \subset L_2(]-L, L[) \longrightarrow L_2(]-L, L[) \\ \qquad\qquad \varphi \longrightarrow T_0^\gamma \varphi = \sum_{n=1}^{\infty} \lambda_n^\gamma \langle \varphi, \varphi_n \rangle \varphi_n \\ \mathscr{D}(T_0^\gamma) = \left\{ \varphi \in L_2(]-L, L[) \text{such that} \sum_{n=1}^{\infty} \lambda_n^{2\gamma} |\langle \varphi, \varphi_n \rangle|^2 < \infty \right\}. \end{cases}$$

13.1.4 Spectral Properties of the Resolvent of the Operator T_0

Theorem 13.1.1 *The resolvent of the operator T_0 belongs to Carleman-class $\mathscr{C}_p$ for any $p > \frac{1}{4}$.* ◇

Proof Due to Lemma 13.1.4 and the fact that the operator T_0 is self-adjoint, we deduce immediately the desired result. Q.E.D.

Lemma 13.1.6 *Let $(\lambda_n)_{n=1}^{\infty}$ be the eigenvalues of T_0. Then, for each $\lambda \neq \lambda_n$, $n \in \mathbb{N}$, we get*

$$(T_0 - \lambda)^{-1} \varphi = \sum_{n=0}^{\infty} \frac{1}{\lambda_n - \lambda} \langle \varphi, \varphi_n \rangle \varphi_n,$$

where $(\varphi_n)_{n=1}^{\infty}$ are the corresponding eigenfunctions of $(\lambda_n)_{n=1}^{\infty}$. Furthermore, if $Im\lambda \neq 0$ and if λ belongs to a ray with origin zero and of angle θ with $\theta \neq 0$ and $\theta \neq \pi$, then

$$\|(T_0 - \lambda)^{-1}\| \leq \frac{1}{|Im\lambda|} = \frac{c(\theta)}{|\lambda|},$$

where $c(\theta)$ is a positive constant that depends on θ. ◇

Proof Lemma 13.1.2(ii) enables us to obtain the desired result. Q.E.D.

Lemma 13.1.7 *There is a sequence of circles $C(O, r_n)$, $n \in \mathbb{N}$, with radii r_n going to infinity such that*

$$\|(T_0 - \lambda)^{-1}\| \leq \frac{c}{|\lambda|^{\frac{3}{4}}} \text{ and } |\lambda| = r_n. \qquad ◇$$

Proof First, we claim that there is $c > 0$ and a sequence $(n_j)_j$ such that

$$|\lambda_{n_j+1}| - |\lambda_{n_j}| \geq c n_j^3.$$

We reason by the absurd. Then, for all $\varepsilon > 0$, there exists n_ε such that for all $n \geq n_\varepsilon$, we get

$$|\lambda_{n+1}| - |\lambda_n| < \varepsilon n^3. \qquad (13.1.19)$$

Now, we may write

$$|\lambda_{n_\varepsilon+p}| - |\lambda_{n_\varepsilon}| = |\lambda_{n_\varepsilon+p}| - |\lambda_{n_\varepsilon+p-1}| + |\lambda_{n_\varepsilon+p-1}| - |\lambda_{n_\varepsilon+p-2}| + \cdots + |\lambda_{n_\varepsilon+1}| - |\lambda_{n_\varepsilon}|$$

and so, by using (13.1.19), we get

$$\begin{aligned} |\lambda_{n_\varepsilon+p}| - |\lambda_{n_\varepsilon}| &< \varepsilon \left(n_\varepsilon^3 + (n_\varepsilon + 1)^3 + \cdots + (n_\varepsilon + p - 1)^3\right) \\ &< \varepsilon p (n_\varepsilon + p)^3 \\ &< \varepsilon (n_\varepsilon + p)^4 \end{aligned}$$

for all $\varepsilon > 0$ and for all $p \in \mathbb{N}^*$. From this inequality, we conclude that

$$\frac{|\lambda_{n_\varepsilon+p}|}{(n_\varepsilon + p)^4} - \frac{|\lambda_{n_\varepsilon}|}{(n_\varepsilon + p)^4} < \varepsilon.$$

When p goes to infinity in the last equation, we get $1 \leq \varepsilon$. This contradicts the fact that $\varepsilon > 0$ and concludes the proof of the claim. We set

$$r_j = \frac{|\lambda_{n_j+1}| + |\lambda_{n_j}|}{2} = |\lambda_{n_j}| + \frac{|\lambda_{n_j+1}| - |\lambda_{n_j}|}{2}.$$

The use of Lemma 13.1.2(ii) allows us to write

$$\|(T_0 - \lambda)^{-1}\| = \frac{1}{\operatorname{dist}(\lambda, \sigma(T_0))}, \tag{13.1.20}$$

where λ belongs to $\rho(T_0)$ and $\operatorname{dist}(\lambda, \sigma(T_0))$ denotes the distance between the point λ and the spectrum of the operator T_0. In view of Eq. (13.1.20), we obtain

$$\begin{aligned} \|(T_0 - \lambda)^{-1}\| &= \frac{1}{\operatorname{dist}(\lambda, \sigma(T_0))} \\ &\leq \frac{2}{|\lambda_{n_j+1}| - |\lambda_{n_j}|}, \end{aligned}$$

where $|\lambda| = r_j$. On the other hand,

$$\begin{aligned} r_j &= \frac{|\lambda_{n_j+1}| + |\lambda_{n_j}|}{2} \\ &\leq (n_j + 1)^4 + n_j^4 \\ &\leq c\, n_j^4. \end{aligned}$$

So,

$$n_j^{-4} \leq c\, r_j^{-1}.$$

Now, the use of

$$\begin{aligned}(|\lambda_{n_j+1}| - |\lambda_{n_j}|)^{-1} &\le c\, n_j^{-3} \\ &= c\,(n_j^{-4})^{\frac{3}{4}} \\ &\le c\,(r_j^{-1})^{\frac{3}{4}},\end{aligned}$$

gives the desired result. Q.E.D.

13.1.5 Compactness Results

The Hankel operator defined in (13.1.12) plays a crucial role in our subsequent analysis. So, our purpose of this section is to give some compactness results about it.

Theorem 13.1.2 *The Hankel operator K is compact on $L_2(]-L, L[)$.* ◇

Proof The function $H_0(\cdot)$ is equivalent, in some neighborhood, $]x-\varepsilon, x+\varepsilon[\ \varepsilon > 0$, of $x = x'$, to the function $\log(k|x-x'|)$ (cf. [11]). Thus, we can write the operator K, defined in (13.1.12), in the following form

$$K\psi(x) = K_1\psi(x) + K_2\psi(x) + K_3\psi(x),$$

where K_1, K_2, and K_3 denote the following operators defined by

$$\begin{cases} K_1 : L_2(]-L, L[) \longrightarrow L_2(]-L, L[) \\ \qquad\qquad \psi \longrightarrow K_1\psi(x) = \dfrac{i}{2}\displaystyle\int_{-L}^{L} H_0(k|x-x'|)\chi_{(-L,x-\varepsilon)}(x')\psi(x')\mathrm{d}x', \end{cases}$$

$$\begin{cases} K_2 : L_2(]-L, L[) \longrightarrow L_2(]-L, L[) \\ \qquad\qquad \psi \longrightarrow K_2\psi(x) = \dfrac{i}{2}\displaystyle\int_{-L}^{L} H_0(k|x-x'|)\chi_{(x-\varepsilon,x+\varepsilon)}(x')\psi(x')\mathrm{d}x', \end{cases}$$

and

$$\begin{cases} K_3 : L_2(]-L, L[) \longrightarrow L_2(]-L, L[) \\ \qquad\qquad \psi \longrightarrow K_3\psi(x) = \dfrac{i}{2}\displaystyle\int_{-L}^{L} H_0(k|x-x'|)\chi_{(x+\varepsilon,L)}(x')\psi(x')\mathrm{d}x', \end{cases}$$

where $\chi_{(-L,x-\varepsilon)}(.)$, $\chi_{(x-\varepsilon,x+\varepsilon)}(.)$, and $\chi_{(x+\varepsilon,L)}(.)$ denote, respectively, the characteristic functions of the intervals $(-L, x-\varepsilon)$, $(x-\varepsilon, x+\varepsilon)$ and $(x+\varepsilon, L)$. In order to conclude, it is sufficient to establish that the operators $K_i, i = 1, 2, 3$ are compact on the space $L_2(]-L, L[)$. We set $r = \frac{5}{4}, \sigma = 1, p = 2$, and $q = 2$, then in view of both Lemma 2.2.1 and the equality

$$\log(k|x-x'|) = \frac{\left(k|x-x'|)\right)^{\frac{2}{3}} \log(k|x-x'|)}{(k|x-x'|))^{\frac{2}{3}}},$$

we deduce that the Hankel operator K is compact from $L_2(]-L, L[)$ into $L_2(]-L, L[)$. In view of Eq. (13.1.8), we deduce that

$$\int_{-L}^{L}\int_{-L}^{L} \left|H_0(k|x-x'|)\chi_{(-L,x-\varepsilon)}(x')\right|^2 \mathrm{d}x'\mathrm{d}x \leq 4L_2c^2$$

and

$$\int_{-L}^{L}\int_{-L}^{L} \left|H_0(k|x-x'|)\chi_{(x+\varepsilon,L)}(x')\right|^2 \mathrm{d}x'\mathrm{d}x \leq 4L_2c^2.$$

Thus, K_1 and K_3 are two Hilbert-Schmidt operators. This achieves the proof. Q.E.D.

13.1.6 Completeness of the System of Root Vectors

In this section, we pove that the system of generalized eigenvectors of the operator $(I + \varepsilon K)^{-1}T_0$ is dense in the space $L_2(]-L, L[)$. Based on the result obtained by M. V. Keldysh (see Theorem 6.1.3), we consider the following eigenvalue problem.

$$\begin{cases} T_0\varphi = \lambda(\varepsilon)(I + \varepsilon K)\varphi \\ \varphi(-L) = \varphi(L) = \varphi'(-L) = \varphi'(L) = 0, \end{cases} \tag{13.1.21}$$

and we state the following theorem.

Theorem 13.1.3 *For $\varepsilon \neq \frac{1}{\|K\|}$, the system of generalized eigenvectors of (13.1.21) is dense in the space $L_2(]-L, L[)$.* ◇

Proof First, the operator $H := T_0^{-1}$ is self-adjoint and belongs to Carleman-class $\mathscr{C}_p$, $p > \frac{1}{4}$ (see Theorem 13.1.1). Second, the operator $S := \varepsilon K$ is compact (see Theorem 13.1.2). Moreover, the operator

$$H(I + S)$$

vanishes only at zero (since $\varepsilon \neq \frac{1}{\|K\|}$). So, using M. V. Keldysh theorem (see Theorem 6.1.3), we deduce the completeness of the system of generalized eigenvectors of the problem (13.1.21) in $L_2(]-L, L[)$. Q.E.D.

13.1.7 On a Riesz Basis in $L_2(-L, L)$

Based on the results obtained by A. Schueller in Theorem 7.6.4, we will show in this section that the system of generalized eigenvectors of the operator

$$T(\varepsilon) := (I + \varepsilon K)^{-1} T_0, \quad \mathscr{D}(T(\varepsilon)) := \mathscr{D}(T_0)$$

forms a Riesz basis in $L_2(]-L, L[)$.

Theorem 13.1.4 *The system of eigenvectors of the operator T_0 forms a basis in $L_2(]-L, L[)$.* ◇

Proof The proof follows from Theorem 7.2.2. Q.E.D.

Theorem 13.1.5 *The system of eigenvectors of T_0 forms a Riesz basis in $L_2(]-L, L[)$.* ◇

Proof Let $u \in L_2(]-L, L[)$. Using Theorem 13.1.4, we can write the function u in the following form

$$u = \sum_{n=0}^{\infty} \langle u, \varphi_n \rangle \varphi_n.$$

Hence,

$$\|u\|^2 = \left\langle \sum_{n=0}^{\infty} \langle u, \varphi_n \rangle \varphi_n, \sum_{n=0}^{\infty} \langle u, \varphi_n \rangle \varphi_n \right\rangle.$$

Thus,

$$\|u\|^2 = \sum_{n,m=0}^{\infty} \langle u, \varphi_n \rangle \overline{\langle u, \varphi_m \rangle} \langle \varphi_n, \varphi_m \rangle.$$

So,

$$\|u\|^2 = \sum_{n=0}^{\infty} |\langle u, \varphi_n \rangle|^2.$$

Q.E.D.

Now, Proposition 7.6.1 allows us to get the desired results.

Let $\varepsilon < \frac{1}{\|K\|}$ for which the operator $I + \varepsilon K$ is invertible. So, the problem (13.1.21) becomes:

Finding the values $\lambda(\varepsilon) \in \mathbb{C}$ for which there exists a solution $\varphi \in H_0^2(]-L, L[) \bigcap H^4(]-L, L[)$, $\varphi \neq 0$ for the equation

$$(I + \varepsilon K)^{-1} \frac{\mathrm{d}^4\varphi}{\mathrm{d}x^4} = \lambda(\varepsilon)\varphi.$$

This is equivalent to:

Finding the values $\lambda(\varepsilon) \in \mathbb{C}$ for which there exists a solution $\varphi \in H_0^2(]-L, L[) \bigcap H^4(]-L, L[)$, $\varphi \neq 0$ for the equation

$$\left(\frac{\mathrm{d}^4}{\mathrm{d}x^4} - \varepsilon K \frac{\mathrm{d}^4}{\mathrm{d}x^4} + \varepsilon^2 K^2 \frac{\mathrm{d}^4}{\mathrm{d}x^4} - \cdots + (-1)^n \varepsilon^n K^n \frac{\mathrm{d}^4}{\mathrm{d}x^4} + \cdots\right) \varphi = \lambda(\varepsilon)\varphi. \tag{13.1.22}$$

We denote by $T_0 := \frac{\mathrm{d}^4}{\mathrm{d}x^4}$, $T_n := (-1)^n K^n \frac{\mathrm{d}^4}{\mathrm{d}x^4}$ and $\mathscr{D}(T_n) := H_0^2(]-L, L[) \bigcap H^4(]-L, L[)$ for $n \geq 1$.

Lemma 13.1.8 *(i) For $\varepsilon < \frac{1}{\|K\|}$, the series*

$$T(\varepsilon)\psi = T_0\psi + \varepsilon T_1\psi + \varepsilon^2 T_2\psi + \cdots + \varepsilon^n T_n\psi + \cdots$$

converges for all $\psi \in H_0^2(]-L, L[) \bigcap H^4(]-L, L[)$.
(ii) For $\varepsilon < \frac{1}{2\|K\|}$, the transformation $T(\varepsilon)$ is closed. ◇

Proof (i) Let $n \in \mathbb{N}$, we get

$$\|T_n\psi\| \leq \|K\|^n \|T_0\psi\|.$$

Clearly, the hypotheses of Corollary 9.2.1 are satisfied by taking $b = \|K\|$, $p = \|K\|$ and $a = 0$. So, the series

$$T(\varepsilon)\psi = T_0\psi + \varepsilon T_1\psi + \varepsilon^2 T_2\psi + \cdots + \varepsilon^n T_n\psi + \cdots$$

converges for all $\psi \in H_0^2(]-L, L[) \bigcap H^4(]-L, L[)$ and for all $\varepsilon < \frac{1}{\|K\|}$.
The assertion (ii) follows immediately from Corollary 9.2.1. Q.E.D.

Theorem 13.1.6 *For $\varepsilon < \frac{1}{2\|K\|}$, $T(\varepsilon)$ is an operator with a compact resolvent and its spectrum consists only of eigenvalues with nonzero imaginary parts and can be ordered as a sequence going to $+\infty$ in modulus and the correspondent eigeinspaces are finite dimensional.* ◇

Proof We claim that $0 \in \rho(T(\varepsilon))$. Indeed, let $\psi \in N(T(\varepsilon))$ i.e., $T(\varepsilon)\psi = 0$. The fact that $(I + \varepsilon K)^{-1} T_0 \psi = 0$, then

$$T_0 \psi = 0$$

(since $\varepsilon < \frac{1}{2\|K\|}$). Hence, $\psi = 0$ because $N(T_0) = \{0\}$. So,

$$N(T(\varepsilon)) = \{0\}.$$

Let $\psi \in N(T(\varepsilon)^*)$ i.e., $T(\varepsilon)^* \psi = 0$. The fact that $T_0(I + \varepsilon K^*)^{-1} \psi = 0$, then

$$(I + \varepsilon K^*)^{-1} \psi = 0$$

because $N(T_0) = \{0\}$. Hence, $\psi = 0$ (since $\varepsilon < \frac{1}{2\|K\|}$). So,

$$N(T(\varepsilon)^*) = \{0\}.$$

This proves that $0 \in \rho(T(\varepsilon))$ and achieves the proof of the claim. Using the fact that the canonical injection i from $\mathscr{D}(T(\varepsilon))$ into $L_2(]-L, L[)$ is compact, we conclude $T(\varepsilon)$ has a compact resolvent. Since T_0 is an accretive operator and K has a nonvanishing imaginary part, we infer that the imaginary parts of the eigenvalues of the problem (13.1.22) are nonvanishing. This completes the proof. Q.E.D.

By using Lemma 13.1.4, we get

$$\lim_{n \to +\infty} (\lambda_{n+1} - \lambda_n) = +\infty.$$

Let $\delta > 0$, there exists $n_0 > 0$ such that for all $n > n_0$, we get

$$\lambda_{n+1} - \lambda_n > \delta.$$

Let

$$r = \min\left(\min_{i \in \{1..n_0\}} \left(\frac{\lambda_{i+1} - \lambda_i}{2} \right), \frac{\delta}{2} \right).$$

We note $\sigma_0 = \{\lambda_n\}$ and $\sigma_1 = \sigma(T_0) \backslash \sigma_0$, where λ_n is the nth eigenvalue of the operator T_0. Let's consider $\mathscr{C}_n = \mathscr{C}(\lambda_n, r)$ the closed circle with center λ_n and with radii r, lying in the resolvent set of T_0, and having σ_0 inside and σ_1 outside. Since $R_\lambda = (\lambda - T_0)^{-1}$ is a regular analytic function of $\lambda \in \rho(T_0)$, then $\|R_\lambda\|$ is a continuous function. Hence, we denote by

$$N_n = \sup_{\lambda \in \mathscr{C}_n} \|T_0 R_\lambda\| = \sup_{\lambda \in \mathscr{C}_n} \| - I + \lambda R_\lambda\|.$$

In view of Corollary 9.3.2, for $\varepsilon < \frac{1}{(1+2N_n)\|K\|}$, the perturbed transformation $T(\varepsilon)$ will have a unique point of the spectrum inside the circle $\mathscr{C}_n = \mathscr{C}(\lambda_n, r)$ and this point $\lambda_n(\varepsilon)$ will also be with principal multiplicity 1. $\lambda_n(\varepsilon)$ is the corresponding eigenvalue to $\varphi_n(\varepsilon)$ of $T(\varepsilon)$. The eigenvalue $\lambda_n(\varepsilon)$ and the eigenvector $\varphi_n(\varepsilon)$ can be developed as an entire series of ε and, we obtain

$$\lambda_n(\varepsilon) = \lambda_n + \varepsilon\lambda_n^1 + \varepsilon^2\lambda_n^2 + \cdots$$

$$\varphi_n(\varepsilon) = \varphi_n + \varepsilon\varphi_n^1 + \varepsilon^2\varphi_n^2 + \cdots,$$

and for all $i \geq 1$,

$$\|\varphi_n^i\| \leq (1+2N_n)^i \|K\|^i \tag{13.1.23}$$

and

$$|\lambda_n^i| \leq r N_n (1+2N_n)^{i-1} \|K\|^i.$$

Let $n \in \mathbb{N}^*$. For each eigenvalue λ_n of T_0, we fix an

$$\varepsilon_n \in \left]0, \frac{1}{\alpha(n+1)(1+2N_n)\|K\|}\right[,$$

where

$$\alpha^2 = \sum_{p=1}^{\infty} \frac{1}{p^2}.$$

Since

$$\left]0, \frac{1}{\alpha(n+1)(1+2N_n)\|K\|}\right[\subset \left]0, \frac{1}{(1+2N_n)\|K\|}\right[,$$

then the eigenvalue $\lambda_n(\varepsilon_n)$ and the corresponding eigenvector $\varphi_n(\varepsilon_n)$ can be developed as an entire series of ε_n and, we get

$$\lambda_n(\varepsilon_n) = \lambda_n + \varepsilon_n\lambda_n^1 + \varepsilon_n^2\lambda_n^2 + \cdots$$

$$\varphi_n(\varepsilon_n) = \varphi_n + \varepsilon_n\varphi_n^1 + \varepsilon_n^2\varphi_n^2 + \cdots.$$

We obtain the following:

Theorem 13.1.7 *The vectors $(\varphi_n(\varepsilon_n))_{n=1}^{\infty}$ form a Riesz basis of $L_2(-L, L)$.* ◇

Proof Let $n \in \mathbb{N}^*$. In view of Eq. (13.1.23), we get

$$
\begin{aligned}
\|\varphi_n(\varepsilon_n) - \varphi_n\| &\leq \sum_{p\geq 1} \varepsilon_n^p \, \|\varphi_n^p\| \\
&\leq \varepsilon_n \sum_{p\geq 0} \varepsilon_n^p \|\varphi_n^{p+1}\| \\
&\leq \varepsilon_n \sum_{p\geq 0} \varepsilon_n^p \, (1+2N_n)^{p+1} \|K\|^{p+1} \\
&\leq \varepsilon_n \, (1+2N_n) \, \|K\| \sum_{p\geq 0} (\varepsilon_n \, (1+2N_n) \, \|K\|)^p \\
&\leq \frac{\varepsilon_n (1+2N_n)\|K\|}{1-\varepsilon_n (1+2N_n)\|K\|} \\
&< \frac{1}{\alpha(1+n)(1-\varepsilon_n(1+2N_n)\|K\|)} \\
&< \frac{1}{\alpha n}.
\end{aligned}
$$

We conclude that

$$
\sum_{n\geq 1} \|\varphi_n(\varepsilon_n) - \varphi_n\|^2 < \sum_{n\geq 1} \frac{1}{\alpha^2 n^2} = 1.
$$

Consequently, combining both Theorems 7.6.4 and 13.1.4, we conclude that the family of functions $(\varphi_n(\varepsilon_n))_{n=1}^{\infty}$ form a Riesz basis of $L_2(-L, L)$. This completes the proof. Q.E.D.

13.2 Vibrating Plate in a Light Fluid

An elastic membrane is excited by a harmonic force $F(x)e^{-i\omega t}$. It occupies the domain $-L < x < L$ of the plane $z = 0$. The two half-spaces $z < 0$ and $z > 0$ are filled with gas. We denote by u the displacement of the membrane and, we consider the following boundary value problem:

$$
\left(\frac{\mathrm{d}^4}{\mathrm{d}x^4} - \frac{m\omega^2}{D}\right) u(x) -
$$

$$
i\rho_0 \int_{-L}^{L} H_0(k|x-x'|) \left(\frac{\omega^2}{D} - \frac{1}{m}\left(\frac{\mathrm{d}^4}{\mathrm{d}x^4} - \frac{\mathrm{d}^2}{\mathrm{d}x^2}\right)\right) u(x')\mathrm{d}x' = \frac{F(x)}{D}, \qquad (13.2.1)
$$

for all $x \in]-L, L[$ such that

$$
u(x) = \frac{\partial u(x)}{\partial x} = 0
$$

for $x = -L$ and $x = L$ and $H_0(\cdot)$ is the Hankel function of the first kind and order 0; while the mechanical parameters of the membrane are E the Young modulus, ν the Poisson ratio, m the surface density, h the thickness of the membrane and

$$D := \frac{Eh^3}{12(1-\nu^2)}$$

the rigidity. The fluid is characterized by ρ_0 the density, c the sound speed and

$$k := \frac{\omega}{c}$$

the wave number.

The problem (13.2.1) satisfy the following system:

$$\left(\frac{\mathrm{d}^4}{\mathrm{d}x^4} - \frac{m\omega^2}{D}\right) u(x) = \frac{1}{D}(F(x) - P(x)) \text{ for all} x \in]-L, L[, \tag{13.2.2}$$

where

$$u(x) = \frac{\partial u(x)}{\partial x} = 0 \text{ for } x = -L \text{ and } x = L, \tag{13.2.3}$$

$$P(x) = \lim_{z \to 0^+} (p(x,z) - p(x,-z)) \tag{13.2.4}$$

and

$$p(x,z) = -sgn(z)\, i \frac{\rho_0}{2} \int_{-L}^{L} H_0\left(k\sqrt{(x-x')^2 + z^2}\right) \left(\omega^2 - \frac{D}{m}\left(\frac{\mathrm{d}^4}{\mathrm{d}x^4} - \frac{\mathrm{d}^2}{\mathrm{d}x^2}\right)\right) u(x')\mathrm{d}x', \tag{13.2.5}$$

for $z < 0$ or $z > 0$. Expressions (13.2.2), (13.2.3), (13.2.4) and (13.2.5) lead to the boundary value problem

$$\left(\frac{\mathrm{d}^4}{\mathrm{d}x^4} - \frac{m\omega^2}{D}\right) u(x) - i\rho_0 \int_{-L}^{L} H_0\left(k|x-x'|\right) \left(\frac{\omega^2}{D} - \frac{1}{m}\left(\frac{\mathrm{d}^4}{\mathrm{d}x^4} - \frac{\mathrm{d}^2}{\mathrm{d}x^2}\right)\right) u(x')\mathrm{d}x' = \frac{F(x)}{D},$$

for all $x \in]-L, L[$ with

$$u(x) = \frac{\partial u(x)}{\partial x} = 0$$

for $x = -L$ and $x = L$.

In the sequel, we consider the following operators:

$$\begin{cases} B = T_0^{\frac{1}{2}} : \mathscr{D}(B) \subset L_2(]-L,L[) \longrightarrow L_2(]-L,L[) \\ \qquad\qquad\qquad \varphi \longrightarrow B\varphi(x) = \dfrac{d^2\varphi}{\mathrm{d}x^2} \\ \mathscr{D}(B) = \left\{\varphi \in L_2(]-L,L[) \text{such that} \displaystyle\sum_{n=1}^{\infty} \lambda_n |\langle \varphi, \varphi_n \rangle|^2 < \infty \right\} \end{cases}$$

and the following eigenvalue problem:

Finding the values $\lambda(\varepsilon) \in \mathbb{C}$ for which there is a solution $u \in H_0^2(]-L,L[) \bigcap H^4(]-L,L[)$, $u \neq 0$, for the equation

$$T_0 u + \varepsilon K(T_0 - B)u = \lambda(I + \varepsilon K)u, \tag{13.2.6}$$

where

$$\lambda = \frac{m\omega^2}{D}$$

and

$$\varepsilon = \frac{2\rho_0}{m}.$$

According to the definition given in reference [10, Chap. 9, Sect. 4], λ is the eigenvalue and u is the eigenmode. Note that λ and u, each one of them depends on the value of ε. So, we denote $\lambda := \lambda(\varepsilon)$ and $u := u(\varepsilon)$. For $|\varepsilon| < \frac{1}{\|K\|}$, the operator $I + \varepsilon K$ is invertible. So, the problem (13.2.6) becomes:

Finding the values $\lambda(\varepsilon) \in \mathbb{C}$ for which there is a solution $\varphi \in H_0^2(]-L,L[) \bigcap H^4(]-L,L[)$, $\varphi \neq 0$, for the equation

$$(I + \varepsilon K)^{-1}\frac{\mathrm{d}^4\varphi}{\mathrm{d}x^4} + \varepsilon(I + \varepsilon K)^{-1}K(T_0 - B)\varphi = \lambda(\varepsilon)\varphi. \tag{13.2.7}$$

The problem (13.2.7) is equivalent to:

Finding the values $\lambda(\varepsilon) \in \mathbb{C}$ for which there exists a solution $\varphi \in H_0^2(]-L,L[) \bigcap H^4(]-L,L[)$, $\varphi \neq 0$, for the equation

$$\left(\frac{\mathrm{d}^4}{\mathrm{d}x^4} - \varepsilon K\frac{\mathrm{d}^2}{\mathrm{d}x^2} + \varepsilon^2 K^2\frac{\mathrm{d}^2}{\mathrm{d}x^2} + \cdots + (-1)^n\varepsilon^n K^n\frac{\mathrm{d}^2}{\mathrm{d}x^2} + \cdots\right)\varphi = \lambda(\varepsilon)\varphi.$$

13.2.1 Elementary Results

For all $n \geq 1$, we denote by

$$T_n := (-1)^n K^n \frac{\mathrm{d}^2}{\mathrm{d}x^2},$$

with domain

$$\mathscr{D}(T_n) = H^2\big(]-L, L[\big).$$

Lemma 13.2.1 *B is $\frac{1}{2}$-subordinate to T_0.* ◇

Proof By using both $\mathscr{D}(T_0) \subset \mathscr{D}(B)$ and (13.1.18), we have for every $\varphi \in \mathscr{D}(T_0)$

$$\begin{aligned}\|B\varphi\|^2 &= \sum_{n=1}^{\infty} \lambda_n |\langle \varphi, \varphi_n \rangle|^2 \\ &= \sum_{n=1}^{\infty} \kappa n^4 |\langle \varphi, \varphi_n \rangle|^2 \\ &= \kappa \sum_{n=1}^{\infty} n^{8\frac{1}{2}} |\langle \varphi, \varphi_n \rangle|^{2\frac{1}{2}} |\langle \varphi, \varphi_n \rangle|^{2(1-\frac{1}{2})}.\end{aligned}$$

Using Hölder's inequality, we get

$$\begin{aligned}\|B\varphi\|^2 &\leq \left(\sum_{n=1}^{\infty} \kappa^2 n^8 |\langle \varphi, \varphi_n \rangle|^2\right)^{\frac{1}{2}} \left(\sum_{n=1}^{\infty} |\langle \varphi, \varphi_n \rangle|^2\right)^{1-\frac{1}{2}} \\ &\leq \|T_0\varphi\| \, \|\varphi\|.\end{aligned}$$

Hence,

$$\|B\varphi\| \leq \|T_0\varphi\|^{\frac{1}{2}} \|\varphi\|^{\frac{1}{2}}$$

for all $\varphi \in \mathscr{D}(T_0)$, which ends the proof of lemma. Q.E.D.

Proposition 13.2.1 *The following properties hold.*

(i) There exist constants $a, b, q > 0$ and $\beta \in \left[\frac{1}{2}, \frac{3}{4}\right]$ such that

$$\|T_k\varphi\| \leq q^{k-1}(a\|\varphi\| + b\|T_0\varphi\|^{\beta} \|\varphi\|^{1-\beta})$$

for all $\varphi \in \mathscr{D}(T_0)$ and for all $k \geq 1$.

(ii) For $|\varepsilon| < \frac{1}{\|K\|}$, the series

$$T(\varepsilon)\varphi := T_0\varphi + \varepsilon T_1\varphi + \varepsilon^2 T_2\varphi + \cdots + \varepsilon^k T_k\varphi + \cdots$$

converges for all $\varphi \in \mathscr{D}(T_0)$. ◇

Proof (i) Let $\beta \in \left[\frac{1}{2}, \frac{3}{4}\right]$ and $\varphi \in \mathscr{D}(T_0)$. Writing $\|T_0\varphi\|^{\frac{1}{2}}\|\varphi\|^{1-\frac{1}{2}}$ as

$$\|T_0\varphi\|^{\frac{1}{2}}\|\varphi\|^{1-\frac{1}{2}} = \|\varphi\|^{1-\beta}\|T_0\varphi\|^{\frac{1}{2}}\|\varphi\|^{\beta-\frac{1}{2}},$$

and since $\beta \geq \frac{1}{2}$, we can apply the foregoing interpolation inequality (see Lemma 1.3.4) to $\|T_0\varphi\|^{\frac{1}{2}}$ and $\|\varphi\|^{\beta-\frac{1}{2}}$ with $\delta = \frac{1}{2}$ and $\theta = \beta$, to deduce that

$$\|T_0\varphi\|^{\frac{1}{2}}\|\varphi\|^{\beta-\frac{1}{2}} \leq \|T_0\varphi\|^{\beta} + \|\varphi\|^{\beta},$$

and

$$\|T_0\varphi\|^{\frac{1}{2}}\|\varphi\|^{\frac{1}{2}} \leq \|T_0\varphi\|^{\beta}\|\varphi\|^{1-\beta} + \|\varphi\|. \tag{13.2.8}$$

Consequently, by using Eq. (13.2.8) and taking into account Lemma 13.2.1, we deduce that

$$\begin{aligned}
\|T_k\varphi\| &= \|K^k B\varphi\| \\
&\leq \|K^k\|\|B\varphi\| \\
&\leq \|K\|^k \left(\|T_0\varphi\|^{\beta}\|\varphi\|^{1-\beta} + \|\varphi\|\right) \\
&\leq \|K\|^{k-1} \left(\|K\|\|T_0\varphi\|^{\beta}\|\varphi\|^{1-\beta} + \|K\|\|\varphi\|\right),
\end{aligned}$$

for all $\varphi \in \mathscr{D}(T_0)$ and $k \geq 1$. Hence, it suffices to take $a = b = q = \|K\|$, which ends the proof of the first assertion.
(ii) The result follows immediately from (i). Q.E.D.

13.2.2 Completeness Results

In the rest of this section, we consider $a = b = q = \|K\|$.

Proposition 13.2.2 *For* $|\varepsilon| < \frac{1}{\|K\|}$*, the series*

$$\sum_{k \geq 0} \varepsilon^k T_k \varphi$$

converges for all $\varphi \in \mathscr{D}(T_0)$*. If we designate its sum by* $T(\varepsilon)\varphi$*, we define a linear operator* $T(\varepsilon)$ *with domain* $\mathscr{D}(T_0)$*. For* $|\varepsilon| < \frac{1}{\|K\|(1+\beta)}$*, the operator* $T(\varepsilon)$ *is closed.* ◇

Proof The result follows from both Proposition 13.2.1 and Corollary 9.2.1. Q.E.D.

In [3], the authors showed that, for $|\varepsilon| < \frac{1}{\|K\|}$, the following Inequality:

$$\left\| \sum_{k=1}^{\infty} (-1)^k \varepsilon^k K^k \frac{\mathrm{d}^2 \varphi_n}{\mathrm{d}x^2} \right\| \leq \frac{|\varepsilon|}{1 - |\varepsilon| \|K\|} \|K\| \kappa n^4,$$

holds. Hence, it is quite difficult to verify Eq. (11.4.2) and consequently, [13, Theorem 1] could not be applied. We begin this subsection by the following:

Theorem 13.2.1 *For $|\varepsilon|$ small enough and $\beta \in [\frac{1}{2}, \frac{7}{8}[$, we have*
$n(\overline{\mathbb{D}}(0, r), T(\varepsilon)) - n(\overline{\mathbb{D}}(0, r), T_0) = O(1).$ ◇

Proof Let λ_n be the nth eigenvalue of T_0 and

$$B(\varepsilon) = \sum_{k=1}^{\infty} \varepsilon^k T_k.$$

For $\alpha = \frac{3}{4}$ in Eq. (11.2.6) and $p \geq \frac{1}{\kappa^{\frac{1}{4}}}$, we have

$$\lambda_{n+p}^{\frac{1}{4}} - \lambda_n^{\frac{1}{4}} = \kappa^{\frac{1}{4}}((n + p) - n) \geq 1. \tag{13.2.9}$$

Further, it follows from Proposition 13.2.1(i), that for all $k \geq 1$, we have

$$\|T_k \varphi\| \leq \|K\|^k \left(\|\varphi\| + \|T_0 \varphi\|^{\beta} \|\varphi\|^{1-\beta}\right), \text{ for all } \varphi \in \mathscr{D}(T_0), \tag{13.2.10}$$

where $\beta \in [\frac{1}{2}, \frac{7}{8}[$. Consequently, in view of Corollary 11.2.1, Eq. (13.1.18), Lemma 13.1.4, Proposition 13.2.2 and Eqs. (13.2.9) and (13.2.10), the spectrum of $T(\varepsilon)$ is discrete for $|\varepsilon| < \frac{1}{\|K\|(1+\beta)}$. On the other hand, Eq. (13.2.10) implies that for $|\varepsilon| < \frac{1}{\|K\|}$, we have

$$\|B(\varepsilon)\varphi_n\|^2 \leq \frac{|\varepsilon|^2}{(1 - |\varepsilon| \|K\|)^2} \|K\|^2 \left(1 + \lambda_n^{\beta}\right)^2.$$

So, for $|\varepsilon| < \frac{1}{\|K\|}$, we get

$$\frac{\|B(\varepsilon)\varphi_n\|^2}{|\lambda - \nu_n|^2} \leq \frac{|\varepsilon|^2}{(1 - |\varepsilon| \|K\|)^2} \frac{\|K\|^2 \left(1 + \lambda_n^{\beta}\right)^2}{|\lambda - \nu_n|^2}.$$

Hence, taking into account that

$$\lambda_n = \kappa n^4,$$

Equation (11.2.10) yields

$$\sum_{n=1}^{\infty}\frac{\lambda_n^{2\beta}}{|\lambda-\nu_n|^2}<\frac{2c}{C_3^2}+\sum_{n<k-1}\frac{c\kappa^{2(\beta-1)}}{\left((k-1)^{4(1-\beta)}-n^{4(1-\beta)}\right)^2}+\sum_{n>k}\frac{c\kappa^{2(\beta-1)}}{\left(n^{4(1-\beta)}-k^{4(1-\beta)}\right)^2}. \tag{13.2.11}$$

Now, if $\beta\in[\frac{1}{2},\frac{3}{4}]$, then

$$4(1-\beta)\geq 1.$$

So, it follows from Eq. (13.2.11) that

$$\begin{aligned}\sum_{n=1}^{\infty}\frac{\lambda_n^{2\beta}}{|\nu_n-\sigma|^2}&<\frac{2c}{C_3^2}+\sum_{n<k-1}\frac{c}{\kappa^{2(1-\beta)}\,(k-1-n)^{8(1-\beta)}}+\sum_{n>k}\frac{c}{\kappa^{2(1-\beta)}\,(n-k)^{8(1-\beta)}}\\&\leq\frac{2c}{C_3^2}+\sum_{n<k-1}\frac{c}{\kappa^{2(1-\beta)}\,(k-1-n)^{8(1-\beta)}}+\sum_{m=1}^{\infty}\frac{c}{\kappa^{2(1-\beta)}m^{8(1-\beta)}}\\&<\frac{2c}{C_3^2}+\frac{2c}{\kappa^{2(1-\beta)}}\sum_{m=1}^{\infty}\frac{1}{m^{8(1-\beta)}}=:\xi_1<\infty.\end{aligned}$$

If $\beta\in]\frac{3}{4},\frac{7}{8}[$, then

$$0<-3+4\beta<\frac{1}{2}$$

and

$$\frac{1}{2}<4-4\beta<1.$$

Hence, Eqs. (11.2.11), (11.2.12) and (13.2.11) imply that

$$\begin{aligned}&\sum_{n=1}^{\infty}\frac{\lambda_n^{2\beta}}{|\lambda-\nu_n|^2}\\&<\frac{2c}{C_3^2}+\sum_{n<k-1}\frac{c\kappa^{2(\beta-1)}}{c_1^2(4-4\beta)^2\,(k-1-n)^{8(1-\beta)}}+\sum_{n>k}\frac{c\kappa^{2(\beta-1)}}{c_2^2(4-4\beta)^2\,(n-k)^{8(1-\beta)}}\\&\leq\frac{2c}{C_3^2}+\sum_{n<k-1}\frac{c\kappa^{2(\beta-1)}}{c'^2(4-4\beta)^2\,(k-1-n)^{8(1-\beta)}}+\sum_{m=1}^{\infty}\frac{c\kappa^{2(\beta-1)}}{c'^2(4-4\beta)^2m^{8(1-\beta)}}\\&<\frac{2c}{C_3^2}+\frac{2c\kappa^{2(\beta-1)}}{c'^2(4-4\beta)^2}\sum_{m=1}^{\infty}\frac{1}{m^{8(1-\beta)}}=:\xi_1'<\infty,\end{aligned}$$

where $c':=\min\{c_1,c_2\}$. Consequence, for $\beta\in[\frac{1}{2},\frac{7}{8}[$, we get

$$\sum_{n=1}^{\infty}\frac{\lambda_n^{2\beta}}{|\lambda-\nu_n|^2}<\max\{\xi_1,\xi_1'\}=:\xi_1'.$$

Moreover, since $\beta \in [\frac{1}{2}, \frac{7}{8}[$, we have

$$4(1 - \frac{\beta}{2}) > 1.$$

Therefore, we obtain

$$\sum_{n=1}^{\infty} \frac{\lambda_n^{\beta}}{|\lambda - \nu_n|^2} < \frac{2c}{C_3^2} + \frac{2c}{\kappa^{(2-\beta)}} \sum_{m=1}^{\infty} \frac{1}{m^{4(2-\beta)}} =: \xi_2' < \infty.$$

In fact, it suffices to use the same reasoning as the above with $\frac{\beta}{2}$. Hence, the series

$$\sum_{n} \frac{2\|K\|^2 \lambda_n^{\beta}}{|\lambda - \nu_n|^2}$$

and

$$\sum_{n} \frac{\|K\|^2 \lambda_n^{2\beta}}{|\lambda - \nu_n|^2}$$

are convergent. So, let ξ be a positive constant such that

$$\|K\|^2 \xi_1' + 2\|K\|^2 \xi_2' = \xi.$$

Then, we obtain

$$\sum_{n=1}^{\infty} \frac{2\|K\|^2 \lambda_n^{\beta}}{|\lambda - \nu_n|^2} + \sum_{n=1}^{\infty} \frac{\|K\|^2 \lambda_n^{2\beta}}{|\lambda - \nu_n|^2} < \xi.$$

Consequently, for $|\varepsilon| < W$, Corollary 12.2.1, Eq. (13.1.18), Lemma 13.1.4, Proposition 13.2.2 and Eqs. (13.2.9) and (13.2.10) yield

$$n(\overline{\mathbb{D}}(0, r), T(\varepsilon)) - n(\overline{\mathbb{D}}(0, r), T_0) = O(1),$$

where

$$W := \min \left\{ \frac{1}{\|K\| + \sqrt{\xi + 4pc_3\|K\|^2}}, \frac{1}{\|K\| + c_4\sqrt{\xi + 2pc_5\|K\|^2}} \right\},$$

with $0 < c_3 < 1$, $c_4 > 1$ and $c_5 > 0$. Q.E.D.

Using the results described above, we have the following:

Proposition 13.2.3 *For $|\varepsilon| < \frac{1}{\|K\|(1+\beta)}$ and $\beta \in [\frac{1}{2}, \frac{7}{8}[$, the system of root vectors of the operator $T(\varepsilon)$ is complete in $L_2(]-L, L[)$.* ◇

Proof In view of Corollary 9.3.10, Lemma 13.1.2 and Eq. (13.2.10), the system of root vectors of the operator $T(\varepsilon)$ is complete in $L_2(]-L, L[)$ for $|\varepsilon| < \frac{1}{\|K\|(1+\beta)}$. This completes the proof. Q.E.D.

13.2.3 Basis with Parentheses

Theorem 13.2.2 *For $|\varepsilon|$ small enough and $\beta \in [\frac{1}{2}, \frac{7}{8}[$, the system of root vectors of the operator $T(\varepsilon)$ forms a basis with parentheses in $L_2(]-L, L[)$.* ◇

Proof In virtue of Proposition 13.2.1(i), we have for all $\varphi \in \mathscr{D}(T_0)$ and for all $k \geq 1$

$$\|T_k\varphi\| \leq \|K\|^k \left(\|\varphi\| + \|T_0\varphi\|^{\beta}\|\varphi\|^{1-\beta}\right), \tag{13.2.12}$$

where $\beta \in [\frac{1}{2}, \frac{7}{8}[$. The result follows immediately from Corollary 12.2.1, Lemma 13.1.2 and Eq. (13.2.12). Q.E.D.

13.3 Notes and Remarks

- In order to obtain the desired results, A. Jeribi and A. Intissar [9] (see also [8]) studied the spectral properties of the operator T_0 and its resolvent. They also derived some compactness results, proved the completeness of the system of root vectors and the existence of the Riesz or unconditional bases formed by the eigenvectors of the perturbed operator.
- **Open problem**: Let Ω be a bounded domain of the $z = 0$ plane, with boundary $\partial\Omega$. It is occupied by a thin plate, while its complement $\widetilde{\Omega}$ is perfectly rigid. The half-spaces $z > 0$ and $z < 0$ are occupied by a perfect fluid at rest. The mechanical parameters of the plate are E: Young modulus; ν: Poisson ratio; h: thickness; $D = \frac{Eh^3}{12(1-\nu^2)}$: rigidity; μ: mass per unit area. The mechanical parameters of the fluid are c_0: sound speed; μ: density. The plate is excited by a harmonic e^{-iwt} source of density F. Let u stand for the plate displacement and p for the radiated sound pressure in the fluid. They are solutions of the following boundary value problem

$$\begin{aligned} D(\Delta^2 - \lambda^4)u + P &= F \text{ in } \Omega \\ (\Delta + k^2)p &= 0 \text{ in } z > 0 \text{ and } z < 0 \\ \partial_z p &= \mu_0 w^2 u \text{ on } z = 0, \text{ in } \Omega \\ &= 0 \text{ on } z = 0, \text{ in } \widetilde{\Omega} \end{aligned} \tag{13.3.1}$$

$$l_1 u = l_2 u = 0 \text{ on } \partial\Omega \tag{13.3.2}$$

$$\text{Condition at infinity} \quad \text{for } p \tag{13.3.3}$$

$$P(x, y) = \lim_{\eta \to 0} [p(x, y, \eta) - p(x, y, -\eta)],\ \eta > 0$$

$$\lambda^4 = \frac{\mu w^2}{D} \quad k^2 = \frac{w^2}{c_0^2}.$$

Condition (13.3.2) expresses the continuity between the normal particle velocity in the fluid and the normal velocity of the structure; condition (13.3.3) is the Sommerfeld condition, or any other equivalent condition. Let G_k be the Green's function of the Helmholtz equation in $z > 0$, which satisfies the Neumann boundary condition on $z = 0$ and the condition (13.3.3) at infinity

$$(\Delta + k^2)G_k = \delta \text{ in } z > 0$$
$$\partial_z G_k = 0 \text{ on } z = 0.$$

Then, the sound pressure field is written as

$$p(x, y, z) = sgn(z) w^2 \mu_0 \int_\Omega u(M') G_k(M, M') dM'$$
$$M = (x, y, z)$$

which leads to

$$P(x, y) = 2w^2 \mu_0 \int_\Omega u(M') G_k(M, M') dM',\ M = (x, y, 0). \tag{13.3.4}$$

Introducing Eq. (13.3.4) into Eq. (13.3.1), one is left with the following system of integro-differential equations

$$\begin{cases} (D\Delta^2 - \mu w^2)u + P = F \text{ in } \Omega \\ 2w^2\mu_0 \int_\Omega u(M')G_k(M, M')dM' - P = 0 \text{ in } \Omega \\ l_1 u = l_2 u = 0 \text{ on } \partial\Omega. \end{cases} \tag{13.3.5}$$

The system (13.3.5) on which the perturbation method will develop. Let $\varepsilon = \frac{2\mu_0}{\mu}$. The angular eigenfrequencies $\Omega_n = \sqrt{\frac{\Lambda_n}{\mu}}$ and eigenmodes (U_n, P_n) of the operator are solution of

$$\begin{cases} (D\Delta^2 - \mu w^2)U_n + P_n = F \text{ in } \Omega \\ \Lambda_n \varepsilon \int_\Omega U_n(M')G_k(M, M')dM' - P_n = 0 \text{ in } \Omega \\ l_1 U_n = l_2 U_n = 0 \text{ on } \partial\Omega. \end{cases} \tag{13.3.6}$$

The authors of the paper [6] suppose the following hypotheses:

(i) There exists an infinite sequence of eigenvalues with a finite order of multiplicity, this sequence being denoted by $\Lambda_n(\varepsilon)$, an eigenvalue of order p appearing p times with different indices.
(ii) To each eigenvalue corresponds to an eigenmode $U_n(\varepsilon)$.
(iii) The set $U_n(\varepsilon)$, $n = 1, 2, \ldots, \infty$ is a basis of the space $H^4(\Omega)$.

In addition, note that the authors of the paper [6] suppose that the eigenvalue $\Lambda_n(\varepsilon)$ and the eigenvector $U_n(\varepsilon)$ of the problem (13.3.6) can be developed as an entire series of ε and so, we can write them in the form
cm

$$\Lambda_n(\varepsilon) = \Lambda_n + \varepsilon \Lambda_n^1 + \varepsilon^2 \Lambda_n^2 + \cdots$$

$$U_n(\varepsilon) = U_n + \varepsilon U_n^1 + \varepsilon^2 U_n^2 + \cdots ,$$

where $\Lambda_n(0) := \Lambda_n$ (respectively, $U_n(0) := U_n$) is the eigenvalue (respectively, eigenvector) of the operator Δ^2.
Is the system (13.3.6) satisfying the assumptions (i), (ii), and (iii)? □

References

1. D. Colton, P. Monk, On a class of integral equations of the first kind in inverse scattering theory. SIAM J. Appl. Math. **53**, 847–860 (1993)
2. D. Colton, R. Kress, Eigenvalues of the far field operator for the Helmholtz equation in an absorbing medium. SIAM J. Appl. Math. **55**, 1724–1735 (1995)
3. H. Ellouz, I. Feki, A. Jeribi, On the asymptotic behavior of the eigenvalues of an analytic operator in the sense of Kato and applications. Serdica Math. J. **45**(1), 55–88 (2019)
4. P.J.T. Filippi, Rayonnement d'une structure vibrante en fluide léger: application des techniques classiques des pertubations. J. Acoustique **2**, 39–45 (1989)
5. P.J.T. Filippi, O. Lagarrigue, P.O. Mattei, Perturbation method for sound radiation by a vibrating plate in a light fluid: comparison with the exact solution. J. Sound Vib. **177**, 259–275 (1994)
6. P.J.T. Filippi, P.O. Mattei, Sound radiation by a thin baffled plate embedded in a "light" fluid: boundary integral equations deduced from a perturbation method. In: Computer Mechnical Southampton, XIV, vol. 1, Seville, pp. 279–291 (1992)
7. D. Givoli, Non-reflecting boundary conditions. J. Comput. Phys. **94**, 1–29 (1991)
8. A. Jeribi, *Denseness, Bases and Frames in Banach Spaces and Applications* (de Gruyter, Berlin, 2018)
9. A. Jeribi, A. Intissar, On an Riesz basis of generalized eigenvectors of the non-self-adjoint problem deduced from a perturbation method for sound radiation by a vibrating plate in a light fluid. J. Math. Anal. Appl. **292**, 1–16 (2004)
10. P.M. Morse, H. Feshbach, *Methods of Theorical Physics* (Book co. Inc, New-York-Toronto-London, 1953)

11. J.C. Nedelec, *Acoustic and Electromagnetic Equations, Integral Representations for Harmonic Problems* (Springer-Verlag, New-York, 2001)
12. B. Noble, *Methods Based on the Wiener-Hopf Technique for the Solution of Partial Differential Equations* (Pergamon Press, New-York, 1958)
13. A.A. Shkalikov, On the basis property of root vectors of a perturbed self-adjoint operator. Proc. Steklov Inst. Math. **269**, 284–298 (2010)
14. A. Sommerfeld, *Lectures on Theoretical Physics* (Academic Press, New-York, 1964)

Chapter 14
Gribov Operator in Bargmann Space

This chapter concens a perturbation method for the Gribov operator in Bargmann space. We treat the Gribov operator in Bargmann space in the cases of finite and infinite sum on null transverse dimension and we confirm the existence of Riesz basis of subspaces, Schauder basis, and Basis with parentheses. It is worth mentioning that each section has its own equations, notations, and symbols. In other words, the reader should remember that the same symbol doesn't have the same meaning or significance from one section to another.

14.1 Finite Sum of Gribov Operators on Null Transverse Dimension ($n = 1$)

In this part, we consider the following Gribov operator

$$H_{\lambda'',\lambda',\mu,\lambda} := \lambda''(A^*A)^3 + \lambda'(A^*A)^2 + \mu A^*A + i\lambda A^*(A + A^*)A,$$

where λ'' is the magic coupling of Pomeron, λ' is the four coupling of Pomeron, μ is the Pomeron intercept, and λ is the triple coupling of Pomeron, the annihilation operator A, and the creation operator A^*, are given by

$$\begin{cases} A : \mathscr{D}(A) \subset E \longrightarrow E \\ \qquad\qquad \varphi \longrightarrow A\varphi(z) = \dfrac{d\varphi}{dz}(z) \\ \mathscr{D}(A) = \{\varphi \in E \text{ such that } A\varphi \in E\} \end{cases}$$

and

A. Jeribi, *Perturbation Theory for Linear Operators*,
https://doi.org/10.1007/978-981-16-2528-2_14

$$\begin{cases} A^* : \mathscr{D}(A^*) \subset E \longrightarrow E \\ \qquad\qquad \varphi \longrightarrow A^*\varphi(z) = z\varphi(z) \\ \mathscr{D}(A^*) = \{\varphi \in E \text{ such that } A^*\varphi \in E\}, \end{cases}$$

and, the Bargmann space E is defined by

$$E := \left\{ \varphi : \mathbb{C} \longrightarrow \mathbb{C} \text{ entire such that } \int_{\mathbb{C}} e^{-|z|^2} |\varphi(z)|^2 dxdy < \infty \text{ and } \varphi(0) = 0 \right\},$$

which is a Hilbert space for the following scalar product

$$\begin{cases} \langle \cdot, \cdot \rangle : E \times E \longrightarrow \mathbb{C} \\ \qquad (\varphi, \psi) \longrightarrow \langle \varphi, \psi \rangle = \frac{1}{\pi} \int_{\mathbb{C}} e^{-|z|^2} \varphi(z) \overline{\psi}(z) dxdy. \end{cases}$$

The associated norm may be denoted by $\| \cdot \|$. The operators A and A^* satisfy the relation

$$[A, A^*] := AA^* - A^*A = I.$$

The expression of $H_{\lambda'',\lambda',\mu,\lambda}$ then becomes

$$H_{\lambda'',\lambda',\mu,\lambda} := \lambda''(A^*A)^3 + \lambda' S + \mu H_0 + i\lambda H_1,$$

where S, H_0 and H_1 are the following operators:

$$\begin{cases} S : \mathscr{D}(S) \subset E \longrightarrow E \\ \qquad\qquad \varphi \longrightarrow S\varphi(z) = (A^*A)^2 \varphi(z) \\ \\ \mathscr{D}(S) = \{\varphi \in E \text{ such that } S\varphi \in E\}, \end{cases}$$

$$\begin{cases} H_0 : \mathscr{D}(H_0) \subset E \longrightarrow E \\ \qquad\qquad \varphi \longrightarrow H_0\varphi(z) = A^*A\varphi(z) \\ \\ \mathscr{D}(H_0) = \{\varphi \in E \text{ such that } H_0\varphi \in E\}, \end{cases}$$

and

$$\begin{cases} H_1 : \mathscr{D}(H_1) \subset E \longrightarrow E \\ \qquad\qquad \varphi \longrightarrow H_1\varphi(z) = A^*(A + A^*)A\varphi(z) \\ \\ \mathscr{D}(H_1) = \{\varphi \in E \text{ such that } H_1\varphi \in E\}. \end{cases}$$

The operator $(A^*A)^3$ is defined by

$$\begin{cases} (A^*A)^3 : \mathscr{D}((A^*A)^3) \subset E \longrightarrow E \\ \qquad \varphi \longrightarrow G\varphi = \sum_{n=1}^{\infty} n^3 \langle \varphi, e_n \rangle e_n \\ \mathscr{D}((A^*A)^3) = \left\{ \varphi \in E \text{ such that } \sum_{n=1}^{\infty} n^6 |\langle \varphi, e_n \rangle|^2 < \infty \right\}. \end{cases}$$

We can see the following.

Proposition 14.1.1 (Intissar [12]) *We have the following items:*

(i) *T_0 is a closed linear operator with dense domain,*

(ii) *the resolvent set of T_0 is not empty. In fact, $0 \in \rho(T_0)$,*

(iii) *T_0 is a self-adjoint operator with compact resolvent,*

(iv) *the eigenvalues of T_0 are simple and isolated,*

(v) *the eigenvectors of T_0 form a Riesz basis in E, and*

(vi) *$\{e_n(z) = \frac{z^n}{\sqrt{n!}}\}_1^{\infty}$ is a system of eigenvectors associated to the eigenvalues $\{n^3\}_{n=1}^{\infty}$ of T_0.* ◇

Let G be the operator defined by

$$\begin{cases} G : \mathscr{D}(G) \subset E \longrightarrow E \\ \qquad \varphi \longrightarrow G\varphi(z) = A^{*3} A^3 \varphi(z) \\ \mathscr{D}(G) = \{\varphi \in E \text{ such that } G\varphi \in E\}, \end{cases}$$

Now, we recall the following result:

Proposition 14.1.2 (Intissar [12])

(i) *The set of each polynomials P_0 which vanish at zero is dense in E.*
(ii) *The family $\{e_n := \frac{z^n}{\sqrt{n!}}\}_{n=1}^{\infty}$ is an orthonormal basis of the space E.*
(iii) *The operators G and H_0 are self-adjoint and with compact resolvents.*
(iv) *The families $\{e_n := \frac{z^n}{\sqrt{n!}}\}_{n=1}^{\infty}$ is a system of eigenvectors for the operator H_0 associated with the eigenvalues $\{n\}_{n=1}^{\infty}$.*
(v) *The families $\{e_n := \frac{z^n}{\sqrt{n!}}\}_{n=2}^{\infty}$ is a system of eigenvectors for the operator G associated with the eigenvalues $\{n(n-1)(n-2)\}_{n=2}^{\infty}$.*
(vi) *The families $\{e_n := \frac{z^n}{\sqrt{n!}}\}_{n=1}^{\infty}$ is a system of eigenvectors for the operator S associated with the eigenvalues $\{n(n-1)\}_{n=1}^{\infty}$.* ◇

Lemma 14.1.1 *(i) There is* $c_1 > 0$ *satisfying*

$$\|H_1\varphi\| \leq c_1 \left(\sum_{n=1}^{\infty} n^3 |\langle \varphi, e_n\rangle|^2\right)^{\frac{1}{2}} \quad \textit{for all } \varphi \in \mathscr{D}(H_1).$$

(ii) We have the following inequality

$$\|S\varphi\| \leq \left(\sum_{n=1}^{\infty} n^6 |\langle \varphi, e_n\rangle|^2\right)^{\frac{1}{3}} \left(\sum_{n=1}^{\infty} |\langle \varphi, e_n\rangle|^2\right)^{\frac{1}{6}} \quad \textit{for all } \varphi \in \mathscr{D}(S). \qquad \diamond$$

Proof (i) Let $\varphi \in \mathscr{D}(H_1)$, then according to Notes and Remarks 14.3(ii), we have

$$\|H_1\varphi\| = \left(\sum_{n=1}^{\infty} \left| n\sqrt{n+1}\langle \varphi, e_{n+1}\rangle + (n-1)\sqrt{n}\langle \varphi, e_{n-1}\rangle \right|^2\right)^{\frac{1}{2}}.$$

In view of Minkowski's inequality (see Lemma 1.3.3), we get

$$\begin{aligned}
\|H_1\varphi\| &\leq \left(\sum_{n=1}^{\infty} \left| n\sqrt{n+1}\langle \varphi, e_{n+1}\rangle \right|^2\right)^{\frac{1}{2}} + \left(\sum_{n=1}^{\infty} \left| (n-1)\sqrt{n}\langle \varphi, e_{n-1}\rangle \right|^2\right)^{\frac{1}{2}} \\
&\leq \left(\sum_{n=1}^{\infty} n^2(n+1)\, |\langle \varphi, e_{n+1}\rangle|^2\right)^{\frac{1}{2}} + \left(\sum_{n=1}^{\infty} (n-1)^2 n\, |\langle \varphi, e_{n-1}\rangle|^2\right)^{\frac{1}{2}}.
\end{aligned}$$

• On the one hand, we have

$$\begin{aligned}
\sum_{n=1}^{\infty} n^2(n+1)|\langle \varphi, e_{n+1}\rangle|^2 &= \sum_{n=2}^{\infty} (n-1)^2 n |\langle \varphi, e_n\rangle|^2 \\
&= \sum_{n=1}^{\infty} (n-1)^2 n |\langle \varphi, e_n\rangle|^2.
\end{aligned}$$

The fact that, for all $n \geq 1$,

$$(n-1)^2 n \leq n^3,$$

then

$$\sum_{n=1}^{\infty} (n-1)^2 n |\langle \varphi, e_n\rangle|^2 \leq \sum_{n=1}^{\infty} n^3 |\langle \varphi, e_n\rangle|^2.$$

Hence,

$$\left(\sum_{n=1}^{\infty} (n-1)^2 n |\langle \varphi, e_n\rangle|^2\right)^{\frac{1}{2}} \leq \left(\sum_{n=1}^{\infty} n^3 |\langle \varphi, e_n\rangle|^2\right)^{\frac{1}{2}}.$$

• On the other hand, we have

$$\begin{aligned}\sum_{n=1}^{\infty} n(n-1)^2 |\langle \varphi, e_{n-1}\rangle|^2 &= \sum_{n=2}^{\infty} n(n-1)^2 |\langle \varphi, e_{n-1}\rangle|^2 \\ &= \sum_{n=1}^{\infty} (n+1)n^2 |\langle \varphi, e_n\rangle|^2.\end{aligned}$$

The fact that, for all $n \geq 1$,

$$(n+1)n^2 \leq (n+1)^3,$$

and

$$\begin{aligned}(n+1)^3 &= n^3 + 1 + 3n^2 + 3n \\ &\leq 8n^3,\end{aligned}$$

we obtain, for all $n \geq 1$,

$$(n+1)n^2 \leq 8n^3.$$

So,

$$\sum_{n=1}^{\infty} (n+1)n^2 |\langle \varphi, e_n\rangle|^2 \leq 8 \sum_{n=1}^{\infty} n^3 |\langle \varphi, e_n\rangle|^2.$$

Hence,

$$\left(\sum_{n=1}^{\infty} (n+1)n^2 |\langle \varphi, e_n\rangle|^2\right)^{\frac{1}{2}} \leq 2\sqrt{2} \left(\sum_{n=1}^{\infty} n^3 |\langle \varphi, e_n\rangle|^2\right)^{\frac{1}{2}}.$$

Thus,

$$\|H_1 \varphi\| \leq c_1 \left(\sum_{n=1}^{\infty} n^3 |\langle \varphi, e_n\rangle|^2\right)^{\frac{1}{2}}$$

for all $\varphi \in \mathscr{D}(H_1)$, where $c_1 = 1 + 2\sqrt{2}$.

(ii) Let $\varphi \in \mathscr{D}(S)$. In view of Notes and Remarks 14.3(ii), we get

$$\begin{aligned}\|S\varphi\| &= \left(\sum_{n=1}^{\infty} |n(n-1)\langle \varphi, e_n\rangle|^2\right)^{\frac{1}{2}} \\ &\leq \left(\sum_{n=1}^{\infty} n^4 \,|\langle \varphi, e_n\rangle|^2\right)^{\frac{1}{2}} \\ &\leq \left(\sum_{n=1}^{\infty} n^4 \,|\langle \varphi, e_n\rangle|^{\frac{4}{3}} \,|\langle \varphi, e_n\rangle|^{\frac{2}{3}}\right)^{\frac{1}{2}}.\end{aligned}$$

By applying Hölder's inequality (see Lemma 1.3.2), we get

$$\|S\varphi\| \leq \left(\sum_{n=1}^{\infty} n^6 |\langle\varphi, e_n\rangle|^2\right)^{\frac{1}{3}} \left(\sum_{n=1}^{\infty} |\langle\varphi, e_n\rangle|^2\right)^{\frac{1}{6}} \quad \text{for all } \varphi \in \mathscr{D}(S).$$

This completes the proof. Q.E.D.

Proposition 14.1.3 *The operator $H_{0,\lambda',\mu,\lambda}$ is G-bounded with order $\frac{2}{3}$.* ◇

Proof It is easy to see that $\mathscr{D}(H_0^{\frac{3}{2}}) \subset \mathscr{D}(H_{0,0,\mu,\lambda})$. Furthermore, in view of Lemma 14.1.1(i), we deduce that there is $c > 0$ such that

$$\|H_{0,0,\mu,\lambda}\varphi\| \leq c\|H_0^{\frac{3}{2}}\varphi\|, \quad \text{for all } \varphi \in \mathscr{D}(H_0^{\frac{3}{2}}). \tag{14.1.1}$$

Let $\varphi \in \mathscr{D}(H_0^3)$. Taking into account of the fact that

$$\mathscr{D}(H_0^3) \subset \mathscr{D}(H_0^{\frac{3}{2}})$$

and in view of the Cauchy-Schwartz's inequality, we get

$$\begin{aligned} \|H_0^{\frac{3}{2}}\varphi\|^2 &= \langle H_0^{\frac{3}{2}}\varphi, H_0^{\frac{3}{2}}\varphi\rangle \\ &= \langle H_0^3\varphi, \varphi\rangle \\ &\leq \|H_0^3\varphi\|\|\varphi\|. \end{aligned} \tag{14.1.2}$$

Hence, in view of both (14.1.1) and (14.1.2) that

$$\|H_{0,0,\mu,\lambda}\varphi\| \leq c\|H_0^3\varphi\|^{\frac{1}{2}}\|\varphi\|^{\frac{1}{2}}.$$

Combining the last inequality together with the fact that $0 \in \rho(H_0)$, we get

$$\begin{aligned} \|H_{0,0,\mu,\lambda}\varphi\| &\leq c\|(H_0^{-1})^3 H_0^3\varphi\|^{\frac{1}{6}}\|H_0^3\varphi\|^{\frac{1}{2}}\|\varphi\|^{\frac{1}{3}} \\ &\leq c\|(H_0^{-1})^3\|^{\frac{1}{6}}\|H_0^3\varphi\|^{\frac{2}{3}}\|\varphi\|^{\frac{1}{3}}. \end{aligned} \tag{14.1.3}$$

So, the operator $H_{0,0,\mu,\lambda}$ is $\frac{2}{3}$ subordinate to H_0^3. Moreover, in view of Lemma 14.1.1(ii) we may write

$$\|S\varphi\| \leq \|H_0^3\varphi\|^{\frac{2}{3}}\|\varphi\|^{\frac{1}{3}}. \tag{14.1.4}$$

Thus, S is also $\frac{2}{3}$ subordinate to the operator H_0^3. Now, by using Proposition 14.1.2(v), we obtain the following estimate:

$$\|H_0^3\varphi\| \leq d_0\|G\varphi\| + d_1\|\varphi\| \tag{14.1.5}$$

holds for some positive constants d_0 and d_1. Finally, Eqs. (14.1.3), (14.1.4) and (14.1.5) guarantee the existence of positive constants a and $b > 0$ satisfying

$$\|H_{0,\lambda',\mu,\lambda}\varphi\| \leq b\|G\varphi\|^{\frac{2}{3}}\|\varphi\|^{1-\frac{2}{3}} + a\|\varphi\|. \qquad \text{Q.E.D.}$$

Let $T_{00} := \lambda'' T_0 = \lambda''(A^*A)^3$, $\mathscr{D} = \mathscr{D}(S)\bigcap\mathscr{D}(H_0)\bigcap\mathscr{D}(H_1)$ and T_1 (respectively, T_2 and T_3) the restriction of $\lambda'' S$ (respectively, $\frac{\lambda''}{c}H_0$ and $\frac{\lambda''}{(1+2\sqrt{2})c}H_1$) to $\mathscr{D}$, where $c = \|(H_0^{-1})^3\|^{\frac{1}{6}}$. Thus, the operators $(T_k)_{1\leq k\leq 3}$ have the same domain $\mathscr{D}$ and, we obtain

$$\mathscr{D}(T_{00}) \subset \mathscr{D}.$$

Moreover, the operator $H_{\lambda'',\lambda',\mu,\lambda}$ can be written as

$$H_{\lambda'',\lambda',\mu,\lambda} := T_{00} + \xi_1 T_1 + \xi_2 T_2 + \xi_3 T_3,$$

where $\xi_1 = \frac{\lambda'}{\lambda''}$, $\xi_2 = \frac{c\mu}{\lambda''}$ and $\xi_3 = \frac{(1+2\sqrt{2})c\lambda}{\lambda''}$. The first result of this section is formulated in the following proposition.

Proposition 14.1.4 *The operators $(T_k)_{1\leq k\leq 3}$ satisfy the following estimation:*

$$\|T_k\varphi\| \leq a\|\varphi\| + b\|T_{00}\varphi\|, \quad \textit{for all} \ \ \varphi \in \mathscr{D}(T_{00}),$$

where $a = \frac{\lambda''}{3}$ and $b = \frac{2}{3}$. ◇

Proof Due to Eq. (14.1.4), we have

$$\|S\varphi\| \leq \|(A^*A)^3\varphi\|^{\frac{2}{3}}\ \|\varphi\|^{\frac{1}{3}},$$

for all $\varphi \in \mathscr{D}((A^*A)^3)$. Using Young's inequality, we obtain

$$\|S\varphi\| \leq \frac{1}{3}\ \|\varphi\| + \frac{2}{3}\ \|(A^*A)^3\varphi\|,$$

for all $\varphi \in \mathscr{D}((A^*A)^3)$. Hence,

$$\|T_1\varphi\| \leq \frac{\lambda''}{3}\ \|\varphi\| + \frac{2}{3}\ \|T_{00}\varphi\|,$$

for all $\varphi \in \mathscr{D}(T_{00})$. Moreover, for all $\varphi \in \mathscr{D}\left(H_0^{\frac{3}{2}}\right)$, we have

$$\|H_0\varphi\| = \left(\sum_{n=1}^{\infty} n^2\,|\langle\varphi, e_n\rangle|^2\right)^{\frac{1}{2}} \leq \left(\sum_{n=1}^{\infty} n^3\,|\langle\varphi, e_n\rangle|^2\right)^{\frac{1}{2}} \leq \left\|H_0^{\frac{3}{2}}\varphi\right\|. \qquad (14.1.6)$$

Taking into account Cauchy-Schwartz's inequality, we get

$$\begin{aligned}\left\| H_0^{\frac{3}{2}}\varphi \right\|^2 &= \left\langle H_0^{\frac{3}{2}}\varphi, H_0^{\frac{3}{2}}\varphi \right\rangle \\ &= \left\langle H_0^3\varphi, \varphi \right\rangle \\ &\leq \|H_0^3\varphi\| \, \|\varphi\|,\end{aligned}$$

for all $\varphi \in \mathscr{D}\left(H_0^3\right)$. So, Eq. (14.1.6) together with the fact that $\mathscr{D}\left(H_0^3\right) \subset \mathscr{D}\left(H_0^{\frac{3}{2}}\right)$ yield

$$\|H_0\varphi\| \leq \|H_0^3\varphi\|^{\frac{1}{2}} \, \|\varphi\|^{\frac{1}{2}}, \quad \text{for all } \varphi \in \mathscr{D}\left(H_0^3\right). \tag{14.1.7}$$

On the other hand, since $0 \in \rho(H_0)$, we get for $\varphi \in \mathscr{D}\left(H_0^3\right)$

$$\begin{aligned}\left\| H_0^3\varphi \right\|^{\frac{1}{2}} \, \|\varphi\|^{\frac{1}{2}} &= \left\| \left(H_0^{-1}\right)^3 H_0^3\varphi \right\|^{\frac{1}{6}} \left\| H_0^3\varphi \right\|^{\frac{1}{2}} \, \|\varphi\|^{\frac{1}{3}} \\ &\leq \left\| \left(H_0^{-1}\right)^3 \right\|^{\frac{1}{6}} \left\| H_0^3\varphi \right\|^{\frac{2}{3}} \, \|\varphi\|^{\frac{1}{3}}.\end{aligned} \tag{14.1.8}$$

Consequently, Eqs. (14.1.7) and (14.1.8) imply that

$$\|H_0\varphi\| \leq c \left\| H_0^3\varphi \right\|^{\frac{2}{3}} \|\varphi\|^{\frac{1}{3}},$$

for all $\varphi \in \mathscr{D}\left(H_0^3\right)$, where

$$c = \left\| \left(H_0^{-1}\right)^3 \right\|^{\frac{1}{6}}.$$

Now, using Young's inequality, we obtain

$$\|H_0\varphi\| \leq \frac{c}{3}\|\varphi\| + \frac{2}{3}\, c \, \|(A^*A)^3\varphi\|,$$

for all $\varphi \in \mathscr{D}\left((A^*A)^3\right)$. Hence,

$$\|T_2\varphi\| \leq \frac{\lambda''}{3} \, \|\varphi\| + \frac{2}{3} \, \|T_{00}\varphi\|,$$

for all $\varphi \in \mathscr{D}(T_{00})$. Now, let us prove the third inequality. In fact, let $\varphi \in \mathscr{D}\left(H_0^3\right)$, we have

$$\|H_1\varphi\| \leq \left(1 + 2\sqrt{2}\right) \, \|H_0^{\frac{3}{2}}\|.$$

Making the same reasoning as above and by using Young's inequality, we infer that for $\varphi \in \mathscr{D}\left(H_0^3\right)$

$$\begin{aligned}\|H_1\varphi\| &\le \left(1+2\sqrt{2}\right)\, c\, \|H_0^3\varphi\|^{\frac{2}{3}}\|\varphi\|^{\frac{1}{3}} \\ &\le \frac{1}{3}\left(1+2\sqrt{2}\right)c\|\varphi\| + \frac{2}{3}\left(1+2\sqrt{2}\right)c\|H_0^3\varphi\|.\end{aligned}$$

Consequently, we have

$$\|T_3\varphi\| \le \frac{\lambda''}{3}\,\|\varphi\| + \frac{2}{3}\,\|T_{00}\varphi\|,$$

for all $\varphi \in \mathscr{D}(T_{00})$. Q.E.D.

Proposition 14.1.5 *If $\xi < \frac{3}{2}$, then the operator $H_{\lambda'',\lambda',\mu,\lambda}$ is closed.* ◇

Proof The result is an immediate consequence of Theorem 9.2.2, Sect. 4.1.5.2 and Proposition 14.1.4. Q.E.D.

Now, we are in position to state the objective of this section.

Theorem 14.1.1 *(i) For $\tau(\xi)$ small enough, the operator $H_{\lambda'',\lambda',\mu,\lambda}$ has a unique point $\lambda_n(\xi)$ of its spectrum in the neighborhood of $\lambda_n = n^3$ and this point is also with multiplicity one. Moreover, $\lambda_n(\xi)$ can be developed into a series*

$$\lambda_n(\xi) := n^3 + \lambda_{n,1} + \lambda_{n,2} + \cdots + \lambda_{n,i} + \cdots .$$

(ii) Let φ_n be an eigenvector of T_{00} associated to the eigenvalue λ_n. Then, setting $\varphi_n(\xi)$ an eigenvector of $H_{\lambda'',\lambda',\mu,\lambda}$ associated to the eigenvalue $\lambda_n(\xi)$, we have for ξ small enough $\varphi_n(\xi)$ can be developed into a series

$$\varphi_n(\xi) := \varphi_n + \varphi_{n,1} + \varphi_{n,2} + \cdots + \varphi_{n,i} + \cdots .$$ ◇

Proof The result follows from Theorem 9.4.1, Sect. 4.1.5.2, Propositions 14.1.4 and 14.1.5. Q.E.D.

14.2 Infinite Sum of Gribov Operators on Null Transverse Dimension ($n = 1$)

We are interested by the Gribov operator, given by

$$\gamma A^{*3}A^3 + (1-\gamma)(A^*A)^3 + \varepsilon A^*(A+A^*)A + \varepsilon^2(A^*A)^{3u_2} + \cdots + \varepsilon^k(A^*A)^{3u_k} + \cdots, \tag{14.2.1}$$

where the parameters $\gamma \in \{0, 1\}$, $\varepsilon \in \mathbb{C}$, and $(u_k)_{k=0}^{\infty}$ is a strictly decreasing sequence with strictly positive terms such that $u_0 = 1$ and $u_1 = \frac{1}{2}$.
We introduce the operators T_0 which is expressed by

$$\begin{cases} T_0 : \mathscr{D}(T_0) \subset E \longrightarrow E \\ \qquad\qquad \varphi \longrightarrow T_0\varphi(z) = (\gamma A^{*3}A^3 + (1-\gamma)(A^*A)^3)\varphi(z) \\ \\ \mathscr{D}(T_0) = \{\varphi \in E \text{ such that } T_0\varphi \in E\}. \end{cases}$$

14.2.1 Riesz Basis of Subspaces in the Case Where $\gamma = 1$

In this part, we consider the following Gribov operator (14.2.1) with $\gamma = 1$:

$$A^{*3}A^3 + \varepsilon A^*(A + A^*)A + \varepsilon^2(A^*A)^{3u_2} + \cdots + \varepsilon^k(A^*A)^{3u_k} + \cdots,$$

where the parameter $\varepsilon \in \mathbb{C}$ and $(u_k)_{k=0}^{\infty}$ is a strictly decreasing sequence with strictly positive terms such that $u_0 = 1$ and $u_1 = \frac{1}{2}$.
We introduce the operators T_0 which is expressed by

$$\begin{cases} T_0 : \mathscr{D}(T_0) \subset E \longrightarrow E \\ \qquad\qquad \varphi \longrightarrow T_0\varphi(z) = A^{*3}A^3\varphi(z) \\ \\ \mathscr{D}(T_0) = \{\varphi \in E \text{ such that } T_0\varphi \in E\}. \end{cases}$$

It is easy to see that for each $k \geq 0$, $\mathscr{D}(H_0^{3u_k}) \subset \mathscr{D}(H_0^{3u_{k+1}})$. Hence,

$$\bigcap_{k\geq 2} \mathscr{D}(H_0^{3u_k}) = \mathscr{D}(H_0^{3u_2}).$$

Consider $\mathscr{D} := \mathscr{D}(H_0^{3u_2}) \bigcap \mathscr{D}(H_1)$ and T_1 (respectively, $(T_k)_{k\geq 2}$) the restriction of H_1 (respectively, $(H_0^{3u_k})_{k\geq 2}$) on $\mathscr{D}$. Thus, the operators $(T_k)_{k\geq 1}$ have the same domain $\mathscr{D}$ and $\mathscr{D}(T_0) \subset \mathscr{D}$. Next, we will prove that the perturbation operators $(T_k)_{k\geq 1}$ satisfy the condition $(H7)$.

Condider the operators $(T_0^{u_k})_{k\geq 0}$

$$\begin{cases} T_0^{u_k} : \mathscr{D}(T_0^{u_k}) \subset E \longrightarrow E \\ \qquad\qquad \varphi \longrightarrow T_0^{u_k}\varphi = \displaystyle\sum_{n=1}^{\infty} n^{3u_k}\langle\varphi, e_n\rangle e_n \\ \mathscr{D}(T_0^{u_k}) = \left\{\varphi \in E \text{ such that } \displaystyle\sum_{n=1}^{\infty} n^{6u_k}|\langle\varphi, e_n\rangle|^2 < \infty\right\}. \end{cases}$$

It is easy to see that for each $k \geq 0$, $\mathscr{D}(T_0^{u_k}) \subset \mathscr{D}(T_0^{u_{k+1}})$. Thus,

$$\bigcap_{k\geq 2} \mathscr{D}(T_0^{u_k}) = \mathscr{D}(T_0^{u_2}).$$

Consider $\mathscr{D} = \mathscr{D}(T_0^{u_2}) \bigcap \mathscr{D}(H_1)$ and T_1 (respectively, $(T_k)_{k\geq 2}$) the restriction of H_1 (respectively, $T_0^{u_k}$) to $\mathscr{D}$. So, the operators $(T_k)_{k\geq 1}$ have the same domain $\mathscr{D}$ and, we obtain $\mathscr{D}(T_0) \subset \mathscr{D}$.

Theorem 14.2.1 *There are constants a, b, $q > 0$ and $\beta \in \left[\frac{1}{2}, 1\right]$ satisfying*

$$\|T_k\varphi\| \leq q^{k-1}(a\|\varphi\| + b\|T_0\varphi\|^{\beta}\|\varphi\|^{1-\beta})$$

for all $k \geq 1$ and for all $\varphi \in \mathscr{D}(T_0)$. ◇

Proof First, we claim that there is $c > 0$ such that

$$\|T_k\varphi\| \leq c\|T_0\varphi\|^{u_k}\|\varphi\|^{1-u_k}$$

for each $k \geq 1$ and for every $\varphi \in \mathscr{D}(T_0)$. Indeed,
• for $k = 1$. Let $\varphi \in \mathscr{D}(T_0)$. Due to Notes and Remarks 14.3(ii), given at the end of this chapter, we obtain

$$T_1\varphi = \sum_{n=1}^{\infty}\left(n\sqrt{n+1}\langle\varphi, e_{n+1}\rangle + (n-1)\sqrt{n}\langle\varphi, e_{n-1}\rangle\right)e_n.$$

Hence,

$$\|T_1\varphi\| = \left(\sum_{n=1}^{\infty}\left|n\sqrt{n+1}\langle\varphi, e_{n+1}\rangle + (n-1)\sqrt{n}\langle\varphi, e_{n-1}\rangle\right|^2\right)^{\frac{1}{2}}.$$

In view of Minkowski's inequality (see Lemma 1.3.3), we get

$$\begin{aligned}
\|T_1\varphi\| &\leq \left(\sum_{n=1}^{\infty}\left|n\sqrt{n+1}\langle\varphi, e_{n+1}\rangle\right|^2\right)^{\frac{1}{2}} + \left(\sum_{n=1}^{\infty}\left|(n-1)\sqrt{n}\langle\varphi, e_{n-1}\rangle\right|^2\right)^{\frac{1}{2}} \\
&\leq \left(\sum_{n=1}^{\infty}n^2(n+1)|\langle\varphi, e_{n+1}\rangle|^2\right)^{\frac{1}{2}} + \left(\sum_{n=1}^{\infty}(n-1)^2 n|\langle\varphi, e_{n-1}\rangle|^2\right)^{\frac{1}{2}}.
\end{aligned}$$

Thus,

$$\|T_1\varphi\| \leq c\left(\sum_{n=1}^{\infty}n^3|\langle\varphi, e_n\rangle|^2\right)^{\frac{1}{2}}, \text{ for all } \varphi \in \mathscr{D}(T_0),$$

where $c = 1 + 2\sqrt{2}$. Hence,

$$\|T_1\varphi\|^2 \leq c^2\|T_0^{\frac{1}{2}}\varphi\|^2, \text{ for all } \varphi \in \mathscr{D}(T_0). \tag{14.2.2}$$

Moreover, we get

$$\begin{aligned}\|T_0^{\frac{1}{2}}\varphi\|^2 &= \langle T_0^{\frac{1}{2}}\varphi, T_0^{\frac{1}{2}}\varphi\rangle \\ &= \langle T_0^{\frac{1}{2}}T_0^{\frac{1}{2}}\varphi, \varphi\rangle \\ &= \langle T_0\varphi, \varphi\rangle \\ &\leq \|T_0\varphi\|\|\varphi\|.\end{aligned}$$

So, Eq. (14.2.2) implies, for all $\varphi \in \mathscr{D}(T_0)$,

$$\|T_1\varphi\| \leq c\|T_0\varphi\|^{\frac{1}{2}}\|\varphi\|^{\frac{1}{2}}. \tag{14.2.3}$$

• Now, for $k \geq 2$ and $\varphi \in \mathscr{D}(T_0)$. We get

$$T_k\varphi = \sum_{n=1}^{\infty} n^{3u_k}\langle\varphi, e_n\rangle e_n.$$

So,

$$\begin{aligned}\|T_k\varphi\|^2 &= \sum_{n=1}^{\infty} n^{6u_k}|\langle\varphi, e_n\rangle|^2 \\ &= \sum_{n=1}^{\infty} n^{6u_k}|\langle\varphi, e_n\rangle|^{2u_k}|\langle\varphi, e_n\rangle|^{2(1-u_k)}.\end{aligned}$$

In view of Hölder's inequality, we have

$$\begin{aligned}\|T_k\varphi\|^2 &\leq \left(\sum_{n=1}^{\infty} n^6|\langle\varphi, e_n\rangle|^2\right)^{u_k}\left(\sum_{n=1}^{\infty}|\langle\varphi, e_n\rangle|^2\right)^{1-u_k} \\ &\leq \|T_0\varphi\|^{2u_k}\|\varphi\|^{2(1-u_k)}.\end{aligned}$$

Hence, for each $\varphi \in \mathscr{D}(T_0)$,

$$\|T_k\varphi\| \leq \|T_0\varphi\|^{u_k}\|\varphi\|^{1-u_k}. \tag{14.2.4}$$

Since $c > 1$, both Eqs. (14.2.3) and (14.2.4) imply that for each $k \geq 1$

$$\|T_k\varphi\| \leq c\|T_0\varphi\|^{u_k}\|\varphi\|^{1-u_k}, \text{ for all } \varphi \in \mathscr{D}(T_0). \tag{14.2.5}$$

This completes the proof of the claim. Second, for $\beta \in [u_1, u_0]$ and $\varphi \in \mathscr{D}(T_0)$, writing $\|T_0\varphi\|^{u_k}\|\varphi\|^{1-u_k}$ as the form

$$\|T_0\varphi\|^{u_k}\|\varphi\|^{1-u_k} = \|\varphi\|^{1-\beta}\|T_0\varphi\|^{u_k}\|\varphi\|^{\beta-u_k},$$

and keeping in mind that for all $k \geq 1$, $0 < u_k < \beta$, we can apply the foregoing interpolation inequality (see Lemma 1.3.4) to $\|T_0\varphi\|^{u_k}$ and $\|\varphi\|^{\beta-u_k}$ with $\theta = \beta$ and $\delta = u_k$, to deduce that

$$\|T_0\varphi\|^{u_k}\|\varphi\|^{\beta-u_k} \leq \|T_0\varphi\|^{\beta} + \|\varphi\|^{\beta},$$

and

$$\|T_0\varphi\|^{u_k}\|\varphi\|^{1-u_k} \leq \|\varphi\| + \|T_0\varphi\|^{\beta}\|\varphi\|^{1-\beta}.$$

Hence, Eq. (14.2.5) implies that for all $k \geq 1$,

$$\|T_k\varphi\| \leq c\|\varphi\| + c\|T_0\varphi\|^{\beta}\|\varphi\|^{1-\beta}$$

for all $\varphi \in \mathscr{D}(T_0)$. Thus, it suffices to choose $a = b = c$ and $q = 1$ and this completes the proof. Q.E.D.

Lemma 14.2.1 *There exists a positive constant c and $\beta \in \left]\frac{1}{2}, \frac{2}{3}\right]$ such that, for all $\varphi \in \mathscr{D}(T_0)$ and all $k \geq 1$,*

$$\|T_k\varphi\| \leq c\|H_0^{3\beta}\varphi\|^{\frac{u_k}{\beta}}\|\varphi\|^{1-\frac{u_k}{\beta}}.$$ ◇

Proof On the one hand, according to Eq. (14.2.5) we claim that, for all $k \geq 1$,

$$\|T_k\varphi\| \leq c\|H_0^3\varphi\|^{u_k}\|\varphi\|^{1-u_k}.$$

On the other hand, since H_0^3 is a self-adjoint positive operator and $u_k < \beta$ for all $k \geq 1$, the first item of Lemma 5.2.2 leads us to deduce that, for all $k \geq 1$,

$$\|T_k\varphi\| \leq c\|(H_0^3)^{\beta}\varphi\|^{\frac{u_k}{\beta}}\|\varphi\|^{1-\frac{u_k}{\beta}}.$$

This achieves the proof of lemma. Q.E.D.

Proposition 14.2.1 *There exists a positive constant q and $\beta \in \left]\frac{1}{2}, \frac{2}{3}\right]$ such that for every sufficiently small η, there exists a continuous and sublinear functional Φ_η satisfying*

$$\|T_k\varphi\| \leq q^{k-1}(\eta\|T_0\varphi\|^{\beta}\|\varphi\|^{1-\beta} + \Phi_\eta(\varphi)),$$

for all $\varphi \in \mathscr{D}(T_0)$ and all $k \geq 1$. ◇

Proof Let $\eta > 0$, combining Lemma 14.2.1 and the second item of Lemma 5.2.2, we obtain

$$\|T_k\varphi\| \leq \eta\|(H_0^3)^{\beta}\varphi\| + \Psi_\eta(\varphi),$$

for every $\varphi \in \mathscr{D}((H_0^3)^{\beta})$, where $\Psi_\eta(\cdot)$ is continuous and sublinear functional. Now, Lemma 5.2.1 allows us to deduce that

$$\|T_k\varphi\| \leq \eta\|H_0^3\varphi\|^{\beta}\|\varphi\|^{1-\beta} + \Psi_\eta(\varphi),$$

for every $\varphi \in \mathscr{D}(H_0^3)$. By virtue of Proposition 14.1.3, there exist two positive constants d_0 and d_1 such that

$$\|H_0^3\varphi\| \leq d_0\|T_0\varphi\| + d_1\|\varphi\|,$$

for every $\varphi \in \mathscr{D}(T_0)$. Hence,

$$\|T_k\varphi\| \leq \eta\|T_0\varphi\|^{\beta}\|\varphi\|^{1-\beta} + \Psi_\eta(\varphi) + d_1^{\beta}\|\varphi\|, \quad \text{for every } \varphi \in \mathscr{D}(T_0).$$

So, it suffices to choose $q = 1$ and

$$\Phi_\eta(\varphi) = \Psi_\eta(\varphi) + d_1^{\beta}\|\varphi\|.$$

This achieves the proof of proposition. Q.E.D.

Now, combining Proposition 14.2.1, Theorems 10.3.1 and 10.3.2, we obtain

Theorem 14.2.2 *For* $\beta \in]\frac{1}{2}, \frac{2}{3}]$, $|\varepsilon| < 1$ *and* $\eta > 0$, *the series*

$$\sum_{k\geq 0} \varepsilon^k T_k\varphi$$

is convergent and for $|\varepsilon| < (1+\beta\eta)^{-1}$, *its limit is a closed linear operator with domain* $\mathscr{D}$. ◇

Let us denote by

$$T(\varepsilon) := A^{*3}A^3 + \varepsilon A^*(A+A^*)A + \varepsilon^2(A^*A)^{3u_2} + \cdots + \varepsilon^k(A^*A)^{3u_k} + \cdots. \tag{14.2.6}$$

The description of the spectrum of the Gribov operator $T(\varepsilon)$ is given by the following theorem:

Theorem 14.2.3 *For every* $\alpha > \frac{\eta|\varepsilon|}{1-|\varepsilon|}$, *there exists* $r_0 > 0$ *such that the spectrum,* $\sigma(T(\varepsilon))$, *of* $T(\varepsilon)$, *given in (14.2.6), satisfies*

$$\sigma(T(\varepsilon)) \subset B_{r_0}(0) \bigcup \{(x+iy);\, x \geq 0 \text{ and } |y| \leq \alpha x^{\beta}\},$$

where $\beta \in]\frac{1}{2}, \frac{2}{3}]$. *Moreover,* $T(\varepsilon)$ *has compact resolvent.* ◇

Proof It follows from Proposition 14.1.2(ii), that the spectrum of T_0 lies in the ray $e^{i\theta_1}$, where $\theta_1 = 0$. Hence, applying Theorem 10.3.3, we obtain the desired result. Q.E.D.

Theorem 14.2.4 *For* $|\varepsilon|$ *small enough, the operator* $T(\varepsilon)$, *given in (14.2.6), has a finitely spectral Riesz basis of subspaces* $(V_k)_{k\in\mathbb{N}}$. ◇

Proof Let $n(\overline{\mathbb{D}}(0, r_n), T_0)$ denotes the sum of multiplicities of all eigenvalues of T_0 which are included in the circle $\mathscr{C}(O, r_n)$ with center O and radius

$$r_n = \frac{\lambda_n + \lambda_{n+1}}{2}.$$

Then, in view of Proposition 14.1.2(ii),

$$\lambda_n = n(n-1)(n-2),$$

and, we get

$$\frac{n(\overline{\mathbb{D}}(0, r_n), G)}{r_n^{1-\frac{2}{3}}} = \frac{2^{\frac{1}{3}} n}{(n(n-1)(2n-1))^{1-\beta}} \underset{+\infty}{\sim} \frac{1}{n^{2-3\beta}}.$$

Using the fact that $\beta \in]\frac{1}{2}, \frac{2}{3}]$, we get

$$\begin{aligned}\liminf_{n \longrightarrow \infty} \frac{n(\overline{\mathbb{D}}(0, r_n), T_0)}{r_n^{1-\beta}} &= \lim_{n \to \infty} \frac{1}{n^{2-3\beta}} \\ &= \begin{cases} 1 \text{ if } \beta = \frac{2}{3} \\ 0 \text{ if } \frac{1}{2} < \beta < \frac{2}{3}. \end{cases}\end{aligned}$$

Finally, using Theorem 10.3.4, we obtain the desired result. Q.E.D.

14.2.2 On the Asymptotic Behavior of the Eigenvalue of Gribov Operator in the Case Where $\gamma = 0$

Let $T_0 = H_0^3$ be the operator, given by

$$\begin{cases} T_0 : \mathscr{D}(T_0) \subset E \longrightarrow E \\ \qquad \varphi \longrightarrow T_0\varphi = \sum_{n=1}^{\infty} n^3 \langle \varphi, e_n \rangle e_n \\ \mathscr{D}(T_0) = \left\{ \varphi \in E \text{ such that } \sum_{n=1}^{\infty} n^6 |\langle \varphi, e_n \rangle|^2 < \infty \right\}. \end{cases}$$

For $n \in \mathbb{N}^*$ and let λ_n be the nth eigenvalue of the operator T_0. If $r_n = \frac{\lambda_{n+1} - \lambda_n}{2}$, then

$$\{z \in \mathbb{C}, \ |z - \lambda_n| \leq r_n\} \bigcap \sigma(T_0) = \{\lambda_n\}.$$

Consider $\mathscr{C}_n = \mathscr{C}(\lambda_n, r_n)$ the closed circle with center λ_n and radius r_n. For $z \in \mathscr{C}_n$, and in view of the fact that T_0 has compact resolvent, then

$$\begin{aligned}\|R_z\| &= \|(T_0 - zI)^{-1}\| \\ &= \frac{1}{\text{dist}(z, \sigma(T_0))} \\ &= \frac{1}{r_n}.\end{aligned}$$

Hence,

$$\begin{aligned}M_n &= \max_{z \in \mathscr{C}_n} \|R_z\| \\ &= \max_{n \geq 1} \frac{1}{r_n}.\end{aligned}$$

We obtain

$$\begin{aligned}\omega_n r_n M_n &= \omega_n r_n \max_{n \geq 1} \frac{1}{r_n} \\ &\leq \omega_n.\end{aligned}$$

Thus,

$$(\omega_n r_n M_n)^2 \leq \omega_n.$$

By using the inequalities

$$\begin{aligned}\sum_{n=1}^{\infty} \omega_n^2 &= \sum_{n=1}^{\infty} \left(\frac{1}{\sqrt{n!}} + \alpha \right)^2 \\ &< \sum_{n=1}^{\infty} \frac{4}{n!} < \infty,\end{aligned}$$

we infer that

$$\sum_{n=1}^{\infty} (\omega_n r_n M_n)^2 < \infty.$$

We obtain

$$\begin{aligned}\|T_0 R_z\| &= \|I + zR_z\| \\ &\leq 1 + |z| \|R_z\|.\end{aligned}$$

Since $z \in \mathscr{C}_n$, we can write

$$z = \lambda_n + r_n e^{it}$$

for $t \in [0, 2\pi[$. Hence,

$$|z| \leq \lambda_n + r_n.$$

So,

$$\begin{aligned}\|T_0 R_z\| &\leq 1 + \frac{\lambda_n + r_n}{r_n} \\ &\leq 2 + \frac{\lambda_n}{r_n}.\end{aligned}$$

Thus,

$$\begin{aligned} N_n &= \max_{z \in \mathscr{C}_n} \|T_0 R_z\| \\ &\le \max_{n \ge 1} \left(2 + \frac{\lambda_n}{r_n}\right). \end{aligned}$$

As a consequence,

$$\begin{aligned} \alpha_n &= a M_n + b N_n^{\beta} M_n^{1-\beta} \\ &\le a \max_{n \ge 1} \frac{1}{r_n} + b \left(\max_{n \ge 1} \left(2 + \frac{\lambda_n}{r_n}\right)\right)^{\beta} \left(\max_{n \ge 1} \frac{1}{r_n} \right)^{1-\beta} \\ &\le \max_{n \ge 1} \left(\frac{a}{r_n} + b \left(2 + \frac{\lambda_n}{r_n}\right)^{\beta} \left(\frac{1}{r_n}\right)^{1-\beta} \right). \end{aligned}$$

Moreover,

$$\begin{aligned} r_n &= \frac{(n+1)^3 - n^3}{2} \\ &= \frac{3n^2 + 3n + 1}{2} \\ &> \frac{3n^2}{2} \\ &> n^2. \end{aligned}$$

Hence, $\frac{1}{r_n} < \frac{1}{n^2}$. Thus,

$$\begin{aligned} \frac{a}{r_n} + b\left(2 + \frac{\lambda_n}{r_n}\right)^{\beta} \left(\frac{1}{r_n}\right)^{1-\beta} &\le \frac{a}{n^2} + b\left(2 + \frac{n^3}{n^2}\right)^{\beta} \left(\frac{1}{n^2}\right)^{1-\beta} \\ &\le \frac{a}{n^2} + b(3n)^{\beta} n^{2(\beta-1)} \\ &\le \frac{a}{n^2} + \frac{3b}{n^{2-3\beta}} \\ &\le \frac{a + 3b}{n^{2-3\beta}}. \end{aligned}$$

Then,

$$\alpha_n \le \max_{n \ge 1} \left(\frac{a + 3b}{n^{2-3\beta}} \right).$$

The fact that $\beta \in \left[\frac{1}{2}, \frac{2}{3}\right]$, we get

$$\alpha_n < d,$$

where $d = a + 3b$. Hence,

$$
\begin{aligned}
q + \alpha_n + \omega_n^2 r_n M_n \alpha_n &\leq 1 + d + \frac{d}{n!} \\
&\leq 1 + 2d.
\end{aligned}
$$

Thus,

$$
\sup_{n \geq 1} (q + \alpha_n + \omega_n^2 r_n M_n) < 1 + 2d < \infty.
$$

Consider $(T_0^{u_k})_{k \geq 0}$ the operators

$$
\begin{cases}
T_0^{u_k} : \mathscr{D}(T_0^{u_k}) \subset E \longrightarrow E \\
\qquad\qquad \varphi \longrightarrow T_0^{u_k} \varphi = \displaystyle\sum_{n=1}^{\infty} n^{3u_k} \langle \varphi, e_n \rangle e_n \\
\mathscr{D}(T_0^{u_k}) = \left\{ \varphi \in E \text{ such that } \displaystyle\sum_{n=1}^{\infty} n^{6u_k} |\langle \varphi, e_n \rangle|^2 < \infty \right\},
\end{cases}
$$

where $(u_k)_{k \in \mathbb{N}}$ is a strictly decreasing sequence with strictly positive terms such that $u_0 = 1$ and $u_1 = \frac{1}{2}$. It is easy to see that for each $k \geq 0$,

$$
\mathscr{D}(T_0^{u_k}) \subset \mathscr{D}(T_0^{u_{k+1}}).
$$

Hence,

$$
\bigcap_{k \geq 2} \mathscr{D}(T_0^{u_k}) = \mathscr{D}(T_0^{u_2}).
$$

Consider $\mathscr{D} = \mathscr{D}(T_0^{u_2}) \bigcap \mathscr{D}(H_1)$ and T_1 (respectively, $(T_k)_{k \geq 2}$) the restriction of H_1 (respectively, $T_0^{u_k}$) to $\mathscr{D}$. Then, the operators $(T_k)_{k \geq 1}$ have the same domain $\mathscr{D}$ and, we obtain

$$
\mathscr{D}(T_0) \subset \mathscr{D}.
$$

Let us consider the Gribov operator (14.2.1) with $\gamma = 0$

$$
T(\varepsilon) := (A^* A)^3 + \varepsilon A^* (A + A^*) A + \varepsilon^2 (A^* A)^{3u_2} + \cdots + \varepsilon^k (A^* A)^{3u_k} + \cdots, \tag{14.2.7}
$$

where $\varepsilon \in \mathbb{C}$.

Proposition 14.2.2 *Let $|\varepsilon| < 1$. Then, the series*

$$
\sum_{k \geq 0} \varepsilon^k T_k \varphi,
$$

given in (14.2.7), converges for all $\varphi \in \mathscr{D}(T_0)$. If we denote its sum by $T(\varepsilon)\varphi$, then we can define a linear operator $T(\varepsilon)$ with domain $\mathscr{D}(T_0)$. Furthermore, for $|\varepsilon| < \frac{1}{1+\beta a}$, the operator $T(\varepsilon)$ is closed. ◇

In [6], the authors showed that it is quite hard to verify Eq. (11.4.1) for this Gribov model. Indeed, for $|\varepsilon| < 1$, we get

$$\left\| \left(\varepsilon A^*(A + A^*)A + \sum_{k=2}^{\infty} \varepsilon^k (A^* A)^{3u_k} \right) \varphi_n \right\| \leq \frac{|\varepsilon|}{1 - |\varepsilon|}(1 + 2\sqrt{2})(1 + n^3). \tag{14.2.8}$$

Hence, it is obvious that Eq. (14.2.8) could not obey to the criteria (Eq. (11.4.2)) of Shkalikov [21]. Consequently, we risk to loose [21, Theorem 1]. The objective of this subsection is formulated in the following theorem.

Theorem 14.2.5 *For $|\varepsilon|$ enough small and $\beta \in [\frac{1}{2}, \frac{5}{6}[$, we have*

$$n(\overline{\mathbb{D}}(0, r), T(\varepsilon)) - n(\overline{\mathbb{D}}(0, r), T_0) = O(1).$$

◇

Proof Let λ_n be the nth eigenvalue of $(A^* A)^3$ and

$$B(\varepsilon) = \sum_{k \geq 1} \varepsilon^k T_k.$$

For $\alpha = \frac{2}{3}$ in Eq. (11.2.6), we have

$$\lambda_{n+p}^{\frac{1}{3}} - \lambda_n^{\frac{1}{3}} = (n + p) - n = p \geq 1. \tag{14.2.9}$$

On the other hand, Theorem 14.2.1 implies that for all $k \geq 1$, we have

$$\|T_k \varphi\| \leq a \left(\|\varphi\| + \|T_0 \varphi\|^{\beta} \|\varphi\|^{1-\beta} \right), \text{ for all } \varphi \in \mathscr{D}(T_0) \tag{14.2.10}$$

where $a = 1 + 2\sqrt{2}$ and $\beta \in [\frac{1}{2}, \frac{5}{6}[$. So, it follows from Corollary 9.3.10, Sect. 4.1.5.2, Proposition 14.2.2 and Eqs. (14.2.9) and (14.2.10) that the spectrum of the operator $T(\varepsilon)$ is discrete for $|\varepsilon| < \frac{1}{1+\beta a}$. Moreover, in view of Eq. (14.2.10), we obtain for $|\varepsilon| < 1$

$$\|B(\varepsilon)\varphi_n\|^2 \leq \frac{|\varepsilon|^2}{(1 - |\varepsilon|)^2} \left(a + a\lambda_n^{\beta} \right)^2.$$

Hence, for $|\varepsilon| < 1$, we get

$$\frac{\|B(\varepsilon)\varphi_n\|^2}{|\lambda - \nu_n|^2} \leq \frac{|\varepsilon|^2}{(1 - |\varepsilon|)^2} \frac{\left(a + a\lambda_n^{\beta} \right)^2}{|\lambda - \nu_n|^2}.$$

On the other hand, as $\lambda_n = n^3$ Eq. (11.2.10) yields

$$\sum_{n=1}^{\infty}\frac{\lambda_n^{2\beta}}{|\lambda-\nu_n|^2}<\frac{2c}{C_3^2}+\sum_{n<k-1}\frac{c}{\left((k-1)^{3(1-\beta)}-n^{3(1-\beta)}\right)^2}+\sum_{n>k}\frac{c}{\left(n^{3(1-\beta)}-k^{3(1-\beta)}\right)^2}. \tag{14.2.11}$$

Now, if $\beta \in [\frac{1}{2}, \frac{2}{3}]$, then

$$3(1-\beta)\geq 1.$$

Hence, in view of Eq. (14.2.11), we get

$$\begin{aligned}\sum_{n=1}^{\infty}\frac{\lambda_n^{2\beta}}{|\lambda-\nu_n|^2}&<\frac{2c}{C_3^2}+\sum_{n<k-1}\frac{c}{(k-1-n)^{6(1-\beta)}}+\sum_{n>k}\frac{c}{(n-k)^{6(1-\beta)}}\\&\leq\frac{2c}{C_3^2}+\sum_{n<k-1}\frac{c}{(k-1-n)^{6(1-\beta)}}+\sum_{m=1}^{\infty}\frac{c}{m^{6(1-\beta)}}\\&<\frac{2c}{C_3^2}+2c\sum_{m=1}^{\infty}\frac{1}{m^{6(1-\beta)}}=:\xi_1<\infty.\end{aligned}$$

If $\beta \in]\frac{2}{3}, \frac{5}{6}[$, then $0<-2+3\beta<\frac{1}{2}$ and $\frac{1}{2}<3-3\beta<1$. So, it follows from Eqs. (11.2.11), (11.2.12) and (14.2.11) that

$$\begin{aligned}\sum_{n=1}^{\infty}\frac{\lambda_n^{2\beta}}{|\lambda-\nu_n|^2}&<\frac{2c}{C_3^2}+\sum_{n<k-1}\frac{c}{c_1^2(3-3\beta)^2\,(k-1-n)^{6(1-\beta)}}+\sum_{n>k}\frac{c}{c_2^2(3-3\beta)^2\,(n-k)^{6(1-\beta)}}\\&\leq\frac{2c}{C_3^2}+\sum_{n<k-1}\frac{c}{c'^2(3-3\beta)^2\,(k-1-n)^{6(1-\beta)}}+\sum_{m=1}^{\infty}\frac{c}{c'^2(3-3\beta)^2 m^{6(1-\beta)}}\\&<\frac{2c}{C_3^2}+\frac{2c}{c'^2(3-3\beta)^2}\sum_{m=1}^{\infty}\frac{1}{m^{6(1-\beta)}}=:\xi_1'<\infty,\end{aligned}$$

where $c':=\min\{c_1,c_2\}$. As a consequence, for $\beta\in[\frac{1}{2},\frac{5}{6}[$, we get

$$\sum_{n=1}^{\infty}\frac{\lambda_n^{2\beta}}{|\lambda-\nu_n|^2}<\max\{\xi_1,\xi_1'\}=:\xi_1'.$$

On the other hand, for $\beta\in[\frac{1}{2},\frac{5}{6}[$, we have $3(1-\frac{\beta}{2})>1$. Thus, by a similar reasoning as the above with $\frac{\beta}{2}$, we obtain

$$\sum_{n=1}^{\infty}\frac{\lambda_n^{\beta}}{|\lambda-\nu_n|^2}<\frac{2c}{C_3^2}+2c\sum_{m=1}^{\infty}\frac{1}{m^{3(2-\beta)}}=:\xi_2'<\infty.$$

Then, the series

$$\sum_n \frac{2a^2\lambda_n^\beta}{|\lambda - \nu_n|^2}$$

and

$$\sum_n \frac{a^2\lambda_n^{2\beta}}{|\lambda - \nu_n|^2}$$

are convergent. Hence, let ξ be a positive constant such that

$$a^2\xi_1' + 2a^2\xi_2' = \xi.$$

So, we get

$$\sum_{n=1}^{\infty} \frac{2a^2\lambda_n^\beta}{|\lambda - \nu_n|^2} + \sum_{n=1}^{\infty} \frac{a^2\lambda_n^{2\beta}}{|\lambda - \nu_n|^2} < \xi.$$

Consequently, in view of Corollary 12.2.1, Sect. 4.1.5.2, Proposition 14.2.2 and Eqs. (14.2.9) and (14.2.10) imply that for $|\varepsilon| < W$

$$n(\overline{\mathbb{D}}(0, r), T(\varepsilon)) - n(\overline{\mathbb{D}}(0, r), T_0) = O(1),$$

where

$$W := \min\left\{\frac{1}{1 + \sqrt{\xi + 4pc_3a^2}}, \frac{1}{1 + c_4\sqrt{\xi + 2pc_5a^2}}\right\},$$

with $0 < c_3 < 1$, $c_4 > 1$ and $c_5 > 0$. Q.E.D.

14.2.3 Basis with Parentheses of Gribov Operator in the Bargmann Space in the Case Where $\gamma = 0$

In Proposition 14.2.1, we take $q = 1$ and $a = b = 1 + 2\sqrt{2}$. Thus, we have the following.

Proposition 14.2.3 *Let $|\varepsilon| < 1$. Then, the series*

$$\sum_{k\geq 0} \varepsilon^k T_k\varphi$$

converges for all $\varphi \in \mathscr{D}(T_0)$. If we denote its sum by $T(\varepsilon)\varphi$, then we can define a linear operator $T(\varepsilon)$ with domain $\mathscr{D}(T_0)$. Furthermore, for $|\varepsilon| < \frac{1}{1+\beta a}$, the operator $T(\varepsilon)$ is closed. ◇

The main results of this subsection are formulated as follows:

Proposition 14.2.4 *For $|\varepsilon| < \frac{1}{1+\beta a}$ and $\beta \in [\frac{1}{2}, \frac{5}{6}[$, the system of root vectors of the operator $T(\varepsilon)$ is complete in E.* ◇

Proof Let λ_n be the nth eigenvalue of $(A^*A)^3$. Then,

$$\lambda_{n+p}^{\frac{1}{3}} - \lambda_n^{\frac{1}{3}} = (n+p) - n = p \geq 1, \quad (\text{where } \alpha = \frac{2}{3}).$$

Furthermore, in view of Theorem 14.2.1, we have for all $\varphi \in \mathscr{D}(T_0)$ and for all $k \geq 1$, we obtain

$$\|T_k\varphi\| \leq a(\|\varphi\| + \|T_0\varphi\|^{\beta}\|\varphi\|^{1-\beta}). \tag{14.2.12}$$

Consequently, in view of Corollary 12.1.1, Sect. 4.1.5.2, Propositions 14.2.2 and 14.2.3 and Eq. (14.2.12) imply that the system of root vectors of the operator $T(\varepsilon)$ is complete in E for $|\varepsilon| < \frac{1}{1+\beta a}$. Q.E.D.

We have proved above that the system of root vectors of the operator $T(\varepsilon)$ is complete in E. Now, it remains to prove that it forms a basis with parentheses in E.

Theorem 14.2.6 *For $|\varepsilon|$ small enough and $\beta \in [\frac{1}{2}, \frac{5}{6}[$, the system of root vectors of the Gribov operator forms a basis with parentheses in E.* ◇

Proof Let λ_n be the nth eigenvalue of $(A^*A)^3$ and

$$B(\varepsilon) = \sum_{k\geq 1} \varepsilon^k T_k.$$

Then,

$$\lambda_{n+p}^{\frac{1}{3}} - \lambda_n^{\frac{1}{3}} = (n+p) - n = p \geq 1, \quad (\text{where } \alpha = \frac{2}{3}).$$

Furthermore, for $|\varepsilon| < 1$, Eq. (14.2.12) yield

$$\frac{\|B(\varepsilon)\varphi_n\|^2}{|\lambda - \lambda_n|^2} \leq \frac{|\varepsilon|^2}{(1-|\varepsilon|)^2}\frac{(a + a\lambda_n^{\beta})^2}{|\lambda - \lambda_n|^2}.$$

Now, let $\sigma = Re\lambda$. Since $\alpha = \frac{2}{3}$ and $\lambda_n = n^3 \geq \left(\frac{n-1}{p} + 1\right)^3$, then for $\beta \in [\frac{1}{2}, \frac{2}{3}]$, we obtain

$$3(1-\beta) \geq 1.$$

Consequently, it follows from Eq. (12.2.27) that

$$\begin{aligned}\sum_{n=1}^{\infty}\frac{\lambda_n^{2\beta}}{|\lambda_n-\sigma|^2} &< \frac{2}{C^2}+\sum_{n<k-1}\frac{p^{6(1-\beta)}}{(k-1-n)^{6(1-\beta)}}+\sum_{n>k}^{\infty}\frac{p^{6(1-\beta)}}{(n-k)^{6(1-\beta)}}\\ &\le \frac{2}{C^2}+\sum_{n<k-1}\frac{p^{6(1-\beta)}}{(k-1-n)^{6(1-\beta)}}+\sum_{m=1}^{\infty}\frac{p^{6(1-\beta)}}{m^{6(1-\beta)}}\\ &< \frac{2}{C^2}+\sum_{m=1}^{\infty}\frac{2p^{6(1-\beta)}}{m^{6(1-\beta)}} =: \xi_1' < \infty.\end{aligned}$$

If $\beta \in]\frac{2}{3}, \frac{5}{6}[$, then

$$0 < -2+3\beta < \frac{1}{2}$$

and

$$\frac{1}{2} < 3-3\beta < 1.$$

Hence, Eq. (12.2.30) yields

$$\begin{aligned}\sum_{n=1}^{\infty}\frac{\lambda_n^{2\beta}}{|\lambda_n-\sigma|^2} &< \frac{2}{C^2}+\frac{p^{6(1-\beta)}}{\gamma_1'^2(3-3\beta)^2}\left(\sum_{n<k-1}\frac{1}{(k-1-n)^{6(1-\beta)}}+\sum_{n>k}^{\infty}\frac{1}{(n-k)^{6(1-\beta)}}\right)\\ &\le \frac{2}{C^2}+\frac{p^{6(1-\beta)}}{\gamma_1'^2(3-3\beta)^2}\left(\sum_{n<k-1}\frac{1}{(k-1-n)^{6(1-\beta)}}+\sum_{m=1}^{\infty}\frac{1}{m^{6(1-\beta)}}\right)\\ &< \frac{2}{C^2}+\frac{2p^{6(1-\beta)}}{\gamma_1'^2(3-3\beta)^2}\sum_{m=1}^{\infty}\frac{1}{m^{6(1-\beta)}} =: \xi_2' < \infty.\end{aligned}$$

So, for $\beta \in [\frac{1}{2}, \frac{5}{6}[$, we get

$$\sum_{n=1}^{\infty}\frac{\lambda_n^{2\beta}}{|\lambda_n-\lambda|^2} < \sum_{n=1}^{\infty}\frac{\lambda_n^{2\beta}}{|\lambda_n-\sigma|^2} < \max\{\xi_1', \xi_2'\} =: \xi_2'.$$

Furthermore, for $\beta \in [\frac{1}{2}, \frac{5}{6}[$, we obtain

$$3\left(1-\frac{\beta}{2}\right) > 1.$$

Moreover, it is easy to check that

$$\sum_{n=1}^{\infty}\frac{\lambda_n^{\beta}}{|\lambda-\lambda_n|^2} < \frac{2}{C^2}+\sum_{m=1}^{\infty}\frac{2p^{3(2-\beta)}}{m^{3(2-\beta)}} =: \xi_3' < \infty.$$

So, the series

$$\sum_n \frac{2a^2\lambda_n^\beta}{|\lambda - \lambda_n|^2}$$

and

$$\sum_n \frac{a^2\lambda_n^{2\beta}}{|\lambda - \lambda_n|^2}$$

are convergent. Hence, let ξ' be a positive constant such that

$$a^2\xi_2' + 2a^2\xi_3' = \xi'.$$

Thus, we get

$$\sum_{n=1}^{\infty} \frac{2a^2\lambda_n^\beta}{|\lambda - \lambda_n|^2} + \sum_{n=1}^{\infty} \frac{a^2\lambda_n^{2\beta}}{|\lambda - \lambda_n|^2} < \xi'.$$

Consequently, the result follows immediately from Corollary 12.2.1, Sect. 4.1.5.2, Proposition 14.2.3 and Eq. (14.2.12). Q.E.D.

14.3 Notes and Remarks

- Specialists in high-energy physics were brought to use the operator

$$H_{\mu,\lambda} = \mu A^* A + i\lambda A^*(A^* + A)A. \tag{14.3.1}$$

  This operator intervenes as a Hamiltonian of the quantum mechanics of the Reggeons. In this theory, it is the simplest Hamiltonian which can describe a field with interaction and it is plausible that it describes the dominant interaction between two hadrons.
- Usually, quantum Hamiltonians are constructed as self-adjoint operators; for certain situations, however, non-self-adjoint Hamiltonians are also of importance. In particular, the reggeon field theory (as invented by Gribov [9] in 1967) for the high-energy behavior of soft processes is governed in zero transverse dimension by the non-self-adjoint operator (14.3.1), where λ and μ are real numbers (μ is the Pomeron intercept and λ is the triple coupling of Pomeron) and $i^2 = -1$.
- Ando and Zerner in [4] and Intissar in [1–3, 10, 11, 13–19] have given a complete spectral theory for the operator $H_{\mu,\lambda}$. In particular, one can consult the list of spectral properties of this operator summarized in [15].
- **Open problem**: For $\mu \neq 0$, the resolvent of $H_{\mu,\lambda}$ belongs to Carleman-class $C_{1+\varepsilon}$ for any $\varepsilon > 0$. However, the completeness of the system of generalized eigenvectors of $H_{\mu,\lambda}$ and the diagonalization of the semigroup $e^{-tH_{\mu,\lambda}}$ are open problems. □

- **Open problem**: For $\mu = 0$, in [13], the boundary conditions at infinity are used in a description of all maximal dissipative extensions in Bargmann space of the minimal operator

$$A^*(A + A^*)A.$$

The characteristic functions of the dissipative extensions are computed and completeness theorems are obtained for the system of generalized eigenvectors. □

- Infinite sum of Gribov operators on null transverse dimension was also well studied by [5, 7, 8] (see also [20]), where the authors considered the following Gribov operator:

$$\gamma A^{*3}A^3 + (1-\gamma)\gamma(A^*A)^3 + \varepsilon A^*(A + A^*)A + \varepsilon^2(A^*A)^{3u_2} + \cdots + \varepsilon^k(A^*A)^{3u_k} + \cdots,$$

where $\gamma \in \{0, 1\}$, $\varepsilon \in \mathbb{C}$ and $(u_k)_{k\in\mathbb{N}}$ is a strictly decreasing sequence with strictly positive terms such that $u_0 = 1$ and $u_1 = \frac{1}{2}$. The case where $\gamma = 1$ was well studied in [8] to investigate the existence of Riesz basis of subspaces and to give a description of the spectrum of the perturbed operator. Some obtained results in [7] are also applied to the Gribov operator, described above, when $\gamma = 0$ to study the basis with parentheses.

(i) H_0 is a self-adjoint operator with compact resolvent. Moreover,

$$\left\{e_n(z) = \frac{z^n}{\sqrt{n!}}\right\}_1^\infty$$

is a system of eigenvectors associated to the eigenvalues $\{n\}$. So, the spectral decomposition of H_0 is given by

$$H_0 = \sum_{n=1}^{\infty} n\langle\cdot, e_n\rangle e_n.$$

Let $T_0 = H_0^3$. So, T_0 is defined by

$$\begin{cases} T_0 : \mathscr{D}(T_0) \subset E \longrightarrow E \\ \qquad\qquad \varphi \longrightarrow T_0\varphi = \displaystyle\sum_{n=1}^{\infty} n^3\langle\varphi, e_n\rangle e_n \\ \mathscr{D}(T_0) = \left\{\varphi \in E \text{ such that } \displaystyle\sum_{n=1}^{\infty} n^6|\langle\varphi, e_n\rangle|^2 < \infty\right\}. \end{cases}$$

(ii) The operators A, A^* A^2, and A^{*2} acting on the Hilbert basis $(e_k)_k$ by

$$Ae_k = \begin{cases} 0 & \text{if } k = 0 \\ \sqrt{k}e_{k-1} & \text{if not,} \end{cases}$$

$$A^*e_k = \sqrt{k+1}e_{k+1},$$

$$A^2 e_k = \sqrt{k(k-1)} e_{k-2},$$

and

$$A^{*2} e_k = \sqrt{(k+1)(k+2)} e_{k+2}.$$

So, we have the following:

$$A[(a_k)_{k=0}^{\infty}] = (\sqrt{k+1} a_{k+1})_{k=0}^{\infty}$$
$$A^*[(a_k)_{k=0}^{\infty}] = (\sqrt{k} a_{k-1})_{k=0}^{\infty}.$$

(iii) The situation that we have checked in this application that 0 is an eigenvalue of T_0 and

$$S = T(\varepsilon) - T_0$$

is β-fully subordinate to T_0 but not β-subordinate to T_0. In this case, the obtained results are of great importance since they allow us to construct a finitely spectral Riesz basis of subspaces.

References

1. M.T. Aimar, A. Intissar, J.M. Paoli, Densité des vecteurs propres généralisés d'une classe d'opérateurs compacts non auto-adjoints et applications. Commun. Math. Phys. **156**, 169–177 (1993)
2. M.T. Aimar, A. Intissar, A. Jeribi, On an unconditional basis of generalized eigenvectors of the non-self-adjoint Gribov operator in Bargmann space. J. Math. Anal. Appl. **231**, 588–602 (1999)
3. M.T. Aimar, A. Intissar, J.M. Paoli, Critères de complétude des vecteurs propres généralisés d'une classe d'opérateurs non auto-adjoints compacts ou à résolvante compacte et applications. Publ. Res. Inst. Math. Sci. **32**(2), 191–205 (1996)
4. T. Ando, M. Zerner, Sur une valeur propre d'un opérateur. Commun. Math. Phys. **93** (1984)
5. S. Charfi, A. Damergi, A. Jeribi, On a Riesz basis of finite-dimensional invariant subspaces and application to Gribov operator in Bargmann space. Line. Mult. Alge. **61**(11), 1577–1591 (2013)
6. H. Ellouz, I. Feki, A. Jeribi, On the asymptotic behavior of the eigenvalues of an analytic operator in the sense of Kato and applications. Serdica Math. J. **45**(1), 55–88 (2019)
7. I. Feki, A. Jeribi, R. Sfaxi, On a Schauder basis related to the eigenvectors of a family of non-self-adjoint analytic operators and applications. Anal. Math. Phys. **3**(4), 311–331 (2013)
8. I. Feki, A. Jeribi, R. Sfaxi, On a Riesz basis of eigenvectors of a non-self-adjoint analytic operator and applications. Line. Mult. Alge. **62**(8), 1049–1068 (2014)
9. V.N. Gribov, A Reggon diagram technique. J. E. T. P., Sov. Phys. **26**, 414–423 (1968)
10. A. Intissar, Etude spectrale d'une famille d'opérateurs non-symétriques intervenant dans la théorie des champs de reggeons. Commun. Math. Phys. **113**, 263–297 (1987)
11. A. Intissar, Quelques nouvelles propriétés spectrales de l'hamiltonien de la théorie des champs de reggeons, C. R. Acad. Sci. Paris, t. 308, Série I, 209–214 (1989)
12. A. Intissar, Analyse fonctionnelle et théorie spectrale pour les opérateurs compacts non-autoadjoints, Editions CEPADUES (1997)

13. A. Intissar, Analyse de Scattering d'un opérateur cubique de Heun dans l'espace de Bargmann. Commun. Math. Phys. **199**, 243–256 (1998)
14. A. Intissar, Diagonalization of non-self-adjoint analytic semigroups and application to the shape memory alloys operator. J. Math. Anal. Appl. **257**, 1–20 (2001)
15. A. Intissar, Approximation of the semigroup generated by the Hamiltonian of Reggeon field theory in Bargmann space. J. Math. Anal. Appl. **305**(2), 669–689 (2005)
16. A. Intissar, Spectral analysis of non-self-adjoint Jacobi-Gribov operator and asymptotic analysis of its generalized eigenvectors. Adv. Math. (China), **44**, 3 (2015)
17. A. Intissar, Regularized trace formula of magic Gribov operator on Bargmann space. J. Math. Anal. Appl. **437**(1), 59–70 (2016)
18. A. Intissar, On spectral approximation of unbounded Gribov-Intissar operators in Bargmann space. Adv. Math., (China) **46**(1), 13–33 (2017)
19. A. Intissar, M. Le Bellac, M. Zerner, Properties of the hamiltonian of reggeon field theory. Phys. Lett. B **113**, 487–489 (1982)
20. A. Jeribi, *Denseness, Bases and Frames in Banach Spaces and Applications* (de Gruyter, Berlin, 2018)
21. A.A. Shkalikov, On the basis property of root vectors of a perturbed self-adjoint operator. Proc Steklov Inst Math **269**, 284–298 (2010)

Chapter 15
Applications in Mathematical Physics and Mechanics

This chapter concentrate on a selection of applications in mathematical physics and mechanics to which the results of the preceding chapters are applied. This chapter contains some applications in mathematical physics and mechanics to investigate the expansion of solution in terms of generalized eigenvectors for a rectilinear transport equation and the behavior of the resolvent in the case of the Lamé system. In this chapter, we apply the results of Chap. 5 to the problem of Mathematical Physics and Mechanics.

15.1 Time-Dependent Rectilinear Transport Equation

The spectral theory of transport equations with general boundary conditions has received a lot of attention in recent years (see, for example, the works [1–60] and the references therein). The model of the time-dependent rectilinear transport equation which was first studied in [74], consists of a segment, say, $-1 \leq x \leq 1$, called the sample, neutrons move along with it in the positive or negative sense, and undergo collisions, with total cross section σ. Time unit is here chosen in such a way that the neutron speed is equal to one, moreover, we assume that the c neutron emerging from a collision, half will take the positive sense of motion and the other half the negative one (isotropic scattering), neutrons going out of the segment are lost. According to the classical approach, let $N^{+}(x, t)$ and $N^{-}(x, t)$ be the density of the neutrons moving in the positive and negative directions of the x-axis, ($x \in [-1, 1]$), at time t, respectively, the rectilinear Boltzmann equation, which is widely developed by [79], can be written as

$$n^{\pm}(x, t) = N^{\pm}(x, t)e^{(\sigma - \frac{c\sigma}{2})t}.$$

This equation can be transformed into a system of two coupled first-order partial differential equations

A. Jeribi, *Perturbation Theory for Linear Operators*,
https://doi.org/10.1007/978-981-16-2528-2_15

$$\begin{cases} \dfrac{\partial n^+}{\partial t} + \dfrac{\partial n^+}{\partial x} = \dfrac{s}{2} n^- \\ \dfrac{\partial n^-}{\partial t} - \dfrac{\partial n^-}{\partial x} = \dfrac{s}{2} n^+, \end{cases} \tag{15.1.1}$$

where $s = c\sigma$. The boundary conditions of no incoming neutrons are expressed by

$$n^+(-1, t) = 0, \quad t > 0$$

$$n^-(1, t) = 0, \quad t > 0$$

while the initial condition is

$$\lim_{t \to 0^+} n^\pm(x, t) = n_0^\pm(x),$$

where $n_0^\pm(x)$ are known functions. The abstract formulation of the problem is obtained by considering the Hilbert space $\mathscr{H}$ of the vectors

$$f = \begin{pmatrix} f^+(x) \\ f^-(x) \end{pmatrix},$$

where $f^+(x)$, $f^-(x)$ belong to the class $L_2(-1, 1)$ of the functions which are square integrable on $(-1, 1)$. The scalar product of two elements $f, g \in \mathscr{H}$ is defined by

$$\langle f, g \rangle = \int_{-1}^{1} \left[f^+(x) g^+(x) + f^-(x) g^-(x) \right] \mathrm{d}x$$

while the norm is

$$\| f \| = \langle f, f \rangle^{\frac{1}{2}}.$$

Let A be the following operator

$$A := \begin{pmatrix} -\dfrac{\partial}{\partial x} & \dfrac{s}{2} \\ \dfrac{s}{2} & \dfrac{\partial}{\partial x} \end{pmatrix} \tag{15.1.2}$$

with domain

$$\mathscr{D}(A) := \left\{ f = \begin{pmatrix} f^+(x) \\ f^-(x) \end{pmatrix} \in \mathscr{H} \text{ such that, } \begin{array}{l} \text{1) } f^+ \text{ and } f^- \text{ are absolutely continuous,} \\ \text{2) } f^+(-1) = f^-(1) = 0, \text{ and} \\ \text{3) } Af \in \mathscr{H} \end{array} \right\}.$$

With the help of these notations, we can write the system (15.1.1) into an evolutionary equation in $\mathscr{H}$

$$\begin{cases} \dfrac{\mathrm{d}}{\mathrm{d}t}n(t) = An(t), \quad t > 0 \\ n(0) = n_0, \end{cases} \tag{15.1.3}$$

where $n(t)$ is a time-dependant vector with values in $\mathscr{D}(A)$ and

$$n_0 = \left\{ n_0^+, n_0^- \right\}$$

is the initial condition which can be given by

$$\lim_{t \to 0^+} \|n(t) - n_0\| = 0.$$

The operator A generates a C_0-semigroup $T(t)$ on $\mathscr{H}$. Hence, the solution of (15.1.3) can be written as

$$n(t) = T(t)n_0.$$

In this case, we need to discuss properties of the C_0-semigroup such that differentiability, compactness, and generalized eigenvectors expansion.

15.1.1 Resolvent and Spectrum of A

Proposition 15.1.1 *Let A be the operator defined as in (15.1.2). Then,*

(i) A is a densely defined closed operator in $\mathscr{H}$, and

(ii) $R(\lambda, A)$ exists and is compact on $\mathscr{H}$, i.e., A is a discrete operator in $\mathscr{H}$. Therefore, the spectrum $\sigma(A)$ consists of isolated eigenvalues only. ◇

Proof (i) It is clear that $\mathscr{D}(A)$ is a closed subspace of, the Sobolev space, $H^1(-1, 1) \times H^1(-1, 1)$, which is dense in $\mathscr{H}$. Hence, assertion (i) follows immediately.

(ii) First, to get the resolvent expression of A, we consider the inhomogeneous equation

$$\lambda f - Af = g, \tag{15.1.4}$$

where $f = \begin{pmatrix} f^+ \\ f^- \end{pmatrix} \in \mathscr{D}(A)$ and $g = \begin{pmatrix} g^+ \\ g^- \end{pmatrix}$ is arbitrary in $\mathscr{H}$. The Eq. (15.1.4) may be written as

$$\begin{pmatrix} \frac{\partial f^+}{\partial x} \\ \frac{\partial f^-}{\partial x} \end{pmatrix} = B \begin{pmatrix} f^+ \\ f^- \end{pmatrix} + \begin{pmatrix} g^+ \\ -g^- \end{pmatrix} \tag{15.1.5}$$

where

$$B = \begin{pmatrix} -\lambda & \frac{s}{2} \\ -\frac{s}{2} & \lambda \end{pmatrix}.$$

The eigenvalues of the matrix B are $\rho_1 = i\mu$ and $\rho_2 = -i\mu$, where $\mu = \sqrt{\left(\frac{s}{2}\right)^2 - \lambda^2}$. The eigenvector associated with the eigenvalue ρ_1 (respectively, ρ_2) is $X_1 = \begin{pmatrix} \lambda - i\mu \\ \frac{s}{2} \end{pmatrix}$ (respectively, $X_2 = \begin{pmatrix} \lambda + i\mu \\ \frac{s}{2} \end{pmatrix}$). The transition matrix P is then

$$P = [X_1, X_2] = \begin{pmatrix} \lambda - i\mu & \lambda + i\mu \\ \frac{s}{2} & \frac{s}{2} \end{pmatrix}. \quad (15.1.6)$$

The matrix P^{-1} is then given by

$$P^{-1} = \frac{i}{\mu s} \begin{pmatrix} \frac{s}{2} & -\lambda - i\mu \\ -\frac{s}{2} & \lambda - i\mu \end{pmatrix}.$$

The matrix B is diagonalizable, and we have $D = P^{-1} B P$, where

$$D = \begin{pmatrix} \rho_1 & 0 \\ 0 & \rho_2 \end{pmatrix}. \quad (15.1.7)$$

Hence, Eq. (15.1.5) leads to the following system:

$$\begin{pmatrix} \frac{\partial f^+}{\partial x} \\ \frac{\partial f^-}{\partial x} \end{pmatrix} = P D P^{-1} \begin{pmatrix} f^+ \\ f^- \end{pmatrix} + \begin{pmatrix} g^+ \\ -g^- \end{pmatrix}.$$

Or,

$$P^{-1} \begin{pmatrix} \frac{\partial f^+}{\partial x} \\ \frac{\partial f^-}{\partial x} \end{pmatrix} = D P^{-1} \begin{pmatrix} f^+ \\ f^- \end{pmatrix} + P^{-1} \begin{pmatrix} g^+ \\ -g^- \end{pmatrix}.$$

Let $U = P^{-1} \begin{pmatrix} f^+ \\ f^- \end{pmatrix}$ and $V = P^{-1} \begin{pmatrix} g^+ \\ -g^- \end{pmatrix}$. We will have

$$\frac{\partial U}{\partial x} = DU + V.$$

This leads to the following system:

$$\begin{aligned} \frac{\partial U_1}{\partial x} &= \rho_1 U_1 + V_1 \\ \frac{\partial U_2}{\partial x} &= \rho_2 U_2 + V_2, \end{aligned} \quad (15.1.8)$$

where $U = \begin{pmatrix} U_1 \\ U_2 \end{pmatrix}$ and $V = \begin{pmatrix} V_1 \\ V_2 \end{pmatrix}$. The equation without a second member

$$\frac{\partial U_1}{\partial x} = \rho_1 U_1$$

admits solution

$$U_1(x) = k(x)e^{\rho_1 x}.$$

Hence,

$$\frac{\partial U_1}{\partial x} = k'(x)e^{\rho_1 x} + \rho_1 U_1(x)$$

and so, Eq. (15.1.8) gives

$$k'(x) = e^{-\rho_1 x} U_1(x).$$

After integrating this equation with respect to x, we obtain

$$k(x) = \int_{-1}^{x} e^{-\rho_1 t} V_1(t)\mathrm{d}t + k_1.$$

Therefore,

$$U_1(x) = \int_{-1}^{x} e^{-i\mu(t-x)} V_1(t)\mathrm{d}t + k_1 e^{i\mu x}. \tag{15.1.9}$$

Similarly,

$$U_2(x) = \int_{-1}^{x} e^{i\mu(t-x)} V_2(t)\mathrm{d}t + k_2 e^{-i\mu x}. \tag{15.1.10}$$

Since $\begin{pmatrix} f^+ \\ f^- \end{pmatrix} = PU$, we have

$$\begin{aligned} f^+(x) &= (\lambda - i\mu)U_1(x) + (\lambda + i\mu)U_2(x) \\ f^-(x) &= \frac{s}{2}(U_1(x) + U_2(x)). \end{aligned} \tag{15.1.11}$$

Now, using the fact that f^+ must satisfy the boundary conditions, Eq. (15.1.11) may be written as

$$f^+(-1) = (\lambda - i\mu)U_1(-1) + (\lambda + i\mu)U_2(-1) = 0.$$

Hence,

$$(\lambda - i\mu)k_1 e^{-i\mu} + (\lambda + i\mu)k_2 e^{i\mu} = 0.$$

So,

$$k_1 = -\frac{\lambda + i\mu}{\lambda - i\mu} e^{2i\mu} k_2.$$

On the other hand,

$$f^-(1) = U_1(1) + U_2(1) = 0$$

which leads to

$$\int_{-1}^{1} e^{-i\mu(t-1)} V_1(t)\mathrm{d}t + k_1 e^{i\mu} + \int_{-1}^{1} e^{i\mu(t-1)} V_2(t)\mathrm{d}t + k_2 e^{-i\mu} = 0.$$

Hence,

$$k_2 e^{i\mu}(\lambda + i\mu)\Delta(\lambda) = \int_{-1}^{1} e^{-i\mu(t-1)} V_1(t)\mathrm{d}t + \int_{-1}^{1} e^{i\mu(t-1)} V_2(t)\mathrm{d}t,$$

where

$$\Delta(\lambda) = \frac{e^{2i\mu}}{\lambda - i\mu} - \frac{e^{-2i\mu}}{\lambda + i\mu}.$$

So,

$$k_2 = \frac{e^{-i\mu}}{(\lambda + i\mu)\Delta(\lambda)} \left[\int_{-1}^{1} e^{-i\mu(t-1)} V_1(t)\mathrm{d}t + \int_{-1}^{1} e^{i\mu(t-1)} V_2(t)\mathrm{d}t \right]$$

and

$$k_1 = -\frac{e^{i\mu}}{(\lambda - i\mu)\Delta(\lambda)} \left[\int_{-1}^{1} e^{-i\mu(t-1)} V_1(t)\mathrm{d}t + \int_{-1}^{1} e^{i\mu(t-1)} V_2(t)\mathrm{d}t \right].$$

Therefore,

$$k_1 = -\frac{1}{(\lambda - i\mu)\Delta(\lambda)} \left[\int_{-1}^{1} e^{i\mu(2-t)} V_1(t)\mathrm{d}t + \int_{-1}^{1} e^{i\mu t} V_2(t)\mathrm{d}t \right],$$

and

$$k_2 = \frac{1}{(\lambda + i\mu)\Delta(\lambda)} \left[\int_{-1}^{1} e^{-i\mu t} V_1(t) \mathrm{d}t + \int_{-1}^{1} e^{-i\mu(2-t)} V_2(t) \mathrm{d}t \right].$$

Now, we can determine $U_1(x)$ and $U_2(x)$. Indeed, by using Eqs. (15.1.9) and (15.1.10), we deduce that

$$U_1(x) = \int_{-1}^{x} e^{-i\mu(t-x)} V_1(t) \mathrm{d}t - \frac{e^{i\mu x}}{(\lambda - i\mu)\Delta(\lambda)} \left[\int_{-1}^{1} e^{i\mu(2-t)} V_1(t) \mathrm{d}t + \int_{-1}^{1} e^{i\mu t} V_2(t) \mathrm{d}t \right],$$

and

$$U_2(x) = \int_{-1}^{x} e^{i\mu(t-x)} V_2(t) \mathrm{d}t + \frac{e^{-i\mu x}}{(\lambda + i\mu)\Delta(\lambda)} \left[\int_{-1}^{1} e^{-i\mu t} V_1(t) \mathrm{d}t + \int_{-1}^{1} e^{-i\mu(2-t)} V_2(t) \mathrm{d}t \right].$$

Eq. (15.1.11) gives

$$\begin{aligned} f^{+}(x) = (\lambda - i\mu) & \int_{-1}^{x} e^{-i\mu(t-x)} V_1(t) \mathrm{d}t + (\lambda + i\mu) \int_{-1}^{x} e^{i\mu(t-x)} V_2(t) \mathrm{d}t - \\ & \frac{e^{i\mu x}}{\Delta(\lambda)} \left[\int_{-1}^{1} e^{i\mu(2-t)} V_1(t) \mathrm{d}t + \int_{-1}^{1} e^{i\mu t} V_2(t) \mathrm{d}t \right] + \\ & \frac{e^{-i\mu x}}{\Delta(\lambda)} \left[\int_{-1}^{1} e^{-i\mu t} V_1(t) \mathrm{d}t + \int_{-1}^{1} e^{-i\mu(2-t)} V_2(t) \mathrm{d}t \right]. \end{aligned} \tag{15.1.12}$$

Moreover,

$$V = \begin{pmatrix} V_1 \\ V_2 \end{pmatrix} = P^{-1} \begin{pmatrix} g^{+} \\ -g^{-} \end{pmatrix} = \frac{i}{\mu s} \begin{pmatrix} \frac{s}{2} & -\lambda - i\mu \\ -\frac{s}{2} & \lambda - i\mu \end{pmatrix} \begin{pmatrix} g^{+} \\ -g^{-} \end{pmatrix}.$$

Hence,

$$V_1(x) = \frac{i}{2\mu} g^{+}(x) + \frac{i}{\mu s} (\lambda + i\mu) g^{-}(x) \tag{15.1.13}$$

$$V_2(x) = -\frac{i}{2\mu} g^{+}(x) - \frac{i}{\mu s} (\lambda - i\mu) g^{-}(x). \tag{15.1.14}$$

Using Eqs. (15.1.13) and (15.1.14), we have the following:

$$(\lambda - i\mu)\int_{-1}^{x} e^{-i\mu(t-x)}V_1(t)\mathrm{d}t =$$

$$i\frac{\lambda - i\mu}{2\mu}\int_{-1}^{x} e^{-i\mu(t-x)}g^+(t)\mathrm{d}t + \frac{is^2}{4\mu s}\int_{-1}^{x} e^{-i\mu(t-x)}g^-(t)\mathrm{d}t, \quad (15.1.15)$$

$$(\lambda + i\mu)\int_{-1}^{x} e^{i\mu(t-x)}V_2(t)\mathrm{d}t =$$

$$-i\frac{\lambda + i\mu}{2\mu}\int_{-1}^{x} e^{i\mu(t-x)}g^+(t)\mathrm{d}t - \frac{is^2}{4\mu s}\int_{-1}^{x} e^{i\mu(t-x)}g^-(t)\mathrm{d}t, \quad (15.1.16)$$

$$\frac{e^{i\mu x}}{\Delta(\lambda)}\int_{-1}^{1} e^{i\mu(2-t)}V_1(t)\mathrm{d}t =$$

$$\frac{ie^{i\mu x}}{2\mu\Delta(\lambda)}\int_{-1}^{1} e^{i\mu(2-t)}g^+(t)\mathrm{d}t + \frac{e^{i\mu x}i(\lambda + i\mu)}{\mu s\Delta(\lambda)}\int_{-1}^{1} e^{i\mu(2-t)}g^-(t)\mathrm{d}t, \quad (15.1.17)$$

$$\frac{e^{i\mu x}}{\Delta(\lambda)}\int_{-1}^{1} e^{i\mu t}V_2(t)\mathrm{d}t =$$

$$-\frac{ie^{i\mu x}}{2\mu\Delta(\lambda)}\int_{-1}^{1} e^{i\mu t}g^+(t)\mathrm{d}t - \frac{e^{i\mu x}i(\lambda - i\mu)}{\mu s\Delta(\lambda)}\int_{-1}^{1} e^{i\mu t}g^-(t)\mathrm{d}t, \quad (15.1.18)$$

$$\frac{e^{-i\mu x}}{\Delta(\lambda)}\int_{-1}^{1} e^{-i\mu t}V_1(t)\mathrm{d}t =$$

$$\frac{ie^{-i\mu x}}{2\mu\Delta(\lambda)}\int_{-1}^{1} e^{-i\mu t}g^+(t)\mathrm{d}t + \frac{e^{-i\mu x}i(\lambda + i\mu)}{\mu s\Delta(\lambda)}\int_{-1}^{1} e^{-i\mu t}g^-(t)\mathrm{d}t, \quad (15.1.19)$$

oindent and

$$\frac{e^{-i\mu x}}{\Delta(\lambda)}\int\limits_{-1}^{1} e^{-i\mu(2-t)}V_2(t)\mathrm{d}t = -\frac{ie^{-i\mu x}}{2\mu\Delta(\lambda)}\int\limits_{-1}^{1} e^{-i\mu(2-t)}g^+(t)\mathrm{d}t - \frac{e^{-i\mu x}i(\lambda-i\mu)}{\mu s\Delta(\lambda)}\int\limits_{-1}^{1} e^{-i\mu(2-t)}g^-(t)\mathrm{d}t. \quad (15.1.20)$$

By using Eqs. (15.1.15), (15.1.16), (15.1.17), (15.1.18), (15.1.19), and (15.1.20), the Eq. (15.1.12) can be written as

$$f^+(x) = f^{++}(x) + f^{+-}(x),$$

where

$$f^{++}(x) = \frac{1}{\Delta(\lambda)}\int\limits_{-1}^{1} R^{++}(x,t,\lambda)g^+(t)\mathrm{d}t$$

and

$$f^{+-}(x) = \frac{1}{\Delta(\lambda)}\int\limits_{-1}^{1} R^{+-}(x,t,\lambda)g^-(t)\mathrm{d}t.$$

Hence,

$$f^{++}(x) = i\frac{\lambda-i\mu}{2\mu}\int\limits_{-1}^{x} e^{-i\mu(t-x)}g^+(t)\mathrm{d}t - i\frac{\lambda+i\mu}{2\mu}\int\limits_{-1}^{x} e^{i\mu(t-x)}g^+(t)\mathrm{d}t - \frac{ie^{i\mu x}}{2\mu\Delta(\lambda)}\int\limits_{-1}^{1} e^{i\mu(2-t)}g^+(t)\mathrm{d}t + \frac{ie^{i\mu x}}{2\mu\Delta(\lambda)}\int\limits_{-1}^{1} e^{i\mu t}g^+(t)\mathrm{d}t + \frac{ie^{-i\mu x}}{2\mu\Delta(\lambda)}\int\limits_{-1}^{1} e^{-i\mu t}g^+(t)\mathrm{d}t - \frac{ie^{-i\mu x}}{2\mu\Delta(\lambda)}\int\limits_{-1}^{1} e^{-i\mu(2-t)}g^+(t)\mathrm{d}t.$$

and

$$f^{+-}(x) = \frac{is^2}{4\mu s}\int_{-1}^{x} e^{-i\mu(t-x)} g^-(t)\mathrm{d}t - \frac{is^2}{4\mu s}\int_{-1}^{x} e^{i\mu(t-x)} g^-(t)\mathrm{d}t$$
$$-\frac{e^{i\mu x} i(\lambda+i\mu)}{\mu s\Delta(\lambda)}\int_{-1}^{1} e^{i\mu(2-t)} g^-(t)\mathrm{d}t + \frac{e^{i\mu x} i(\lambda-i\mu)}{\mu s\Delta(\lambda)}\int_{-1}^{1} e^{i\mu t} g^-(t)\mathrm{d}t$$
$$+\frac{e^{-i\mu x} i(\lambda+i\mu)}{\mu s\Delta(\lambda)}\int_{-1}^{1} e^{-i\mu t} g^-(t)\mathrm{d}t - \frac{e^{-i\mu x} i(\lambda-i\mu)}{\mu s\Delta(\lambda)}\int_{-1}^{1} e^{-i\mu(2-t)} g^-(t)\mathrm{d}t.$$

A tedious but elementary calculation gives

$$R^{++}(x,t,\lambda) =$$
$$\frac{1}{4i\mu}e^{i\mu x}\left[-2e^{i\mu(2-t)}\chi_{[-1,x]}(t) + 2\frac{\lambda-i\mu}{\lambda+i\mu}e^{-i\mu(2+t)}\chi_{[-1,x]}(t) + 2e^{i\mu(2-t)} - 2e^{i\mu t}\right]+$$
$$\frac{1}{4i\mu}e^{-i\mu x}\left[-2e^{-i\mu(2-t)}\chi_{[-1,x]}(t) + 2\frac{\lambda+i\mu}{\lambda-i\mu}e^{i\mu(2+t)}\chi_{[-1,x]}(t) + 2e^{-i\mu(2-t)} - 2e^{-i\mu t}\right],$$

and

$$R^{+-}(x,t,\lambda) = \frac{1}{4i\mu}e^{-i\mu x}\left[\frac{s}{\lambda-i\mu}e^{i\mu(2+t)}\chi_{[-1,x]}(t) - \frac{s}{\lambda+i\mu}e^{-i\mu(2-t)}\chi_{[-1,x]}(t) + \right.$$
$$\left.\frac{s}{\lambda+i\mu}e^{-i\mu(2-t)} - \frac{s}{\lambda-i\mu}e^{-i\mu t}\right]+$$
$$\frac{1}{4i\mu}e^{i\mu x}\left[\left(\frac{s}{\lambda+i\mu}e^{-i\mu(2+t)} - \frac{s}{\lambda-i\mu}e^{i\mu(2-t)}\right)\chi_{[-1,x]}(t)+\right.$$
$$\left.\frac{s}{\lambda-i\mu}e^{i\mu(2-t)} - \frac{s}{\lambda+i\mu}e^{i\mu t}\right].$$

The same way, we can find the expression of $f^-(x)$. Hence, the solution of (15.1.4) is $f = R(\lambda, A)g$ and reveals that

$$\begin{cases} f^+(x) = \dfrac{1}{\Delta(\lambda)}\displaystyle\int_{-1}^{1} R^{++}(x,t,\lambda)g^+(t)\mathrm{d}t + \dfrac{1}{\Delta(\lambda)}\displaystyle\int_{-1}^{1} R^{+-}(x,t,\lambda)g^-(t)\mathrm{d}t \\ f^-(x) = \dfrac{1}{\Delta(\lambda)}\displaystyle\int_{-1}^{1} R^{-+}(x,t,\lambda)g^+(t)\mathrm{d}t + \dfrac{1}{\Delta(\lambda)}\displaystyle\int_{-1}^{1} R^{--}(x,t,\lambda)g^-(t)\mathrm{d}t, \end{cases}$$

where

$$R^{++}(x,t,\lambda) = \frac{e^{-i\mu x}}{4i\mu}\left\{-2e^{-i\mu t} + (1\pm 1)\frac{\lambda+i\mu}{\lambda-i\mu}e^{i\mu(2+t)} + (1\mp 1)e^{-i\mu(2-t)}\right\}$$
$$+\frac{e^{i\mu x}}{4i\mu}\left\{-2e^{-i\mu t} + (1\mp 1)e^{i\mu(2-t)} + (1\pm 1)\frac{\lambda-i\mu}{\lambda+i\mu}e^{-i\mu(2+t)}\right\},$$

$$R^{+-}(x,t,\lambda) = \frac{e^{-i\mu x}}{4i\mu}\left\{-2\frac{\frac{s}{2}}{\lambda - i\mu}e^{-i\mu t} + (1\pm 1)\frac{\frac{s}{2}}{\lambda - i\mu}e^{i\mu(2+t)} + (1\mp 1)\frac{\frac{s}{2}}{\lambda + i\mu}e^{-i\mu(2-t)}\right\}$$
$$+\frac{e^{i\mu x}}{4i\mu}\left\{-2\frac{\frac{s}{2}}{\lambda + i\mu}e^{i\mu t} + (1\mp 1)\frac{\frac{s}{2}}{\lambda - i\mu}e^{i\mu(2-t)} + (1\pm 1)\frac{\frac{s}{2}}{\lambda + i\mu}e^{-i\mu(2+t)}\right\},$$

$$R^{-+}(x,t,\lambda) = \frac{e^{-i\mu x}}{4i\mu}\left\{-2\frac{(\frac{s}{2})e^{-i\mu t}}{\lambda + i\mu} + (1\pm 1)\frac{(\frac{s}{2})e^{i\mu(2+t)}}{\lambda - i\mu} + (1\mp 1)\frac{(\frac{s}{2})e^{-i\mu(2-t)}}{\lambda + i\mu}\right\}$$
$$+\frac{e^{i\mu x}}{4i\mu}\left\{-2\frac{(\frac{s}{2})e^{i\mu t}}{\lambda - i\mu} + (1\mp 1)\frac{(\frac{s}{2})e^{i\mu(2-t)}}{\lambda - i\mu} + (1\pm 1)\frac{(\frac{s}{2})e^{-i\mu(2+t)}}{\lambda + i\mu}\right\},$$

$$R^{--}(x,t,\lambda) = \frac{e^{-i\mu x}}{4i\mu}\left\{-2e^{i\mu t} + (1\mp 1)\frac{(\lambda - i\mu)e^{-i\mu(2-t)}}{\lambda + i\mu} + (1\pm 1)e^{i\mu(2+t)}\right\}$$
$$+\frac{e^{i\mu x}}{4i\mu}\left\{-2e^{i\mu t} + (1\mp 1)\frac{(\lambda + i\mu)e^{i\mu(2-t)}}{\lambda - i\mu} + (1\pm 1)e^{-i\mu(2+t)}\right\}.$$

(The upper signs hold for $t < x$, the lower ones for $t > x$). So, $R(\lambda, A)$ exists and is compact on $\mathscr{H}$ by the Sobolev embedding theorem. Thus, the spectrum of A consists of isolated eigenvalues only and these eigenvalues are exactly the zeros set of $\Delta(\lambda)$. Q.E.D.

Theorem 15.1.1 *The spectrum of A is given by*

$$\sigma(A) = \sigma_p(A) = \left\{z \in \mathbb{C}_{\text{such that}}\, z \sin[s\sqrt{1-z^2}] + \sqrt{1-z^2}\cos[s\sqrt{1-z^2}] = 0\right\}. \diamond$$

Proof Consider the eigenvalue problem

$$\lambda f - Af = 0, \tag{15.1.21}$$

where $f = \begin{pmatrix} f^+ \\ f^- \end{pmatrix} \in \mathscr{D}(A)$. The Eq. (15.1.21) may be written as

$$\begin{pmatrix} \frac{\partial f^+}{\partial x} \\ \frac{\partial f^-}{\partial x} \end{pmatrix} = B\begin{pmatrix} f^+ \\ f^- \end{pmatrix} := \begin{pmatrix} -\lambda & \frac{s}{2} \\ -\frac{s}{2} & \lambda \end{pmatrix}\begin{pmatrix} f^+ \\ f^- \end{pmatrix}.$$

The matrix B is diagonalizable, and we have

$$D = P^{-1}BP,$$

where P and D are given in (15.1.6) and (15.1.7), respectively. Let $U = P^{-1}\begin{pmatrix} f^+ \\ f^- \end{pmatrix}$. We will have

$$\frac{\partial U}{\partial x} = DU.$$

This leads to the following system:

$$\frac{\partial U_1}{\partial x} = \rho_1 U_1$$
$$\frac{\partial U_2}{\partial x} = \rho_2 U_2,$$

where $U = \begin{pmatrix} U_1 \\ U_2 \end{pmatrix}$. Hence,

$$U_1(x) = k_1 e^{i\mu x},$$
$$U_2(x) = k_2 e^{-i\mu x}.$$

Since $\begin{pmatrix} f^+ \\ f^- \end{pmatrix} = PU$, we have

$$f^+(x) = (\lambda - i\mu)U_1(x) + (\lambda + i\mu)U_2(x) \tag{15.1.22}$$
$$f^-(x) = \frac{s}{2}(U_1(x) + U_2(x)).$$

Now, using the fact that f^+ must satisfy the boundary conditions, Eq. (15.1.22) may be written as

$$f^+(-1) = (\lambda - i\mu)U_1(-1) + (\lambda + i\mu)U_2(-1) = 0.$$

Hence,

$$(\lambda - i\mu)k_1 e^{-i\mu} + (\lambda + i\mu)k_2 e^{i\mu} = 0.$$

So,

$$k_1 = -\frac{\lambda + i\mu}{\lambda - i\mu} e^{2i\mu} k_2.$$

On the other hand

$$f^-(1) = U_1(1) + U_2(1) = 0$$

which leads to

$$k_1 e^{i\mu} + k_2 e^{-i\mu} = 0.$$

Therefore,

$$k_1 = -e^{-2i\mu} k_2.$$

Hence

$$-e^{-2i\mu} k_2 = -\frac{\lambda + i\mu}{\lambda - i\mu} e^{2i\mu} k_2.$$

Since k_2 is supposed not zero, we can write

$$e^{-2i\mu} = \frac{\lambda + i\mu}{\lambda - i\mu} e^{2i\mu}.$$

Hence,

$$(\lambda - i\mu)e^{-2i\mu} = (\lambda + i\mu)e^{2i\mu}.$$

So,

$$\lambda \sin(2\mu) = -\mu \cos(2\mu). \qquad (15.1.23)$$

Let $\lambda = \frac{s}{2}\nu$, where $\nu \in \mathbb{C}$. We obtain

$$\mu = \frac{s}{2}\sqrt{1 - \nu^2}.$$

Therefore, by Eq. (15.1.23), we deduce

$$\nu \sin(s\sqrt{1 - \nu^2}) + \sqrt{1 - \nu^2} \cos(s\sqrt{1 - \nu^2}) = 0.$$

Let's define the function g by

$$g(z) = z \sin(s\sqrt{1 - z^2}) + \sqrt{1 - z^2} \cos(s\sqrt{1 - z^2}), \quad z \in \mathbb{C}. \qquad (15.1.24)$$

g is a holomorphic function. The zeros of g are a set, $Z(g)$, at most countable. It follows that the eigenvalues of A are exactly the set $Z(g)$. Moreover, the calculation made above shows that the singular points $\lambda \in \mathbb{C}$ of the resolvent of the operator A, $(\lambda - A)^{-1}$, are exactly the zeros of the function $g(\cdot)$. Thus, it leads to the fact that the spectrum of the operator A consists of isolated eigenvalues which are nothing other than the zeros of the function $g(\cdot)$. Hence,

$$\sigma(A) = \sigma_p(A) = \{z \in \mathbb{C} \text{ such that } g(z) = 0\}. \qquad \text{Q.E.D.}$$

In the following, we will explicitly study the zeros of the function $g(\cdot)$, given in (15.1.24). Let $z \in \mathbb{C}$ such that

$$g(z) = 0.$$

The Eq. (15.1.24) is then written

$$\tan(s\sqrt{1 - z^2}) = -\frac{\sqrt{1 - z^2}}{z}.$$

This leads to the following equations:

$$\sin(s\sqrt{1-z^2}) = \pm\sqrt{1-z^2}$$

$$\cos(s\sqrt{1-z^2}) = \mp z.$$

Write $\sqrt{1-z^2} = \beta + i\gamma$ and $z = \eta + i\tau$, we have

$$\sin(s\beta + is\gamma) = \pm(\beta + i\gamma).$$

Furthermore,

$$\begin{aligned}\sin(s\beta + is\gamma) &= \sin(s\beta)\cos(is\gamma) + \cos(s\beta)\sin(is\gamma)\\ &= \sin(s\beta)\cosh(s\gamma) + i\cos(s\beta)\sinh(\gamma s).\end{aligned}$$

By identification, one obtains

(i) $\sin(s\beta)\cosh(s\gamma) = \pm\beta$
(ii) $\cos(s\beta)\sinh(\gamma s) = \pm\gamma$. On the other hand,

$$\cos(s\beta + is\gamma) = \cos(s\beta)\cosh(s\gamma) - i\sin(s\beta)\sinh(\gamma s).$$

Or, by identification,
(iii) $\cos(s\beta)\cosh(s\gamma) = \mp\eta$
(iv) $\sin(s\beta)\sinh(\gamma s) = \pm\tau$. By dividing (ii) by (iii), one obtains

$$\frac{\tanh(\gamma s)}{\gamma} = -\frac{1}{\eta}. \tag{15.1.25}$$

Note that the function $\tanh(\cdot)$ is an odd function. Therefore, the two quantities $\tanh(\gamma s)$ and γ are of the same sign (since $s \geq 0$). Hence,

$$\frac{\tanh(\gamma s)}{\gamma} \geq 0$$

(since $s \geq 0$). Therefore, the Eq. (15.1.25) is satisfied for $s > 0$ and $\gamma \neq 0$ if $\eta < 0$. This shows that for $\eta \neq 0$ and $\eta > 0$, Eq. (15.1.25) is not satisfied and therefore,

$$g(\eta + i\tau) \neq 0.$$

Hence, the real part of the complex roots of $g(\cdot)$ is necessarily negative. Suppose now that $\tau = 0$ and $z = \eta > 1$, a similar reasoning shows that $g(\cdot) \neq 0$.
It is now assumed that $z \in]0, 1[$. Hence,

$$\beta = \sqrt{1 - z^2}.$$

So,

$$\tan(s\beta) = -\frac{\beta}{\sqrt{1-\beta^2}}. \tag{15.1.26}$$

By plotting function graphs

$$\beta \longrightarrow \mathrm{tg}(s\beta)$$

and

$$\beta \longrightarrow -\frac{\beta}{\sqrt{1-\beta^2}},$$

we can see that the Eq. (15.1.26) admits solutions β if $s > \frac{\pi}{2}$ and no root if $s \geq \frac{\pi}{2}$. Moreover, $z = 1$ is the root of $g(\cdot)$ for all s.

The question that now arises is whether there are roots of $g(\cdot)$ of the form $z = i\tau$, $\tau \neq 0$ (pure imaginary). We suppose that $z = i\tau$, $\tau \neq 0$. Under these assumptions, we arrive at a contradiction because equation (iv) is not verified. Hence, the function $g(\cdot)$ does not admit pure imaginary roots.

If $z = 0$ i.e., $\eta = \tau = 0$, then we get $\cos(s) = 0$ and $\sin(s) = \pm 1$. So,

$$s = \frac{1}{2}(2n+1)\pi, \quad n = 0, 1, 2, \ldots.$$

15.1.2 Distribution of the Eigenvalues of the Operator A

The next theorem gives us information about the distribution of the eigenvalues of the operator A. Indeed, the spectrum of A is located in a left-half plane and there are only finitely a great number of eigenvalues of A in any vertical strip.

Theorem 15.1.2 (Montagnini and Pierpaoli [74, Theorem 2]) *There is an infinite number of complex eigenvalues, distributed in pairs of complex conjugates, the real parts of which are negative. An asymptotic formula for the high-order eigenvalues of the operator A is given by*

$$\lambda_{\pm n} \sim -\frac{1}{2}\log\frac{(2n+1)\pi}{s} \pm i\frac{(2n+1)\pi}{4} \quad (n >> 1). \qquad \diamond$$

Let $\sigma(A) = \{\lambda_n, n \in \mathbb{N}\}$. For each $\lambda_n \in \sigma(A)$, we now find its projection $E(\lambda_n, A)$. It is easy to see that the adjoint operator of A, A^*, is given by

$$A^* := \begin{pmatrix} \frac{\partial}{\partial x} & \frac{s}{2} \\ \frac{s}{2} & -\frac{\partial}{\partial x} \end{pmatrix}$$

with

$$\mathscr{D}(A^*) := \left\{ f = \begin{pmatrix} f^+(x) \\ f^-(x) \end{pmatrix} \in L_2(-1,1) \times L_2(-1,1) \text{ such that } f^+(1) = f^-(-1) = 0 \right\}.$$

For each $\lambda_n \in \sigma(A)$, the corresponding eigenvector of A is

$$\phi(\lambda_n) = \left(\left[k_1(\lambda_n - i\mu_n)e^{i\mu_n x} + k_2(\lambda_n + i\mu_n)e^{-i\mu_n x} \right], \frac{s}{2}\left[k_1 e^{i\mu_n x} + k_2 e^{-i\mu_n x} \right] \right),$$

and the eigenvector of A^* corresponding to $\overline{\lambda_n}$ is

$$\psi(\overline{\lambda_n}) = \left(\left[k_1(\overline{\lambda_n} + i\overline{\mu_n})e^{i\overline{\mu_n} x} + k_2(\overline{\lambda_n} - i\overline{\mu_n})e^{-i\overline{\mu_n} x} \right], \frac{s}{2}\left[k_1 e^{i\overline{\mu_n} x} + k_2 e^{-i\overline{\mu_n} x} \right] \right),$$

where $\mu_n = \sqrt{(\frac{s}{2})^2 - \lambda_n^2}$. For any $F \in \mathscr{H}$, we have

$$E(\lambda_n, A)F = \langle F, \psi(\overline{\lambda_n}) \rangle_{\mathscr{H}} \phi(\lambda_n)$$

and

$$\|E(\lambda_n, A)\| = \|\psi(\overline{\lambda_n})\| \|\phi(\lambda_n)\|.$$

An elementary calculation reveals that

$$\begin{aligned} \|E(\lambda_n, A)\|^2 = {} & 4k_1^2 \sinh(2Im\mu_n) \left[|\lambda_n|^2 + |\mu_n|^2 + \left(\frac{s}{2}\right)^2 + 2Re\lambda_n Im\mu_n - 2Im\lambda_n Re\mu_n \right] \\ & + 4k_2^2 \sinh(2Im\mu_n) \left[|\lambda_n|^2 + |\mu_n|^2 + \left(\frac{s}{2}\right)^2 - 2Re\lambda_n Im\mu_n + 2Im\lambda_n Re\mu_n \right] \\ & + 8k_1k_2 \sin(2Re\mu_n) \left[|\mu_n|^2 - |\lambda_n|^2 - \left(\frac{s}{2}\right)^2 \right]. \end{aligned}$$

Now, observing that

$$\mu_n = \pm i\lambda_n \left\{ 1 + O\left(\frac{1}{|\lambda_n|^2} \right) \right\},$$

we can deduce that

$$\sup_n \|E(\lambda_n, A)\| = \infty.$$

So, the eigenvectors fail to be a basis for $\mathscr{H}$.

15.1.3 Differentiability of the Semigroup Generated by A

Lemma 15.1.1 *The resolvent of the operator A belongs to Carleman-class $\mathscr{C}_{1+\varepsilon}$ for every $\varepsilon > 0$.* ◇

Proof First, from Theorem 15.1.2, we infer that the asymptotic formula for the high-order eigenvalues of the operator A is given by

$$\lambda_{\pm n} \sim -\frac{1}{2}\log\frac{(2n+1)\pi}{s} \pm i\frac{(2n+1)\pi}{4} \quad (n >> 1).$$

Then,

$$|\lambda_n| = \sqrt{\frac{1}{4}\log^2\left(\frac{(2n+1)\pi}{s}\right) + \left(\frac{(2n+1)\pi}{4}\right)^2}$$

and satisfies

$$\frac{(2n+1)\pi}{4} \leq |\lambda_n| \leq \alpha(2n+1), \tag{15.1.27}$$

where $\alpha = \frac{\pi}{2}\sqrt{\frac{1}{s^2}+\frac{1}{4}}$. Obviously, $0 \in \rho(A)$, so let $K = A^{-1}$ and $s_n(K)$ the eigenvalues of the operator $\sqrt{K^*K}$. From Inequality (15.1.27) and Notes and Remarks 15.4(i), it is clear that

$$\sum_{n=1}^{\infty} s_n^{1+\varepsilon}(K) < \infty,$$

for every $\varepsilon > 0$. Then, A^{-1} belongs to Carleman-class $\mathscr{C}_{1+\varepsilon}$ for every $\varepsilon > 0$. Finally, using Notes and Remark 4.6(vii), we deduce the desired result. Q.E.D.

Theorem 15.1.3 (Montagnini and Pierpaoli [74]) *The operator A generates a C_0-semigroup on $\mathscr{H}$.* ◇

Now, applying Lemma 5.1.7, Theorem 5.1.2, together with Corollary 5.1.1, we obtain the following theorem.

Theorem 15.1.4 *Let A be defined as (15.1.2) and $T(t)$ be the semigroups generated by A. Then, for $t > 2$, the series*

$$\sum_{n=1}^{\infty} T(t)E(\lambda_n, A)$$

is compact and converges in the operator norm. Moreover, $T(t)$ is differentiable for $t > 3$. ◇

As a consequence of the spectral mapping theorem for differential semigroups, we have the following result.

Corollary 15.1.1 *System (15.1.3) satisfies the spectrum determined growth assumption, i.e.,*

$$S(A) = w(A),$$

where

$$S(A) := \sup_{\lambda\in\sigma(A)} Re\lambda,$$

and $w(A) := \inf\{w \mid$ *there exists* $C \geq 1$ *such that* $\|e^{At}\| \leq Ce^{wt}$ *for all* $t \geq 0\}$. ◇

15.2 Behavior of Resolvent in the Case of the Lamé System

We want to describe the structure of the solutions of the homogeneous Lamé system in a truncated sector Ω, with homogeneous Dirichlet condition on the rectilinear sides of the sector. More precisely, we assume that

$$\Omega = \{(r\cos\theta, r\sin\theta) \text{ such that } 0 < r < 1,\ 0 < \theta < \omega\}, \tag{15.2.1}$$

where $w \in]0, 2\pi]$ and we assume that $\varphi \in H^1(\Omega)^2$ fulfills

$$L\varphi := \mu\Delta\varphi + (\lambda+\mu)\nabla \mathrm{div}\varphi = 0$$

in Ω, where λ and μ are the Lamé coefficients and $\varphi = 0$ for $\theta = 0$ and $\theta = w$. Of course, we do not assume anything on the traces of φ on the curved part of the boundary of Ω. Our goal is to derive an expansion of φ into a series of functions of the form

$$r^\xi p(\log r; \theta),$$

where p is a polynomial in $\log r$. Let us perform the change of variable $r = e^{-t}$. Then, the problem is to develop φ into a series of exponential polynomials in t. We shall apply the method described in Sect. 5.3.5.

15.2.1 Explicit Expression for the Operator A

In the first step, we have to make the operator A explicit. The notation due to the use of polar coordinates is the following. We denote, respectively, by φ_r and φ_θ the radial and angular components of the vector field φ. We have the well-known formulae

$$\begin{aligned}(L\varphi)_r &= (\lambda+2\mu)D_r E(\varphi) - \mu r^{-1} D_\theta R(\varphi),\\ (L\varphi)_\theta &= (\lambda+2\mu)r^{-1} D_\theta E(\varphi) + \mu D_r R(\varphi),\end{aligned}$$

where

$$E(\varphi) = r^{-1} D_r (r\varphi_r) + r^{-1} D_\theta \varphi_\theta,$$

and

$$R(\varphi) = r^{-1} D_r (r\varphi_\theta) - r^{-1} D_\theta \varphi_r.$$

In order to make easier the change of variable $r = e^{-t}$ (in which $rD_r = -D_t$) it is more convenient to rewrite the above equations in rD_r and D_θ. One gets

$$\begin{aligned} r^2(L\varphi)_r &= (\lambda + 2\mu) r^2 D_r E(\varphi) - \mu D_\theta r R(\varphi) \\ &= (\lambda + 2\mu)(rD_r - 1) r E(\varphi) - \mu D_\theta r R(\varphi), \end{aligned}$$

and

$$\begin{aligned} r^2(L\varphi)_\theta &= (\lambda + 2\mu) D_\theta r E(\varphi) + \mu r^2 D_r R(\varphi) \\ &= (\lambda + 2\mu) D_\theta r E(\varphi) + \mu (rD_r - 1) r R(\varphi), \end{aligned}$$

where

$$rE(\varphi) = (rD_r + 1)\varphi_r + D_\theta \varphi_\theta,$$

and

$$rR(\varphi) = (rD_r + 1)\varphi_\theta - D_\theta \varphi_\theta.$$

If $L\varphi = 0$, one has

$$\begin{aligned} &(\lambda + 2\mu)(-D_t - 1)[(-D_t + 1)\varphi_r + D_\theta \varphi_\theta] - \mu D_\theta [(-D_t + 1)\varphi_\theta - D_\theta \varphi_r] = 0 \\ &(\lambda + 2\mu) D_\theta [(-D_t + 1)\varphi_r + D_\theta \varphi_\theta] + \mu(-D_t - 1)[(-D_t + 1)\varphi_\theta - D_\theta \varphi_r] = 0. \end{aligned}$$

One obtains Eq. (5.3.1) by reducing order in t i.e., setting

$$u = (u_1, u_2, u_3, u_4) = (\varphi_r, \varphi_\theta, D_t\varphi_r, D_t\varphi_\theta).$$

We get

$$\begin{aligned} &(\lambda + 2\mu)[(D_t u_3 - u_1) - D_\theta(u_4 + u_2)] - \mu D_\theta[-u_4 + u_2 - D_\theta u_1] = 0 \\ &\mu[(D_t u_4 - u_2) + D_\theta(u_3 + u_1)] + (\lambda + 2\mu) D_\theta[-u_3 + u_1 + D_\theta u_2] = 0. \end{aligned}$$

Adding the obvious equations

$$D_t u_1 = u_3 \ \text{ and } \ D_t u_2 = u_4,$$

one gets the following 4×4 system

$$\begin{aligned}
&D_t u_1 - u_3 = 0,\\
&D_t u_2 - u_4 = 0,\\
&D_t u_3 - u_1 - D_\theta(u_4 + u_2) - \mu(\lambda + 2\mu)^{-1} D_\theta[-u_4 + u_2 - D_\theta u_1] = 0,\\
&D_t u_4 - u_2 + D_\theta(u_3 + u_1) + (\lambda + 2\mu)\mu^{-1} D_\theta[-u_3 + u_1 + D_\theta u_2] = 0,
\end{aligned}$$

which provides the matrix

$$A := \begin{pmatrix} 0 & 0 & -1 & 0 \\ 0 & 0 & 0 & -1 \\ a_{31} & a_{32} & 0 & a_{34} \\ a_{41} & a_{42} & a_{43} & 0 \end{pmatrix}, \tag{15.2.2}$$

where

$$\begin{aligned}
a_{31} &= -1 + \mu(\lambda + 2\mu)^{-1} D_\theta^2,\\
a_{32} &= -(\lambda + 3\mu)(\lambda + 2\mu)^{-1} D_\theta,\\
a_{34} &= -(\lambda + \mu)(\lambda + 2\mu)^{-1} D_\theta,\\
a_{41} &= (\lambda + 3\mu)\mu^{-1} D_\theta,\\
a_{42} &= -1 + (\lambda + 2\mu)\mu^{-1} D_\theta^2,\\
a_{43} &= -(\lambda + \mu)\mu^{-1} D_\theta.
\end{aligned}$$

We shall view A as operating in the space

$$H = H_0^1(]0, w[)^2 \times L_2(]0, w[)^2,$$

with domain

$$\mathscr{D}(A) = \{H^2(]0, w[) \bigcap H_0^1(]0, w[)\}^2 \times H_0^1(]0, w[)^2.$$

15.2.2 Estimation of the Resolvent

In order to find explicitly the resolvent of A, we solve the following equation, where $f \in H$ is given. We are looking for $u \in \mathscr{D}(A)$ such that

$$Au - \xi u = f \tag{15.2.3}$$

or equivalently

$$-\xi u_1 - u_3 = f_1,$$
$$-\xi u_2 - u_4 = f_2,$$
$$-\xi u_3 - u_1 - D_\theta(u_4 + u_2) - \mu(\lambda + 2\mu)^{-1} D_\theta[-u_4 + u_2 - D_\theta u_1] = f_3,$$
$$-\xi u_4 - u_2 + D_\theta(u_3 + u_1) + (\lambda + 2\mu)\mu^{-1} D_\theta[-u_3 + u_1 + D_\theta u_2] = f_4.$$

Clearly now, following the inverse route that has been followed above, we must eliminate u_3 and u_4. We obtain

$$(\lambda + 2\mu)[-\xi(-\xi u_1 - f_1) - u_1 - D_\theta([-\xi u_2 - f_2] + u_2)] - \mu D_\theta(-[-\xi u_2 - f_2] + u_2 - D_\theta u_1) = (\lambda + 2\mu) f_3$$

$$\mu(-\xi[-\xi u_2 - f_2] - u_2 + D_\theta([-\xi u_1 - f_1] + u_1)) + (\lambda + 2\mu) D_\theta(-[-\xi u_1 - f_1] + u_1 + D_\theta u_2) = \mu f_4,$$

putting the data on the right side,

$$(\lambda + 2\mu)([\xi^2 - 1]u_1 - D_\theta[-\xi + 1]u_2) - \mu D_\theta([\xi + 1]u_2 - D_\theta u_1) = g_1, \tag{15.2.4}$$
$$\mu([\xi^2 - 1]u_2 + D_\theta([-\xi + 1]u_1)) + (\lambda + 2\mu) D_\theta([\xi + 1]u_1 + D_\theta u_2) = g_2, \tag{15.2.5}$$

where

$$g_1 = (\lambda + 2\mu)[f_3 - \xi f_1] + (\lambda + \mu) D_\theta f_2,$$

and

$$g_2 = \mu[f_4 - \xi f_2] - (\lambda + \mu) D_\theta f_1.$$

In what follows it is desirable to group the terms involving μ and $(\lambda + \mu)$ separately. This is consistent with the structure of the Lamé system

$$L\varphi := \mu\Delta\varphi + (\lambda + \mu)\nabla \mathrm{div}\varphi.$$

Thus, one obtains

$$\mu([\xi^2 - 1]u_1 + D_\theta^2 u_1 - 2D_\theta u_2) + (\lambda + \mu)([\xi^2 - 1]u_1 + [\xi - 1]D_\theta u_2) = g_1, \tag{15.2.6}$$
$$\mu([\xi^2 - 1]u_2 + D_\theta^2 u_2 + 2D_\theta u_1) + (\lambda + \mu)([\xi + 1]D_\theta u_1 + D_\theta^2 u_2) = g_2. \tag{15.2.7}$$

Lemma 15.2.1 *Let* $\xi = \zeta + i\eta$. *Then, for*

$$|\zeta| \leq \frac{\sqrt{\mu}}{\sqrt{\lambda + 2\mu}}|\eta|,$$

there is uniqueness of u_1 and u_2, the solutions of (15.2.4)–(15.2.5) which fulfill the boundary conditions $u_1(0) = u_2(0) = u_1(w) = u_2(w) = 0$. ◇

Proof As usual, we shall derive estimates for u_1 and u_2 by integrating the Eqs. (15.2.6) and (15.2.7) against $\{\overline{u}_1, \overline{u}_2\}$. Integrating by parts and noticing that

$$u_1(0) = u_2(0) = u_1(w) = u_2(w) = 0,$$

one gets

$$\begin{aligned}
&\mu\left\{[\xi^2-1]\|u_1\|^2 - \|D_\theta u_1\|^2 - 2\int D_\theta u_2\overline{u}_1 d\theta + [\xi^2-1]\|u_2\|^2\right.\\
&\left.-\|D_\theta u_2\|^2 + 2\int D_\theta u_1\overline{u}_2 d\theta\right\}\\
&+(\lambda+\mu)\left\{[\xi^2-1]\|u_1\|^2 + [\xi-1]\int D_\theta u_2\overline{u}_1 d\theta + [\xi+1]\int D_\theta u_1\overline{u}_2 d\theta - \|D_\theta u_2\|^2\right\}\\
&=\mu\left\{[\xi^2-1]\|u_1\|^2 - \|D_\theta u_1\|^2 + [\xi^2-1]\|u_2\|^2 - \|D_\theta u_2\|^2\right.\\
&\left.-2\int D_\theta u_2\overline{u}_1 d\theta - 2\int u_1 D_\theta\overline{u}_2 d\theta\right\}\\
&+(\lambda+\mu)\left\{[\xi^2-1]\|u_1\|^2 + [\xi-1]\int D_\theta u_2\overline{u}_1 d\theta - [\xi+1]\int u_1 D_\theta\overline{u}_2 d\theta - \|D_\theta u_2\|^2\right\}\\
&=\int g_1\overline{u}_1 d\theta + \int g_2\overline{u}_2 d\theta.
\end{aligned}$$

Now, let us set $\xi = \zeta + i\eta$, we get

$$\begin{aligned}
&\mu\Big\{[\zeta^2-\eta^2+2i\zeta\eta-1]\|u_1\|^2 - \|D_\theta u_1\|^2 + [\zeta^2-\eta^2+2i\zeta\eta-1]\|u_2\|^2\\
&\quad-\|D_\theta u_2\|^2 - 2\int D_\theta u_2\overline{u}_1 d\theta - 2\int u_1 D_\theta\overline{u}_2 d\theta\Big\}\\
&\quad+(\lambda+\mu)\Big\{[\zeta^2-\eta^2+2i\zeta\eta-1]\|u_1\|^2 + [\zeta+i\eta-1]\int D_\theta u_2\overline{u}_1 d\theta\\
&\quad-[\zeta+i\eta+1]\int u_1 D_\theta\overline{u}_2 d\theta - \|D_\theta u_2\|^2\Big\}\\
&\quad=\int g_1\overline{u}_1 d\theta + \int g_2\overline{u}_2 d\theta.
\end{aligned} \tag{15.2.8}$$

The multiplier of $(\lambda + \mu)$ will be rewritten taking into account the following identity:

$$
\begin{aligned}
|D_\theta u_2 + (1+i\eta)u_1|^2 &= [D_\theta u_2 + (1+i\eta)u_1][D_\theta \overline{u}_2 + (1-i\eta)\overline{u}_1] \\
&= |D_\theta u_2|^2 + (1+i\eta)u_1 D_\theta \overline{u}_2 + (1-i\eta)D_\theta u_2 \overline{u}_1 + (1+\eta^2)|u_1|^2.
\end{aligned}
$$

Consequently, one has

$$
\begin{aligned}
&\|D_\theta u_2 + (1+i\eta)u_1\|^2 \\
&\quad = \|D_\theta u_2\|^2 + (1+i\eta)\int u_1 D_\theta \overline{u}_2 d\theta + (1-i\eta)\int D_\theta u_2 \overline{u}_1 d\theta + (1+\eta^2)\|u_1\|^2.
\end{aligned}
$$

It follows that

$$
\begin{aligned}
&[\zeta^2 - \eta^2 + 2i\zeta\eta - 1]\|u_1\|^2 + [\zeta + i\eta - 1]\int D_\theta u_2 \overline{u}_1 d\theta \\
&- [\zeta + i\eta + 1]\int u_1 D_\theta \overline{u}_2 d\theta - \|D_\theta u_2\|^2 \\
&\quad = [\zeta^2 + 2i\zeta\eta]\|u_1\|^2 + \zeta\int D_\theta u_2 \overline{u}_1 d\theta - \zeta\int u_1 D_\theta \overline{u}_2 d\theta - \|D_\theta u_2 + (1+i\eta)u_1\|^2
\end{aligned}
$$

whose real part is

$$
\zeta^2\|u_1\|^2 - \|D_\theta u_2 + (1+i\eta)u_1\|^2.
$$

Changing signs in (15.2.8) and taking the real parts yield

$$
\begin{aligned}
&\mu\left\{(1+\eta^2-\zeta^2)[\|u_1\|^2 + \|u_2\|^2] + [\|D_\theta u_1\|^2 + \|D_\theta u_2\|^2] + 4Re\int D_\theta u_2 \overline{u}_1 d\theta\right\} \\
&\qquad + (\lambda+\mu)[-\zeta^2\|u_1\|^2 + \|D_\theta u_2 + (1+i\eta)u_1\|^2] \\
&\qquad = -Re\left\{\int g_1 \overline{u}_1 d\theta + \int g_2 \overline{u}_2 d\theta\right\}. \qquad (15.2.9)
\end{aligned}
$$

To take advantage of this identity, we observe that

$$
\begin{aligned}
\left|4Re\int D_\theta u_2 \overline{u}_1 d\theta\right| &= 2\left|Re\int D_\theta u_2 \overline{u}_1 d\theta\right| + 2\left|Re\int u_2 D_\theta \overline{u}_1 d\theta\right| \\
&\le \|u_1\|^2 + \|D_\theta u_1\|^2 + \|u_2\|^2 + \|D_\theta u_2\|^2.
\end{aligned}
$$

Then, it follows from (15.2.9) that

$$
\begin{aligned}
&\mu(\eta^2-\zeta^2)[\|u_1\|^2 + \|u_2\|^2] - (\lambda+\mu)\zeta^2\|u_1\|^2 \\
&\quad = (\mu\eta^2 - [\lambda+2\mu]\zeta^2)\|u_1\|^2 + \mu(\eta^2-\zeta^2)\|u_2\|^2 \\
&\quad \le \|g_1\|\|u_1\| + \|g_2\|\|u_2\|.
\end{aligned}
$$

Then, for

$$|\zeta| \leq \frac{\sqrt{\mu}}{\sqrt{\lambda + 2\mu}} |\eta|,$$

there is uniqueness of u_1 and u_2, the solutions of (15.2.4)–(15.2.5) which fulfill the boundary conditions $u_1(0) = u_2(0) = u_1(w) = u_2(w) = 0$. Q.E.D.

By using Lemma 15.2.1 and in addition for every

$$\delta < \arctan\left(\frac{\sqrt{\mu}}{\sqrt{\lambda + 2\mu}}\right),$$

there exists a constant $C(\delta)$ such that

$$\|u_1\| + \|u_2\| \leq C(\delta)|\xi|^{-2}[\|g_1\| + \|g_2\|].$$

We also obtain

$$\|D_\theta u_1\| + \|D_\theta u_2\| \leq C(\delta)|\xi|^{-2}[\|g_1\| + \|g_2\|].$$

Since it is a differential system with boundary conditions, the Fredholm alternative holds and therefore, there is existence and uniqueness of u_1 and u_2 under the same assumptions. Going back to u, the solution of (15.2.3), we have proved the invertibility of $A - \xi$ in the double sector S defined by

$$|\zeta| \leq \frac{\sqrt{\mu}}{\sqrt{\lambda + 2\mu}} |\eta|.$$

Taking into account the form of g_1 and g_2, one gets an estimate of the inverse

$$\|g_1\| + \|g_2\| = O(1 + |\xi|)\|f\|_H.$$

To sum up, we have proven the existence of a constant $C'(\delta)$ such that

$$\|(A - \xi)^{-1}\|_{H \to H} \leq C'(\delta)|\xi|^{-1} O(1 + |\xi|).$$

We see that $\|(A - \xi)^{-1}\|_{H \to H}$ remains bounded in S with the possible exception of a neighborhood of zero. We still have to study the invertibility of A itself since $\xi = 0$ has been excluded in the previous calculations. Back to (15.2.4)–(15.2.5) when $\xi = 0$ and $f = 0$, one gets

$$(\lambda + 2\mu)(-u_1 - D_\theta u_2) - \mu D_\theta(u_2 - D_\theta u_1) = 0,$$
$$\mu(-u_2 + D_\theta u_1) + (\lambda + 2\mu) D_\theta(u_1 + D_\theta u_2) = 0.$$

Integrating by parts against u_1, u_2 (assumed to be real since the coefficients are real) one gets

$$-\mu\left\{\|D_\theta u_1\|^2 + \|u_2\|^2 + 2\int u_2 D_\theta u_1 d\theta\right\} - (\lambda+2\mu)\left\{\|u_1\|^2 + \|D_\theta u_2\|^2 + 2\int u_2 D_\theta u_1 d\theta\right\} = 0.$$

In other words, one has

$$-\mu\|D_\theta u_1 - u_2\|^2 - (\lambda+2\mu)\|u_1 + D_\theta u_2\|^2 = 0.$$

It follows that

$$D_\theta u_1 - u_2 = 0 \text{ and } u_1 + D_\theta u_2 = 0, \tag{15.2.10}$$

or that

$$D_\theta^2 u_1 + u_1 = 0$$

and

$$D_\theta^2 u_2 + u_2 = 0.$$

With the boundary conditions, this implies that $u_1 = 0$ and $u_2 = 0$ unless $\frac{w}{\pi}$ is an integer. In this last case, one has $u_1 = c_1 \sin\theta$, $u_2 = c_2 \sin\theta$, but noting (15.2.10), one sees that $c_1 = c_2 = 0$. In conclusion, A is always invertible. So, the operator A, defined in (15.2.2), fulfills $(A4)$ with $m = 0$ and

$$\delta < \arctan\sqrt{\frac{\mu}{\lambda+2\mu}}.$$

15.2.3 *Study of the Fredholm Determinant in the Case of the Lamé System*

Lemma 15.2.2 *The eigenvalues of A are the numbers ξ such that*

$$\frac{\sin^2(\xi w)}{\xi^2} - \left[(\lambda+\mu)\frac{\sin w}{\lambda+3\mu}\right]^2 = 0. \qquad \diamond$$

Proof We begin by looking for the eigenvalues of A; i.e., the numbers ξ such that there exists $u \neq 0$ solution of

$$Au - \xi u = 0.$$

This amounts to Eq. (15.2.3) in the particular case when $f = 0$. This is also equivalent to (15.2.4)–(15.2.5) when $g_1 = g_2 = 0$, with the homogeneous boundary conditions

$$u_1(0) = u_2(0) = u_1(w) = u_2(w) = 0.$$

Taking into account the calculations in Sect. 15.2, it is equivalent to look for

$$\varphi = r^{\xi}(u_1(\theta), u_2(\theta))$$

solution of

$$L\varphi = \mu\Delta\varphi + (\lambda+\mu)\nabla \mathrm{div}\varphi = 0 \tag{15.2.11}$$

with the same boundary conditions; i.e., $\varphi = 0$ for $\theta = 0$ and $\theta = w$. These calculations are classical (and elementary, though quite tedious). One shows at first that for $\xi \neq 0$ a basis of the space of u such that

$$\varphi = r^{\xi} u$$

is a solution of (15.2.11) with the condition $\varphi(0) = 0$ is constituted of the vector fields

$$\begin{aligned}\varphi(\xi,\theta) &= (\xi[\cos(\xi-2)\theta - \cos\xi\theta, -\xi\sin(\xi-2)\theta + [\xi - t]\sin\xi\theta)^T\\ \psi(\xi,\theta) &= (\xi\sin(\xi-2)\theta - [\xi+t]\sin\xi\theta, \xi[\cos(\xi-2)\theta - \cos\xi\theta])^T,\end{aligned}$$

where

$$t = \frac{\lambda+3\mu}{\lambda+\mu}.$$

One gets the eigenvectors in the form

$$u = x\varphi + y\psi,$$

where x and y are solutions of

$$x\varphi(\xi,w) + y\psi(\xi,w) = 0.$$

The eigenvalues are the numbers ξ such that

$$\det[\varphi(\xi,w);\psi(\xi,w)] = 0;$$

i.e.,

$$W(\xi) = \left[\frac{\sin(\xi w)}{\xi}\right]^2 - \left[\frac{\sin w}{t}\right]^2 = 0, \tag{15.2.12}$$

which completes the proof. Q.E.D.

(i) First, let us assume $w < 2\pi$. It is easy to check that unless for $\xi = \pm\xi_0$, with

$$\xi_0 = \left[\frac{t^2}{\sin^2 w} - \frac{1}{w^2}\right]^{\frac{1}{2}},$$

the roots of W (see (15.2.12)) are simple. It follows that the corresponding poles of $(A - \xi)^{-1}$ are simple and consequently the algebraic and geometric multiplicities are equal. The geometric multiplicity is one since the rank of the matrix

$$[\varphi(\xi, w), \psi(\xi, w)]$$

is one. One concludes that $d(\xi)$ and $W(\xi)$ have the same roots with the same multiplicities with the possible exception of $\pm\xi_0$. The function W obviously has the exponential growth

$$|W(\xi)| = O(e^{2w|\xi|})$$

and consequently, Hadamard's uniqueness theorem shows that

$$W(\xi) = e^{\alpha\xi+\beta} d(\xi) g(\xi),$$

where α and β are complex numbers and g is a polynomial. To conclude, one observes that W is even. Consequently, ξ and $-\xi$ arc simultaneously roots of W. It follows that $d(\cdot)$ is also an even function and this forces α down to 0. Summing up, we have

$$d(\xi) = O(e^{2w|\xi|}).$$

(ii) The case when $w = 2\pi$ is slightly different. In the above notation, one checks that

$$\varphi(\xi, w) = \psi(\xi, w) = 0$$

for every ξ solution of

$$W(\xi) = \frac{\sin^2 2\pi\xi}{\xi^2} = 0;$$

i.e., $\xi = \frac{k}{2}$ with $k \neq 0$ an integer. Thus, the geometric multiplicity is obviously too. An explicit calculation of the resolvent of A shows that it involves the factors

$$\frac{\varphi_i(\xi, w)}{W(\xi)} \text{ and } \frac{\psi_i(\xi, w)}{W(\xi)}, \quad i = 1, 2$$

and consequently, the poles of the resolvent are simple and the algebraic and geometric multiplicities are equal. As before one concludes that

$$d(\xi) = \frac{W(\xi)}{W(0)}$$

and thus,

$$d(\xi) = O(e^{4\pi|\xi|}).$$

So, the growth of the Fredholm determinant of A, defined in (15.2.2), is given by

$$d(\xi) = O(e^{2w|\xi|}). \tag{15.2.13}$$

15.2.4 Conclusion

Let us summarize the properties of the operator A defined in (15.2.2). Obviously this is a closed linear operator with a dense domain $\mathscr{D}(A)$ in H. It fulfills assumption $(A4)$. The injection of $H^{m+1}(]0, w[)$ into $H^m(]0, w[)$ belongs to the Carleman-class $\mathscr{C}_p$ for every $p > 1$. Since A is invertible, it follows that A has an almost nuclear resolvent and fulfills $(A5)$ for every $p > 1$. Finally, the explicit calculations of (15.2.13) show that $d(\xi)$ has the exponential growth

$$|d(\xi)| = O(e^{2w|\xi|}).$$

In order to make it possible to apply Theorem 5.3.10, we just have to find the real numbers r_j such that (5.3.11) hold.

There exists a constant $C > 0$ such that

$$\xi^2 d(\xi) = C[\sin^2(\xi w) - \gamma^2\xi^2],$$

where one has set

$$\gamma = (\lambda + \mu)\sin w/(\lambda + 3\mu).$$

Therefore, one has

$$|\xi^2 d(\xi)|^2 = C^2|\sin(\xi w) - \gamma\xi|^2|\sin\xi w + \gamma\xi|^2.$$

Let us consider separately the two factors $\sin\xi w \pm \gamma\xi$. Setting $\xi = x + iy$ one has

$$Re[\sin\xi w \pm \gamma\xi] = \sin xw\cosh yw \pm \gamma x,$$
$$Im[\sin\xi w \pm \gamma\xi] = \sinh yw\cos xw \pm \gamma y,$$

and thus,

$$|\sin\xi w \pm \gamma\xi|^2 = [\sin xw\cosh yw \pm \gamma x]^2 + [\sinh yw\cos xw \pm \gamma y]^2.$$

In particular, for

$$x = \frac{\pi}{2w} + \frac{j\pi}{w},$$

one has

$$|\sin \xi w \pm \gamma \xi|^2 = [\pm \cosh wy \pm \gamma x]^2 + [\gamma y]^2$$

and

$$|\xi^2 d(\xi)|^2 = C^2([\pm \cosh wy \pm \gamma x]^2 + [\gamma y]^2)^2.$$

So, there exists a constant $a > 0$ such that

$$|d(\xi)| \geq a$$

for $Re\xi = \frac{\pi}{2w} + \frac{j\pi}{w}$, $j = 1, 2, \ldots$. The application of Theorem 5.3.10 with

$$r_j = \frac{\pi}{2w} + \frac{j\pi}{w}$$

yields the following result.

Theorem 15.2.1 (Geymonat and Grisvard [63, Theorem 7.2]) *Let $\varphi \in H^1(\Omega)^2$ be a solution of (15.2.11) in Ω (see (15.2.1) and $\varphi = 0$ for $\theta = 0$ and $\theta = w$). There exists $R_0 > 0$ such that*

$$\varphi = \sum_j \left\{ \sum r^{\xi_k} p_k(\log(r, \theta) \right\},$$

where the interior summation is over the roots ξ_k of

$$\frac{\sin^2(\xi w)}{\xi^2} - \left[(\lambda + \mu) \frac{\sin w}{\lambda + 3\mu} \right]^2 = 0$$

such that

$$\frac{\pi}{2w} + \frac{j\pi}{w} < Re\xi_k < \frac{\pi}{2w} + \frac{(j+1)\pi}{w}$$

and the p_k do not depend on $\log r$ (with the possible exception of the value)

$$\xi_k = \left\{ \frac{t^2}{\sin^2 w} - \frac{1}{w^2} \right\}^{\frac{1}{2}}$$

when $w < 2\pi$, in which case p_k is affine in $\log r$) and are infinitely differentiable functions in θ, and in addition the series in j converge uniformly for $r \leq R_0$. ◇

Proof One checks easily with the help of the boundary conditions that $r^{-1}\varphi$, $\nabla\varphi$ are square integrable in Ω. By the change of variable $r = e^{-t}$, it follows that

$$u = (\varphi_r, \varphi_\theta, D_t\varphi_r, D_t\varphi_\theta) \in L_2(\mathbb{R}_{\geq 0}, H).$$

This is a weak solution of (5.3.1) with the operator A defined in Eq. (15.2.2). The desired result is nothing but (5.4.1). Q.E.D.

15.3 Differential Operator with Dirichlet Boundary Condition

Consider the operator T on $L_2([0, 1])$ defined by

$$Tu = -u'' + g_1u' + g_0u,$$

with domain

$$\mathscr{D}(T) = \{u \in W^{2,2}([0, 1]), \text{ such that } u(0) = u(1) = 0\},$$

where $g_0 \in L_2([0, 1])$, $g_1 \in L^\infty([0, 1])$. In this case, we consider the operators

$$T_0u = -u'',$$

$$T_1u = g_1u' + g_0u$$

and $\xi_1 = 1$ with $\mathscr{D}(T_0) = \mathscr{D}(T_1) = \mathscr{D}(T)$.

Direct calculations show that T_0 is self-adjoint with compact resolvent and its eigenvalues are

$$\lambda_n = \pi^2n^2, \ k = 1, 2, \ldots,$$

which are all simple.

Lemma 15.3.1 *The operator T_1 is fully subordinate to T_0.* ◇

Proof Integrating by parts and using both the boundary condition and the Cauchy-Schwarz inequality, we obtain for $u \in \mathscr{D}(T_0)$

$$\begin{aligned}\int_0^1 |u'(x)|^2 \mathrm{d}x &= \int_0^1 u'(x)\overline{u'(x)}\mathrm{d}x \\ &= -\int_0^1 u(x)\overline{u''(x)}\mathrm{d}x \\ &\leq \|u\|_{L_2}\|u''\|_{L_2}.\end{aligned}$$

Hence,

$$\|u'\|_{L_2} \leq \|u\|_{L_2}^{\frac{1}{2}} \|T_0 u\|_{L_2}^{\frac{1}{2}} \text{ for all } u \in \mathscr{D}(T_0).$$

Moreover, for $u \in \mathscr{D}(T_0)$, the identity

$$u(x) = \int_0^x u'(t)\mathrm{d}t$$

yields

$$\begin{aligned}|u(x)| &\leq \int_0^1 |u'(t)|\mathrm{d}t \\ &\leq \left(\int_0^1 \mathrm{d}t\right)^{\frac{1}{2}} \left(\int_0^1 |u'(t)|^2 \mathrm{d}t\right)^{\frac{1}{2}} \\ &= \|u'\|_{L_2}\end{aligned}$$

and thus,

$$\|u\| \leq \|u'\|_{L_2}.$$

We obtain the estimate

$$\begin{aligned}\|T_1 u\|_{L_2} &\leq \|g_1\|_\infty \|u'\|_{L_2} + \|g_0\|_{L_2} \|u\|_\infty \\ &\leq (\|g_1\|_\infty + \|g_0\|_{L_2}) \|u'\|_{L_2} \\ &\leq (\|g_1\|_\infty + \|g_0\|_{L_2}) \|u\|_{L_2}^{\frac{1}{2}} \|T_0 u\|_{L_2}^{\frac{1}{2}}\end{aligned}$$

for all $u \in \mathscr{D}(T_0)$. Hence, T_1 is $\frac{1}{2}$-subordinate to T_0 with bound $b \leq \|g_1\|_\infty + \|g_0\|_{L_2}$. Since T_0 is a positive definite operator, then the use of Lemma 5.2.2 makes us conclude that T_1 is fully subordinate to T_0. Q.E.D.

Lemma 15.3.2 *The family $\{e^{i\sqrt{\lambda_n}t}\}$ forms a Riesz basis in $L_2(0, 2)$.* ◇

Proof The solution of the following problem

$$\begin{cases} T_0\varphi = \lambda\varphi \\ \varphi \in \mathscr{D}(T_0) \end{cases}$$

is

$$\varphi(x) = \alpha e^{i\sqrt{\lambda_n}x} + \beta e^{-i\sqrt{\lambda_n}x}$$

and satisfies the boundary conditions

$$\varphi(0) = \varphi(1) = 0.$$

So, we find ourselves with a linear system of two equations with two unknowns and which are vanish. To avoid a trivial solution, it suffices that the determinant of the following system be zero

$$\det\begin{pmatrix} e^{i\sqrt{\lambda_n}} & e^{-i\sqrt{\lambda_n}} \\ 1 & 1 \end{pmatrix} = 0.$$

Hence, the eigenvalue $\lambda \in \sigma(T_0)$ if, and only if, it is the zero of the function g defined as well

$$g(\lambda) := \sin(\sqrt{\lambda}).$$

Then, the zeros $\{\lambda_n\}$ (respectively, $\{\mu_n\}$) according to the function of $g(\lambda)$ (respectively, $f(\lambda) := g(\lambda^2)$) are related through by

$$\mu_n = \sqrt{\lambda_n}.$$

Now, it remains to check the Riesz basis property of the exponential family $\{e^{i\mu_n t}\}$. First, we claim that f is of exponential type 1 and also of sine-type function. In fact, a simple computation gives

$$Ce^{|y|} \leq |f(z)| = |\sin(z)| \leq De^{|y|}, \quad z := x + iy \in \mathbb{C} \tag{15.3.1}$$

for some positive constants C and D and for all $y \in \mathbb{R}$ whenever y is sufficiently large. Obviously, f is uniformly bounded on the real axis, hence it belongs to the Cartwright-class. Hence, the indicator diagram is an interval. Second, it follows from Eq. (15.3.1) that

$$h_f(\phi) = |\sin\phi|,$$

and

$$G_f = [-i, i].$$

Hence, we conclude that the width of indicator diagram of f equals $T = 2$. Then, f is the generating function of the family $\{e^{i\mu_n t}\}$. Finally, it is easy to check that the zeros $\{\mu_n\}$ are separated. Indeed, let $n \in \mathbb{N}^*$, we get

$$|\mu_{n+1} - \mu_n| = \Big|(n+1)\pi - n\pi\Big| = \pi. \qquad \text{Q.E.D.}$$

Now, taking account of Theorem 10.4.3, Lemmas 15.3.1 and 15.3.2, we obtain the following theorem:

Theorem 15.3.1 *The exponential family $\{e^{i\sqrt{|\lambda_n(\xi)|}t}\}_1^\infty$, associated to the perturbed operator T_ξ, forms a Riesz basis in $L_2(0, 2)$.* ◇

15.4 Notes and Remarks

- It was shown in [74] that the operator A, given in (15.1.2), generates a C_0-semigroup $T(t)$ on $\mathscr{H}$, hence the solution of (15.1.3) can be written as

$$n(t) = T(t)n_0.$$

In this case, it is interesting to discuss properties of the C_0-semigroup such that differentiability, compactness, and generalized eigenvectors expansion. Since very frequently one merely had information about the infinitesimal generator A, it is important to know how to deduce these properties from those of A. It is along this line of thoughts that the authors in [27] used spectral information of A to investigate properties of the semigroups.
In fact, to show that the system (15.1.3) satisfies the spectrum determined growth condition is an interesting but difficult problem. More precisely, if the system operator generates an eventually norm continuous semigroup, then the spectrum determined growth condition holds. However, verifying this condition is difficult because the semigroup usually may not have a workable expression. So, using the spectrum of the system operator becomes an attractive alternative. In fact, the above transport operator is very particular. Indeed, it was shown in [74] that the operator A has a compact resolvent. Hence, its spectrum is discrete, i.e., $\sigma(A)$ consists of isolated eigenvalues $\{\lambda_n\}_{n=1}^{\infty}$. In this case, a natural question is to seek if every ϕ belonging to $\mathscr{H}$ (or to a subspace of $\mathscr{H}$) can be expanded in terms of generalized eigenvectors of A. The exact question is to know whether the formal Fourier series

$$\sum_n E(\lambda_n, A)\phi$$

of Riesz projections, $E(\lambda_n, A)$, of A (or at least a subsequence of its partial sums) converges in the norm for every ϕ in $\mathscr{H}$ (or in a subspace of $\mathscr{H}$) and its sum are equal to ϕ. This question of summability of the Riesz projections gave rise to major difficulties and answering this question is important in this transport model. In literature, much progress has been made for a different kind of problems, for example in [54, 61, 64–69, 71–73, 75–78, 80], the authors showed that the partial sum

$$\sum_{k=1}^{r} E(\lambda_k, A)$$

is uniformly bounded and unconditionally convergent in the strong topology. More precisely, the authors in literatures [65, 75, 76, 80] proved the Riesz basis property of the eigenvectors of the studied system operator to show that the spectrum determined growth condition is well satisfied. Unfortunately, for the time-dependent rectilinear transport equation given in (15.1.3), the eigenvectors of the system operator fail to form a basis because

$$\sup_n \|E(\lambda_n, A)\| = \infty.$$

In the spirit of [81], Charfi et al. [27] gave a unified treatment and provided an answer to a long time unsolved problem about the summability of Riesz projections and differentiability of the semigroups generated by this transport operator.

- To expand the solutions of an operational differential equation

$$u'(t) + Au(t) = 0, \quad t > 0,$$

into sums of exponential-polynomial solutions when A is an unbounded and non-normal operator in a Hilbert space. The assumptions on A are motivated by some examples in the theory of plane elasticity, where one wishes to separate variables in polar coordinates. One of the basic tools is the behavior of the Fredholm determinant of A which can be calculated in the case of the homogeneous Lamé system in a truncated sector with homogeneous Dirichlet condition on the rectilinear sides of the sector for which we describe the structure of its solutions.

(i) Since the operator A has a compact resolvent, so the adjoint A^* has the same property [62, p. 2354]. On the other hand, it follows that the eigenvalues of A are symmetric about the real axis, then A^* will have the same eigenvalues as A [73, p. 26] with the same algebraic multiplicity for the conjugate eigenvalues ([62, p. 2354]).

(ii) It follows from Theorem 4.4.1 and Lemma 15.1.1 that the resolvent of the operator A is a meromorphic function of finite exponential type at most $h = 1$.

(iii) We infer from the norm expression of $\|E(\lambda_n, A)\|$ that

$$\rho_1 = \lim_{n\to\infty} \frac{\log \|E(\lambda_n, A)\|}{Re\lambda_n} = 1$$

and

$$\rho_2 = \lim_{n\to\infty} \frac{\log |\lambda_n|}{Re\lambda_n} = 1.$$

(iv) Since the series

$$\sum_{n=1}^{\infty} \frac{1}{|\lambda_n|^{1+\varepsilon}} < \infty,$$

for all $\varepsilon > 0$, then the series

$$\sum_{n=1}^{\infty} e^{Re\lambda_n \tau_0}$$

converges for $\tau_0 = 1 + \varepsilon$.

(v) An elementary study of the asymptotic behavior of the roots of

$$\frac{\sin^2(\xi w)}{\xi^2} - \left[(\lambda + \mu)\frac{\sin w}{\lambda + 3\mu}\right]^2 = 0$$

shows that for large values of j, there is only one pair of conjugated simple roots in the strip

$$\frac{\pi}{2w} + \frac{j\pi}{w} < Re\xi_k < \frac{\pi}{2w} + \frac{(j+1)\pi}{w}.$$

If, in addition, we assume that φ is real (which is physically plausible) the grouping by pairs of the terms in the above series corresponds to the real parts of

$$r^{\xi_k} p_k(\log r, \theta).$$

(vi) One can derive similar expansions in the case of Neumann boundary conditions (i.e., φ is traction-free for ($\theta = 0$ and w) since in that case the corresponding Wronskian is simply

$$W(\xi) = \frac{\sin^2(\xi w)}{\xi^2} - \sin^2 w.$$

(vii) In order to write more explicitly the above expansion in the case $w = 2\pi$, one takes advantage of the fact that the eigenvalues ξ_k are the numbers $\frac{k}{2}$ with integral k and that they are semi-simple of multiplicity two. There is only one eigenvalue such that

$$\frac{\pi}{2w} + \frac{j\pi}{w} = \frac{1}{4} + \frac{j}{2} < Re\xi_k < \frac{\pi}{2w} + \frac{(j+1)\pi}{w} = \frac{1}{4} + \frac{j+1}{2},$$

and the expansion (5.3.9) reduces to

$$u(t) = \sum_k e^{-t\frac{k}{2}} \sum_{1 \le l \le 2} \langle u(0), \psi_{k,l}\rangle \frac{\varphi_{k,l}}{\langle \varphi_{k,l}, \psi_{k,l}\rangle}.$$

It remains, therefore, to calculate $\varphi_{k,j}$ and $\psi_{k,j}$.

References

1. A. Jeribi, *Spectral Theory and Applications of Linear Operators and Block Operator Matrices* (Springer, Cham, 2015), p. 599. (ISBN 978-3-319-17565-2; 978-3-319-17566-9). xvi
2. A. Jeribi, *Denseness, Bases and Frames in Banach Spaces and Applications* (de Gruyter, Berlin, 2018), p. 406. (ISBN 978-3-11-048488-5; 978-3-11-049240-8). xv
3. A. Ammar, A. Jeribi, K. Mahfoudhi, Global bifurcation from the real leading eigenvalue of the transport operator. J. Comput. Theor. Transp. **46**(4), 229–241 (2017)
4. F. Abdmouleh, A. Ammar, A. Jeribi, A Characterization of the pseudo-Browder essential spectra of linear operators and application to a transport equation. J. Comput. Theor. Transp. **44**(3), 141–153 (2015)

5. A. Jeribi, N. Moalla, S. Yengui, S-essential spectra and application to an example of transport operators. Math. Methods Appl. Sci. **37**(16), 2341–2353 (2014)
6. H. Baloudi, A. Jeribi, Left-Right Fredholm and Weyl spectra of the sum of two bounded operators and applications. Mediterr. J. Math. **11**, 939–953 (2014)
7. S. Charfi, A. Jeribi, R. Moalla, Essential spectra of operator matrices and applications. Math. Methods Appl. Sci. **37**(4), 597–608 (2014)
8. B. Abdelmoumen, O. Jedidi, A. Jeribi, Time asymptotic description of an abstract Cauchy problem solution and application to transport equation. Appl. Math. **59**(1), 53–67 (2014)
9. A. Ammar, A. Jeribi, A characterization of the essential pseudospectra and application to a transport equation. Extr. Math. **28**(1), 95–112 (2013)
10. A. Ammar, A. Jeribi, A characterization of the essential pseudospectra on a Banach space. Arab. J. Math. (Springer) **2**(2), 139–145 (2013)
11. F. Abdmouleh, A. Ammar, A. Jeribi, Stability of the S-essential spectra on a Banach space. Math. Slovaca **63**(2), 299–320 (2013)
12. A. Ammar, A. Jeribi, N. Moalla, A note on the spectra of a 3×3 operator matrix and application. Ann. Funct. Anal. **4**(2), 153–170 (2013)
13. S. Charfi, A. Jeribi, N. Moalla, Time asymptotic behavior of the solution of an abstract Cauchy problem given by a one-velocity transport operator with Maxwell boundary condition. Collect. Math. **64**(1), 97–109 (2013)
14. B. Abdelmoumen, A. Jeribi, M. Mnif, Measure of weak noncompactness, some new properties in Fredholm theory, characterization of the Schechter essential spectrum and application to transport operators. Ricerche Mat. **61**(2), 321–340 (2012)
15. B. Abdelmoumen, A. Jeribi, M. Mnif, On graph measures in Banach spaces and description of essential spectra of multidimensional transport equation. Acta Math. Sci. Ser. B Engl. Ed. **32**(5), 2050–2064 (2012)
16. F. Abdmouleh, S. Charfi, A. Jeribi, On a characterization of the essential spectra of the sum and the product of two operators. J. Math. Anal. Appl. **386**(1), 83–90 (2012)
17. B. Abdelmoumen, A. Jeribi, M. Mnif, Invariance of the Schechter essential spectrum under polynomially compact operators perturabation. Extr. Math. **26**(1), 61–73 (2011)
18. B. Abdelmoumen, A. Jeribi, M. Mnif, Time asymptotic description of the solution to an abstract Cauchy problem and application to transport equation. Math. Z. **268**(3–4), 837–869 (2011)
19. F. Abdmouleh, A. Jeribi, Gustafson, Weidman, Kato, Wolf, Schechter, Browder, Rakocevic and Schmoeger essential spectra of the sum of two bounded operators and application to a transport operator. Math. Nachr. **284**(2–3), 166–176 (2011)
20. A. Jeribi, I. Walha, Gustafson, Weidmann, Kato, Wolf, Schechter and Browder essential spectra of some matrix operator and application to two-group transport equation. Math. Nachr. **284**(1), 67–86 (2011)
21. N. Ben Ali, A. Jeribi, N. Moalla, On some subsets of Schechter's essential spectrum of a matrix operator and application to transport operator. Math. Nachr. **283**(9), 1245–1256 (2010)
22. A. Ben Amar, A. Jeribi, B. Krichen, Essential spectra of a 3×3 operator matrix and application to three-group transport equation. Inte. Equa. Oper. Theo. **68**(1), 1–21 (2010)
23. A. Ben Amar, S. Chouayekh, A. Jeribi, New fixed point theorems in Banach algebras under weak topology features and applications to nonlinear integral equations. J. Funct. Anal. **259**(9), 2215–2237 (2010)
24. S. Charfi, A. Jeribi, I. Walha, Essential spectra, matrix operator and applications. Acta Appl. Math. **111**(3), 319–337 (2010)
25. A. Ben Amar, A. Jeribi, M. Mnif, Some applications of the regularity and irreducibility on transport theory. Acta Appl. Math. **110**(1), 431–448 (2010)
26. F. Abdmouleh, A. Jeribi, Symmetric family of Fredholm operators of indices zero, stability of essential spectra and application to transport operators. J. Math. Anal. Appl. **364**(2), 414–423 (2010)
27. S. Charfi, A. Intissar, A. Jeribi, Expansion of solution in terms of generalized eigenvectors for a rectilinear transport equation. Trans. Theo. Stati. Phys. **38**(6), 330–345 (2009)

28. A. Jeribi, N. Moalla, A characterization of some subsets of Schechter's essential spectrum and application to singular transport equation. J. Math. Anal. Appl. **358**(2), 434–444 (2009)
29. S. Charfi, A. Jeribi, On a characterization of the essential spectra of some matrix operators and applications to two-group transport operators. Math. Z. **262**(4), 775–794 (2009)
30. A. Jeribi, N. Moalla, I. Walha, Spectra of some block operator matrices and application to transport operators. J. Math. Anal. Appl. **351**(1), 315–325 (2009)
31. B. Abdelmoumen, A. Dehici, A. Jeribi, M. Mnif, Some new properties in Fredholm theory, Schechter essential spectrum, and application to transport theory. J. Inequal. Appl. Art. ID 852676, 1–14 (2008)
32. A. Ben Amar, A. Jeribi, M. Mnif, Some fixed point theorems and application to biological model. Numer. Funct. Anal. Optim. **29**(1–2), 1–23 (2008)
33. A. Jeribi, S.A. Ould Ahmed Mahmoud, R. Sfaxi, Time asymptotic behaviour for a one-velocity transport operator with Maxwell boundary condition. Acta Appl. Math. **98**(3), 163–179 (2007)
34. M. Damak, A. Jeribi, On the essential spectra of matrix operators and applications. Electron. J. Differ. Equ. **11**, 16 (2007)
35. A. Dehici, A. Jeribi, K. Latrach, Spectral analysis of a transport operator arising in growing cell populations. Acta Appl. Math. **92**(1), 37–62 (2006)
36. N. Moalla, M. Damak, A. Jeribi, Essential spectra of some matrix operators and application to two-group transport operators with general boundary conditions. J. Math. Anal. Appl. **323**(2), 1071–1090 (2006)
37. A. Jeribi, N. Moalla, Fredholm operators and Riesz theory for polynomially compact operators. Acta Appl. Math. **90**(3), 227–245 (2006)
38. A. Jeribi, M. Mnif, Fredholm operators, essential spectra and application to transport equations. Acta Appl. Math. **89**(1–3), 155–176 (2006)
39. A. Jeribi, K. Latrach, H. Megdiche, Time asymptotic behavior of the solution to a Cauchy problem governed by a transport operator. J. Integr. Eqn. Appl. **17**(2), 121–139 (2005)
40. A. Ben Amar, A. Jeribi, M. Mnif, On a generalization of the Schauder and Krasnosel'skii fixed points theorems on Dunford-Pettis spaces and applications. Math. Methods Appl. Sci. **28**(14), 1737–1756 (2005)
41. A. Jeribi, H. Megdiche, N. Moalla, On a transport operator arising in growing cell populations. II. Cauchy problem. Math. Methods Appl. Sci. **28**(2), 121–145 (2005)
42. A. Jeribi, Fredholm operators and essential spectra. Arch. Ineq. Appl. **2**(2–3), 123–140 (2004) (Reviewer: Harro Heuser) 47A53 (45B05 47A10 47A55 47N20 47N55) Review PDF Clipboard Journal Article
43. A. Jeribi, A. Intissar, On an Riesz basis of generalized eigenvectors of the non-self-adjoint problem deduced from a perturbation method for sound radiation by a vibrating plate in a light fluid. J. Math. Anal. Appl. **292**(1), 1–16 (2004)
44. A. Jeribi, Time asymptotic behaviour for unbounded linear operator arising in growing cell populations. Nonlinear Anal. Real World Appl. **4**(5), 667–688 (2003) (Reviewer: Hassan Emamirad) 47G20 (35B40 35R10 47N20 92D25) Review PDF Clipboard Journal Article
45. A. Dehici, K. Latrach, A. Jeribi, On a transport operator arising in growing cell populations. Spectral analysis. Advances in mathematics research, Vol. 1 (Nova Sci. Publ., Hauppauge, New-York, 2002), pp. 159–185
46. A. Jeribi, On the Schechter essential spectrum on Banach spaces and application. Facta Univ. Ser. Math. Inform. **17**, 35–55 (2002)
47. A. Jeribi, A nonlinear problem arising in the theory of growing cell populations. Nonlinear Anal. Real World Appl. **3**(1), 85–105 (2002)
48. A. Jeribi, Some remarks on the Schechter essential spectrum and applications to transport equations. J. Math. Anal. Appl. **275**(1), 222–237 (2002)
49. A. Jeribi, A characterization of the essential spectrum and applications. Boll. dell. Unio. Mate. Ital. Sez. B Artic. Ric. Mat. **8**(3, 5), 805–825 (2002)
50. A. Jeribi, A characterization of the Schechter essential spectrum on Banach spaces and applications. J. Math. Anal. Appl. **271**(2), 343–358 (2002)

51. A. Jeribi, Une nouvelle caractérisation du spectre essentiel et application. (French) [A new characterization of the essential spectrum and application] C. R. Acad. Sci. Paris, t. **331**(7), Sér. I, 525–530 (2000)
52. A. Jeribi, Un probléme non linéaire intervenant en dynamique des populations. (French) [A nonlinear problem arising in growing cell populations] C. R. Acad. Sci. Paris, t. **330**(9), Sér. I, 795–800 (2000)
53. K. Latrach, A. Jeribi, A nonlinear boundary value problem arising in growing cell populations. Nonlinear Anal. **36**(7), Ser. A: Theory Methods, 843–862 (1999)
54. M.T. Aimar, A. Intissar, A. Jeribi, On an unconditional basis of generalized eigenvectors of the non-self-adjoint Gribov operator in Bargmann space. J. Math. Anal. Appl. **231**(2), 588–602 (1999)
55. A. Jeribi, Quelques remarques sur le spectre de Weyl et applications. (French) [Some remarks on the Weyl spectrum and applications] C. R. Acad. Sci., Paris Sér. I Math. **327**(5), 485–490 (1998) (Reviewer: Hassan Emamirad) 47A10 (45J05 47G20 47N55 82C70) Review PDF Clipboard Journal Article
56. K. Latrach, A. Jeribi, Some results on Fredholm operators, essential spectra and application. J. Math. Anal. Appl. **225**(2), 461–485 (1998)
57. K. Latrach, A. Jeribi, Sur une équation de transport intervenant en dynamique des populations. (French) [On a transport equation arising in growing cell populations] C. R. Acad. Sci. Paris, t. **325**(10), Sér. I, 1087–1090 (1997)
58. A. Jeribi, Quelques remarques sur les opérateurs de Fredholm et application l'équation de transport. (French) [Some remarks on Fredholm operators and application to the transport equation] C. R. Acad. Sci. Paris Sér. I **325**(1), 43–48 (1997)
59. K. Latrach, A. Jeribi, On the essential spectrum of transport operators on L_1-spaces. J. Math. Phys. **37**(12), 6486–6494 (1996)
60. A. Jeribi, K. Latrach, Quelques remarques sur le spectre essentiel et application l'équation de transport. (French) [Some remarks on the essential spectrum and application to the transport equation] C. R. Acad. Sci. Paris, t. **323**(5), Sér. I, 469–474 (1996)
61. N. Ben Ali and A. Jeribi, On the Riesz basis of a family of analytic operators in the sens of Kato and application to the problem of radiation of a vibriting structure in a light fluid, J. Math. Anal. Appl., 320, no.1, 78–94 (2006)
62. N. Dunford, J.T. Schwartz, *Linear Operators* (Part III, Wiley-Interscience, New-York, 1971)
63. G. Geymonat, P. Grisvard, Expansions on generalized eigenvectors of operators arising in the theory of elasticity. Diff. Inte. Equa. **4**(3), 459–481 (1991)
64. B.Z. Guo, Riesz basis property and exponential stability of controlled Euler-Bernoulli beam equations with variable coefficients. SIAM J. Control Optim. **40**, 1905–1923 (2002)
65. B.Z. Guo, K.Y. Chan, Riesz basis generation, eigenvalues distribution and exponential stability for Euler-Bernoulli beam with joint feedback control. Rev. Mat. Complut. **14**, 205–229 (2001)
66. B.Z. Guo, Y.H. Luo, Riesz basis property of a second order hyperbolic system with collocated scalar input/output. IEEE Trans. Automat. Control **47**, 693–698 (2002)
67. B.Z. Guo, G.Q. Xu, Expansion of solution in terms of generalied eigenfunctions for a hyperbolic system with static boundary condition. J. Func. Anal. **231**, 245–268 (2006)
68. B.Z. Guo, R. Yu, The Riesz basis property of discrete operators and application to a Euler-Bernoulli beam equation with boundary linear feedback control. IMA J. Math. Control Inform. **18**, 241–251 (2001)
69. A. Intissar, Analyse fonctionnelle et théorie spectrale pour les opérateurs compacts non-autoadjoints, Editions CEPADUES, (1997)
70. A. Jeribi, *Denseness, Bases and Frames in Banach Spaces and Applications* (de Gruyter, Berlin, 2018)
71. P. Lang, J. Locker, Spectral decomposition of a Hilbert space by a Fredholm operators. J. Funct. Anal. **79**, 9–17 (1988)
72. P. Lang, J. Locker, Spectral representation of the resolvent of a discrete operator. J. Funct. Anal. **79**, 18–31 (1988)

73. J. Locker, *Spectral Theory of Non-Self-Adjoint Two-Point Differential Operators, Mathematical Surveys and Monographs, 73* (Amer. Math. Soci, Providence, RI, 2000)
74. B. Montagnini, V. Pierpaoli, The time-dependent rectilinear transport equation. Tran. Theo. Stat. Phys. **1**(1), 59–75 (1971)
75. B. P. Rao, Optimal energy decay rate in a damped Rayleigh beam, in: S. Cox, I. Lasiecka (Eds.), Optimisation Methods in Partial Differential Equations, in: Contemp. Math., 209, 221–229 (1997)
76. M.A. Shubov, Basis property of eigenfunctions of non-self-adjoint operator pencils generated by the equation of nonhomogeneous damped string. Inte. Equa. Oper. Theo. **25**, 289–328 (1996)
77. M.A. Shubov, Asymptotic and spectral analysis of the spatially nonhomogeneous Timoshenko beam model. Math. Nachr. **241**, 125–162 (2002)
78. M.A. Shubov, Spectral operators generated by damped hyperbolic equations. Inte. Equa. Oper. Theo. **28**, 358–372 (1997)
79. G.M. Wing, *An Introduction to Transport Theory* (John Wiley and Sons, Inc., New-York., 1962)
80. G.Q. Xu, D.X. Feng, Riesz basis property of a Timoshenko beam with boundary feeback and application. IMA J. Appl. Math. **67**(4), 357–370 (2002)
81. G.Q. Xu, S.P. Yung, Properties of a class of $\mathscr{C}_0$-semigroups on a Banach spaces and their applications. J. Math. Anal. Appl. **328**, 245–256 (2007)

Index

A. Jeribi, *Perturbation Theory for Linear Operators*,
https://doi.org/10.1007/978-981-16-2528-2

GPSR Compliance
The European Union's (EU) General Product Safety Regulation (GPSR) is a set of rules that requires consumer products to be safe and our obligations to ensure this.

If you have any concerns about our products, you can contact us on

ProductSafety@springernature.com

In case Publisher is established outside the EU, the EU authorized representative is:

Springer Nature Customer Service Center GmbH
Europaplatz 3
69115 Heidelberg, Germany

www.ingramcontent.com/pod-product-compliance
Ingram Content Group UK Ltd.
Pitfield, Milton Keynes, MK11 3LW, UK
UKHW021913280526
471597UK00003B/161
* 9 7 8 9 8 1 1 6 2 5 2 7 5 *